Statistics and Computing

Series Editors:
J. Chambers
D. Hand
W. Härdle

For other titles published in this series, go to
www.springer.com/series/3022

Mervyn G. Marasinghe · William J. Kennedy

SAS for Data Analysis

Intermediate Statistical Methods

With 100 SAS Programs

Mervyn G. Marasinghe
Iowa State University
Department of Statistics
Ames, IA 50011
USA
mervyn@iastate.edu

William J. Kennedy
Iowa State University
Department of Statistics
Ames, IA 50011
USA
wjk@iastate.edu

Series Editors:
John Chambers
Department of Statistics-Sequoia
  Hall
390 Serra Mall
Stanford University
Stanford, CA 94305-4065
USA

W. Härdle
Institut für Statistik und
  Ökonometrie
Humboldt-Universität zu
  Berlin
Spandauer Str. 1
D-10178 Berlin
Germany

David Hand
Department of Mathematics
South Kensington Campus
Imperial College London
London, SW7 2AZ
United Kingdom

ISSN: 1431-8784
ISBN: 978-0-387-77371-1   e-ISBN: 978-0-387-77372-8
DOI: 10.1007/978-0-387-77372-8

Library of Congress Control Number: 2008932955

The program code and output for this book was generated using SAS software, Version 9.1 of the SAS System for Windows. ©2002–2003 SAS Institute Inc. SAS and all other SAS Institute Inc. product or service names are registered trademarks or trademarks of SAS Institute Inc., Cary, NC, USA.

Figures 1.24, 3.1, A.1–A.4, A.12, and Table A.1 were created with SAS® software. ©2002–2007, SAS Institute Inc., Cary, NC, USA. All Rights Reserved. Reproduced with permission of SAS Institute Inc., Cary, NC.

© 2008 Springer Science+Business Media, LLC
All rights reserved. This work may not be translated or copied in whole or in part without the written permission of the publisher (Springer Science+Business Media, LLC, 233 Spring Street, New York, NY 10013, USA), except for brief excerpts in connection with reviews or scholarly analysis. Use in connection with any form of information storage and retrieval, electronic adaptation, computer software, or by similar or dissimilar methodology now known or hereafter developed is forbidden.
The use in this publication of trade names, trademarks, service marks, and similar terms, even if they are not identified as such, is not to be taken as an expression of opinion as to whether or not they are subject to proprietary rights.

Printed on acid-free paper

springer.com

This work is gratefully dedicated

to our parents

Preface

This book is intended for use as the textbook in a second course in applied statistics that covers topics in multiple regression and analysis of variance at an intermediate level. Generally, students enrolled in such courses are primarily graduate majors or advanced undergraduate students from a variety of disciplines. These students typically have taken an introductory-level statistical methods course that requires the use a software system such as SAS for performing statistical analysis. Thus students are expected to have an understanding of basic concepts of statistical inference such as estimation and hypothesis testing.

Understandably, adequate time is not available in a first course in statistical methods to cover the use of a software system adequately in the amount of time available for instruction. The aim of this book is to teach how to use the SAS system for data analysis. The SAS language is introduced at a level of sophistication not found in most introductory SAS books. Important features such as SAS data step programming, pointers, and line-hold specifiers are described in detail. The powerful graphics support available in SAS is emphasized throughout, and many worked SAS program examples contain graphic components.

The basic theory of those statistical methods covered in the text is discussed briefly and then is extended beyond the elementary level. Particular attention has been given to topics that are usually not included in introductory courses. These include models involving random effects, covariance analysis, variable subset selection in regression methods, categorical data analysis, and graphical tools for residual diagnostics. However, a thorough knowledge of advanced theoretical material such as linear model theory will not be assumed or required to assimilate the material presented.

SAS programs and SAS program outputs are used extensively to supplement the description of the analysis methods. Example data sets are from the biological and physical sciences and engineering. Exercises are included in each chapter. Most exercises involve constructing SAS programs for the analysis of given observational or experimental data. Complete pdf files of all

SAS examples used in the book can be downloaded from the Springer website www.springer.com/978-0-387-77371-1. The text versions of all data sets used in examples and exercises are available from the website. Statistical tables are not reprinted in the book. These are also on the website.

The first author has taught a one-semester course based on material from this book for several years. The coverage depends on the preparation and maturity level of students enrolled in a particular semester. In a class mainly composed of graduate students from disciplines other than statistics, with adequate knowledge of statistical methods and the use of SAS, most of the book is covered. Otherwise, in a mixed class of undergraduate and graduate students with little experience using SAS, the coverage is usually 5 weeks of introduction to SAS, 5 weeks on regression and graphics, and 5 weeks of anova applications. This amounts to approximately 60% of the material in the textbook. The structure of sections in the chapters facilitates this kind of selective coverage.

SAS for Data Analysis: Intermediate Statistical Methods, although intended to be used as a textbook, may also be useful as a reference to researchers and data analysts both in the academic setting and in industry.

The first author wishes to thank his former instructors at Kansas State University, Professor George Milliken and Professor Dallas Johnson, for introducing him to the SAS language as well as to the fundamentals of statistical data analysis. He is also grateful to Professor Kenneth Koehler, Chair of the Department of Statistics at Iowa State University for steady encouragement and allowing him the time and resources to complete this work. The students in his data analysis course were responsible for locating numerous errors and typos in earlier versions of the manuscript. The authors also wish to thank Professor Dan Nettleton, Laurence H. Baker Endowed Chair in Biological Statistics and Professor of Statistics at Iowa State University for valuable discussions concerning inference from the mixed model and providing a nice exercise problem and Dr. Grace H. Liu of Pioneer Hi-Bred for reading an earlier version of the manuscript and offering constructive suggestions.

MERVYN G. MARASINGHE
Associate Professor
Department of Statistics
Iowa State University, Ames, IA 50011,
USA

WILLIAM J. KENNEDY
Professor Emeritus
Department of Statistics
Iowa State University, Ames, IA 50011,
USA

Contents

Preface VII

1 Introduction to the SAS Language 1
   1.1 Introduction 1
   1.2 Basic Language: Rules and Syntax 5
   1.3 Creating SAS Data Sets 8
   1.4 The INPUT Statement 11
   1.5 SAS Data Step Programming Statements and Their Uses 16
   1.6 Data Step Processing 24
   1.7 More on INPUT Statement 31
      1.7.1 Use of pointer controls 31
      1.7.2 The `trailing @` line-hold specifier 33
      1.7.3 The `trailing @@` line-hold specifier 35
      1.7.4 Use of RETAIN statement 36
      1.7.5 The use of line pointer controls 38
   1.8 Using SAS Procedures 40
   1.9 Exercises 48

2 More on SAS Programming and Some Applications 55
   2.1 More on the DATA and PROC Steps 55
      2.1.1 Reading data from files 56
      2.1.2 Combining SAS data sets 58
      2.1.3 Saving and retrieving permanent SAS data sets 64
      2.1.4 User-defined informats and formats 69
      2.1.5 Creating SAS data sets in procedure steps ... 74
   2.2 SAS Procedures for Computing Statistics 79
      2.2.1 The UNIVARIATE procedure 81
      2.2.2 The FREQ procedure 88
   2.3 Some Useful Base SAS Procedures 103
      2.3.1 The PLOT procedure 104
      2.3.2 The CHART procedure 113

		2.3.3 The TABULATE procedure 119

- 2.4 Exercises 122

3 Statistical Graphics Using SAS/GRAPH 129
- 3.1 Introduction 129
- 3.2 An Introduction to SAS/GRAPH........................... 129
  - 3.2.1 Useful SAS/GRAPH procedures...................... 130
    - GPLOT procedure 130
    - GCHART procedure............................... 133
  - 3.2.2 Writing SAS/GRAPH programs..................... 136
- 3.3 Quantile Plots 146
- 3.4 Empirical Quantile-Quantile Plots 151
- 3.5 Theoretical Quantile-Quantile Plots or Probability Plots..... 154
- 3.6 Profile Plots of Means or Interaction Plots 159
- 3.7 Two-Dimensional Scatter Plots and Scatter Plot Matrices 163
  - 3.7.1 Two-Dimensional Scatter Plots...................... 163
  - 3.7.2 Scatter Plot Matrices 166
- 3.8 Histograms, Bar Charts, and Pie Charts.................... 169
- 3.9 Other SAS Procedures for High-Resolution Graphics 175
- 3.10 Exercises 181

4 Statistical Analysis of Regression Models 187
- 4.1 An Introduction to Simple Linear Regression................ 187
  - 4.1.1 Simple linear regression using PROC REG 189
  - 4.1.2 Lack of fit test using PROC ANOVA 195
  - 4.1.3 Diagnostic use of case statistics 197
  - 4.1.4 Prediction of new y values using regression 204
- 4.2 An Introduction to Multiple Regression Analysis 208
  - 4.2.1 Multiple regression analysis using PROC REG 211
  - 4.2.2 Case statistics and residual analysis.................. 217
  - 4.2.3 Residual plots..................................... 222
  - 4.2.4 Examining relationships among regression variables 225
- 4.3 Types of Sums of Squares Computed in PROC REG and PROC GLM 231
  - 4.3.1 Model comparison technique and extra sum of squares . 231
  - 4.3.2 Types of sums of squares in SAS 233
- 4.4 Subset Selection Methods in Multiple Regression 235
  - 4.4.1 Subset selection using PROC REG 241
  - 4.4.2 Other options available in PROC REG for model selection 249
- 4.5 Inclusion of Squared Terms and Product Terms in Regression Models 251
  - 4.5.1 Including interaction terms in the model 252
  - 4.5.2 Comparing slopes of regression lines using interaction .. 253

        4.5.3 Analysis of models with higher-order terms with
              PROC REG 254
   4.6 Exercises 261

5 **Analysis of Variance Models** 275
   5.1 Introduction 275
        5.1.1 Treatment Structure 278
        5.1.2 Experimental Designs 279
        5.1.3 Linear Models 280
   5.2 One-Way Classification 282
        5.2.1 Using PROC ANOVA to analyze one-way
              classifications 291
        5.2.2 Making preplanned (or a priori) comparisons using
              PROC GLM 297
        5.2.3 Testing orthogonal polynomials using contrasts 302
   5.3 One-Way Analysis of Covariance 309
        5.3.1 Using PROC GLM to perform one-way covariance
              analysis 312
        5.3.2 One-way covariance analysis: Testing for equal slopes .. 321
   5.4 A Two-Way Factorial in a Completely Randomized Design ... 328
        5.4.1 Analysis of a two-way factorial using PROC GLM 331
        5.4.2 Residual analysis and transformations 336
   5.5 Two-Way Factorial: Analysis of Interaction 338
   5.6 Two-Way Factorial: Unequal Sample Sizes 346
   5.7 Two-Way Classification: Randomized Complete Block Design . 358
        5.7.1 Using PROC GLM to analyze a RCBD 361
        5.7.2 Using PROC GLM to test for nonadditivity 367
   5.8 Exercises 369

6 **Analysis of Variance: Random and Mixed Effects Models** .. 389
   6.1 Introduction 389
   6.2 One-Way Random Effects Model 393
        6.2.1 Using PROC GLM to analyze one-way random
              effects models 396
        6.2.2 Using PROC MIXED to analyze one-way random
              effects models 400
   6.3 Two-Way Crossed Random Effects Model 407
        6.3.1 Using PROC GLM and PROC MIXED to analyze
              two-way crossed random effects models 410
        6.3.2 Randomized complete block design: Blocking when
              treatment factors are random 417
   6.4 Two-Way Nested Random Effects Model 418
        6.4.1 Using PROC GLM to analyze two-way nested
              random effects models 421

        6.4.2 Using PROC MIXED to analyze two-way nested
              random effects models 425
  6.5 Two-Way Mixed Effects Model 427
        6.5.1 Two-way mixed effects model: Randomized complete
              blocks design 430
        6.5.2 Two-way mixed effects model: Crossed classification ... 441
        6.5.3 Two-way mixed effects model: Nested classification 453
  6.6 Models with Random and Nested Effects for More Complex
      Experiments 465
        6.6.1 Models for nested factorials 466
        6.6.2 Models for split-plot experiments 472
        6.6.3 Analysis of split-plot experiments using PROC GLM . 474
        6.6.4 Analysis of split-plot experiments using
              PROC MIXED 481
  6.7 Exercises 488

APPENDICES

A **SAS/GRAPH** 503
  A.1 Introduction 503
  A.2 SAS/GRAPH Statements 513
        A.2.1 Goptions statement 516
        A.2.2 SAS/GRAPH global statements 516
  A.3 Printing and Exporting Graphics Output 526

B **Tables** 529

References 549

Index 553

1
Introduction to the SAS Language

1.1 Introduction

The Statistical Analysis System (SAS) is a computer package program for performing statistical analysis of data. The system incorporates data manipulation and input/output capabilities as well as an extensive collection of routines for analysis of data. The SAS system achieves its versatility by providing users with the ability to write their own program statements to manipulate data as well as call up SAS routines called *procedures* for performing major statistical analysis on specified *data sets*. The user-written program statements usually perform data modifications such as transforming values of existing variables, creating new variables using values of existing variables, or selecting subsets of observations. Once data sets have thus been prepared, they are used as input to the particular statistical procedure that performs the desired analysis of the data. SAS will perform any statistical analysis that the user correctly specifies using a SAS procedure.

Most statistical analyses do not require knowledge of a great many features available in the SAS system. However, even a simple analysis will involve the use of some of the extensive capabilities of the language. Thus, to be able to write SAS programs effectively, it is necessary to learn at least a few SAS statement structures and how they work. The following SAS program contains features that are common to many SAS programs.

SAS Example A1

The data to be analyzed in this program consist of gross income, tax, age, and state of individuals in a group of people. The only analysis required is a *listing* of all observations in the data set. The statements necessary to accomplish this task are given in the program for SAS Example A1 shown in Fig. 1.1. In this program, note that each SAS statement ends with a semicolon. The statements that follow the `data first;` statement up to and including the first `run;` statement cause a *SAS data set* to be created. Names for the SAS

1 Introduction to the SAS Language

```
data first ; [2]
input (income tax age state)(@4 2*5.2 2. $2.);
datalines ; [1]
123546750346535IA
234765480895645IA
348578650595431IA
345786780576541NB
543567511268532IA
231785870678528NB
356985650756543NB
765745630789525IA
865345670256823NB
786567340897534NB
895651120504545IA
785650750654529NB
458595650456834IA
345678560912728NB
346685960675138IA
546825750562527IA
;
run;
proc print ; [3]
title 'SAS Listing of Tax data';
run;
```

Fig. 1.1. SAS Example A1: Program

variables to be saved in the data set and the location of their values on each line of data are given in the `input` statement. The raw data are embedded in the input stream (i.e., physically inserted within the SAS program) preceded by a `datalines;` statement [1]. The `proc print;` performs the requested analysis of the SAS data set created; namely to print a listing of the entire SAS data set.

As observed in the SAS Example A1 program, SAS programs are usually made up of *two kinds of statements*:

- Statements that lead to the creation of SAS data sets
- Statements that lead to the analysis of SAS data sets

There may be several groups of each kind of these statements in a SAS program. The occurrence of a group of statements used for creating a SAS data set (called a *SAS data step*) can be recognized because it begins with a `data` statement [2], and a group of statements used for analyzing a SAS data set (called a *SAS proc step*) can be recognized because it begins with a `proc` statement [3].

SAS interprets and executes these steps in their order of appearance in a program. Therefore, the user must make sure that there is a logical progression in the operations carried out. Thus, a `proc` step must follow the `data` step that creates the SAS data set to be analyzed by the `proc` step. Moreover, statements within each group must also satisfy this requirement, in general, except for certain *declarative* or *nonexecutable* statements. For example, an input statement that names variables must precede any SAS statement that references these variable names.

One very important characteristic of the execution of a SAS data step is that the statements in a data step are executed and an observation written to the output SAS data set, repeatedly for every line of data input, until every line is processed. A detailed discussion of *data step processing* is given in Section 1.6.

The first statement following the data statement **2** in the data step usually (but not always) is an input statement. The input statement used here is a moderately complex example of *a formatted input* statement, described in detail in Section 1.4. The *symbols* and *informats* used to read the data values for the variables income, tax, age, and state from the data lines in SAS Example A1 and their effects are itemized as follows:

- @4 causes SAS to begin reading each data line *at* column 4.
- 2*5.2 reads data values for income and tax from columns 4-8 and 9-13, respectively, using the informat 5.2 twice; that is, two decimal places are assumed for each value.
- 2. reads the data value for age from columns 14 and 15 as a whole number (i.e., a number without a fraction portion) using the informat 2.
- $2. reads the data value for state from columns 16 and 17 as a character string of length 2, using the informat $2.

A semicolon symbol ";" appearing by itself in the first column in a data line signals the end of the lines of raw data supplied instream in the current data step. On its encounter, SAS proceeds to finish the creation of the SAS data set named first. The proc print; **3** that follows the data step signals the beginning of a proc step. The SAS data set processed in this proc step is, by default, the data set created immediately preceding it (in this program the SAS data set first, was the only one created). Again, by default, all variables and observations in the SAS data set will be processed in this proc step.

The output from execution of the SAS program consists of two parts: the *SAS Log* (see Fig. 1.2), which is a running commentary on the results of executing each *step* of the entire program, and the *SAS Output* (see Fig. 1.3), which is the output produced as a result of the statistical analysis. In interactive mode (on a PC, say), SAS will display these in separate windows called the log and output windows. When the results of a program executed in the batch mode are printed, the SAS log and the SAS output will begin on new pages.

The *SAS log* contains error messages and warnings and provides other useful information via NOTES **4**. For example, the first NOTE in Fig. 1.2 indicates that a work file containing the SAS data set created is saved on the hard disk and is named WORK.FIRST. This file is a *temporary* file because it will be discarded at the end of the current SAS session.

The printed output produced by the proc print; statement appears in Fig. 1.3. It contains a listing of data for all 16 observations and 4 variables in

```
 2    data first ;
 3    input (income tax age state)(@4 2*5.2 2. $2.);
 4    datalines;

NOTE: The data set WORK.FIRST has 16 observations and 4 variables.
NOTE: DATA statement used (Total process time): 4
      real time           0.12 seconds
      cpu time            0.01 seconds

21    ;
22    run;
23    proc print ;
24    title 'SAS Listing of Tax data';
25    run;

NOTE: There were 16 observations read from the data set WORK.FIRST.
NOTE: PROCEDURE PRINT used (Total process time):
      real time           0.10 seconds
      cpu time            0.03 seconds
```

Fig. 1.2. SAS Example A1: Log

the data set. By default, *variable names* are used in the SAS output to identify the data values for each variable and an observation number is automatically generated that identifies each observation. Note also that the data values are also automatically *formatted* for printing using default format specifications. For example, values of both the `income` and `tax` variables are printed correct to two decimal places, those of the variable `age` as whole numbers, and those of the variable `state` as a string of two characters. These are default formats because it was not specified in the program how these values must appear in the output.

```
              SAS listing of Tax data                1

         Obs    income      tax    age   state

          1     546.75     34.65    35    IA
          2     765.48     89.56    45    IA
          3     578.65     59.54    31    IA
          4     786.78     57.65    41    NB
          5     567.51    126.85    32    IA
          6     785.87     67.85    28    NB
          7     985.65     75.65    43    NB
          8     745.63     78.95    25    IA
          9     345.67     25.68    23    NB
         10     567.34     89.75    34    NB
         11     651.12     50.45    45    IA
         12     650.75     65.45    29    NB
         13     595.65     45.68    34    IA
         14     678.56     91.27    28    NB
         15     685.96     67.51    38    IA
         16     825.75     56.25    27    IA
```

Fig. 1.3. SAS Example A1: Output

1.2 Basic Language: Rules and Syntax

Data Values

Data values are classified as either *character* values or *numeric* values. A character value may consist of as many as 32,767 characters. It may include letters, numbers, blanks, and special characters. Some examples of character values are

$$\text{MIG7, D'Arcy, 5678, South Dakota}$$

A standard numeric value is a number with or without a decimal point that may be preceded by a plus or minus sign but may not contain commas. Some examples are

$$71, \ .0038, \ -4., \ 8214.7221, \ 8.546E{-}2$$

Data that are not one of these standard types (such as dates with slashes or numbers with embedded commas) may be accessed using special *informats*. They are stored then in SAS data sets as character or numeric values as appropriate.

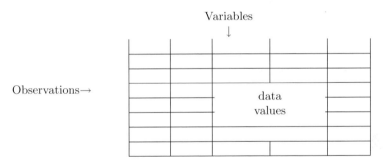

Fig. 1.4. Structure of a SAS data set

SAS Data Sets

SAS data sets consist of *data values* arranged in a rectangular array as displayed in Fig. 1.4. Data values in a column represents a *variable* and those in a row comprise an *observation*. In addition to the data values, *attributes* associated with each variable, such as the name and type of a variable, are also kept in the SAS data set. Internally, SAS data sets have a special organization that is different from that of data sets created using simple editing (e.g., ASCII files). SAS data sets are ordinarily created in a SAS data step and may be stored as *temporary* or *permanent* files. SAS procedures can access data only from SAS data sets. Some procedures are also capable of creating SAS data sets to save information computed as results of an analysis.

Variables

Each column of data values in a SAS data set represents a SAS variable. Variables are of two types: *numeric* or *character*. Values of a numeric variable must be numeric data values and those of a character variable must be character data values. A character variable can include values that are numbers, but they are treated like any other sequence of characters. SAS cannot perform arithmetic operations on values of a character variable.

SAS variables have several *attributes* associated with them. The *name* of the variable and its *type* are two examples of variable attributes. The other attributes of a SAS variable are *length* (in bytes), *relative position* in the data set, *informat, format,* and *label*. Attribute information of SAS variables is also saved in a SAS data set in addition to data values.

Observations

An observation is a group of data values that represent different measurements on the same individual. "Individual" here can mean a person, an experimental animal, a geographic region, a particular year, and so forth. Each row of data values in a SAS data set represent an observation. However, each observation in a SAS data set may be formed using one or more lines of input data.

SAS Names

SAS users select names for many elements in a SAS program, including variables, data sets, and statement labels, etc. Many SAS names can be up to 32 characters long; others are limited to a length of eight characters. The first character in a SAS name must be an alphabetic character. Embedded blanks are not allowed. Characters after the first can be alphabetic, numeric, or underscores. Some examples are H22A, rep_no, and yield.

SAS Variable Lists

A list of SAS variables consists of the names of the variables separated by one or more blanks. For example H22A rep_no yield. A user may **define** or **reference** a sequence of variable names in SAS statements by using an abbreviated list of the form

```
charsxx-charsyy
```

where "chars" is a set of characters and the "xx" and "yy" indicate a sequence of numbers. Thus, the list of indexed variables q2 through q9 may appear in a SAS statement as

```
q2 q3 q4 q5 q6 q7 q8 q9
```

or equivalently as q2-q9.

Using this form in an **input** statement implies that a variable corresponding to each intermediate number in the sequence will be created in the SAS data set and values for them therefore must be available in the lines of data. For example, `var1-var4` implies that `var2` and `var3` are also to be defined as SAS variables.

Any subset of variables already in a SAS data set may be *referenced*, whether the variable names are numbered sequentially or not, by giving the first and last names in the subset separated by two dashes (e.g., `id--grade`). To be able to do this, the user must make sure that the list of variables referenced appears in the SAS data set in the required sequence.

SAS Statements

In every SAS manual or online documentation describing syntax of particular SAS statements, the general form of the statement is given. In these descriptions, words in boldface letters are *SAS keywords*. Keywords must be used exactly as they appear in the description. SAS keywords may not be used as SAS names. Words in lowercase letters specified in the general form of a SAS statement describe the information a user must provide in those positions.

For example, the general form of the **drop** statement is specified as

$$\text{DROP } variable\text{-}list;$$

To use this statement, the keyword *drop* must be followed by the names of the variables that are to be omitted from a SAS data set. The variable list may contain one or more variable names; for example,

```
drop x y2 age;
```

The individual statement descriptions indicate what information is optional, usually by enclosing them in angled brackets < >; several choices are indicated by the term <options>. Some examples are

OUTPUT <*data-set-name(s)*>;

FILENAME *fileref* <*device-type*><*options*>
<*operating-environment-options*>;

PROC CHART <option(s)>;
   HBAR variable(s) </option(s >) ;
   VBAR variable(s) </option(s >) ;

Syntax of SAS Statements

Some general rules for writing SAS statements are as follows:

- SAS statements can begin and end in any column.
- SAS statements end with a semicolon.
- More than one SAS statement can appear on a line.

- SAS statements can begin anywhere on one line and continue onto any number of lines.
- Items in SAS statements should be separated from neighboring items by one or more blanks. If items in a statement are connected by special symbols such as +, -, /, *, or =, blanks are unnecessary. For example, in the statement x=y; no blanks are needed. However, the statement could also be written in one of the forms x = y; or x= y; or x =y;, all of which are acceptable.

Missing Values

A missing numeric value on the input data line can be represented by blanks or a single period, depending on how the values on a data line are input (i.e., what type of *input* statement is used; see below). Once SAS determines a value to be missing in the current observation, the value of the variable for that observation will be set to the SAS missing value indicator, which is also the period symbol. A missing character value in SAS data is represented by a blank character.

A missing value can be used in comparison operations. For example, to check whether a value of a numeric variable, say **age**, is missing for a particular observation and then to remove the entire observation from the data set, the following SAS *programming statement* may be used:

```
if age=. then delete;
```

SAS Programming Statements

SAS programming statements are *executable* statements used in **data** step programming and are discussed in Section 1.5. Other SAS statements such as the drop statement discussed earlier are declarative (i.e., they are used to assign various attributes to variables) and thus are nonexecutable statements.

1.3 Creating SAS Data Sets

Creating a SAS data set suitable for subsequent analysis in a **proc** step involves the following three actions by the user:

a. Use the **data** statement to indicate the beginning of the **data** step and, optionally, name the data set.
b. Use one of the statements **input**, **set**, **merge**, or **update** to specify the location of the information to be included in the data set.
c. Optionally, modify the data before inclusion in the data set by means of user written **data** step programmming statements. Some of the statements that could be used to do this are described in Section 1.5.

In this section the use of the SAS **data step** for the creation of SAS data sets is illustrated by means of some examples. Each example is also used to introduce some variations in the use of different SAS statements.

```
data first ; 1
input (income tax age state)(@4 2*5.2 2. $2.);
datalines;
123546750346535IA
234765480895645IA
348578650595431IA
345786780576541NB
543567511268532IA
231785870678528NB
356985650756543NB
765745630789525IA
865345670256823NB
786567340897534NB
895651120504545IA
785650750654529NB
458595650456834IA
345678560912728NB
346685960675138IA
546825750562527IA
;
run;
data second; 2
set first;
if age<35 & state='IA';
run;
proc print; 3
title 'Selected observations from the Tax data set';
run;
```

Fig. 1.5. SAS Example A2: Program

SAS Example A2

In the program for SAS Example A2, shown in Fig. 1.5, two SAS data sets are created in separate `data` steps. The first data set (named `first` 1) uses data included instream preceded by a `datalines;` statement, as in SAS Example A1. The second data set (named `second` 2) is created by extracting a subset of observations from `first`. This is done in the second step of the SAS program.

In the second data step, observations in the SAS data set `first` that satisfy the condition(s) in the `if` data modification statement that follows the `set` statement are used to create the new data set named `second`. The input data for this `data step` are already available in a SAS data set `first` that is named in the `set` statement. Note that the if statement used here is of the form if (expression);, where the *expression* is a logical expression. As will be discussed in detail in a later section, such expressions may have one of two possible values TRUE or FALSE. In this form of the if statement, the resulting action is to write the current observation to the output SAS data set if the expression evaluates to a TRUE value. The `if` statement, if present, must follow the `set` statement. (As a rule, SAS programming statements follow the `input` or the `set` statement in data steps.) Clearly, two *data* steps and one *proc* step 3 can be identified in this SAS program.

```
1    options ls=80 nodate pageno=1;
2    data first ;
3    input (income tax age state)(@4 2*5.2 2. $2.);
4    datalines;

NOTE: The data set WORK.FIRST has 16 observations and 4 variables.  4
NOTE: DATA statement used:
      real time          0.00 seconds
      cpu time           0.00 seconds

21   ;run;
22   data second;
23   set first;
24   if age<35 & state='IA';
25   run;

NOTE: There were 16 observations read from the data set WORK.FIRST.
NOTE: The data set WORK.SECOND has 5 observations and 4 variables.  5
NOTE: DATA statement used:
      real time          0.00 seconds
      cpu time           0.00 seconds

26   proc print;
27   title 'Selected observations from the Tax data set';
28   run;

NOTE: There were 5 observations read from the data set WORK.SECOND.
NOTE: PROCEDURE PRINT used:
      real time          0.01 seconds
      cpu time           0.01 seconds
```

Fig. 1.6. SAS Example A2: Log

The SAS log obtained from executing the SAS Example A2 program is reproduced in Fig. 1.6. Note carefully that this indicates the creation of two temporary data sets: WORK.FIRST 4 and WORK.SECOND 5. The output from executing the SAS Example A2 program, shown in Fig. 1.7, displays the listing of the observations in the SAS data set named second because the proc print; step, by default, processes the most recently created SAS data set. It can be verified that these constitute the subset of the observations in the SAS data set named first for which the values for the variable age are less than 35 and those for state are equal to the character string IA.

```
        Selected observations from the Tax data set           1

            Obs     income      tax     age    state

             1      578.65    59.54      31     IA
             2      567.51   126.85      32     IA
             3      745.63    78.95      25     IA
             4      595.65    45.68      34     IA
             5      825.75    56.25      27     IA
```

Fig. 1.7. SAS Example A2: Output

SAS Example A3

```
data first ;
input (income tax age state)(@4 2*5.2 2. $2.);
datalines;
1235467503465351A
2347654808956451A
3485786505954311A
345786780576541NB
5435675112685321A
231785870678528NB
356985650756543NB
7657456307895251A
865345670256823NB
786567340897534NB
8956511205045451A
785650750654529NB
4585956504568341A
345678560912728NB
3466859606751381A
5468257505625271A
;
run;
proc print;
where age<35 & state='IA'; 1
title 'Selected observations from the Tax data set';
run;
```

Fig. 1.8. SAS Example A3: Program

The SAS Example A3 program, shown in Fig. 1.8, illustrates how the `proc step` in SAS Example A2 can be modified to obtain the listing of the same subset of observations without the creation of a new SAS data set. This is achieved by the use of the `where` statement in the `proc step`. The `where` statement **1** is an example of a *procedure information statement* described in Section 1.8.

1.4 The INPUT Statement

The *input statement* describes the arrangement of data values in each data line. SAS uses the information supplied in the input statement to produce observations in a SAS data set being created by reading in data values for each of the variables listed in the input statement. There are several methods to input values for variables to form a data set; three of these are summarized below.

List Input

When the data values are separated from one another by one or more blanks, a user may describe the data line to SAS with

INPUT *variable_name_list*;

The variable names are the names chosen to be given for the variables that are to be created in the new SAS data set. These names follow the rules for valid SAS names. Examples of the use of `list input` are:

```
input age weight height;
```

```
input score1-score10;
```

SAS assigns the first value of each data line to the first variable, the second value to the second variable, and so on. Note that the second statement is a convenient shortened form to read data values into 10 variables named `score1, score2,...,score10`, respectively.

List input can be used for reading data values for either numeric or character variables. To describe character variables with *list input*, the $ symbol is entered after each character variable name in the list of variables in the input statement. For example, when

```
input state $ pop income;
```

is used, SAS infers that the variable `state` will contain character values and pop and `income` will contain numeric values. SAS gives character variables described in this way a maximum length of eight characters (bytes) by default. If a value has fewer than eight characters, then it is filled on the right with blanks up to eight characters total. If a value is longer than eight characters, it is truncated on the right to eight characters. Character variables expected to contain values of length more than eight characters can be read using the formatted input method discussed below.

If SAS does not find a value it is looking for on the current data line when using list input, it will go to the next data line and continue to look for the values. For this reason, when using the list input method, if there are any missing data values they must be indicated on the data line by entering a period (the SAS missing value indicator as described previously) separated from other data values by at least one blank on either side of the period.

Formatted Input

For many instream data sets, or those accessed from recording media such as disks or CDs , `list input` may be inappropriate. This is because, in order to save space, the data values contiguous to one another may have been prepared with no spaces or, other characters such as commas, separating them. In such cases, `SAS informats` must be used to input the data.

In general, `informats` can be used to read data lines present in almost any form. They provide information to SAS such as how many columns are occupied by a data value, how to read the data value, and how to store the data value in the SAS data set. The two most commonly used `informats` are those available for the purpose of inputting numeric and character data values.

To read a data value from a data line, the user must specify in which column the data value begins, how many columns to read, whether the data value is numeric or character, and where, if needed, a decimal point should be placed in the case of a numeric value.

To indicate the column to begin reading a data value, the character "@" followed by the column number, both placed before the name of a variable, is used. For example,

> input @26 store @45 sales;

tells SAS that a value for the variable `store` is to be read beginning in column 26 and a value for `sales` beginning in column 45. Here it is assumed that the values in each data line are separated by blanks; otherwise, informats are required to read these values, as described below. When the data values appear in consecutive columns, the use of "@" symbol is not necessary for indicating the position to begin reading the next value, because the next value is read beginning at the column number immediately following the columns from which the previous value was read.

For a numeric variable, the `informat` "w." specifies that the next w columns beginning at the current column be read as the variable's value. The w must be a positive integer. For example,

> input @25 weight 3.;

tells SAS to go to column 25 and read the next three columns (i.e, columns 25, 26, and 27) and store the numeric value (in internal floating point form) as a value for the variable `weight`.

The `informat` "w.d" tells SAS to read the variable's value as above and then insert a decimal point before the last d digits. For example,

> input @10 price 6.2;

tells SAS to begin at column 10, read the next six columns as a value of `price`, and insert a decimal point before the last two digits. If a data value already has a decimal point entered, SAS leaves it in place, overriding the specification given in the informat. In the latter case, the w in "w.d" must also include a column for the decimal point.

For a character variable, the `informat` "$w." tells SAS to begin in the current column and read the next w columns as a character value. Leading and trailing blanks are removed. For example,

> input @30 name $45.;

tells SAS to read columns 30–74 as a value of the character variable `name`. To retain leading and trailing blanks if they appear in the data line, a user may use the `$CHARw.` `informat` instead of $w. Some examples below illustrate the use of `informats` in practice. Suppose a data line contains

0001IA005040891349

where 0001 is the I.D. number of a survey response, IA is the state in which the respondent resides, 5.04 is the number of tons of fertilizer sold in February 1985, 0.89 is the percentage of sales to members, and 1349 is the number of members for this responding farmers' cooperative. Let `id`, `state`, `fert`, `percent`, and `members` be the names assigned by the user to the corresponding variables. An appropriate `input` statement would be

```
input id 4.   state $2.   fert 5.2   percent 3.2   members 4.;
```

Note that no "@" symbol is needed here because data values are read beginning in column 1 and all data values are read consecutively with no blank columns appearing in between. Thus an "@" symbol is not needed for skipping to any position at the beginning or in the interior of the line of data.

Suppose, instead, that the data line has the following appearance:

0001xxxxIA00504x089xxxxxx1349

where the x's represent columns of data that are not of interest for the current problem; these columns may or may not be blanks. Instead of reading these columns, it is possible to skip over to the appropriate column using the "@" symbol or the "+" symbol. For example, after reading a value for `id`, the value for `state` is read beginning in column 9, using "@9," and after reading values for `state` and `fert`, one column is skipped using "+1." The input statement thus could be of the form

```
input id 4. @9 state $2. fert 5.2 +1 percent 3.2 @26 members 4.;
```

Symbols, such as "@" and "+" that could be used on input statements are called *pointer control* symbols. The use of the *pointer* and *pointer controls* in reading data from an input data line is described in detail in Section 1.7.

Finally, the variable names and informats (including pointer controls) that occur on an input statement can be grouped into two separate *lists* enclosed in parentheses. For example, the above statement could also be written as

```
input (id state fert percent members)(4. @9 $2 5.2 +1 3.2 @26 4.);
```

Here, each informat or pointer control-informat combination is associated with a variable name in the list sequentially. If the informat list is shorter, then the entire informat list is reapplied to the remaining variables as required.

Column INPUT

Column input is another alternative to list input when the data values are not separated by blanks or other separators but the user prefers not to use informats. In this case, the values must occupy the same columns on all data lines, a requirement that is also necessary for using formatted input. In this case, however, in the input statement the variable name is followed by the range of columns that the data value occupies in the data instead of an informat. The column numbers are specified in the form `begin-end` and are optionally followed by an integer preceded by a decimal point to indicate the number of decimal places to be assumed for the data value. For inputting character strings, the "$" symbol must follow the variable name but before the column specification. Blanks occurring both *before* and *after* the data value are ignored. For example, if the data line has the appearance

```
0001IA   5.04 891349
```

then it could be read, using column input as

```
input id 1-4 state $ 5-6 fert 7-12 percent 13-15 .2 members 16-19;
```

This reads the value for `id` from columns 1 through 4 as an integer and the value for `state` as a character string from the next two columns. The .2 following 13–15 indicates where the decimal point must be assumed when reading the value for `percent`. The value for `percent` will thus be read as .89 and the value for `members` as 1349 from the above data line.

Combining INPUT Styles

An input statement may contain a combination of the above styles of input. For example, as in the previous example, if the data line has the appearance:

```
0001IA   5.04 891349
```

then it could be read, using a combination of column, formatted, and list input styles as

```
input id 1-4 state $2. fert percent 2.2 members 16-19;
```

Here, SAS uses column input to read the value for `id`, formatted input to read the value for `state` and switches to list input to read the value for `fert`. As discussed later in Section 1.7, this causes the *pointer* to move to column 14 after reading the value for `fert`. Thus, when using an informat to read the value for `percent`, the width of field must be 2 instead of 3 (i.e., no leading blank). Consequently, the informat 2.2 is used instead of 3.2, as in the previous example. Then the value for `members` is read using column input again. Thus, a knowledge of how the *pointer* is handled by the three styles of input is necessary to use them correctly in a single statement.

1.5 SAS Data Step Programming Statements and Their Uses

SAS allows the user to perform various kinds of modification to the variables and observations in the data set as it is being created in the data step. The use of the `if age<35 & state='IA';` statement to obtain a subset of observations in SAS Example A2 is an example of a typical SAS *programming statement*. SAS programming statements are generally used to modify the data during the process of creating a new SAS data set, either from raw data or from data already available in a SAS data set; hence, they must follow an `input`, `set`, `merge`, or `update` statement. The syntax and usage of several statements available for SAS *data step programming* are discussed below.

Assignment Statements

Assignment statements are used to create new variables and change the values of existing ones. The general form of the `assignment` statement is

variable_name= *expression;*

New variables can be created by combining one or more existing variables in an *arithmetic expression*. This may involve combining *arithmetic operators*, *SAS functions*, and other arithmetic expressions enclosed in *parentheses*, and assigning the value of that expression to a new variable name. For example, in the SAS data step

Example 1.5.1

```
data sample;
input(x1-x7) (@5 3*5.1 4*6.2);
y1 = x1+x2**2;
y2 = abs(x3)
y3 = sqrt(x4+4.0*x5**2)-x6;
x7 = 3.14156*log(x7);
datalines;
   ⋮
;
```

three new variables `y1`, `y2`, and `y3` are created. The value of `y1` for each observation in the data set, for example, will be the sum of the value of `x1` and the square of the value of `x2` in that observation. The `variable_name` in an assignment statement may be a new variable to be created and assigned the value of the expression; or it may be an already existing variable, in which case the original value of the variable is replaced by the value of the expression. Thus, in the above data set, each value of the variable `x7` that is read-in, is replaced by the logarithm of the original value of `x7` multiplied by 3.14156.

Arithmetic expressions are normally evaluated beginning from the left and proceeding to the right, but applying the Rules 1, 2, and 3, given in Fig. 1.9, may change the order of evaluation. The result of an arithmetic expression containing a missing value is a missing value. The SAS system incorporates a large number of mathematical functions that can be used in the expressions, as shown in the above example. Some of the commonly used functions are `abs`, `log`, and `sqrt`.

Arithmetic expressions are evaluated according to a set of rules called *precedence rules*. These rules, summarized in Fig. 1.9, specify the order of evaluation of entities within an expression. It is good programming practice to follow these rules when writing expressions.

Rule 1. Expressions within parenthesis are evaluated first.

Rule 2. An operator in a higher ranking group below has higher priority and therefore is evaluated before an operator in lower ranking group.

Group I	**, +(prefix), −(prefix), (NOT), ><, <>
Group II	*, /
Group III	+(infix), −(infix)
Group IV	||
Group V	<, <=, =, =, >=, >, >
Group VI	&(AND)
Group VII	|(OR)

Rule 3. Operators with the same priority (same group) are evaluated from left to right of the expression (except for Group I operators, which are evaluated right to left).

Fig. 1.9. Order of evaluating expressions

The assignment statements used in Example 1.5.1 contain arithmetic expressions only. However, variable names may be combined using comparison operators to form *logical expressions* as described in the paragraph below. Both arithmetic and logical expressions may be combined using *logical oper-*

ators such as the **and** operator ("&") or the **or** operator ("|") to form more complex expressions.

Conditional Execution

As in any programming language, several constructs for altering the normal top-down flow of a program are available in SAS. The `if-then` and `else` statements allow the execution of SAS programming statements depending on the value of an *expression*. The syntax of the statements are

$$\text{IF } expression \text{ THEN } statement;$$
$$< \text{ELSE } statement; >$$

The *expression*, in many cases, is a *logical expression* that evaluates to a *one* if the expression is true or a *zero* if the expression is false. A *logical expression* consists of numerical or character comparisons made using *comparison operators*. These may be combined using *logical operators* such as the **and** operator (&) or the **or** operator (|) to form more complex logical expressions. The *statement* in the above syntax is any executable SAS statement. Several SAS statements enclosed in a `do-end` group may be substituted for a single statement.

The following examples illustrate typical uses of `if-then/else` statements.

Example 1.5.2

```
if score < 80 then weight=.67;
else weight=.75;
```

In this example, the expression `score < 80` evaluates to a one if the current value of the variable `score` is less than 80 and, thus, the assignment statement `weight=.67` will be executed; otherwise, the expression evaluates to a zero and the statement `weight=.75` will be executed. The following statement illustrates a more advanced method for obtaining the same result using the numerical values of the comparisons `score < 80` and `score >= 80`:

```
weight=(score < 80) *.67 + (score >= 80) *.75;
```

Example 1.5.3

```
if state= 'CA' | state= 'OR' then region='Pacific Coast';
```

This is an example of the use of an `if-then` statement without the `else` statement. The expression here is a *logical expression* that will evaluate to a one if at least one of the comparisons `state= 'CA'` or `state= 'OR'` is true or to a zero otherwise. Thus, the current value of the SAS variable `region` will be set to the character string `'Pacific Coast'` if the current value of the

1.5 SAS Data Step Programming Statements and Their Uses

SAS variable `state` is either 'CA' or 'OR'. If this is not so, then the current value of `region` will be determined by other `if-then` statements in the SAS data step or will be left blank.

Example 1.5.4

```
if income= . then delete;
```

A special SAS program statement, `delete`, stops the current observation from being processed further. It is not written to the SAS data set being created, and control returns to the beginning of the data step to process the next observation. In this example, if the current value of the variable `income` is found to be a SAS missing value, then the observation is not written into the data set as a new observation.

In SAS Example A2 (see program in Fig. 1.5), the *subsetting if* statement used was of the form

IF *expression*;

This statement is equivalent to the statement

IF not *expression* THEN delete;

The result is that if the computed value of the expression is FALSE, then the current observation is not written to the output SAS data set. On the other hand, it will be written to the output SAS data set if the expression evaluates to TRUE.

Example 1.5.5

```
if 6.5<=rate<=7.5 then go to useit;
           ⋮
    ··· SAS program statements ···
       ··· to calculate new rate ···
           ⋮
useit: cost= hours*rate;
           ⋮
```

Sometimes it may be required to jump over a few SAS program statements depending on the value of an expression. For this purpose, SAS program statements could be *labeled* using the `label:` notation. In the above example, `useit:` is the label that identifies the SAS statement `cost= hours*rate;`, and if `6.5<=rate<=7.5` is true, then control transfers to this statement. Note that the `if 6.5<=rate<=7.5` statement is a condensed version of the equivalent statement `rate>=6.5 & rate<=7.5`, which will evaluate to a one only if both of the comparisons `rate>=6.5` AND `rate<=7.5` are true or to a zero otherwise.

Example 1.5.6

```
if score < 80 then do;
    weight=.67;
    rate=5.70;
    end;
else do;
    weight=.75;
    rate=6.50;
    end;
```

A `do-end` group can be used to extend the conditional evaluation of single SAS statements to conditionally executing groups of SAS statements. The above example is a straightforward extension of Example 1.5.2.

```
data group1;  1
    input age @@;
datalines;
1 3 7 9 12 17   21 26 30 32 36 42 45 51
;
run;

data group2;  2
set group1;
    if 0<=age<10 then agegroup=0;
    else if 10<=age<20 then agegroup=10;
    else if 20<=age<30 then agegroup=20;
    else if 30<=age<40 then agegroup=30;
    else if 40<=age<50 then agegroup=40;
    else if age >=50 then agegroup=50;
run;

proc print;run;

data group3;  3
set group1;
agegroup=int(age/10)*10;
run;

proc print; run;
```

Fig. 1.10. SAS Example A4: Program

SAS Example A4

The extended example shown in Fig. 1.10 illustrates how `if-then/else` statements can be use to create values for a new variable, as well as how they may be avoided using a convenient transformation.

1.5 SAS Data Step Programming Statements and Their Uses

In the SAS Example A4 program, there are three different data steps and they create SAS data sets named **group1**, **group2**, and **group3**, respectively. In the first data step **1**, data are read using list input with the statement `input age @@;`. The `@@` pointer control symbol causes the `input` statement to be repeatedly executed for the data line. Thus, the data set named **group1** will have 14 observations.

In the second data step **2**, the SAS data set **group2** will be formed using the observations in **group1**, with an additional variable named **agegroup**. The variable **agegroup** will be assigned a value for each observation determined by the value of **age** in the current observation, by executing the series of `if-then/else` statements.

In the third data step **3**, the SAS data set **group3** will be formed using the observations in **group1**. The values for the additional variable **agegroup** this time are determined simply by executing the arithmetic expression $\text{int}(\text{age}/10) * 10$ that converts the value of **age** to the required values of **agegroup**.

The two `proc print;` statements constitute two proc steps that list two of these data sets **group2** and **group3**, which are identical in content. One of the two data sets is displayed in Fig. 1.11.

```
                 The SAS System                    1

            Obs      age     agegroup

             1        1         0
             2        3         0
             3        7         0
             4        9         0
             5       12        10
             6       17        10
             7       21        20
             8       26        20
             9       30        30
            10       32        30
            11       36        30
            12       42        40
            13       45        40
            14       51        50
```

Fig. 1.11. SAS Example A4: Output

Repetitive Computation

Repetitive computation is achieved through the use of *do loops* or *for loops* respectively in commonly known low-level languages Fortran and C. In the SAS data step language, several forms of `do` statements, in addition to the `do-end` groups discussed earlier are available. The statements iterative `do`, `do while`, and `do until` are very flexible, allow a variety of uses, and can

1 Introduction to the SAS Language

be combined. The use of iterative do loops in the data step is illustrated in Examples 1.5.7-1.5.9.

Example 1.5.7

```
data scores;
input quiz1-quiz5 test1-test3;
array scores {8} quiz1-quiz5 test1-test3;
do i= 1 to 8;
    if scores{i}= . then scores{i}= 0;
end;
datalines;
    :
```

An *iterative do loop*, in general, is used to perform the same operation on a sequence of variables. This requires the sequence of variables to be defined as *elements of an array*, using the array statement. This statement, being nonexecutable, may appear anywhere in the data step, but in practice, it is inserted immediately after the variables are defined (usually in the input statement). The array definition allows the user to reference the variables using the corresponding array elements. This is achieved by the use of *subscripts*.

In Example 1.5.7, the variables quiz1,...,quiz5, test1,...,test3 are defined as elements of the array named scores, and they are referenced in the do loop as scores{1},...,scores{8}, respectively, where the values $1,\ldots,8$ are the subscripts. Within the do loop, the subscripting is achieved through the use of an *index variable*, here i, that is used as a *counting variable* in the do statement. During the execution of the loop (do through the end statements), the value of i runs through the values $1,\ldots,8$. The task performed by the do loop in Example 1.5.7 is to set a missing value read from any data line for any of the above variables to zero in the corresponding observation written to the data set created.

Example 1.5.8

```
data load;
input d1-d7;
array day {7} d1-d7;
array hour {7} h1-h7;
    do i= 1 to 7;
    if day{i}= 999 then day{i}=.;
    hour{i}= day{i}*12;
end;
datalines;
    :
```

Variables defined in two different arrays may be processed in a single do loop if the two arrays are of the same length. In this example, two arrays, day and hour are defined – the first corresponding to the variables d1-d7 and the second corresponding to a new set of variables h1-h7. In the loop, first the value of each of the variables d1-d7 is converted to a missing value if the current value is 999. Then the current value of each of the variables h1-h7 is set to 12 times the value of each of the corresponding variables d1-d7, respectively.

Example 1.5.9

```
data index;
do a= 1 to 4;
    do b= 3,6,9;
        c=(a-1)*10+b;
        output;
    end;
end;
proc print data=index;
title 'Creating indices';
run;
```

In this SAS program, a *nested do loop* is illustrated using an example where counter variables a and b of the do statements are manipulated to create the values of a new variable c. This technique is often used for generating factor levels of combinations of factors or interactions in factorial experiments. The output statement inside the loop forces the writing of a new observation containing current values of the variables a, b, and c to the data set, each pass through the loop. Thus at the end of the processing of the loop the SAS data set index will contain 12 observations. The printed listing of this data set is

```
                Creating indices

        Obs     a       b       c

         1      1       3       3
         2      1       6       6
         3      1       9       9
         4      2       3       13
         5      2       6       16
         6      2       9       19
         7      3       3       23
         8      3       6       26
         9      3       9       29
        10      4       3       33
        11      4       6       36
        12      4       9       39
```

The do while statement repeatedly executes statements in a do loop repetitively while a condition, checked before each iteration, is true. The do until statement executes statements similarly but checks the condition at the end of the loop.

1.6 Data Step Processing

A basic understanding of the operations in the SAS data step is necessary to effectively use the capabilities, such as data step programming, available in the data step. The discussion here is kept to a minimum technical level by making use of illustrations and examples. When SAS begins execution of a data step, the statements are first syntax checked and compiled into machine code. At this stage, SAS has sufficient information to create the following:

- an *input buffer*, an area in the memory where the current line of data can be temporarily stored
- a *program data vector* (PDV), an area in the memory where SAS builds an observation to be written to a SAS data set

The PDV is a temporary *place holder* for *a single value* of each of the variables in a list recognized by SAS to exist at this stage. If some of these variables are not given a value either by reading in a value from the input buffer or as a result of a calculation by executing a SAS programmming statement, they will remain as missing values until the end of the data step processing. At the discretion of the user, some or all of the variables in the PDV may form the observation written to the SAS data set at the end of the data step. The SAS data set is a file in which each observation is written as a separate record and thus will contain the entire set of observations the user opts to include in the data set. On the other hand, the PDV contains only those values of the variables obtained from the *current data line* (or new values calculated using them) at any point in the execution of the data step.

The basic SAS data step begins at the *data* statement. Values for the variables in the PDV are initialized to SAS missing values, a line of data is read into the *input buffer*, and data values transferred into the PDV from the input buffer, replacing the missing values in the PDV. The pointer control symbols and informats in the `input` statement facilitate the conversion of the columns in the input buffer into data values for the variables in the PDV. SAS programming statements are then executed using the *current data values* in the PDV, the current values in the PDV are written as a new observation in the SAS data set, and then control is returned back to the beginning of the data step. This is an *iteration* of a SAS data step and the automatically generated SAS variable _N_ keeps track of the the current iteration number. The user may make use of this variable in any programming statement in the data step.

In this description it has been assumed that the data step is operating under its default behavior. It is possible for the user to alter the flow of operations described above by various actions, implemented via the inclusion of one or more *executable* SAS programming statement at different points in the data step. For example, if an `output` statement is inserted among the SAS programmming statements, instead of writing an observation to the SAS data set at the end of an iteration of the SAS data step, SAS will do so at the point

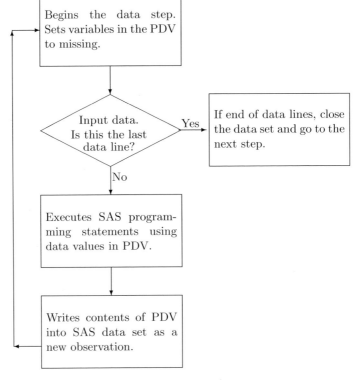

Fig. 1.12. Flow of operations in a data step

the `output` statement is encountered. The flow of operations in a data step is summarized in the chart shown in Fig. 1.12.

SAS Example A5

A simple example is used to illustrate the flow of operations in a data step described above. Consider the data step in the SAS program shown in Fig.1.13. This data step creates the SAS data set named `four`. Four data lines with data values for three variables `x1`, `x2`, and `x3` are read instream. The values of variable `x3` are transformed **1**, and a new variable `x4` **2** is created. Further, only variables `x3` and `x4` are written to the data set.

At the beginning of each iteration of the data step, variables `x1`, `x2`, `x3`, and `x4` are initialized to missing values in the PDV because SAS has detected their presence in the data step during the compile stage. The data step execution proceeds as follows:

- The first line of data is transferred to the input buffer.

```
data four;
input x1-x3;
x3= 3*x3-x1**2; ■1
x4=sqrt(x2); ■2
drop x1 x2; ■3
datalines;
3 4 5
-2 9 3
. 16 8
-3 1 4
;
run;
proc print data=four;
title 'Flow of operations in a data step';
run;
```

Fig. 1.13. SAS Example A5: Program

- The values 3, 4, and 5 are accessed using list input from the input buffer and transferred to the PDV as the new values of variables x1, x2, and x3; the value for x4 still remains a missing value.
- A value of 6 is computed by substituting the values of x1=3 and x3=5 in the expression $3*x3 - x1**2$. This replaces the current value 5 of the variable x3 in the PDV.

```
2     data four;
3     input x1-x3;
4     x3= 3*x3-x1**2;
5     x4=sqrt(x2);
6     drop x1 x2;
7     datalines;
NOTE: Missing values were generated as a result of performing an
      operation on missing values.
      Each place is given by: (Number of times) at (Line):(Column).
      1 at 4:9      1 at 4:12
NOTE: The data set WORK.FOUR has 4 observations and 2 variables.
NOTE: DATA statement used (Total process time):
      real time           0.00 seconds
      cpu time            0.01 seconds

12    ;
13    run;
14    proc print data=four;
15    title 'Flow of operations in a data step';
16    run;

NOTE: There were 4 observations read from the data set WORK.FOUR.
NOTE: PROCEDURE PRINT used (Total process time):
      real time           0.00 seconds
      cpu time            0.01 seconds
```

Fig. 1.14. SAS Example A5: Log

- The square root of 4, the value of x2 in the PDV, replaces the missing value of the variable x4.
- SAS ascertains that the end of the data step has been reached and writes the observation into the data set (named `four`) using the current values of the variables in the PDV. Variables x1 and x2 are excluded from the data set because they appear in the `drop` statement ■3■.
- Next, SAS goes back to the beginning of the data step (the `input` statement) and reinitializes the PDV to missing values, and the next line of data is transferred to the input buffer.

The appearance of the PDV (just before writing the first observation to the SAS data set) is

```
x1   x2   x3   x4    _N_   _ERROR_
 3    4    6    2     1       0
```

The sequence of operations described above continues until end-of-file is detected (i.e., end of the data is encountered) by the `input` statement. SAS then closes the data set and proceeds to the next step. The `drop` statement is a nonexecutable SAS statement and thus may appear anywhere in the data step. It results in the variables listed in the statement being marked so that those variables will be omitted from the observations written to the SAS data set. A listing of the data set created in the data step described above is produced as the output from the next step in the program and is shown in Fig. 1.15. The SAS Log is shown in Fig. 1.14.

```
Flow of operations in a data step          1

           Obs    x3    x4

            1      6     2
            2      5     3
            3      .     4
            4      3     1
```

Fig. 1.15. SAS Example A5: Output

SAS Example A6

The SAS program shown in Fig. 1.16 uses the `array`, `do`, and `output` statements to create a SAS data set that is markedly different in appearance from the instream data set used to create it. This technique, called *transposing*, is useful for preparing SAS data sets for analysis of data obtained from statistically designed experiments (e.g., factorial experiments).

The data for this example, displayed in Fig. 1.16, are scores received by students for five quizzes. The name of the student and the five scores are entered with at least one blank as a separator so that the data can be read using list input. It is important to note that when using list input, missing values must be indicated by a period. If a blank is entered as the missing value, the input statement will mistakenly read the next data value available as the value for the variable for which a value is actually missing in the current line of data.

```
data quizzes;
input name $ quiz1-quiz5;
array qz {5} quiz1-quiz5; 1
drop quiz1-quiz5;
do test= 1 to 5;
   if qz{test}=. then qz{test}= 0 ;
   score = qz{test};
   output; 2
end;
datalines;
Smith 8 7 9 . 3
Jones 4 5 10 8 4
;
run;
proc print data=quizzes;
run;
```

Fig. 1.16. SAS Example A6: Program

The five variable names for the quiz scores quiz1,...,quiz5 are declared in an array named qz **1** and then used in the do loop with a counter variable named test. At the beginning of the data step, variables name, quiz1,...,quiz5, test, and score are all initialized to missing values in the program data vector. Thus, the appearance of the PDV at the beginning of the data step is

```
name test quiz1 quiz2 quiz3 quiz4 quiz5  score    _N_  _ERROR_
 .    .     .     .     .     .     .      .       1      0
```

The first line of data Smith 8 7 9 . 3 is transferred to the input buffer. The input statement reads these values from the input buffer using the list input style and assigns them as new values of variables quiz1,...,quiz5 in the PDV. Thus, the appearance of the PDV at this stage is

```
name  test quiz1 quiz2 quiz3 quiz4 quiz5 score  _N_  _ERROR_
Smith  .    8     7     9     .     3     .     1      0
```

The statements in the do loop are executed with the counter variable test taking values 1 through 5, incremented by +1. With the value of test set to 1, if qz{test}=. then qz{test}= 0; determines whether the value for quiz1 in the PDV is a missing value, and if so, replaces it with a zero. Here qz{test}=. is false so SAS proceeds to execute the next statement.

```
              Obs    name    test    score

               1     Smith    1       8
               2     Smith    2       7
               3     Smith    3       9
               4     Smith    4       0
               5     Smith    5       3
               6     Jones    1       4
               7     Jones    2       5
               8     Jones    3      10
               9     Jones    4       8
              10     Jones    5       4
```

Fig. 1.17. SAS Example A6: Output

The next statement `score= qz{test};` causes the value of `score` to be set to the value of `quiz1` since `qz{1}` refers to the variable `quiz1`. Thus, the appearance of the PDV at this point is

```
name   test  quiz1 quiz2 quiz3 quiz4 quiz5  score  _N_  _ERROR_
Smith   1     8     7     9     .     3      8      1     0
```

A new observation containing current values of the variables `name`, `test`, and `score` is written to the SAS data set `quizzes` at this time because the output statement **2** is encountered. The observation written to the SAS data set is

```
           Obs     name     test     score

            1      Smith      1        8
```

because the variables `quiz1,...,quiz5` are excluded from the SAS data set as they are named in a `drop` statement.

The above steps are repeated for each pass through the *do loop* for the set of data values already in the PDV (i.e., without reading in a new data line). Thus, for each line of data input, five observations are written to the SAS data set, each observation in the data set corresponding to a quiz score for a student.

The printed listing of this data set will thus have the appearance shown in Fig. 1.17.

SAS Example A7

Figure 1.18 displays an example of construction of a SAS data set intended to be used as input to a SAS analysis of variance procedure. This program uses nested do loops. The experiment involves two factors: `amount` with levels `0.9`, `0.8`, `0.7`, `0.6`, and `concentration` with levels `1%`, `1.5%`, `2%`, `2.5%`, `3%`. The data values, consisting of reaction times measured for each combination of `amount` and `concentration`, are available as a table, with the columns corresponding to the levels of `concentration` and the rows to the levels of `amount`. Since data from factorial experiments are typically tabulated in this form, entering the data with each row in the table as a line of data is convenient.

```
data reaction;
length conc $4;
do amount =.9 to .6 by -.1;  ❷
    do conc = '1%' , '1.5%' , '2%' , '2.5%' , '3%' ;
    input time @@;  ❶
    output;  ❸
    end;
end;
datalines;
10.9 11.5 9.8 12.7 10.6
 9.2 10.3 9.0 10.6  9.4
 8.7  9.7 8.2  9.4  8.5
 7.2  8.6 7.5  9.7  7.7
;
run;
proc print;
title 'Reaction times for biological substrate';
run;
```

Fig. 1.18. SAS Example A7: Program

The data are entered instream, as shown in Fig. 1.18. The `input time @@;` ❶ statement inside two nested do loops with index variables `conc` and `amount` is used to read the data values one at a time. Note carefully that `conc` runs through the set of *character* values 1%, 1.5%, 2%, 2.5%, and 3% for each value of `amount` and that `amount` runs through the values 0.9, 0.8, 0.7, and 0.6, in that order ❷. The `@@` pointer control symbol ❶ causes the `input` statement to read values from the same line of data, until the end of that data line is reached. This enables values for the variable `time` to be read, one at a time, from each line of data (See Section 1.7 for more about the use the `@@` pointer control.)

The `output` statement ❸ will cause an observation containing the current values in the PDV for the variables `conc`, `amount`, and `time` to be written to the SAS data set named `reaction`. This will be repeated for all combinations of the index variables `conc` and `amount`; that is, 20 observations will be written, one for each combination of values for these variables. The printed listing of this data set, shown in Fig. 1.19, displays the values of these variables for each observation.

```
         Reaction times for biological substrate              1

              Obs     conc    amount    time

               1      1%       0.9     10.9
               2      1.5%     0.9     11.5
               3      2%       0.9      9.8
               4      2.5%     0.9     12.7
               5      3%       0.9     10.6
               6      1%       0.8      9.2
               7      1.5%     0.8     10.3
               8      2%       0.8      9.0
               9      2.5%     0.8     10.6
              10      3%       0.8      9.4
              11      1%       0.7      8.7
              12      1.5%     0.7      9.7
              13      2%       0.7      8.2
              14      2.5%     0.7      9.4
              15      3%       0.7      8.5
              16      1%       0.6      7.2
              17      1.5%     0.6      8.6
              18      2%       0.6      7.5
              19      2.5%     0.6      9.7
              20      3%       0.6      7.7
```

Fig. 1.19. SAS Example A7: Output

1.7 More on INPUT Statement

In this section, the column pointer controls @ and +, line-hold specifiers trailing @ and trailing @@, and the line pointer control #n are discussed.

1.7.1 Use of pointer controls

The SAS input statement uses a *pointer* to track the position in the input buffer where it begins reading a data value for each of the variables in the PDV. At the start of the execution of the input statement, the pointer is positioned at the beginning of the input buffer (position one) and then moves along the buffer as each successive informat or pointer control in the input statement is encountered. As the pointer moves along the input buffer, the input statement reads data values from the input buffer beginning at the current pointer position and converts them to values for successive variables in the PDV. This conversion is done using the informats supplied by the user in the input statement (or using a default informat if one is not supplied, as in the case of list input). For example, the following input statement

input id 4. @9 state $2. fert 5.2 +1 percent 3.2 @26 members 4.;

was used in Section 1.4 to read the data line

    0001xxxxIA00504x089xxxxxx1349

Suppose that the data line has been moved to the input buffer and that the pointer is positioned at the beginning of the buffer as follows:

```
0001xxxxIA00504x089xxxxxx1349
↑
```

The SAS numeric informat 4. reads the value 0001 for the variable id (and inserts it in the PDV) and the pointer is repositioned at column 5 of the input buffer:

```
0001xxxxIA00504x089xxxxxx1349
    ↑
```

The pointer control @9 then causes the pointer to move to column 9:

```
0001xxxxIA00504x089xxxxxx1349
        ↑
```

The SAS character informat $2 reads the value IA as the value for state and inserts it in the PDV, and the pointer moves over those two columns to column 11 of the input buffer:

```
0001xxxxIA00504x089xxxxxx1349
          ↑
```

Next, the SAS numeric informat 5.2 is used to read the value 00504 as the current value 5.04 of the variable fert in the PDV and the pointer moves over five columns:

```
0001xxxxIA00504x089xxxxxx1349
               ↑
```

The pointer control +1 the moves the pointer one more column:

```
0001xxxxIA00504x089xxxxxx1349
                ↑
```

Three columns are read using the informat 3.2 to obtain the current value of the variable percent in the PDV, and the pointer moves a further three columns:

```
0001xxxxIA00504x089xxxxxx1349
                   ↑
```

The pointer control @26 moves the pointer to column 26:

```
0001xxxxIA00504x089xxxxxx1349
                         ↑
```

Notice that it would have been more convenient to use +6 to move the pointer to column 26 than determining the required number, 26, of columns to move the pointer to the current position from position one. At this stage the value 1349 for the variable members is read from the input buffer, using the informat 4., and inserted in the PDV. The pointer moves to column 31:

```
0001xxxxIA00504x089xxxxxx1349
                              ↑
```

At this point, SAS recognizes that the end of the input statement has been reached and the execution of the SAS programming statements using the values in the PDV begins.

Note that the pointer will move *backward* along the input buffer if the @ pointer control is used with a value that points to a position to the left of the current pointer position. There are other variations of the use of the pointer controls @ and + available. For example, the form *@numeric-variable* or *@numeric-expression* can be used to read subsequent data by positioning the pointer at the value of a *numeric-variable* or a *numeric-expression*, respectively. Thus, moving of the pointer can be made dependent on a value of a variable or an expression. The pointer control + can also be used to move the pointer backward. For example, +(-3) or +num, where the value of num is set to -3, moves the pointer back three columns from the current position.

1.7.2 The trailing @ line-hold specifier

So far, it is understood that when the end of the input statement is reached (the SAS pointer is positioned at the end of the input buffer), SAS proceeds to execute the programming statements that follow using the values in the PDV. In addition, the input buffer will be replaced with the next line of data when this occurs.

Sometimes it may become necessary to execute SAS programming statements after reading only some of the data values from the input buffer. One situation of this kind occurs when reading the rest of the data values depends on the value(s) of variable(s) read so far. Obviously, it is necessary to use a second input statement to read the rest of the data values from the input buffer. However, this not possible because, ordinarily, completing the execution of an input statement will cause the input buffer to be replaced with the next line of data. Thus, the values yet to be read from the previous line of data will become unavailable.

The use of an @ symbol appearing by itself as the last item on an input statement (i.e., just before the semicolon), called a trailing @, is one solution to this problem. The trailing @ forces SAS to hold the pointer at the current position on the input buffer and allows SAS to execute another input statement before the contents of the current input buffer are replaced.

SAS Example A8

Figure 1.20 shows an example where the trailing @ is used twice to hold the same data line (in the input buffer). First, it is used on the statement input store : $13. count @; **1** to hold the data line after reading values for the variables store and count. The variable count contains the number of pairs of values for the variables item and price to be read from the same data

```
data garden;
input store : $13. count  @; ▮1
    do i=1 to count;
        input item : $10. price : 5.2 @; ▮2
        output;
    end;
drop i;
datalines;
JMart 4 rake 1250 sprinkler 875 bench 12000 chair 3525
Woodsons 3 edging 750 planter 1365 basket 870
Home'nGarden 5 sweeper 1185 gloves 350 shears 2100
          spade 3450 trimmer 7640
;
run;
proc print data=garden;
title 'Gardening materials purchased Spring 2004 ';
run;
```

Fig. 1.20. SAS Example A8: Program

line. These pairs of values are read using the statement `input item : $10. price : 5.2 @;`. This statement appears within a do loop that executes a number of times equal to the value of the `count` variable, read previously from the same data line.

The second use of `trailing @` in this example occurs in the `input item : $10. price : 5.2 @;` ▮2. It holds the line after each pair of values for `item` and `price` are read, leaving the pointer at the correct position for the next execution of the same `input` statement. After a pair of values for `item` and `price` are read, the `output` statement causes the values in the PDV for the variables `store`, `count`, `item`, and `price` to be written as an observation in the SAS data set named `garden` created in this data step.

```
                Gardening materials purchased Spring 2004                   1

        Obs    store           count    item           price

         1     JMart             4      rake           12.50
         2     JMart             4      sprinkler       8.75
         3     JMart             4      bench         120.00
         4     JMart             4      chair          35.25
         5     Woodsons          3      edging          7.50
         6     Woodsons          3      planter        13.65
         7     Woodsons          3      basket          8.70
         8     Home'nGarden      5      sweeper        11.85
         9     Home'nGarden      5      gloves          3.50
        10     Home'nGarden      5      shears         21.00
        11     Home'nGarden      5      spade          34.50
        12     Home'nGarden      5      trimmer        76.40
```

Fig. 1.21. SAS Example A8: Output

Note that the data values for the last observation in the input stream continues on to a second data line. These data values are processed correctly because a value of 5 read for the variable `count` results in the statement `input item : $10. price : 5.2 @;` being repeatedly executed five times. This causes SAS to encounter the end of a data line after reading the third pair of values `item` and `price` from the input buffer and thus move the second data line into the input buffer. The next two pairs of values for `item` and `price` are then read using the same input statement.

Additionally, this program illustrates combining the : *modifier* with both character and numeric informats in the list input style. First, `store : $13.` allows the reading of character strings shorter than the specified width of the data value to be read (i.e., 13 bytes) **1**. The : modifier causes `$13.` to recognize the first blank encountered in the data field as a delimiter, as is the case when using list input with simply a `$` symbol without specifying a length. Thus, data values of shorter lengths than 13 characters are read correctly as values of `store`. Second, `price : 5.2` **2** allows the reading of numeric data values of varying widths delimited by blanks using the numeric informat `5.2`. The listing of the data set produced by the SAS Example A8 is displayed in Fig. 1.21.

1.7.3 The `trailing @@` line-hold specifier

It was stated in Section 1.7.2 that SAS assumes implicitly that the processing of a data line is over when the end of the input statement is reached and automatically goes to read a new line of data. There, `trailing @` pointer control was used to hold the current data line for further processing.

Another situation in which it is necessary that SAS does not assume that processing of a data line is complete when the end of an input statement is reached occurs when information for multiple observations are to be read from the same line of data. The trailing `@@` causes the `input` statement to be repeatedly executed for the same data line, and each time the `input` statement is executed, an iteration of the data step is also executed (as if a completely new data line has been read).

Example 1.7.1

```
data sat;
input name $ verbal math @@;
total= verbal + math;
datalines;
Sue 610 560 John 720 640 Mary 580 590
Jim 650 760 Bernard 690 670 Gary 570 680 Kathy 720 780
Sherry 640 720
;
run;
proc print;
run;
```

In Example 1.7.1, several sets of data values, consisting of the values for the variables `name`, `verbal`, and `math` are entered into several data lines in the input stream. Three values at a time are read from each line using list input from the input buffer and transferred to the PDV, with the pointer maintaining its current position while the three values are being processed. The program statement `total= verbal + math;` is then executed and an observation is written to the data set, as SAS has reached the end of the data step. The above actions describe a single iteration of the data step.

Instead of returning to read a new data line, the next set of values will now be read from the input buffer beginning from the current position of the pointer. The presence of the `trailing @@` caused the data line to be *held* in the input buffer. Note, however, that using a `trailing @` instead of the `trailing @@` will not work in this case. This is because the input buffer would have been reinitialized to missing values at the end of each iteration of the data step, thus wiping out the data values that are yet to be transferred to the PDV. The printed output from `proc print;` in Example 1.7.1 is shown as follows:

Obs	name	verbal	math	total
1	Sue	610	560	1170
2	John	720	640	1360
3	Mary	580	590	1170
4	Jim	650	760	1410
5	Bernard	690	670	1360
6	Gary	570	680	1250
7	Kathy	720	780	1500
8	Sherry	640	720	1360

1.7.4 Use of RETAIN statement

Recall that each time SAS returns to the top of the data step, every variable value in the PDV is initialized to missing values. The `retain` statement is a declarative statement that causes the value of each variable listed in the statement to be retained in the PDV from one iteration of the data step to the next. The general form of the `retain` statement is

$$\text{RETAIN } \textit{variable-list} < (\textit{initial-values}) >;$$

By default, the *initial-values* assigned to the variables in the list are missing values; however, the retain statement allows the user to specify the values to be used for initializing as well. The SAS Example A9 program displayed in Fig. 1.22 illustrates the use of the `retain` statement.

The `retain` statement is most useful when multiple types of data lines are to be processed in a data step. It is necessary to retain data values read

```
data ledger;
retain store region month;
input type $1. @;
if type='S' then input @3 store 4. region $10. month : $8. ; 2
else do;
     input @4 date ddmmyy8. sales 7.2; 3
     output;
     end;
drop type;
datalines;
S 0021 Southeast   March 1
     10/05/04 134510
     12/05/04  23675
     21/05/04  96860
     28/05/04 265036
S 0173 Northwest   January
     15/05/04  67200
     18/05/04 158325
     29/05/04 127950
     30/05/04  45845
     02/06/04 304730
;
run;
proc print data=ledger;
 id store;
 format store z4. date ddmmyy8. sales dollar10.2 ;
 title 'Sales Analysis for Martin & Co.';
run;
```

Fig. 1.22. SAS Example A9: Program

in one type of a data line in the PDV, so that they can be combined with information read from other types of data lines to form a single observation to be output to the SAS data set. Usually, values retained in the PDV remain there until they are overwritten by new values read from a data line of the same type.

SAS Example A9

In the SAS Example A9 program, there are two kinds of data lines: one kind, identified by an 'S' entered in the first column 1, specifies values for the variables store, region, and month, and the other containing a blank in the first column specifies values for the variables date and sales. Notice that in the second input statement 2, the : modifier is used to read the value for the variable month, and the informat ddmmyy8. is used in the third input statement 3 to read the date value.

As seen from the listing of the output data set displayed in Fig. 1.23, the values for date and sales have been combined with those of store, region, and month to form each individual observation in the data set. It is important to recognize that the values for date and sales are read from a new data line using the statement input @4 date ddmmyy8. sales 7.2;, following which an observation is written to the SAS data set (named ledger). Then SAS

		Sales Analysis for Martin & Co.		
store	region	month	date	sales
0021	Southeast	March	10/05/04	$1,345.10
0021	Southeast	March	12/05/04	$236.75
0021	Southeast	March	21/05/04	$968.60
0021	Southeast	March	28/05/04	$2,650.36
0173	Northwest	January	15/05/04	$672.00
0173	Northwest	January	18/05/04	$1,583.25
0173	Northwest	January	29/05/04	$1,279.50
0173	Northwest	January	30/05/04	$458.45
0173	Northwest	January	02/06/04	$3,047.30

Fig. 1.23. SAS Example A9: Output

returns to the top of the data step and variables in the PDV are all initialized to missing values *except* for `store`, `region`, and `month`. The values for these variables in the PDV remain the same as those that were previously read from the last type 'S' data line. A new set of values for `date` and `sales` are read from the next data line, unless the next data line is of type 'S'. Note that the data lines are required to be arranged precisely in the sequence they appear in the SAS program for the example to work as described.

1.7.5 The use of line pointer controls

The pointer controls discussed in Section 1.7.1 are called *column pointer controls* because they facilitate the movement of the pointer along the columns of a data line (in the input buffer). The pointer control `#n` moves the pointer to the first column of the nth data line in the input buffer. This implies that it is possible for the input buffer to contain multiple data lines. The largest value of n that is used in an input statement is used by SAS to determine how many data lines will be read into the input buffer at a time. The user may specifically state the number of lines to be read using the `n=` option in the `infile` statement. Once several lines are in the input buffer, `#n` or one of its other forms `#numeric-variable` or `#numeric-expression` may be used in the input statement, to move the pointer among these lines of data to read data values into the PDV. The pointer may move either forward or backward among these lines depending on the value of n, the `numeric-variable`, or the `numeric-expression`.

Data lines continued onto several lines may also occur due to reasons different from the situation described in Section 1.7.4. An observation may constitute information entered on several data lines simply because the data values are too numerous to be entered on a single line. Such data may arise as a result of surveys, longitudinal studies in which many variables are measured over time on each experimental subject, or experiments involving repeated measures. In Example 1.7.2, the data set is the result of a questionnaire in which demographic data are entered in the first data line and the responses to 20 questions are recorded in the second data line as a series of single-digit numbers corresponding to responses made by the subject (identified by the I.D. number on the first line).

Example 1.7.2

```
data survey;
input #1 id 1-4 gender $ bdate ddmmyy6. (race marital educ) (1.)
      #2 @5 (q1-q20) (1.) ;
format bdate ddmmyy8. ;
datalines;
3241 F 100287012
     13431043321110310022
5673 M 211178124
     11031002231134310433
4702 M 170780025
     31134310433211103100
2496 F 030979013
     22311542102231152111
6543 M 090885124
     03100343104332111031
;
run;
proc print;
run;
```

Note that the informat 1. is applied repeatedly to each of the variables in the list (race marital educ) to read the responses to these variables, which are single-digit numbers entered in adjoining columns in the first line of data. Similarly, the informat 1. is applied to each variable in the list (q1-q20). The printed output from proc print; in Example 1.7.2 is as follows:

The SAS System 1

O b s	i d	g e n d e r	b d a t e	r a c e	m a r i t a l	e d u c	q 1	q 2	q 3	q 4	q 5	q 6	q 7	q 8	q 9	q 1 0	q 1 1	q 1 2	q 1 3	q 1 4	q 1 5	q 1 6	q 1 7	q 1 8	q 1 9	q 2 0
1	3241	F	10/02/87	0	1	2	1	3	4	3	1	0	4	3	3	2	1	1	1	0	3	1	0	0	2	2
2	5673	M	21/11/78	1	2	4	1	1	0	3	1	0	0	2	2	3	1	1	3	4	3	1	0	4	3	3
3	4702	M	17/07/80	0	2	5	3	1	1	3	4	3	1	0	4	3	3	2	1	1	1	0	3	1	0	0
4	2496	F	03/09/79	0	1	3	2	2	3	1	1	5	4	2	1	0	2	2	3	1	1	5	2	1	1	1
5	6543	M	09/08/85	1	2	4	0	3	1	0	0	3	4	3	1	0	4	3	3	2	1	1	1	0	3	1

1.8 Using SAS Procedures

The Proc Step

It has previously been noted that the group of SAS statements used to invoke a procedure for performing a desired statistical analysis of a SAS data set is designated as a `proc step` and that the group begins with a statement of the form

<p align="center">PROC <i>procedure_name;</i></p>

One may use options and parameters in the `proc` statement to provide additional information to the procedure. Some procedures also allow optional *procedure information statements*, which usually follow the `proc` statement, to be included in the proc step. Thus the most general form of a proc step is:

<p align="center">PROC <i>proc_name options_list;</i>
<<i>procedure information statements;</i>>
< <i>variable attribute statements;</i>></p>

If the user only requires that

- the most recently created data set will be analyzed,
- all variables in the data set are to be processed, and
- the entire data set is to be processed instead of subsets of observations,

then most SAS procedures can be invoked by using a simple `proc` statement as in the SAS Example A1 program. For example, in the code

```
data new;
input x y z;
datalines;
   ⋮
;
run;
proc print;
run;
```

`proc print` will produce a listing of the data values in the entire SAS data set created in the `data step` immediately preceding the `proc print` statement.

Specifying Options in the PROC *Statement*

On the other hand, if a user intends to analyze a data set that is not the most recently created one or if a user wishes to specify additional information to the procedure, these may be specified as `options` in the `proc` statement. For example,

<p align="center"><code>proc print data=one;</code></p>

specifies that the **print** procedure use the data from the data set named **one**, irrespective of whether it is the most recent SAS data set created in the current job. Thus, the data set named **one** may have been created in any one of the several data steps preceding the current **proc step**.

```
proc corr kendall;
```

provides an example of using a keyword option to specify a type of computation to be performed. Here the keyword **kendall** specifies that Kendall's tau-b correlation coefficients be computed when procedure **corr** is executed, instead of the Pearson correlations that would have been computed by default.

Procedure Information Statements

Certain statements may be optionally included in a **proc** step to provide additional information to be used by the procedure in its execution. Some statements of this type are **var, by, output**, and **title**. For example, the requirement that the analysis is to be performed only on some of the variables in the data set can be specified by using the procedure information statement **var**. In the example

```
proc means data=store mean std;
var bolts nuts screws;
```

the **var** statement requires that procedure **means** compute the mean and standard deviation only of the variables named **bolts, nuts,** and **screws** in the data set named **store**. These variables are thus identified as the *analysis variables*.

A special procedure information statement, the **by** statement, allows many SAS procedures to process subsets of a specified data set based on the values of the variable (or variables) listed in the **by** statement. This in effect means that SAS executes the procedure repeatedly on each subset of data separately. The form of the **by** statement is

$$\text{BY } \textit{variables\_list};$$

When a **by** statement appears, the SAS procedure expects the data set to be arranged in the order of values of the variable(s) listed in the **by** statement. The essential requirement is that those observations with identical values for each of these variables occur together in the input data set. This is required so that these observations can be analyzed by the procedure as subsets called *by groups*. See Example 1.8.1.

This requirement is most conveniently achieved by using the SAS **sort** procedure to rearrange the data set prior to analyzing it. When used in the **proc sort** step, a **by** statement specifies the keys to be used for sorting. In the **sort** procedure, observations are first arranged in the increasing order of the values of the first variable specified in the **by** statement. Within each of the resulting groups, observations are arranged in the increasing order of the values of the second variable specified and so on.

For numeric sort keys, the signed value of a variable is used to determine the ordering with SAS missing value assigned the lowest rank. For character variables, ordering is determined using the ASCII sequence in UNIX and Windows operating environments. The main features of the ASCII sequence are that digits are sorted before uppercase letters, and uppercase letters are sorted before lowercase letters. The blank is the smallest displayable character. Thus, the string 'South ' is larger than the string 'North ' but is smaller than the string 'Southern'.

By default, `proc sort` will overwrite the input SAS data set with the data rearranged as requested in the by statement. However, if needed, the sorted output can be written to a new SAS data set using an `out=` option on the `proc sort` statement to name the new data set.

Example 1.8.1

As an example, suppose that it is required to analyze the variables in a data set by *gender* and *income class* where the respondents were assigned to one of, say, 3 income classes 1, 2, or 3. Once the SAS data set is created, a sequence of SAS statements comparable to those shown below may be used to obtain a required analysis. First, `proc sort` uses the SAS data set as input, rearranges it as specified in the by statement, and overwrites the input data set. The `proc print` uses this data set to produce a listing of the rearranged data.

```
        ⋮
   proc sort;
   by gender income;
   proc print;
   by gender;
        ⋮
```

The output from `proc print` will result in a listing of observations that are grouped by `gender` first, and within each group arranged in the increasing values of `income`, as shown in Output 1.

Output 1

```
              GENDER = F

         listing of observations in ascending order of
         INCOME values and a value of F for GENDER

              GENDER = M

         listing of observations in ascending order of
         INCOME values and a value of M for GENDER
```

1.8 Using SAS Procedures

If, in addition, it is required to find the mean and variance of the variables in the data set by both gender and income class, the following statements may be added:

```
proc means mean var;
by gender income;
var age food rent;
    ⋮
```

producing the table of means and variances for the variables age, rent, and food for each subgroup defined by values of gender and income, as shown in Output 2.

Output 2

        GENDER = F   INCOME = 1

Means and Variances computed for values of variables AGE etc.

        GENDER = F   INCOME = 2

    . . . . .

        GENDER = F   INCOME = 3

    . . . . .

        GENDER = M   INCOME = 1

    . . . . .

        GENDER = M   INCOME = 2

    . . . . .

        GENDER = M   INCOME = 3

    . . . . .

The `where` statement used in SAS Example A3 (see Fig. 1.8) is another example of a procedure information statement.

```
data prindata;
input region : $8. state $2. +1 month monyy5. headcnt revenue
      expenses; 1
format month monyy5. revenue dollar12.2; 2
label region='Sales Region' headcnt='Sales Personnel';
datalines;
SOUTHERN FL JAN78 10 10000 8000
SOUTHERN FL FEB78 10 11000 8500
SOUTHERN FL MAR78 9 13500 9800
SOUTHERN GA JAN78 5 8000 2000
SOUTHERN GA FEB78 7 6000 1200
PLAINS NM MAR78 2 500 1350
NORTHERN MA MAR78 3 1000 1500
NORTHERN NY FEB78 4 2000 4000
NORTHERN NY MAR78 5 5000 6000
EASTERN NC JAN78 12 20000 9000
EASTERN NC FEB78 12 21000 8990
EASTERN NC MAR78 12 20500 9750
EASTERN VA JAN78 10 15000 7500
EASTERN VA FEB78 10 15500 7800
EASTERN VA MAR78 11 16600 8200
CENTRAL OH JAN78 13 21000 12000
CENTRAL OH FEB78 14 22000 13000
CENTRAL OH MAR78 14 22500 13200
CENTRAL MI JAN78 10 10000 8000
CENTRAL MI FEB78 9 11000 8200
CENTRAL MI MAR78 10 12000 8900
CENTRAL IL JAN78 4 6000 2000
CENTRAL IL FEB78 4 6100 2000
CENTRAL IL MAR78 4 6050 2100
;
run;
proc sort;
by region state month; 3
run;
proc print;
run;
proc print label;
  by region state ; 4
  format expenses dollar10.2 ; 5
  label state= State month= Month revenue='Sales Revenue'
        expenses='Overhead Expenses';
  id region state;
  sum revenue expenses; 6
  sumby region;
  title ' Sales report by state and region';
run;
```

Fig. 1.24. SAS Example A10: Program

Variable Attribute Statements

These statements allow SAS users to specify the `format`, `informat`, `label`, and `length` of the variables in a `proc step`. Such specifications are associated with the variables only during the execution of a `proc step` if specified in that `proc step`. On the other hand, these statements may also be used in a `data step` to specify attributes of SAS variables, in which case they would be permanently associated with the variables in the data set created by the `data step`. Thus, these attributes will be available subsequently to any SAS procedure for use within a `proc step`. The `format` and `label` statements are two variable attribute statements frequently used in `proc` steps.

The FORMAT statement: The `format` statement may be used both in the data step and the proc step to specify *formats*. SAS `formats` are used for converting data values stored in a SAS data set to the form desired in printed output. They provide information to SAS such as how many character positions are to be used by a data value, in what form the data value must appear in the printed output, and what additional symbols, such as decimal points, commas, dollar signs, etc., must appear in the printed form of the data value. For example, a date value stored internally as a binary value may be printed in one of several date formats provided by SAS for printing dates. The two most commonly used `formats` are those available for the purpose of printing numeric and character data values.

The LABEL statement: The `label` statement is also used in both the data and the proc steps to give more descriptive *labels* than the variable names to identify the data values (or statistics computed on the data values) in the output.

SAS Example A10

The SAS Example A10 program, displayed in Fig. 1.24, illustrates several features of the `data` and `proc` steps discussed so far. The two different `proc print` steps provide listings of the same SAS data set; the first step is a simple invocation of the procedure, whereas in the second step, several procedure information and variable attribute statements are used to produce more complete annotation.

In the `input` statement **1**, the : modifier is used to read the value for the variable `region` and the date informat `monyy5.` to read the `date` value. The `format` statement **2** used in this data step specifies formats for the variables `month` and `revenue`. Thus, these formats were used in the first `proc print` for printing the values of these variables, as illustrated in Output 1 shown in Fig. 1.25.

The statement `by region state month;` **3** was used with `proc sort`; thus, as expected, in the listing produced by the first `proc print` step (Output 1), observations appear arranged in the increasing order of `month` values within groups with the same `region` and `state` values. These appear in the increasing

Sales report by state and region 1

Obs	region	state	month	headcnt	revenue	expenses
1	CENTRAL	IL	JAN78	4	$6,000.00	2000
2	CENTRAL	IL	FEB78	4	$6,100.00	2000
3	CENTRAL	IL	MAR78	4	$6,050.00	2100
4	CENTRAL	MI	JAN78	10	$10,000.00	8000
5	CENTRAL	MI	FEB78	9	$11,000.00	8200
6	CENTRAL	MI	MAR78	10	$12,000.00	8900
7	CENTRAL	OH	JAN78	13	$21,000.00	12000
8	CENTRAL	OH	FEB78	14	$22,000.00	13000
9	CENTRAL	OH	MAR78	14	$22,500.00	13200
10	EASTERN	NC	JAN78	12	$20,000.00	9000
11	EASTERN	NC	FEB78	12	$21,000.00	8990
12	EASTERN	NC	MAR78	12	$20,500.00	9750
13	EASTERN	VA	JAN78	10	$15,000.00	7500
14	EASTERN	VA	FEB78	10	$15,500.00	7800
15	EASTERN	VA	MAR78	11	$16,600.00	8200
16	NORTHERN	MA	MAR78	3	$1,000.00	1500
17	NORTHERN	NY	FEB78	4	$2,000.00	4000
18	NORTHERN	NY	MAR78	5	$5,000.00	6000
19	PLAINS	NM	MAR78	2	$500.00	1350
20	SOUTHERN	FL	JAN78	10	$10,000.00	8000
21	SOUTHERN	FL	FEB78	10	$11,000.00	8500
22	SOUTHERN	FL	MAR78	9	$13,500.00	9800
23	SOUTHERN	GA	JAN78	5	$8,000.00	2000
24	SOUTHERN	GA	FEB78	7	$6,000.00	1200

Fig. 1.25. SAS Example A10: Output 1

order of state values within groups with the same region values and, finally, in the increasing order of values of the region variable. However, these by groups are not clearly identifiable in the output from the first proc print step. This is because a by statement is not used in the proc step.

In the second proc print step, the statement by region state ❹ is used, causing separate listing for each by group as defined by identical values for state within groups with the same region values, as seen in Fig. 1.26. The format ❺ statement in this proc step provides a format for printing values of the variable expenses. In addition, the statements sum revenue expenses; ❻ and sumby region; in the second proc print step, illustrate how totals are calculated for each of the numeric variables revenue and expenses and are displayed for each group defined by the same value for the by variable region, respectively.

```
                    Sales report by state and region                    2

        Sales                        Sales         Sales       Overhead
        Region       State   Month   Personnel ■7  Revenue     Expenses

        CENTRAL      IL      JAN78   4             $6,000.00   $2,000.00 ■8
                             FEB78   4             $6,100.00   $2,000.00
                             MAR78   4             $6,050.00   $2,100.00

        CENTRAL      MI      JAN78   10            $10,000.00  $8,000.00
                             FEB78   9             $11,000.00  $8,200.00
                             MAR78   10            $12,000.00  $8,900.00

        CENTRAL      OH      JAN78   13            $21,000.00  $12,000.00
                             FEB78   14            $22,000.00  $13,000.00
                             MAR78   14            $22,500.00  $13,200.00
        --------     -----                         ------------ ----------
        CENTRAL                                    $116,650.00 $69,400.00

        EASTERN      NC      JAN78   12            $20,000.00  $9,000.00
                             FEB78   12            $21,000.00  $8,990.00
                             MAR78   12            $20,500.00  $9,750.00

        EASTERN      VA      JAN78   10            $15,000.00  $7,500.00
                             FEB78   10            $15,500.00  $7,800.00
                             MAR78   11            $16,600.00  $8,200.00
        --------     -----                         ------------ ----------
        EASTERN                                    $108,600.00 $51,240.00

        NORTHERN     MA      MAR78   3             $1,000.00   $1,500.00

        NORTHERN     NY      FEB78   4             $2,000.00   $4,000.00
                             MAR78   5             $5,000.00   $6,000.00
        --------     -----                         ------------ ----------
        NORTHERN                                   $8,000.00   $11,500.00

        PLAINS       NM      MAR78   2             $500.00     $1,350.00

        SOUTHERN     FL      JAN78   10            $10,000.00  $8,000.00
                             FEB78   10            $11,000.00  $8,500.00
                             MAR78   9             $13,500.00  $9,800.00

        SOUTHERN     GA      JAN78   5             $8,000.00   $2,000.00
                             FEB78   7             $6,000.00   $1,200.00
        --------     -----                         ------------ ----------
        SOUTHERN                                   $48,500.00  $29,500.00
                                                   ============ ==========
                                                   $282,250.00 $162990.00
```

Fig. 1.26. SAS Example A10: Output 2

Although labels for the variables `region` and `headcnt` were also specified in a `label` statement in the data step, they do not appear in the printed output from the first `proc print`. This is because attributes available in a data set will not be used by some proc steps, unless they are specifically requested by the use of an option. The *label* option in `proc print label;` in the second `proc print` step is an example. The output from this step, displayed in Output 2, shows the labels for `region` and `headcnt` ■7 being used (see Fig. 1.26). Note also that a format for printing values of `expenses` was not available in the data set, so the default format is used in the first `proc print`. In the second `proc print`, a format specifically for printing values of expenses was included. 8

1.9 Exercises

1.1 In each of the following cases show the observations written to the SAS data set (variable names and corresponding data values) when the given lines of data are read by the given `input` statement.
   a. `input id gender $ age height weight;`
      101␣M␣23␣68␣155
      102␣F␣.␣61␣␣102
      103␣␣M␣␣55␣␣␣70␣␣␣202
   b. `input id dob : mmddyy8. dx;`
      1␣10/21/46␣256.20
      2␣9/15/44␣232.4
   c. `input id $1-4 @8 pulse 3. +2 weight 4.1 pushups;`
      L4WP236␣85␣92517␣28
      M5XQ␣47␣␣␣␣␣1423␣32
      ␣␣␣␣158␣79␣81145␣19
   d. `input name :$12. age 2. income 6.2;`
      Peters␣26␣53740
      Hefflefinger␣34276520

   Note: The symbol ␣ denotes one blank column.
1.2 Display the output resulting from executing the following SAS program. Describe in your own words the flow of operations in the data step in creating this data set.

```
data two;
input score @@;
if score > 50 then do;
   result = 'P';
   addon = (score - 50)*10;
   end;
else if score < 50 then do;
   result = 'F';
   end;
datalines;
47 49 50 52 55 .
;
run;
proc print data=two; run;
```

1.3 The following data lines are input in a data step:

$$21 \quad 50.2 \; 17 \; 47.5 \; 54 \; 32.1$$
$$12. \; 54.3$$
$$23.$$
$$45.6$$

What would be the contents of the SAS data set if the input statement used was each of the following? Write a brief explanation of what takes place in each data step.
a. input id   score1;
b. input id   score1 @@;
c. input id;
d. input id   score1   score2;

1.4 The program data vector in a SAS data step has variables with values as follows:

$$\text{code} = \text{'VLC'}$$
$$\text{size} = \text{'M'}$$
$$v1 = 2$$
$$v2 = 3$$
$$v3 = 7$$
$$v4 = .$$

Determine the results of the following SAS expressions:
a. (v1 + v2 - v3)/3
b. v3 - v2/v1
c. v1*v2 - v3
d. v2*v3/v1
e. v1**2 + v2**2
f. code = 'VLC'
g. code = 'VLC' & size = 'M'
h. code = 'VLC'|size = 'M'
i. code = 'VLC' & v4^=.
j. (v3=.) + (v2=3)
k. v1 + v2 + v3 ^= 12
l. code = 'VLC' | (size = 'M' & v1 = 3)
m. 3 < v2 < 5

Hint: Recall that logical expressions evaluate to numeric values 1 (for 'TRUE') or 0 (for 'FALSE').

1.5 Show the output produced by executing the following SAS program. Display the contents of the program data vector at the point the first observation is to be written to the SAS data set.

```
data one;
input x1-x3 @@;
x3=3*x3-x1**2;
x4=sqrt(x2);
drop x1 x2;
```

```
datalines;
3    4    5    -2    9    3    -3    1    4    .    16    8
;
run;
proc print data=one; run;
```

1.6 Show the values for the variable miles that will be stored in the SAS data set distance:

```
data distance;
input miles 5.2;
datalines;
3
34
345
3451
34512
3.
3.4
34.
345.1
;
```

1.7 Sketch the printed output produced by executing the following SAS program. Display the contents of the program data vector immediately after processing the first line of data (just before it is written to the SAS data set).

```
data two;
input team1-team5;
array tms98{5} team1-team5;
drop i;
do i=1 to 5;
    if tms98{i}=. then tms98{i}=0;
    tms98{i} = tms98{i}**2;
    tmstot + tms98{i};
end;
datalines;
4    6    0    -1    .
3    -2   8    -9    12
5    .    -4   7     6
7    5    10   4     5
;
run;
proc print; run;
```

1.8 Write a SAS data step to create a data set named corn with variables variety and yield using input data lines entered with varying numbers of pairs of values for the two variables as shown in the following:

```
A 24.2  B 31.5  B 32.0  C 43.9
C 45.2  A 21.8
B 36.1  A 27.2  C 34.6
```

1.9 Consider the following SAS data step:

```
data result;
input type c1 c2 ;
datalines;
5 0 2
7 3 1
. 0 0
;
proc print;
```

Display what the output from the print looks like if each of the following sets of statements appeared between input and datalines; when the program is executed:

a. index = (2*c1) + c2;
b. if type <= 6 then do;
          index = (2*c1) + c2;
          output;
          end;
c. if type <= 6 then do;
          index = (2*c1) + c2;
          end;
     else delete;
d. if type > 6 then delete;
e. if type > 6 then delete;
      index = (2*c1) + c2;

1.10 Study the the following program:

```
data tests;
input name $ score1 score2 score3 team $ ;
datalines;
Peter 12 42 86 red
Michael 14 29 72 blue
Susan 15 27 94 green
;
run;
proc print; run;
```

a. Sketch the printed output produced from executing this program.
b. What would be the printed output if the **input** statement is changed to the following:

    input name $ score1 score2 score3;

c. What would you do to modify the above program if the data value for the variable **score2** was missing for Michael?
d. Would the above **input** statement still work if the datalines were of the form given below. Explain why or why not.

    Peter
    12 42 86
    red
    Michael
    ...

e. Use the SAS function **sum()** in a single SAS assignment statement to create a new variable called **total**. Where would you insert this statement in the above program?

1.11 A researcher had five subjects in a placebo group and five in a drug group. A score variable is measured on each subject. She would, for example, like to structure the SAS data set to appear as follows:

GROUP	SCORE
P	77
P	76
P	74
P	72
P	78
D	80
D	84
D	88
D	87
D	90

a. Write an **input** statement to read the data values entered in data lines (i.e., with one or more blanks between the group designation and the score and one data line for each pair of values).
b. Suppose it is preferred to enter the data values on two lines as follows:

    P 77 P 76 P 74 P 72 P 78
    D 80 D 84 D 88 D 87 D 90

Write an **input** statement for this arrangement.
c. Instead, consider that the five scores for the placebo group are placed on the first line and the five scores for the drug group on the next as follows:

```
P 77 76 74 72 78
D 80 84 88 87 90
```

Write a **data** step to read these data. Make sure that the data set contains a **group** as well as a **score** variable.

d. Modify the part (c) program so that the five subjects in each group has a subject number from 1 to 5.

1.12 A research project at a college department has collected data on athletes. A subset of the data as follows:

Id	Age	Race	Systolic Blood Pressure	Diastolic Blood Pressure	Heart Rate
4101	18	W	130	80	60
4102	18	W	140	90	70
4103	19	B	120	70	64
4104	17	B	150	90	76
4105	18	B	124	86	72
4106	19	W	145	94	70
4107	23	B	125	78	68
4108	21	W	140	85	74
4109	18	W	150	82	65
4110	20	W	145	95	75

Start up a SAS Windows session and enter statements necessary in the editor window to accomplish the following:

a. Write SAS statements necessary to create a SAS data set named **athlete** containing these data. Name the variables **id, age, race, spb, dbp,** and **hr**, respectively. Include a label statement for the purpose of describing the variables **spb, dbp,** and **hr**. [This would be the first data step.]

b. Submit this data step for execution. Use the SAS Explorer to look in the **work** folder for the temporary SAS data set named **work.athlete**. Open this file by clicking on the name and use the **file** menu to obtain a printed copy of the window thus opened.

c. Add a **proc** step to obtain a SAS listing of the data set **athlete** and submit it. [This would be the first proc step.]

d. *Average* blood pressure is defined as a weighted average of systolic blood pressure and diastolic blood pressure. Since the heart spends more time in its relaxed state (diastole), the diastolic pressure is weighted two-thirds and the systolic blood pressure is weighted one-third. Modify the above program to add a new variable named **abp**, which contains values of average blood pressure computed for each athlete. Submit the modified SAS program to provide a SAS listing of the changed data set. [This would now complete the first data step and the first proc step.]

e. Add SAS statements to create a new data set of a subset of observations named `project` from the above data set containing only those athletes with a value greater than or equal to 100 for average blood pressure and a heart rate greater than 70. Provide a SAS listing of this data set. Suppress the Observation number appearing in this listing, instead identifying the athletes in the output by their Id number. [It will require adding a second data step and a second proc step to do this part.]

f. Obtain the same listing as in part (e) but without creating a new SAS data set to do it. Instead, use the SAS statement `where` within the `print` procedure step to select the subset of observations to be processed. [This will require a third proc step.]

1.13 Ms. Anderson wants to use a SAS program to compute the total score, assign letter grades, and compute summary statistics for her college Stat 101 class. A subset of the data appears as follows:

Id	Major	Year	Quiz	Exam1	Exam2	Lab	Final
5109	Psych	3	75	78	90	87	87
7391	Stat	1	87	75	80	85	75
.	.	.	.	.	.	.	.
.	.	.	.	.	.	.	.
2962	Econ	2	93	68	60	75	93

Note: `Year` records the year in school. `Quiz` and `Lab` are the totals for 10 quizzes and 10 labs, respectively.

a. Write SAS statements to create a SAS dataset named `stat101`. Name your variables as given above and assume that data will be included instream with data values entered delimited by blanks.

b. A maximum of 100 points each could be earned for the quizzes, each midterm exam, the labs, and the final exam. Write a SAS statement to be added to the above data step to include a new variable `total` containing values of the course percentage, calculated weighting the points obtained for the quizzes by 10%, each of the two midterms and the labs by 20%, and the final by 30%.

c. Write SAS statement(s) to be added to the above data step to create a new variable `grade` containing letter grades A, B, C, D, and F, using 90%, 80%, 70%, 60% cutoffs, respectively. You may use the variable `total` from part (b) in your statements.

2

More on SAS Programming and Some Applications

Although several approaches are possible for introducing the SAS language, in presenting the material in Chapter 1 in this book, the authors have consciously avoided a cookbook approach. The earlier students encountered the concepts of pointers and program data vectors, for example, the better their understanding of the working of the SAS data step. Without a basic understanding of the flow of operations in the data step, they will be not be prepared to use the data step effectively. From experience, it has been observed that this technique is more effective in getting the students to a higher point in the learning curve much earlier than using a cookbook approach. Having mastered the material in Chapter 1, readers will be ready to examine the use of SAS for data analysis in greater detail. In this chapter, several SAS procedures are used to illustrate the use of some statements common to many SAS procedures as well as a few that are specific to each procedure. To begin Chapter 2, some useful SAS statements available in both data and proc steps not introduced previously are discussed in detail.

2.1 More on the DATA and PROC Steps

In the previous chapter, presentation of many important aspects of both the data and proc steps were deferred in order to keep the material presented in those sections, to some degree, less cluttered. Some of these topics are covered in detail in this section. Many readers who are already familiar with SAS may have observed that in previous examples, the raw data were input as instream data when the content of the data sets called for the data to be read from external text files. This choice was made because it has been the experience of the authors that the introduction of the `infile` statement, the primary tool for accessing data from external files, at an early stage impedes the beginning SAS user from understanding the basic data step operations. This is due to the fact it makes it more difficult to follow the data step processing clearly when the raw data lines are not in direct view of the user. Also, some users have a tendency to confuse the external data file with the SAS data set being created. Thus, SAS data sets and their creation was introduced early.

2.1.1 Reading data from files

The `infile` statement is primarily used to specify the external file containing the raw data, but it includes options to allow the user more control during the process of transferring data values from the raw data file into a SAS data set. For example, the user may use an option available in the `infile` statement to change the "delimiter", used by the *list input style* for reading data with the input statement, from a blank space to another character such as a comma. Another option allows the user to be given control when end-of-file is reached when reading external data so that other actions may be initiated before closing the new SAS data set.

The INFILE Statement

In previous SAS examples presented in Chapter 1, the data lines were inserted instream preceded by a `datalines` statement to identify the beginning of the data lines (see SAS Example A1 program in Fig. 1.1 in Chapter 1). The `infile` statement is an executable statement required to access data from an external file. In a SAS data step, it must obviously be present before the input statement because the execution of input statement requires the knowledge of the source of the raw data. The general form of the `infile` statement is

INFILE *file-spec* <*options*> ;

where *file-spec* represents a file specification. In the Windows environment, the file specification is easiest to be given directly as a path name to a file inserted within quotes; for example,

```
infile 'C:\Documents and Settings\...\demogr.data';
```

However, this may become cumbersome if some options are also to be included in the `infile` statement. Thus, it may be convenient to use a *fileref*.

The FILENAME Statement

The nonexecutable `filename` statement associates the physical name and location of an external file with a *fileref*, which is an alias for the file. The fileref is then available for use within the current SAS program. Under the Windows environment, a fileref is synonymous with the path name of the file. Text files previously saved in a folder can be given a fileref by including a `filename` statement in the SAS program. The following statements assign a fileref to the file named `demogr.data` and uses it in an infile statement:

```
filename mydata 'C:\Documents and Settings\...\demogr.data';
infile mydata;
```

When a `datalines` statement is used to process instream data, SAS automatically assumes an `infile` statement with the file specification `datalines`; thus the infile statement is not required, unless the user wants to use an option available on the infile statement. In that case, the user must include an infile statement even if the data are included instream. An example of the use of an option while reading instream data is

```
filename datalines eof=last;
```

The above option specifies that once the last data line has been processed, the data step is to be continued by transferring control to the SAS statement labeled `last` instead of closing the SAS data set and terminating the data step. For instance, if the last observation has not yet been written to the SAS data set when end-of-file is encountered (for some reason, such as the last data line being incomplete), this allows the user to define how that situation should be handled.

Example 2.1.1

This is a simple conversion of SAS Example A1 displayed in Fig. 1.1 in Chapter 1 to read the raw data set from a file instead from data entered instream. First, suppose that the data set is available as a text file prepared by entering the data lines into a simple text editor such as Notepad (if a word processor is used to enter the data, the user must make sure that the file is saved as a simple text file). Assume the file is named, say, `wages.txt` and is saved in a folder under the Windows environment. The SAS Example A1 program must be modified to access the data from this file as follows:

```
data first ;
infile 'C:\Documents and Settings\...\wages.txt';
input (income tax age state)(@4 2*5.2 2. $2.);
run;
proc print ;
title 'SAS Listing of Tax data';
run;
```

Here the file specification is a quoted string giving the path of the file containing the raw data.

Some Infile Statement Options

There are several infile statement options that may be useful for managing the conversion of information in data lines to an observation, such as the `eof=` option discussed earlier or the `n=` option discussed in the Section 1.7.5. They are too numerous to be discussed in detail in this book; however, a few are sufficiently important to be briefly mentioned here. The `delimiter=` or the `dlm=` option allows the user to change the default value of the separator of data values, when using the the list input style to read data, from a space to the character specified. To read data separated by commas, use

```
infile datalines delimiter=',';
```

If any of the data values contain an embedded comma, this option will not work; instead, the dsd must be used:

```
infile datalines dsd;
```

With this option in force, a missing value is assumed if two consecutive commas are detected. When using a list input style if a line contains fewer data values than indicated in the input statement, use the missover option to prevent SAS from moving the input pointer to the next line to read the needed values:

```
infile datalines missover;
```

The missover option sets the remaining input statement variables to missing values. The option flowover is the default. The default causes the pointer to move to the next input data line if the current input line is not complete. The options such as firstobs= and obs= allow the user to access a specified number of data lines beginning from a specified line of data in the external data set. For example, the following processes data lines 20 through 50:

```
infile datalines firstobs=20 obs=50;
```

If firstobs= is omitted, SAS will access the first 50 data lines. The n= option specifies the number of lines that the pointer can move to in the input buffer using the # pointer control in a single execution of the input statement. The default value is 1. See Section 1.7.5 for more details.

2.1.2 Combining SAS data sets

When several data sets are created using multiple sources, they must be combined before a meaningful statistical analysis can be performed. Depending on the structure and the format of the input data sets and those required of the output data set, a variety of methods are available in SAS to form a combined data set. The SAS data step statements set, merge, and update are the primary tools available for combining data sets in a SAS data step. In SAS Example A2 (see Fig. 1.5 in Chapter 1 for the program), the set statement was used to illustrate how a SAS data set containing a subset of another SAS data set may be created as follows:

```
data second;
set first;
if age<35 & state='IA';
run;
```

Here an `if` statement was used to select those observations that satisfy the condition specified.

SAS Example B1

In this section an example is used to illustrate the use of the `set` statement to combine two SAS data sets by appending observations from one data set to those of the other. This process is called *concatenation* and allows the combination of several data sets. It is usually practicable when the data sets contain data from similar studies. This implies that the input data sets are expected to contain exactly the same variables (i.e., variables with identical names). It is possible that a few variables are different among some data sets due to decisions taken during the data collection process. If one or more of the data sets contain variables that are not common to all, the combined data set will contain those variables, but with missing values in the observations formed from the data sets that do not contain those variables.

```
data third;
input w 1-2 x 3-5 y 6;
datalines;
211023
312034
413045
;
run;
data fourth;
input x y z;
datalines;
14 5 7862
15 6 6517
16 7 8173
;
run;
data fifth;
set third fourth;
run;
proc print;
title 'Combining SAS data sets end-to-end ';
run;
```

Fig. 2.1. SAS Example B1: Program

Three SAS data sets named `third`, `fourth`, and `fifth` are created in the SAS Example B1 program (see Fig. 2.1), the first two using external data and the other by combining the two SAS data sets previously created. The first data step uses the column input style to create the data set `third` and the second uses the list input style to create the data set `fourth`, both containing **three** observations and **three** variables, respectively (see the abbreviated SAS log in Fig. 2.2). By observing the program, it can be determined that the variable `z` is not present in the data set `third` and the variable `w` is not present in the data set `fourth`, whereas the variables `x` and `y` are common to both data sets. The data set `fifth` is formed using the data step

>   data fifth;
>   set third fourth;

The data set `fifth` is formed by concatenating the observations in the two data sets `third` and `fourth` and so will contain *six* observations and *four* variables `w`, `x`, `y`, and `z`. In the simplest use of the set statement illustrated here, SAS reads observations from the first data set in the list, `third`, transfers data values to the PDV, and the writes them sequentially to the new data set `fifth`. For example, the PDV following reading the first observation is

$$
\begin{array}{cccccc}
w & x & y & z & \_N\_ & \_ERROR\_ \\
21 & 102 & 3 & . & 1 & 0
\end{array}
$$

Although, only the variables `w`, `x`, and `y` are in data set `third`, SAS has detected the presence of the variable `z` in the data step during the compiling stage. Thus, a slot for `z` is created in the PDV and a missing value is inserted at initialization. When the observation is written to the output data set `fifth`, it will contain the values for the *four* variables `w`, `x`, `y`, and `z` as given above.

```
2     data third ;
3     input w 1-2 x 3-5 y 6;
4     datalines;

NOTE: The data set WORK.THIRD has 3 observations and 3 variables.
NOTE: DATA statement used (Total process time):

8     ;
9     run;
10    data fourth;
11    input x y z;
12    datalines;

NOTE: The data set WORK.FOURTH has 3 observations and 3 variables.
NOTE: DATA statement used (Total process time):

16    ;
17    run;
18    data fifth;
19    set third fourth;
20    run;

NOTE: There were 3 observations read from the data set WORK.THIRD.
NOTE: There were 3 observations read from the data set WORK.FOURTH.
NOTE: The data set WORK.FIFTH has 6 observations and 4 variables.

21    proc print;
22    title 'Combining SAS data sets end-to-end ';
23    run;

NOTE: There were 6 observations read from the data set WORK.FIFTH.
NOTE: PROCEDURE PRINT used (Total process time):
```

Fig. 2.2. SAS Example B1: Log

Once the data in the first data set listed is exhausted, SAS begins reading data from the second data set `fourth` and transfers data values to the PDV. The PDV following reading the first observation from the data set `fourth` is

```
 w   x   y    z   _N_  _ERROR_
 .  14   5  7862   1      0
```

Again, an observation containing values for the variables `w`, `x`, `y`, and `z` is written to the output data set `fifth`. This process continues until data set `fourth` reaches end-of-file. Then the data step comes to an end and SAS closes the output data set `fifth` and exits. The number of observations in the new data set is the total number of observations in the two input data sets, and the order of appearance is all observations from the first data set followed by all observations from the second data set, with missing values set appropriately for `z` and `w`, respectively. The output from `proc print` (shown in Fig. 2.3) displays a listing of the data set `fifth`.

```
                Combining SAS data sets end-to-end                    1

                 Obs    w     x    y     z

                  1    21   102   3      .
                  2    31   203   4      .
                  3    41   304   5      .
                  4     .    14   5    7862
                  5     .    15   6    6517
                  6     .    16   7    8173
```

Fig. 2.3. SAS Example B1: Output

The SET Statement

The general form of the `set` statement is

SET <*SAS-data-set(s)* <(*data-set-option(s)*)>> <options> ;

where *data-set-options* are those options that may be specified in parentheses after a SAS data set name, whether it is an input data set (as in a `set` statement) or an output data set (as in an `input` statement). More commonly used options such as `firstobs=`, `obs=`, or `where=` specify observations; those such as `drop=`, `keep=`, and `rename=` have variable names as arguments. When a set statement is used, it is more efficient to use an option to access only those variables required:

```
data fifth;
set third(keep=x y) fourth(drop=z);
run;
```

This will result in a SAS data set named `fifth` without any missing values:

Obs	x	y
1	102	3
2	203	4
3	304	5
4	14	5
5	15	6
6	16	7

If the variable z in the SAS data set `fourth` is renamed to be w, it will also result in a SAS data set with no missing values (albeit one different from the above):

```
data fourth;
set third fourth(rename=(z=w));
run;
```

This would be an option if variables measuring the same quantity have been assigned different names in the two data sets. By renaming z to be w, a variable that already exists in the data set `third`, the user is in fact recognizing this fact. The resulting data set is

Obs	w	x	y
1	21	102	3
2	31	203	4
3	41	304	5
4	7862	14	5
5	6517	15	6
6	8173	16	7

In the SAS Example A2 program (see Fig. 1.5), the data set option `where=` could have been used to select the required subset of observations; thus,

```
data second;
set first(where=(age<35 & state='IA'));
run;
```

There are several options that are unique to the `set` statement; among them are those that enable accessing observations nonsequentially according to a value given in the `key=` option or according to the observation number in the `point=` option.

Programming statements other than the *subsetting if*, such as assignment statements, may be used following the `set` statement, just as one would use following an `input` statement. In particular, one could use the `output` statement to create multiple observations in the output data set from a single observation in the input data set, similar to its use in SAS Example A7 (see

Fig. 1.18). The use of a by following the set statement allows interleaving of observations in several data sets. The observations in the output data set are arranged by the values of the by variable(s), in the order of the data sets in which they occur. Consider the two data sets AAA and BBB containing information for identical subjects:

Data set AAA		Data set BBB	
Id	Height	Id	Weight
111	65	111	145
222	70	222	156
333	58	333	148
444	71	444	166
555	69	555	175
777	70	666	136

The following SAS data step results in the formation of an interleaved SAS data set:

```
data CCC;
  set AAA BBB;
  by id;
run;
```

The resulting data set CCC (a listing is shown below) has 12 observations, which is the total number of observations from both data sets. The new data set contains all variables from both data sets. The values of variables found in one data set but not in the other are set to a missing value, and the observations are arranged in the order of the values of the variable id. In particular, note that the observation with id equal to 666 occurs before that with the id equal to 777 in the output data set, although the second observation came from the data set AAA listed first in the set statement. Note that observations in each of the original data sets were already arranged in the increasing order of the values of id. Thus, it is required for interleaving to ensure that the observations are sorted or grouped in each input data set by the variable or variables that are in the by statement.

id	height	weight
111	65	.
111	.	145
222	70	.
222	.	156
333	58	.
333	.	148
444	71	.
444	.	166
555	69	.
555	.	175
666	.	136
777	70	.

Instead of a set statement, a merge statement may be used to combine these two data sets:

```
data CCC;
  merge AAA BBB;
  by id;
run;
```

A listing of the resulting data set is

Obs	id	height	weight
1	111	65	145
2	222	70	156
3	333	58	148
4	444	71	166
5	555	69	175
6	666	.	136
7	777	70	.

Note that missing values are generated for the variables height and weight for those observations with no common id values in both data sets.

2.1.3 Saving and retrieving permanent SAS data sets

In SAS examples discussed so far in this chapter, raw data, input either instream or from text files, were used to create temporary SAS data sets. As discussed in Section 1.2, SAS data sets contain not only the rectangular array of data but also other information such as variable attributes. In practice, the creation of a SAS data set requires substantial effort so that a user may want to save it permanently for future analysis using SAS procedures for performing different statistical applications. The availability of a carefully constructed permanent SAS data set allows the user to bypass the data set creation step at least for the duration of a research project. In addition, SAS data sets have become a convenient vehicle for transfer of large data sets to other users.

Two SAS examples are used in this subsection to illustrate how to use raw data to create a permanent data set and how to retrieve data for analysis from a previously saved SAS data set. The concept of a SAS *library* is easily understood in the context of running SAS programs under the Windows environment. Recall that the complete path name of a file was used with the filename statement to associate a fileref with the physical name and location of a file. Similarly, the libname statement associates the physical name and location of an external folder (directory) with a *libref*, which is an alias for the complete path name of the folder (directory). The following statement assigns the libref mylib to the folder named projectA:

```
libname mylib 'C:\Documents and Settings\...\projectA\';
```

2.1 More on the DATA and PROC Steps 65

To save a SAS data set in a folder given in a libref as above, the user must specify a *two-level* SAS data set name, where the first level is the libref and the second level is the actual data set name. A two-level SAS data set name, in general, is a name that contain two parts separated by a period of the form libref.membername and is used to refer to *members* of a library libref. The *membername* is the name of a SAS data set when the members stored in the library are SAS data sets. Under the Windows operating system, a library is synonymous with a folder (or a directory). Thus, SAS data sets can be saved in a folder directly by executing statements in SAS programs giving two-level names to the data sets to be saved.

For example, mylib.survey refers to a SAS data set named survey to be saved in the above folder projectA. The libref defined in a SAS program is available for use only within the current SAS program. Many SAS data sets may be saved in the same folder (as members of the library) by using the libref mylib as the first-level name as many times as needed in the same program. A different name may be used as a libref to associate the same library in another SAS program, thus allowing the user to access previously stored members or add new members to the library.

SAS Example B2

```
libname mylib1 'G:\stat479\Class\';
data mylib1.first;
input x1-x5;
datalines;
1 2 3 4 5
2 3 4 5 6
6 5 4 3 2
1 2 1 2 1
7 2 55 5 5
;
run;
```

Fig. 2.4. SAS Example B2: Program

The SAS Example B2 program (see Fig. 2.4) is a simple example illustrating the use of the libname statement and two-level names to create and access permanent SAS data sets. Instream raw data lines are used to create a SAS data set using the two-level name mylib1.first. The first part of the two-level name mylib1 refers to a folder in a disk mounted on a zip drive. Thus, the SAS data set named first created in the data step is saved as a permanent file in the specified folder. Thus, the data set first will be a member of this library. The SAS log reproduced in Fig. 2.5 indicates this fact by listing the two-level name MYLIB1.FIRST and identifying the name of the folder as G:\stat479\Class. The actual physical name of the file saved is

```
2    libname mylib1 'G:\stat479\Class\';
NOTE: Libref MYLIB1 was successfully assigned as follows:
      Engine:           V9
      Physical Name: G:\stat479\Class
3    data mylib1.first;
4    input x1-x5;
5    datalines;

NOTE: The data set MYLIB1.FIRST has 5 observations and 5 variables.
NOTE: DATA statement used (Total process time):
      real time           0.03 seconds
      cpu time            0.01 seconds

11   ;
12   run;
```

Fig. 2.5. SAS Example B2: Log page

first.sas7bdat, as can be verified by manually obtaining a listing of the Class folder (see Fig. 2.6). Obviously, if a single-level name, say first, was used instead in the data statement, the SAS data set would have been temporarily saved in the WORK folder (and the SAS data set thus created referred to as WORK.FIRST in the log page).

Name	Size	Type	Date Modified
cities2.sas7bdat	97 KB	SAS Data Set	8/28/2006 11:59 AM
cities.sas7bdat	61 KB	SAS Data Set	8/28/2006 11:23 AM
first.sas7bdat	5 KB	SAS Data Set	9/5/2007 11:41 AM
fuel.data	2 KB	DATA File	11/3/2004 11:45 AM
second.sas7bdat	5 KB	SAS Data Set	12/19/2007 4:23 PM
zany.data	1 KB	DATA File	9/14/2004 9:21 AM

Fig. 2.6. Screen shot of Class folder listing

SAS Example B3

By including a libname statement of the form shown in the SAS program shown in Fig. 2.4 (possibly with a different libref, but the same physical path name of the folder), one or more SAS data sets stored permanently in a library can be accessed for further processing in another SAS program to be executed subsequently.

In the SAS Example B3 program (see Fig. 2.7) the SAS data set first is accessed from the library for processing using this method. The following libname statement in this program associates the libref mydef1 with the same library where the data set first was saved when the SAS program in Fig. 2.4 was executed:

```
libname mydef1 'G:\stat479\Class\';
```

```
libname mydef1 'G:\stat479\Class\';

proc print data=mydef1.first; run;

proc means data=mydef1.first; run;

data mydef1.second;
input y1-y3;
datalines;
31 34 38
43 45 47
10 11 12
908 97 96
;
run;
proc contents data=mydef1.first; run;

proc datasets library=mydef1 memtype=data;
    contents data=first directory details;
run;
```

Fig. 2.7. SAS Example B3: Program

This allows the two-level name mydef1.first to be used as shorthand for accessing the SAS data set first from the library to be analyzed using the SAS procedure proc print by naming it in the data= option. The listing resulting from this statement is shown on page 1 of the output produced by the SAS Example B3 program, displayed in Fig. 2.8.

```
                         The SAS System                                    1

                 Obs    x1    x2    x3    x4    x5

                  1      1     2     3     4     5
                  2      2     3     4     5     6
                  3      6     5     4     3     2
                  4      1     2     1     2     1
                  5      7     2    55     5     5


                         The SAS System                                    2

                       The MEANS Procedure

Variable    N          Mean         Std Dev       Minimum       Maximum

x1          5      3.4000000       2.8809721     1.0000000     7.0000000
x2          5      2.8000000       1.3038405     2.0000000     5.0000000
x3          5     13.4000000      23.2873356     1.0000000    55.0000000
x4          5      3.8000000       1.3038405     2.0000000     5.0000000
x5          5      3.8000000       2.1679483     1.0000000     6.0000000
```

Fig. 2.8. SAS Example B3: Output pages 1 and 2

68   2 More on SAS Programming and Some Applications

The statement `proc means data=mydef1.first;` produces the statistical analysis shown on page 2 of the output from the SAS Example B3 program displayed in Fig. 2.8. Again, `proc means` accesses the SAS data set `first` from the same library and computes the default statistics for all variables in the data set as shown on page 2. There is nothing to preclude the user from adding new SAS data sets to the same library, in the same program or in separate SAS programs. The data step shown in Fig. 2.7 reads instream data using list input as usual and creates the SAS data set named `second` and then saves it permanently in the library identified by the libref `mydef1`.

```
                         The SAS System                                   3

                       The CONTENTS Procedure

Data Set Name        MYDEF1.FIRST                Observations           5
Member Type          DATA                        Variables              5
Engine               V9                          Indexes                0
Created              Wed, Sep 05, 2007 11:41:54 AM   Observation Length     40
Last Modified        Wed, Sep 05, 2007 11:41:54 AM   Deleted Observations   0
Protection                                       Compressed            NO
Data Set Type                                    Sorted                NO
Label
Data Representation  WINDOWS_32
Encoding             wlatin1  Western (Windows)

                    Engine/Host Dependent Information

           Data Set Page Size            4096
           Number of Data Set Pages      1
           First Data Page               1
           Max Obs per Page              101
           Obs in First Data Page        5
           Number of Data Set Repairs    0
           File Name                     G:\stat479\Class\first.sas7bdat
           Release Created               9.0101M3
           Host Created                  XP_PRO

                Alphabetic List of Variables and Attributes

                    #     Variable    Type     Len

                    1     x1          Num      8
                    2     x2          Num      8
                    3     x3          Num      8
                    4     x4          Num      8
                    5     x5          Num      8
```

Fig. 2.9. SAS Example B3: Output page 3

The SAS procedure `contents` enables the user to examine the contents of SAS data sets stored in a library. The output from the first use of `proc contents` in the SAS Example B3 program is displayed in Fig. 2.9. This contains general technical information about the file stored, such as the length of an observation and the version of SAS used to create the data set as well as more specific information such as the number of variables and the number of observations. For a user accessing a SAS data set from another source, the more useful information appears at the bottom of the page. This is an alphabetic list of the variables and their attributes. Note that if the data analyst who originally created this data set has defined such information as formats and labels for the variables in the data step, they would appear here.

A more generally applicable SAS procedure named `datasets` is also available for examining the contents of an existing SAS library. An illustration of the use of this procedure is shown in Fig. 2.7. Here, the procedure statement option `library=` names the library to be examined and the `memtype=` option names the member type of interest. The `data=` option in the `contents` statement can be set to the name of a specific member to be examined or to the SAS keyword `_ALL_` to indicate that all data sets are to be examined. The keyword option `directory` requests that a list of the SAS data sets in the SAS library be printed and the `details` option requests that information contained in the data sets, such as the number of observations and number of variables, be included in the output. The printed output from the above `proc datasets` step is identical to the output from `proc contents` shown in Fig. 2.9.

2.1.4 User-defined informats and formats

Before discussing the use of the `format` procedure for creating user-defined informats and formats, a review of these two variable attributes is informative. Several simple informats such as `$10.` or `5.2` and more complex informats such as `dollar10.2` or `monyy5.` were used in several examples in Chapter 1. Informats determine how raw data values are read and converted to a number or a character string to be stored in memory locations. An informat contains information of the type of data (character or numeric) to be read, the length it occupies in the data field, how to handle leading, trailing, or embedded blanks and zeros, where to place the decimal point, and so forth. For example, the informat `ddmmyy8.0` converts the date value 19/10/07 entered in a data line into the binary number 17458 to be stored as a value of a SAS variable. Similarly, `formats` convert data values from internal form into a form the user wants them to appear in printed output. For example, the format `dollar15.2` prints the value of `cost=2317438.3921`, which is (say) the result of the product of the values of `quantity=2346.678` and `price=987.54`, as $2,317,438.39.

SAS system contains a large number of predefined informats and formats to handle many types of data conversion. However, it is not possible to provide informats or formats for every conceivable need; for example, an informat might be needed to convert the character strings 'YES' and 'NO' to be stored as the numeric values one or zero, respectively; or a numeric value stored internally as a one or a two to indicate gender will be best converted to be output as character strings 'Female' or 'Male', respectively. `Proc format` is a SAS procedure that allows the user to define informats or formats to do these kinds of specialized conversion. In this section, the emphasis will be on the use of `proc format` for creating output formats, although the importance of user-defined informats cannot be overstated. In practical terms, the primary use of user-constructed informats is for data validation and some examples of this type of application appear below. The general structure of a `proc format` step (with three procedure information statements to be illustrated below) is

```
PROC FORMAT <option(s)>;
  INVALUE <$>name <(informat-option(s))> value-range-set(s);
  PICTURE name <(format-option(s))> value-range-set-1
                                    <(picture-1-option(s))>
                <...value-range-set-n <(picture-n-option(s))> >;
  VALUE <$>name <(format-option(s))>value-range-set(s);
```

In the above description, the phrase *value-range-set* refers to an assignment type specification that defines a one-to-one or a many-to-one relationship between value or values to be converted to another value. The specification and action of a `value-range-set` depends on the context of its usage.

In the case of an `invalue` statement, *value-range-set* is of the form

```
value or range = informatted-value|[existing-informat]
```

where `value` is a value such as 1 or 'NB', `range` is a list of values usually specified in the form 100-999 or 'A'-'Z'. The words `low` and `high` may be used to define the end points of any range (numeric or character), implying that the specified range covers the entire range of values below the upper end point or above the lower end point, respectively. For example, the range 100-high covers every value greater than 100. The `informatted-value` (on the right-hand side of the equal sign) specifies the internal value that the raw data value that is equal to the value (or in the range of values) on the left is to be converted.

In the case of a `value` statement, the *value-range-set* is of the form

```
value or range = 'formatted-value'|[existing-format].
```

The definition of value and range is the same as for the `invalue` statement. The `'formatted-value'` specifies a character string to which the value (or the range of values) that appears on the left side of the equal sign is to be converted for printing. The `'formatted-value'` is a character string, regardless of whether the format created is a character or numeric format.

A discussion of several possible options on the `value` and `invalue` statements are omitted here but can be found under the description of `proc format`. For example, the option `fuzz=` allows the user to specify a fuzz factor for matching values to a range. If a value does not exactly match or falls in a range but comes within the fuzz factor, then the format or the informat will consider it to be a match or in the range. This facility is useful especially when the raw data contains fractions that need to be rounded up or down to be exactly in a prespecified range. For example, the value 99.9 may be considered in the range 100-200 if a fuzz factor of .1 has been specified (`fuzz=.1`) and values below 100 are not considered to be in the conversion range.

SAS Example B4

```
proc format;
    invalue $st 'IA'='Iowa'
                'NB'='Nebraska';
run;
data first ;
length state $ 12;
infile 'C:\Documents and Settings\...\wages.txt';
input (income tax age state)(@4 2*5.2 2. $st2.);
run;
proc print noobs;
format income tax dollar8.2  state $12. ;
var income tax age state;
title 'SAS Listing of Tax data';
run;
```

Fig. 2.10. SAS Example B4: Program

In the SAS Example B4 program displayed in Fig. 2.10, the two-character state codes used in the raw data set of SAS Example A1 (see Fig. 1.1) are converted to values that are longer character strings identifying the name of the respective state. Since there are several SAS functions (e.g., `stname1()`) available for state name conversions, this informat (named $st) is created only as an illustration. In the proc format step, an `invalue` statement is used to define the required conversion. Note that only values expected to be in the data for the character variable `state` are used in this definition. If other state values are expected, then they must be included in the format definition. Since character type variables are assigned lengths of 2 bytes by default, a `length` statement (that must appear before the `input` statement) specifies the length of the `state` variable to be 12 bytes. Thus, a format with a width of at least 12 positions is needed to print values of the `state` variable. As seen in the SAS program, the format $12. is associated with the `state` variable and the resulting output is shown in Fig. 2.11.

If the only state codes allowed in the data set are 'NB' and 'IA', the `invalue` statement may be modified to flag any other state code used as an error as follows:

```
invalue $st 'IA'='Iowa' 'NB'='Nebraska' other='Invalid St.';
```

In the above, the word `other` is a SAS keyword that will match any value that is not the strings 'NB' or 'IA'. Thus, if this informat is used to input values for a character variable with values other than 'NB' or 'IA', the respective observations will contain the string 'Invalid St.' as the value of that variable.

| SAS Listing of Tax data | | | 1 |
income	tax	age	state
$546.75	$34.65	35	Iowa
$765.48	$89.56	45	Iowa
$578.65	$59.54	31	Iowa
$786.78	$57.65	41	Nebraska
$567.51	$126.85	32	Iowa
$785.87	$67.85	28	Nebraska
$985.65	$75.65	43	Nebraska
$745.63	$78.95	25	Iowa
$345.67	$25.68	23	Nebraska
$567.34	$89.75	34	Nebraska
$651.12	$50.45	45	Iowa
$650.75	$65.45	29	Nebraska
$595.65	$45.68	34	Iowa
$678.56	$91.27	28	Nebraska
$685.96	$67.51	38	Iowa
$825.75	$56.25	27	Iowa

Fig. 2.11. SAS Example B4: Output

This is an example of the use of an informat for data validation, an important step in data analysis.

It is important to note that user-defined informats read only **character** values, although these can be converted to either character or numeric values. In the above example, if the state FIPS codes were input as numbers (19 for IA and 31 for NB), they must still be accessed as character data by the informat. So the appropriate invalue statement is

> invalue $st '19'='Iowa' '31'='Nebraska';

If the expansion of the state code was required only for the printed output, then it would have been sufficient to create a format (as opposed to an informat) for this purpose. The above program is modified as shown in Fig. 2.12 (the output from this SAS program is not shown).

```
proc format;
    value $st 'IA'='Iowa'
              'NB'='Nebraska';
run;
data first ;
infile 'C:\Documents and Settings\...\wages.txt';
input (income tax age state)(@4 2*5.2 2. $2.);
run;
proc print noobs;
format income tax dollar8.2  state $st10. ;
var income tax age state;
title 'SAS Listing of Tax data';
run;
```

Fig. 2.12. SAS Example B4: Modified program

A `value` statement is used to define a format for printing values of the variable `state`. Note that the new format $st is used in a `format` statement to specify the values of the variable `state`. Note carefully that the values to be stored in `state` were read using the informat $2. and hence will be one of the strings 'IA' or 'NB'. The conversion takes place when they are output using the format $st10., where these values will be printed as 'Iowa' and 'Nebraska', respectively, in using 10 print positions aligned to the left.

Note the difference between the `format` procedure and the `format` statement carefully. As in the above example, `proc format` is used to create user-defined formats or informats. The `format` (or the `informat`) statement associates an existing format (or informat) with one or more variables. Either standard SAS or user-defined formats or informats can be associated with variables this way. For example, the statement

```
format income tax dollar8.2  state $st10.;
```

associates the SAS format `dollar8.2` with the variables `income` and `tax`, whereas the user-defined format `$st10.` is associated with the variable `state`. `Proc format` stores user-defined informats and formats as entries in *SAS catalogs* (specially structured files), either temporarily in the WORK library or permanently in a user-specified library.

Finally, the following example illustrates how an existing SAS informat or a format can be used as an informatted or a formatted value in `value-range-set` definitions in invalue or value statements, respectively. Recall that the definitions of `value-range-sets` in these two statements were

```
value or range = informatted-value|[existing-informat]

value or range = 'formatted-value'|[existing-format].
```

Instead of an informatted or a formatted value, the user can specify an existing SAS informat or a format placed inside box brackets that will be used for the conversion of the value or the range on the left hand side of the `value-range-set` definition.

Example 2.1.2

```
proc format;
  invalue ff 0-high=[4.2]
              -1 = .;
run;

data ex212;
input a 2.0 b ff4.2 ;
datalines;
10 205
20 216
```

```
                    30 237
                    40 257
                    50  -1
                    60 469
                    ;
                    run;
```

The user-defined numeric informat (named **ff**) converts all positive data values using the SAS informat **4.2**. When the data value is -1, it is converted to the SAS missing value for a numeric variable i.e., a period. A scenario for the need to use such an informat may arise if the raw data set has been prepared where a -1 has been entered instead of using SAS missing values or spaces to indicate missing data values where the actual data values are all positive numbers. The output data set is

Obs	a	b
1	10	2.05
2	20	2.16
3	30	2.37
4	40	2.57
5	50	.
6	60	4.69

2.1.5 Creating SAS data sets in procedure steps

Several SAS procedures used for statistical analysis have the capability to let the user specify which statistics, calculated by the procedure, are to be saved in newly created SAS data sets. In some procedures, these data sets are organized in special structures that allow them to be read by another SAS procedures for further analysis by specifying the **type=** attribute of the data set. For example, **proc corr** creates a data set with the attribute **type=corr** containing a correlation matrix, that can be directly input to a procedure such as **proc reg** as an input data set. If the required analysis performed by **proc reg** is solely based on the correlation matrix then much of the overhead spent on recomputing the correlation matrix can be avoided.

In this subsection, the discussion is limited to a description of the use of the **output** statement in several SAS procedures that compute an extensive number of statistics for variable values. In most of these procedure steps, a **class** statement specifies classification variables in the data set that are discrete-valued variables that identify groups, classes, or categories of observations in the data set. They may be numeric or character-valued and may be observed ordinal- or nominal-valued variables or user-constructed variables. In practice, continuous-valued variables may be used to define new grouping variables that can then be used in **class** statements. An example would be the creation of a variable defining income groups with values, say, 1, 2, and 3, or 'Low', 'Medium', and 'High', using the values of the continuous-valued variable `income` to form the groups. A `var` statement (i.e., the variables statement) identifies the analysis variables (that must all be of numeric type). The statistics are computed on the values of analysis variables for subsets of observations defined by the classification variables.

In the SAS Example B5 program, `proc means` is used to introduce the basic use of the output statement. A simplified general form of the output statement used in this example is

`OUTPUT <OUT=SAS-data-set> <output-statistic-specification(s)>;`

where an *output-statistic-specification* is of the general form

`statistic-keyword<(variable-list)>=<name(s)>`

where *statistic-keyword* specifies the statistic to be calculated and is stored as a value of a variable in the output data set. Some of available statistic keywords are `n`, `mean`, `median`, `var`, `cv`, `std`, `stderr`, `max`, `min`, `range`, `cv`, `skewness`, `kurtosis`, `q1`, `q3`, `qrange`, `p1`, `p5`, `p10`, `p90`, `p95`, `p99`, `t`, and `probt`. The optional *variable-list* specifies the names of one or more analysis variables on whose values the specified statistic is to be computed. If this list is omitted, the specified statistic is computed for all the analysis variables. The optional *name(s)* specifies one or more names for the variables in the output data set that will contain the analysis variable statistics in the same sequence that the analysis variables are listed in the `var` statement. The first name contains the statistic for the first analysis variable; the second name contains the statistic for the second analysis variable; and so on. If the names are omitted, the analysis variable names are used to name the variables in the output data set.

SAS Example B5

The SAS Example B5 program (see Fig. 2.13) illustrates the use of `proc means` to calculate and print statistics for an input data set named `biology` and, in addition, save the statistics in a new SAS data set created in the proc step. The simple data set to be analyzed includes a numeric variable `year` (indicating class in college) and a character variable `sex` that will be used as classification variables and two numeric analysis variables `height` and `weight`. Ordinarily, `proc means` produces printed output of five default statistics (n (sample size), mean, standard deviation, minimum, and maximum) calculated for every variable in the `var` statement list, for subsets of observations formed by all combinations of the levels of the `class` variables. The option `maxdec=3` used on the proc statement limits the number of decimal places output when printing all calculated statistics. Page 1 of the SAS output (see Fig. 2.14) displays the printed output in its standard format. As described above, the five default statistics are computed for the variables `height` and `weight` for

```
data biology;
input id sex $ age year height weight;
datalines;
7389   M   24   4   69.2   132.5
3945   F   19   2   58.5   112.0
4721   F   20   2   65.3    98.6
1835   F   24   4   62.8   102.5
9541   M   21   3   72.5   152.3
2957   M   22   3   67.3   145.8
2158   F   21   2   59.8   104.5
4296   F   25   3   62.5   132.5
4824   M   23   4   74.5   184.4
5736   M   22   3   69.1   149.5
8765   F   19   1   67.3   130.5
5734   F   18   1   64.3   110.2
4529   F   19   2   68.3   127.4
8341   F   20   3   66.5   132.6
4672   M   21   3   72.2   150.7
4823   M   22   4   68.8   128.5
5639   M   21   3   67.6   133.6
6547   M   24   2   69.5   155.4
8472   M   21   2   76.5   205.1
6327   M   20   1   70.2   135.4
8472   F   20   4   66.8   142.6
4875   M   20   1   74.2   160.4
;
run;

proc means data=biology fw=8 maxdec=3;
    class year sex;
    var height weight;
    output out=stats mean=av_ht av_wt stderr=se_ht se_wt;
run;

proc print data=stats;
    title 'Biology Class Data Set: Output Statement';
run;
```

Fig. 2.13. SAS Example B5: Program

groups observations defined by the levels 'F' and 'M', respectively, of the sex variable within each value 1, 2, 3, or 4, of the year variable, respectively.

The *output-statistic-specifications* used in the output statement has the basic form of statistic=names. They are mean=av_ht av_wt and stderr=se_ht se_wt. Since the var statement used in the proc step is var height weight;, the above specifications request that the means and standard errors of the variables height and weight are to be computed and stored in the new variables av_ht, av_wt, se_ht, and se_wt, respectively. Page 2 of the SAS output (see Fig. 2.15) displays a listing of the SAS data set named stats produced in the proc means step. It can be observed that there are 15 observations displaying values for the variables year, sex, _TYPE_, _FREQ_, av_ht, av_wt, se_ht, and se_wt. The value of the _TYPE_ variable (0, 1, 2, or 3) indicates which combinations of the class variables are used to define the subgroups of observations used for computing the statistics. For example, there is exactly one observation (Observation 1) with the value _TYPE_=0. In

```
                  Biology Class Data Set: Output Statement                    1

                            The MEANS Procedure

                    N
year   sex         Obs   Variable   N      Mean    Std Dev    Minimum    Maximum

 1     F            2    height     2     65.800     2.121     64.300     67.300
                         weight     2    120.350    14.354    110.200    130.500

       M            2    height     2     72.200     2.828     70.200     74.200
                         weight     2    147.900    17.678    135.400    160.400

 2     F            4    height     4     62.975     4.614     58.500     68.300
                         weight     4    110.625    12.455     98.600    127.400

       M            2    height     2     73.000     4.950     69.500     76.500
                         weight     2    180.250    35.143    155.400    205.100

 3     F            2    height     2     64.500     2.828     62.500     66.500
                         weight     2    132.550     0.071    132.500    132.600

       M            5    height     5     69.740     2.481     67.300     72.500
                         weight     5    146.380     7.535    133.600    152.300

 4     F            2    height     2     64.800     2.828     62.800     66.800
                         weight     2    122.550    28.355    102.500    142.600

       M            3    height     3     70.833     3.182     68.800     74.500
                         weight     3    148.467    31.183    128.500    184.400
```

Fig. 2.14. SAS Example B5: Output page 1

this observation, the variables `year` and `sex` are set to respective missing values, indicating that both of these variables are ignored in determining the sample used to compute the statistics shown for this observation; that is, the subgroup for their computation is the entire data set as evidenced by the value of _FREQ_=22.

Similarly, for _TYPE_=1, there are two subgroups formed for each of the values of the `sex` variable ignoring the `year` variable. Note carefully that `sex` variable appears rightmost in the variable list in the `class` statement; hence, its levels form _TYPE_=1 groups. Observations 2 and 3 list statistics computed based on these groups of observations and note sample sizes given by _FREQ_=10 and _FREQ_=12, respectively.

There are four observations with _TYPE_=2, and these statistics are based on the groups of observations that correspond to each level of `year` ignoring the levels of `sex`. Observations with _TYPE_=3 correspond to subgroups defined by all combinations of levels of `year` and the levels of `sex`. Thus, the complete set is formed by $1+2+4+8=15$; thus, 15 observations are included in `stats`.

Other forms of the *output-statistic-specifications* used in the `output` statement can be used to alter the appearance of the SAS data set created. Several procedure information statements and proc statement options are available

		Biology Class Data Set: Output Statement						
Obs	year	sex	_TYPE_	_FREQ_	av_ht	av_wt	se_ht	se_wt
1	.		0	22	67.8955	137.591	0.97773	5.4643
2	.	F	1	10	64.2100	119.340	1.03489	4.8887
3	.	M	1	12	70.9667	152.800	0.85390	6.4766
4	1		2	4	69.0000	134.125	2.11069	10.3181
5	2		2	6	66.3167	133.833	2.72255	16.4964
6	3		2	7	68.2429	142.429	1.30783	3.4515
7	4		2	5	68.4200	138.100	1.89642	13.3320
8	1	F	3	2	65.8000	120.350	1.50000	10.1500
9	1	M	3	2	72.2000	147.900	2.00000	12.5000
10	2	F	3	4	62.9750	110.625	2.30701	6.2277
11	2	M	3	2	73.0000	180.250	3.50000	24.8500
12	3	F	3	2	64.5000	132.550	2.00000	0.0500
13	3	M	3	5	69.7400	146.380	1.10932	3.3698
14	4	F	3	2	64.8000	122.550	2.00000	20.0500
15	4	M	3	3	70.8333	148.467	1.83697	18.0037

Fig. 2.15. SAS Example B5: Output page 2

for controlling the contents of this data set. The following examples of the ways and types statements illustrate some of these choices. These two statements allow the user to select the set of observations to be included in the output data set as defined by the _TYPE_ variable discussed earlier. The ways statement uses integers to indicate number of class variables to form the combinations; for example, one may specify 2 to request that subgroups are to be formed by combining all possible pairs of class variables in the class variable list. The types statement, on the other hand, allows the user to specify class variables and how they are to be combined directly.

Including the statement ways 1; in the proc step in the above example produces the output data set shown in Fig. 2.16. This requests that subgroups are to be defined by the levels of the class variables taken one at a time. Here, two sets of statistics are produced for the two levels of sex and four sets for the four levels of year. The printed output (not shown) is similarly structured; two tables of statistics are produced for each class variable separately.

		Biology Class Data Set: Output Statement						
Obs	year	sex	_TYPE_	_FREQ_	av_ht	av_wt	se_ht	se_wt
1	.	F	1	10	64.2100	119.340	1.03489	4.8887
2	.	M	1	12	70.9667	152.800	0.85390	6.4766
3	1		2	4	69.0000	134.125	2.11069	10.3181
4	2		2	6	66.3167	133.833	2.72255	16.4964
5	3		2	7	68.2429	142.429	1.30783	3.4515
6	4		2	5	68.4200	138.100	1.89642	13.3320

Fig. 2.16. SAS Example B5: Result using the WAYS statement

```
                    Biology Class Data Set: Output Statement                    2

   Obs    year    sex    _TYPE_    _FREQ_    av_ht     av_wt    se_ht    se_wt

    1      1      F        3         2       65.8000   120.350  1.50000  10.1500
    2      1      M        3         2       72.2000   147.900  2.00000  12.5000
    3      2      F        3         4       62.9750   110.625  2.30701   6.2277
    4      2      M        3         2       73.0000   180.250  3.50000  24.8500
    5      3      F        3         2       64.5000   132.550  2.00000   0.0500
    6      3      M        3         5       69.7400   146.380  1.10932   3.3698
    7      4      F        3         2       64.8000   122.550  2.00000  20.0500
    8      4      M        3         3       70.8333   148.467  1.83697  18.0037
```

Fig. 2.17. SAS Example B5: Result from using TYPES statement

Other possibilities in this example are ways 0; when no subgroups are formed, meaning statistics are computed for the entire data set, and ways 2; when subgroups are formed for all eight combinations of the two class variables.

Including the statement types year sex; in the proc step produces the same data set shown in Fig. 2.16 and the corresponding printed output (not shown). If, instead, types year; is used, then only those statistics for the subgroups defined by the four levels of year will be calculated. The statement types year*sex; produces statistics for subgroups formed for the eight combinations of the two class variables year and sex as shown in Fig. 2.17.

2.2 SAS Procedures for Computing Statistics

The data set shown in Table B.1 of Appendix B appeared in Weisberg (1985) and was extracted from the *American Almanac* and the *World Almanac* for 1974. It lists the values of fuel consumption for each of the 48 contiguous states, in addition to several other measured variables. This data set is used to illustrate several SAS procedures that are classified as Base SAS procedures. As a prelude to the use of several SAS procedures for analysis, a SAS data set that contains user-generated labels, formats, grouping variables (ordinal or nominal variables with values that identify groups of observations belonging to different classes or strata), etc. is created and stored in a library. This data set is then accessed repeatedly in several SAS proc steps.

SAS Example B6

The SAS data set fueldat is created in the SAS Example B6 program shown in Fig. 2.18. The following actions are taken in the data step of this program. The data are input from a text file using an infile statement. The SAS data set is saved in a folder using the two-level name mylib.fueldat to be accessed in other SAS programs later. Mnemonic variable names are used, but label statements are included to provide more descriptive labeling as necessary. In the same data step, five new variables are created as follows:

```
libname mylib 'C:\Documents and Settings\...\stat479\';
data mylib.fueldat;
filename fueldd 'C:\Documents and Settings\...\fuel.txt';
infile fueldd;
input (st pop tax numlic income roads fuelc)
      ( $2.  5. 2.1    5.   4.3  5.3   5.);

label pop='Population(in thousands)'
      tax='Motor Fuel Tax Rate(in cents/gallon)'
      numlic='No. of Licensed Drivers'
      income='Per Capita Income(in thsnds.)'
      roads='Miles of Primary Highways(in thsnds.)' ;

percent=100*numlic/pop;

fuel=1000*fuelc/pop;

if income=<3.8 then incomgrp=1;
else if 3.8<income=<4.4 then incomgrp=2;
else incomgrp=3;

if tax<8.0 then taxgrp='Low ';
else taxgrp='High';

label percent='% of Population with Driving Licenses'
      fuel='Fuel Consumption (in gallons/person)'
      incomgrp='Per capita Income'
      taxgrp='Fuel Tax'
      state='State' ;

format percent 4.1 fuel 7. ;

state=stnamel(st);

drop fuelc st;
run;

proc print label;
title 'Complete Data Set' ;
run;
```

Fig. 2.18. SAS Example B6: Program

a. A numeric variable `percent` that will contain the percent of population with driving licenses in each state
b. A numeric variable `fuel` that measures the per capita motor fuel consumption in gallons in each state
c. An ordinal variable called `incomgrp` assigned the value 1, 2, or 3 according to whether the per capita income (in thousands of dollars) is less than or equal to 3.8 , greater than 3.8 and less than or equal to 4.4, or over 4.4, respectively
d. A nominal variable called `taxgrp` with a value of 'Low ' when the fuel tax is less than 8 cents and a value of 'High' otherwise
e. A character variable named `state` containing the state name in uppercase and lowercase, for example, Kansas

A `format` statement ensures that values of the variable created in (a) are printed rounded to one decimal place and those of the variable created in (b) are printed as whole numbers (i.e., appropriate print formats are associated with `percent` and `fuel` variables). A `drop` statement is used to exclude variables `fuelc` and `st` from the data set created. Printed output from the program, a listing of the data set, is not reproduced here.

2.2.1 The UNIVARIATE procedure

Although there are several SAS procedures that produce descriptive statistics, `proc univariate` is best suited for studying the empirical distributions of variables in a data set. It produces a variety of descriptive statistics such as moments and percentiles and optionally creates output SAS data sets containing selected sample statistics. In addition, `proc univariate` can be used to produce high-resolution graphics such as histograms with overlayed kernel density estimates, quantile-quantile plots, and probability plots supplemented with goodness-of-fit statistics for a variety of distributions. A discussion of the statements that produce high-resolution graphics is deferred until Chapter 3. In this subsection, a brief discussion of several statements available for calculating sample statistics and saving those in a SAS data set are presented. This is followed by an illustrative example. The general structure of a `proc univariate` step (that includes five of the procedure information statements to be illustrated) is

```
PROC UNIVARIATE < options > ;
  BY variables ;
  CLASS variable-1 <(v-options)> < variable-2 <(v-options)> >
            ...< / KEYLEVEL= value1 | ( value1 value2 ) >;
  VAR variables ;
  ID variables ;
  OUTPUT < OUT=SAS-data-set >
      < keyword1=names...keywordk=names > < percentile-options >;
```

A large number of `proc` statement options are available for `proc univariate`. Although some of these are standard options such as the `data=` option for naming the data set to be analyzed or the `noprint` for suppressing printed output, others are more specialized. Some of these special `proc` statement options are summarized below.

Some PROC Statement Options

$\boxed{\text{alpha=}}$ option specifies an α for calculating $(1-\alpha)100\%$ confidence intervals, the default being .05

$\boxed{\text{cibasic} < (< \text{type=} ><\text{alpha=} >) >}$ option calculates $(1-\alpha)100\%$ confidence intervals for the mean, standard deviation, and variance assuming that the data are normally distributed. Optionally, type may be set equal to one of the keywords lower, upper, or two sided. The defaults are type=twosided and alpha value set in the above alpha= proc option or the default value of .05.

boxed{mu0=} option is used to list value(s) (μ_0) stipulated in the hypotheses for tests concerning population means corresponding to the variables listed in the var statement. The tests performed are the Student's t-test, the sign test, and the Wilcoxon signed rank test.

boxed{normal} option requests tests for normality. Computed test statistics and p-values for the Shapiro-Wilk test (for sample sizes less than or equal to 2000), the Kolmogorov-Smirnov test, the Anderson-Darling test, and the Cramér-von Mises test are output.

boxed{pctldef=} gives the user the option of selecting one of five methods (labeled 1, 2, 3, 4, or 5) that proc univariate uses for calculating sample percentiles. These methods depend on the sample size n and the percentile p and are described in the documentation. The default method is 5.

boxed{plots} option requests that low-resolution descriptive graphics (a histogram or a stem-and-leaf plot, a box plot, and a normal probability plot) are to be produced.

boxed{trim=values < (<type= ><alpha= >) >} option requests the computation of trimmed means for each variable in the var list, where the *values* list is the numbers k or the fractions p of observations to be *trimmed* from both ends of the observations ordered smallest to largest. If p are specified, then the numbers trimmed equal np rounded up to the nearest integer, respectively. Confidence intervals for the population means are also calculated based on the trimmed means and estimates of their standard errors; the options type= and alpha= may be used to change their default settings, as described for the cibasic proc option earlier.

boxed{vardef=} specifies the divisor to be used in the calculation of variance and standard deviation. The default value for the divisor is df when the degrees of freedom $n-1$ will be used. Other possible values that may be specified are n, wdf, and weight or wgt, respectively, when the sample size n, the weighted degrees of freedom $\sum w_i - 1$, or the sum of weights $\sum w_i$ (where w_i are the weights specified in weight statement) will be used.

Some CLASS Statement Options

The variables list in the class statement specifies groups into which the observations in a data set are classified into for the purpose of calculating statistics.

The values of these variables can be numeric or character and are called *levels*. For the purpose of displaying output from such an analysis (e.g., tables), procedures such as `univariate` must be provided with a way to determine in what order the statistics calculated for each level of a class variable are to be displayed. In many procedures, the `order=` option is available as a `proc` statement option to be used for this purpose. In `proc univariate`, this option is available as one of the *v-options* in the `class` statement, so each level of each class variable may be separately ordered.

The `class` statement allows the *v-options* `missing` and `order=` to be specified, enclosed in parentheses, for each of the variables in the *class variable list*. For example, using the `order=` option for each variable allows the user to specify the display order of the levels of each of the class variables separately. The default setting for the `order=` option is `internal`, which specifies that the internal unformatted (character or numeric) value of a variable be used for this purpose. The other available choices are `data`, in which case the levels will be displayed in the order they appeared in the input data, `formatted`, which requests that the levels be ordered by their formatted values, and `freq`, which requests that levels be listed in the decreasing order of frequency of observations for each level.

```
libname mylib 'C:\Documents and Settings\...\stat479\';

proc univariate data=mylib.fueldat plots normal;
   var pop income;
   id state;
   title 'Use of Proc Univariate to Examine Distributions:1';
run;

proc univariate data=mylib.fueldat cibasic mu0=4 500 trim=2;
   var income fuel;
   id state;
   title 'Use of Proc Univariate to Compute Statistics:2';
run;

proc univariate data=mylib.fueldat noprint;
   var fuel percent;
   output out=stats pctlpts=33.3 66.7 pctlpre=fuel lic;
   title 'Calculation of User Specified Percentile Points';
run;

proc print data=stats;
run;
```

Fig. 2.19. SAS Example B7: Program

SAS Example B7

Several variables in the SAS data set created in SAS Example B6 on fuel consumption data are analyzed using `proc univariate` in the SAS Example B7 program (see Fig. 2.19) to illustrate the use of the procedure options and statements discussed in this section. The previously saved SAS data set named

```
          Use of Proc Univariate to Examine Distributions:1                    1

                          The UNIVARIATE Procedure
                 Variable:  pop  (Population(in thousands))

                                    Moments

   N                           48    Sum Weights                   48
   Mean                  4296.91667  Sum Observations          206252
   Std Deviation          4441.1087  Variance               19723446.5
   Skewness              2.00893087  Kurtosis               4.32297497
   Uncorrected SS        1813249642  Corrected SS            927001986
   Coeff Variation        103.355709 Std Error Mean          641.018826

                          Basic Statistical Measures

              Location                        Variability

          Mean      4296.917    Std Deviation              4441
          Median    2982.500    Variance               19723447
          Mode             .    Range                     20123
                               Interquartile Range         3894

                             Tests for Normality

          Test                  --Statistic---    -----p Value------

          Shapiro-Wilk          W     0.771583    Pr < W     <0.0001
          Kolmogorov-Smirnov    D     0.208119    Pr > D     <0.0100
          Cramer-von Mises      W-Sq  0.591415    Pr > W-Sq  <0.0050
          Anderson-Darling      A-Sq  3.36823     Pr > A-Sq  <0.0050
```

Fig. 2.20. SAS Example B7: Summary statistics and tests for Normality

fueldat is accessed using the two-level name mylib.fueldat. In the first proc step in this program, the options plots and normal produce line-printer style

```
          Use of Proc Univariate to Examine Distributions:1                    2

                          The UNIVARIATE Procedure
                 Variable:  pop  (Population(in thousands))

                           Quantiles (Definition 5)

                           Quantile      Estimate

                           100% Max       20468.0
                           99%            20468.0
                           95%            11926.0
                           90%            11251.0
                           75% Q3          4989.0
                           50% Median      2982.5
                           25% Q1          1095.5
                           10%             579.0
                           5%              527.0
                           1%              345.0
                           0% Min          345.0
```

Fig. 2.21. SAS Example B7: Quantiles

```
                        Extreme Observations

-----------------Lowest----------------    ----------------Highest---------------

Value   state                    Obs       Value   state                    Obs

 345    Wyoming                   40       11251   Illinois                  12
 462    Vermont                    3       11649   Texas                     37
 527    Nevada                    45       11926   Pennsylvania               9
 565    Delaware                  22       18366   New York                   7
 579    South Dakota              19       20468   California                48
```

Fig. 2.22. SAS Example B7: Extreme values

low-resolution stem-and-leaf plots, box plots, and normal probability plots of the population and income variables, named (`pop` and `income`), descriptive statistics including percentiles and extreme values, and several test statistics for testing normality. Those output for the population variable are reproduced in Figs. 2.20-2.24.

```
            Use of Proc Univariate to Examine Distributions:1           3

                         The UNIVARIATE Procedure
                Variable:  pop  (Population(in thousands))

            Stem Leaf                        #              Boxplot
             20 5                            1                 *
             19
             18 4                            1                 *
             17
             16
             15
             14
             13
             12
             11 369                          3                 0
             10 8                            1                 |
              9 1                            1                 |
              8                                                |
              7 34                           2                 |
              6                                                |
              5 238                          3              +-----+
              4 015788                       6              |  +  |
              3 134579                       6              *-----*
              2 02334679                     8              |     |
              1 0011589                      7              +-----+
              0 355666788                    9                 |
                ----+----+----+----+
            Multiply Stem.Leaf by 10**+3
```

Fig. 2.23. SAS Example B7: Stem-and-leaf and box plots

Note that the percentiles shown in Fig. 2.21 are calculated using `definition 5`. The lowest and highest five extreme values, also output in page 2,

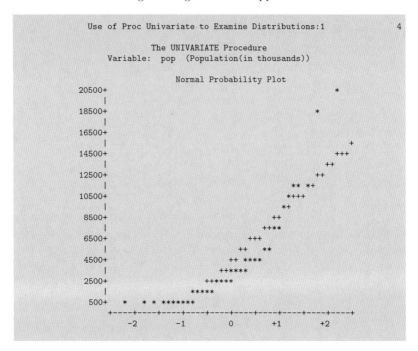

Fig. 2.24. SAS Example B7: Normal probability plot

are shown in Fig. 2.22. The `id state;` statement results in these values being identified by the corresponding state name. The stem-and-leaf plot and the box plot (see Fig. 2.23) show a highly positively skewed population distribution with two extreme values, which were identified as those that correspond to the states of New York and California in Fig. 2.22.

The points (plotted as asterisks) in the normal probability plot (see Fig. 2.24) show a clear bowl-shaped pattern, indicating a right-skewed distribution for the data. The interpretation of normal probability plots is discussed in more detail in Section 3.5 of Appendix A. Using the Shapiro-Wilk statistic, the null hypothesis of normality is rejected, as the p-value is $< .0001$ (see Fig. 2.20), thus substantiating the evidence from the graphical displays. Note, too, that the **skewness** statistic (which is expected to be near zero for symmetrical distributions) is quite large here.

The second proc step analyzes the distribution of the variables `income` and `fuel` and includes the procedure options `cibasic`, `mu0=4 500`, and `trim=2`. By default, the `cibasic` option produces 95% confidence intervals for the population mean μ, the population standard deviation σ, and the population variance σ^2 for each of the variables calculated under the normality assumption for the data. These are shown in Fig. 2.25 for the `income` variable (recall that the data values are per capita income figures in thousands of dollars).

```
                      Basic Statistical Measures                          9

              Location                        Variability

       Mean      4.241833      Std Deviation           0.57362
       Median    4.298000      Variance                0.32904
       Mode      5.126000      Range                   2.27900
                               Interquartile Range     0.85050

              Basic Confidence Limits Assuming Normality

       Parameter          Estimate        95% Confidence Limits

       Mean                4.24183        4.07527      4.40840
       Std Deviation       0.57362        0.47752      0.71851
       Variance            0.32904        0.22803      0.51626

                      Tests for Location: Mu0=4

       Test              -Statistic-      -----p Value------

       Student's t     t   2.920853      Pr > |t|      0.0053
       Sign            M          7      Pr >= |M|     0.0595
       Signed Rank     S      248.5      Pr >= |S|     0.0093
```

Fig. 2.25. SAS Example B7: Confidence intervals and tests for the `income` variable

The confidence interval reported for μ is (4.07527, 4.40840) and the p-value for the two-sided t-test of $H_0 : \mu = 4$ vs. $H_a : \mu \neq 4$ is calculated to be .0053. The p-value for the corresponding one-sided test is, of course, $.0053/2 = .0026$. The tests and confidence intervals are not reproduced here for the `fuel` variable, but they are to be similarly interpreted.

```
              Use of Proc Univariate to Compute Statistics:2              10

                      The UNIVARIATE Procedure
              Variable:  income   (Per Capita Income(in thsnds.))

                            Trimmed Means

  Percent    Number                  Std Error
  Trimmed    Trimmed     Trimmed     Trimmed      95% Confidence
  in Tail    in Tail     Mean        Mean         Limits              DF

   4.17         2       4.239795     0.087055   4.064233  4.415358    43

                            Trimmed Means

                    Percent
                    Trimmed     t for H0:
                    in Tail     Mu0=4.00    Pr > |t|

                     4.17       2.754533     0.0086
```

Fig. 2.26. SAS Example B7: Trimmed means for the `income` variable

```
         Calculation of User Specified Percentile Points              15

   Obs      fuel33_3      fuel66_7      lic33_3      lic66_7

    1        524.994       609.991      54.4421      58.0087
```

Fig. 2.27. SAS Example B7: Calculating user-specified percentiles

The estimate, confidence interval, and associated t-test for the mean μ under trimming for the income variable appear in Fig. 2.26. The option `trim=2` requested that the `trimmed mean` be computed after the *2 smallest* and the *2 largest* observations are deleted from the sample, which is equivalent to approximately 4% trimming from the tails of the distribution of the income variable. Associated confidence intervals and a t-test for the population mean μ is computed based on the standard error of the trimmed mean. For a symmetric distribution, the trimmed mean is an unbiased estimate of the population mean. The results under trimming here indicate that the estimates and test statistics are not significantly different from those statistics calculated from the complete sample (see Figs. 2.25 and 2.26). This is also an indicator of the symmetry of the population distribution of the income variable.

The final proc step requests the calculation of the 33.3 and 66.7 percentiles of the `fuel` (Fuel Consumption (in gallons/person)) and `percent` (% of Population with Driving Licenses) variables. In this example, the printed output is suppressed (as a result of the `noprint` proc option) and the output is written to a new SAS data set named `stats`. The percentiles that cannot be directly requested via the usage of a keyword such as `p1`, `p10`, or `p90` are calculated by the use of the pair of keywords `pctlpts=` and `pctlpre=` used concurrently. For example, the use of `pctlpts=33.3 66.7 pctlpre=fuel lic` generates the 33.3 and 66.7 percentiles for the two analysis variables (variables in the var statement) and adds them to the new data set as values of new variables named `fuel33_3`, `fuel66_7`, `lic33_3`, and `lic66_7`, respectively. The `proc print data=stats;` statement produces the output of these values shown in Fig. 2.27.

2.2.2 The FREQ procedure

The `freq` procedure in SAS computes many statistics and measures related to the analysis of categorical data. The discussion in this subsection is primarily intended to illustrate the use of statements and options to generate these statistics rather than a presentation of statistical methodology involved in the analysis of categorical data. It is recommended that the prospective user of `proc freq` consult references cited to learn more about techniques available for hypothesis testing and measuring association among categorical variables. Moreover, the type of inference depends on many factors such as sampling strategy; thus, a knowledge of how the data are collected is also necessary for making relevant conclusions.

A chi-square goodness-of-fit test can be used to test several types of hypothesis using frequency counts. For example, using a one-way frequency table with k classes, one could compute a chi-square statistic to test whether the counts conform to sampling from a multinomial population with specified probabilities. In this case, the null hypothesis of interest is

$$H_0 : p_i = p_{i0}, \quad i = 1, 2, \ldots, k$$

where the p_{i0}'s are postulated values of the multinomial probabilities. The Pearson chi-square statistic is given by

$$\chi^2 = \sum_{i=1}^{k} \frac{(f_i - e_i)^2}{e_i}$$

where f_i is the observed frequency count in class i and e_i is the expected frequency calculated under the null hypothesis (i.e., $e_i = p_{i0} N$ where N is the total number of responses). Another application of a chi-square test is for testing *homogeneity* of several multinomial populations. In this case, random samples are taken from each population and then classified by a categorical variable. The populations are usually defined by levels of variables such as gender, age group, state, etc., and the levels of the categorical variable form the k categories of the multinomial populations.

For example, suppose samples are drawn from two populations (say, males and females or persons below and above the age of 40) and they are grouped into three categories (say, according to three levels of support for a certain local bond issue). Suppose that the multinomial probabilities for each population are as given in the following table:

	Groups		
Populations	p_{11}	p_{12}	p_{13}
	p_{21}	p_{22}	p_{23}

Then the null hypothesis of homogeneity of populations (i.e., whether random samples were drawn from the same multinomial population) is given by

$$H_0 : p_{11} = p_{21}, \ p_{12} = p_{22}, \ p_{13} = p_{23}$$

Note carefully that the sampling procedure here is different from the process used in the construction of a *contingency table*. In the above situation, random samples are drawn from two different populations and then each sample is classified into three different groups. Contingency tables are constructed by multiple classification of a single random sample. Observations in a sample may be cross-classified by variables with ordinal or nominal data values defining *categorical variables*. These variables may be covariates already present in the data set (e.g., gender, marital status, or region), thus forming natural subsets or strata of the data or may be generated from other quantitative variables such as `population` or `income`. For example, observations in a sample may be categorized into three income groups (say "low," "middle," or "high") by creating a new variable, say `incgrp`, and assigning the above character strings as its values according to whether the value of the income variable is below $30,000, between $30,000 and $70,000, or above $70,000, respectively.

A chi-square statistic can be computed for a two-way $r \times c$ contingency table to test whether the two categorical variables are independent; that is, the null hypothesis tested is of the form

$$H_0: p_{ij} = p_{i.}p_{.j}, \quad i = 1, 2, \ldots, r, \ j = 1, 2, \ldots, c$$

where the p_{ij}, are the probabilities considering that the entire sample is from a multinomial population, and $p_{i.}$ and $p_{.j}$, called the *marginal* probabilities, are probabilities for multinomial populations defined by each categorical variable. The chi-square statistic for the test of this hypothesis is given by

$$\chi^2 = \sum_i \sum_j \frac{(f_{ij} - e_{ij})^2}{e_{ij}}$$

where f_{ij} is the observed frequency in the ijth cell and $e_{ij} = f_{i.}f_{.j}/N$, where $f_{i.}$ and $f_{.j}$ are the observed row and column marginal frequencies, respectively.

When the row and column variables are independent, the above statistic has an asymptotic chi-square distribution with $(r-1)(c-1)$ degrees of freedom. Instead of χ^2, the likelihood ratio chi-square statistic, usually denoted by G^2, that has the same asymptotic null distribution may be computed. If the row and columns are ordinal variables, the Mantel-Haenszel chi-square statistic tests the alternative hypothesis that there is a linear association between them. Fisher's exact test is another test of association between the row and column variables that does not depend on asymptotic theory. It is thus suitable for small sample sizes and for sparse tables. One may compute measures of association between variables that may or may not depend on the chi-square test of independence. Some of these are illustrated in SAS Example B8.

One would use `proc freq` to analyze count data using one-way frequency tables or two-way or higher-order contingency tables. In addition to chi-square statistics for testing whether two categorical variables are independent, for a two-way contingency table, `proc freq` also computes measures of association that estimate the strength of association between the pair of variables. In this subsection, a brief discussion of several statements available in `proc freq` followed by an illustrative example are presented. The general structure of a `proc freq` step is

```
PROC FREQ < options > ;
  BY variables ;
  TABLES requests < / options > ;
```

```
EXACT statistic-options < / computation-options > ;
TEST options ;
OUTPUT < OUT=SAS-data-set > options ;
```

The primary statement in `proc freq` is the `tables` statement for requesting tables having different structures with options for selecting statistics to be included in those tables. Since the computation of frequencies requires that the variables used in the `tables` statement are necessarily discrete-valued (containing either nominal or ordinal data) such as category, classification, or grouping type variables, statements such as `var` or `class` are not available in `proc freq`.

The syntax of the tables statement allow the user to request one-way tables just by listing the variables in the tables statement, and two-way tables by two variable names combined with an asterisk between them. Thus, the statement `tables region;` produces one-way tables with frequency counts of observations for each *level* of `region` and the statement `tables taxgrp*region;` produces a two-way cross-tabulation with the *levels* of the variable `taxgrp` as the rows of the table and the *levels* of `region` as the columns. A combination of a level of `taxgrp` and a level of `region` forms a *cell* in the table. In this case, frequencies of observations are tallied for every possible combination of the two variables and entered in the respective cells; they are called *cell frequencies*. Multiway combinations of variables such as p*q*r*s are possible in which case two-way tables are produced for every combination of levels of each of the variables p and q. The cells in each two-way table are formed by combinations of a level of variables r and s. The proc statement option `page` may be used to force these tables to be output on different pages. The `tables` statement syntax also allows variations such as `tables q*(r s)`, which is equivalent to the specification `tables q*r q*s`, or `tables (p q)*(r s)`, which is equivalent to `tables p*r q*r p*s q*s`.

By default, one-way frequency tables contain the statistics frequency, cumulative frequency, percentage frequency, and cumulative percentage computed for each level of the variable and a two-way or multi-way tables may include cell frequency, cell percentage of the total frequency, cell percentage of row frequency, and cell percentage of column frequency computed for each cell. Many options are available with the `tables` statement to control the statistics that are calculated and output by `proc freq`. While some of these are simple options for suppressing the statistics computed by default, others request additional statistics such as goodness-of-fit statistics and measures of association to be computed. An abbreviated description of these options is provided below.

Some TABLES Statement Options

nocol   suppresses printing the column percentage for each cell.

`nocum` suppresses printing the cumulative frequencies and cumulative percentages in one-way frequency tables and in list format.

`norow` suppresses printing the row percentage for each cell.

`nopercent` suppresses printing the percentage, row percentage, and column percentage in two-way tables, or percentages and cumulative percentages in one-way tables and in list format.

`noprint` suppresses printing the frequency table but displays other statistics.

`list` prints multiway tables in list format.

`binomial` requests binomial proportion, confidence limits and test for one-way tables.

`testf=` specifies expected frequencies for a one-way table chi-square test.

`testp=` specifies expected proportions for a one-way table chi-square test.

`chisq` requests chi-square tests and measures of association based on chi-square.

`cellchi2` prints each cell's contribution to the total Pearson chi-square statistic.

`deviation` prints the deviation of the cell frequency from the expected value for each cell.

`expected` prints the expected cell frequency for each cell under the null.

`fisher` requests Fisher's exact test for tables larger than 2×2.

`measures` requests measures of association and their asymptotic standard errors.

`cl` requests confidence limits for the measures statistics.

`alpha=` sets the confidence level for confidence limits.

`agree` requests tests and measures of classification agreement.

Options such as `binomial`, `testf=`, and `testp=` are used for specifying either the postulated probabilities (p_{i0}'s) where $H_0 : p_i = p_{i0}$, $i = 1, 2, \ldots, k$, or the expected frequencies in a sample of size n classified in a one-way frequency table for performing a chi-square goodness-of-fit test. An application is provided as an exercise at the end of this chapter.

SAS Example B8

In the following contrived example of a two-way table, suppose that subjects are classified according to levels of two variables A and B. The column variable A has three categories, say a_1, a_2 and a_3 and the row variable B has three categories, say b_1, b_2, and b_3. Most often, the row variable is called the *dependent variable* if the categories of the variable are recognized as possible *outcomes* or *responses*. An example would be where factor B is Marital Status and factor A is a response to a question with three possibilities. Consider the table of frequencies:

	b_1	b_2	b_3	Total
a_1	8	16	31	55
a_2	9	18	74	101
a_3	34	23	17	74
Total	51	57	122	230

In this case, the column variable is called the independent variable with categories being *classes*, *groups*, or *strata*. The designation of whether the two types of variable are assigned to rows or columns is usually a matter of choice.

In the above setup, subjects from each of the column categories (independent variable) can be viewed as being classified into one of the row categories (dependent variable). The choice of the independent and dependent variables does not affect the statistical analysis of the data except when part of the inference is measuring the predictability of a response category given that an object belongs to a certain group or class. Otherwise, when variables cannot be clearly identified as independent and dependent variables, statistics and measures unaffected by an arbitrary designation are preferred.

In the SAS Example B8 program (see Fig. 2.28), the cell frequencies are directly input to `proc freq` instead of raw data (which are not available in this example). The use of the the `weight` statement allows SAS to construct the two-way cross-tabulation using the cell counts. In the first part of the output (see Fig. 2.29), only the statistics observed count f_i, the expected frequency e_i, and its contribution to the total chi-squared statistic are displayed in each cell (i.e., percentage of the total frequency, percentage of row frequency, and percentage of column frequency are suppressed using `tables` statement options given earlier). It is clear that the cells (2,1), (2,3), (3,1), and (3,3) provide the largest contributions to the total chi-squared statistic of 52.4. It is observed that at the lowest level of B, the response is smaller than expected for group 2 of A and larger than expected for group 3 of A. This pattern is reversed at the highest level of B.

```
options   formchar="|----|+|---+=|-/\<>*";
data ex8;
input A $ B $  count @@;
datalines;
a1 b1   8 a1 b2 16 a1 b3 31
a2 b1   9 a2 b2 18 a2 b3 74
a3 b1 34 a3 b2 23 a3 b3 17
;
run;
proc freq;
weight count;
tables A*B/chisq expected cellchi2 nocol nopercent norow measures;
title 'Example B8: Illustration of Tables Options';
run;
```

Fig. 2.28. SAS Example B8: Program

For two-way frequency tables, the chi-square test of independence is a test of *general association*, where the null hypothesis is that the row and column variables are independent (no association) and the alternative hypothesis is that an association exists between the two variables with the type of association unspecified. The chi-square statistic and the likelihood ratio statistic are both suitable for testing this hypothesis. `proc freq` computes these statistics in response to `chisq` option in the `tables` statement. For large sample sizes and if the null hypothesis is true, these test statistics have approximately a chi-square distribution. (For small samples, the user may request that Fisher's exact test be computed by specifying the `exact` option in the `tables` statement.) In Fig. 2.29, the *p*-values of both the chi-square statistic and the likelihood ratio statistic are smaller than, say .01. Thus, the null hypothesis that the two variables are independent is rejected, leading to the conclusion that there is some type of association between these two variables.

The three measures `phi coefficient` ϕ, `contingency coefficient` C, and `Cramer's V` displayed next in the output are suitable for measuring the strength of the dependency between nominal variables but are also applicable for ordinal variables. The value of C is zero if there is no association between the two variables but has a value that is less than 1 even with perfect dependence. Its value is dependent on the size of the table with a maximum value of $\sqrt{(r-1)/r}$ for an $r \times r$ table. For a 3×3 table, this value is 0.816. Thus, the value of 0.43 of C appears to indicate a strength midway between no association and a perfect association.

The other statistics also lead to similar conclusions. Cramer's V is a normed measure, so its value is between 0 and 1; thus, a value of 0.34 is approximately in the bottom third of the scale. The range for ϕ is $0 < \phi < \min\{\sqrt{r-1}, \sqrt{c-1}\}$. Since for this table the maximum is 1.4, a strength of association similar to the above measures is indicated by a value of 0.48 for ϕ.

```
                Example B8: Illustration of Tables Options                1
                          The FREQ Procedure
                          Table of A by B

           A            B

                Frequency    |
                Expected     |
                Cell Chi-Square|b1       |b2       |b3       | Total
                ---------------+---------+---------+---------+
                a1             |      8  |    16   |    31   |   55
                               | 12.196  | 13.63   | 29.174  |
                               |  1.4434 |  0.4119 |  0.1143 |
                ---------------+---------+---------+---------+
                a2             |      9  |    18   |    74   |  101
                               | 22.396  | 25.03   | 53.574  |
                               |  8.0124 |  1.9747 |  7.7878 |
                ---------------+---------+---------+---------+
                a3             |     34  |    23   |    17   |   74
                               | 16.409  | 18.339  | 39.252  |
                               | 18.859  |  1.1846 | 12.615  |
                ---------------+---------+---------+---------+
                Total                51       57       122       230


                   Statistics for Table of A by B

           Statistic                      DF       Value      Prob
           ------------------------------------------------------------
           Chi-Square                      4      52.4031    <.0001
           Likelihood Ratio Chi-Square     4      53.1793    <.0001
           Mantel-Haenszel Chi-Square      1      25.0216    <.0001
           Phi Coefficient                         0.4773
           Contingency Coefficient                 0.4308
           Cramer's V                              0.3375
```

Fig. 2.29. SAS Example B8: A × B chi-square test

The above three measures of association are all derived from the Pearson chi-square statistic. There are many other measures of association between two categorical variables available and `proc freq` calculates several of these. Some of these statistics are briefly discussed here. Many of these statistical measures also require the assignment of a dependent variable and an independent variable, as the goal is to predict a rank (category) of an individual on the dependent variable given that the individual belongs to a certain category in the independent variable.

For calculating the following measures for the two variables under consideration, pairs of observations are first classified as concordant or discordant. A pair is concordant if the observation with the larger value for variable one also has the larger value for variable two, and it is discordant if the observation with the larger value for variable one has the smaller value for variable two. Thus, the pair of observations (12, 2.7) and (15, 3.1) are concordant and the pair (12, 2.7) and (10, 3.1) are discordant.

```
                  Example B8: Illustration of Tables Options                 2

                             The FREQ Procedure

                        Statistics for Table of A by B

          Statistic                              Value        ASE
          ---------------------------------------------------------
          Gamma                                -0.4375       0.0828
          Kendall's Tau-b                      -0.2981       0.0586
          Stuart's Tau-c                       -0.2804       0.0555

          Somers' D C|R                        -0.2891       0.0575
          Somers' D R|C                        -0.3074       0.0601

          Pearson Correlation                  -0.3306       0.0626
          Spearman Correlation                 -0.3354       0.0650

          Lambda Asymmetric C|R                 0.1574       0.0607
          Lambda Asymmetric R|C                 0.2326       0.0622
          Lambda Symmetric                      0.1983       0.0540

          Uncertainty Coefficient C|R           0.1138       0.0293
          Uncertainty Coefficient R|C           0.1082       0.0281
          Uncertainty Coefficient Symmetric     0.1109       0.0286

                             Sample Size = 230
```

Fig. 2.30. SAS Example B8: A × B measures of association

Gamma is a normed measure of association based on the numbers of concordant and discordant pairs. If there are no discordant pairs, Gamma is +1 and perfect positive association exists between the two variables and if there are no concordant pairs, Gamma is −1 and perfect negative association exists between the two variables. Values in between −1 and +1 measure the strength of negative or positive association. If the numbers of discordant and concordant pairs are equal, Gamma is zero and the rank of the independent variable cannot be used to predict the rank of the dependent variable. In the SAS Example B8 output (see Fig. 2.30), Gamma= −0.4375 with an estimated asymptotic standard error (ASE) of 0.0828, indicating a negative association.

Kendall's tau-b is the ratio of the difference between the number of concordant and discordant pairs to the total number of pairs. It is scaled to be between −1 and +1 when there are no ties, but not otherwise. The ordinal measure Somers' D on the other hand adjusts for ties by counting pairs where ties occur only on the independent variable so that the value of the statistic lies between −1 and +1 when such ties occur. Usually, two values of this statistic are computed: one when the row variable is considered the independent variable (Somers' D $C|R$) and one when the column is considered the independent variable (Somers' D $R|C$). The values differ because of the way ties are counted. In SAS Example B8, Somers' D $R|C = 0.2326$, showing a moderate positive association.

The nominal measure `asymmetric lambda`, $\lambda(R|C)$, is interpreted as the proportional improvement in predicting the dependent (row) variable given the independent (column) variable. Asymmetric lambda has the range $0 \leq \lambda(R|C) \leq 1$, although values around 0.3 are considered high. The measure $\lambda(C|R)$ may be interpreted similarly. In SAS Example B8, if information about variable B is used to predict A, the *proportional reduction in error* in the prediction is 23.26% compared to not using that information. `Stuart's tau c` makes an adjustment for table size in addition to a correction for ties. Tau-c is appropriate only when both variables lie on an ordinal scale. Tau-c also is in the range $-1 \leq \tau_c \leq 1$.

The `Pearson correlation coefficient` and the `Spearman rank-order correlation coefficient` are also appropriate for ordinal variables. The Pearson correlation describes the strength of the linear association between the row and column variables. It is computed using the row and column scores specified by the `scores=` option in the `tables` statement. By default, the row or column scores are the integers 1, 2,... for character variables and the actual variable values for numeric variables. Consult SAS documentation for other options. The Spearman correlation is computed with rank scores.

SAS Example B9

The analysis of the SAS data set on fuel consumption created in SAS Example B6 is continued in SAS Example B9 (see Fig. 2.31 for the program) using `proc freq` to illustrate the statistics resulting from some of the `tables` statement options discussed in this section. A new SAS data set is created by supplementing the original data set with two category variables, `fuelgrp` and `licgrp`, each with three levels, in the data step. The 33.3 and 66.7 percentiles of the `fuel` and `percent` variables calculated in SAS Example B7 (see Fig. 2.27) aid in the determination of cutoff values for creating the corresponding category variables. Thus, the category variables `fuelgrp` and `licgrp` will have three levels each. In addition, `proc format` described in Section 2.1.4 facilitates the creation of output formats to convert the ordinal levels of the three categorical variables `fuelgrp`, `incomgrp`, and `licgrp` to more descriptive strings when they are printed.

The proc step generates three contingency tables for combinations of the variable `fuelgrp` with each of `taxgrp`,`incomgrp`, and `licgrp`. The output is shown in Figs. 2.32, 2.33, and 2.34. The `fuelgrp` by `taxgrp` table is a 3×2 cross-tabulation where the levels of `fuelgrp` are ordered by their internal (unformatted) values of 1, 2, and 3. However, internal values of the two levels of `taxgrp` are the strings 'Low ' and 'High'; thus, they are ordered by their alphanumeric values. The *p*-values of both the chi-square statistic and the likelihood ratio statistic are smaller than, say, .01. Thus, the null hypothesis that the two variables are independent is rejected.

```
options  formchar="|----|+|---+=|-/\<>*";
libname mylib 'C:\Documents and Settings\mervyn\My Documents\Classwork\stat479\';

data fueldat2;
set mylib.fueldat;

if fuel=<525 then fuelgrp=1;
else if 525<fuel=<610 then fuelgrp=2;
else fuelgrp=3;

if percent=<54 then licgrp=1;
else if 54<percent=<58 then licgrp=2;
else licgrp=3;

label licgrp='% Driving Licenses'
      fuelgrp='Fuel Consumption';
run;

proc format;
   value lg 1='below 54%'
            2='54 to 58%'
            3='above 58%' ;
   value fg  1 = 'Low Fuel Use'
             2 = 'Medium Fuel Use'
             3 = 'High Fuel Use';
   value ing 1 = 'Low Income'
             2 = 'Middle Income'
             3 = 'High Income';
run;

proc freq data=fueldat2;
   tables fuelgrp*(taxgrp incomgrp)/chisq expected
                    cellchi2 nocol nopercent norow;
   tables fuelgrp*licgrp/chisq expected nocol nopercent norow measures;
   tables taxgrp*fuelgrp/list ;
   format fuelgrp fg. licgrp pg. incomgrp ing. ;
   title 'Output from Proc Freq';
run;
```

Fig. 2.31. SAS Example B9: Program

To study the strength of association between the two variables fuelgrp and taxgrp, the statistics in Fig. 2.32 are used here. The three measures phi coefficient ϕ, contingency coefficient C, and Cramer's V displayed next in the output are suitable statistics for measuring the strength of the dependency between nominal variables and are also applicable for ordinal variables, as in this example. The value of C is zero if there is no association between the two variables but has a value that is less than 1 even with perfect dependence. Its value is dependent on the size of the table with a maximum value of $\sqrt{(r-1)/r}$ for an $r \times r$ table. For a 3×3 table, this value is 0.816. Thus, the value of 0.40 for C appears to indicate a strength of about 50% of a perfect association.

The other statistics in Fig. 2.32 also lead to similar conclusions. Cramer's V is a normed measure, so its value is between 0 and 1; thus, a value of 0.44 is about in the middle of the scale. The range for ϕ is $0 < \phi <$

$\min\{\sqrt{r-1}, \sqrt{c-1}\}$. Thus, for this table, the maximum is 1, so again a strength of association similar to the above measures is indicated. The cross-tabulation fuelgrp by incomgrp shown in Fig. 2.33 is a 3×3 table. Again, the chi-square and the likelihood ratio statistic are both significant at the .01 level, indicating dependency. A study of the values for the three measures discussed above, shown in Fig. 2.33, indicates a slightly stronger association between the variables fuelgrp and incomgrp.

The hypothesis of independence between the two variables fuelgrp and licgrp is rejected at .05 (the likelihood ratio statistic has a p-value of .0026; see Fig. 2.34). This is one situation where Fisher's exact test may be performed instead of the chi-square test because conditions for that test are clearly not met due to several small cell frequencies. In this example, the inclusion of the exact option produces the output in Fig. 2.35. The p-value is smaller than .05 so the conclusion is that the independence hypothesis is rejected at $\alpha = .05$

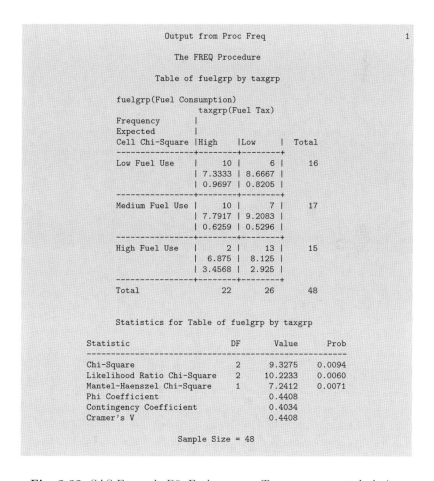

Fig. 2.32. SAS Example B9: Fuel group × Tax group cross-tabulation

```
                    Output from Proc Freq                       2
                      The FREQ Procedure

                 Table of fuelgrp by incomgrp

fuelgrp(Fuel Consumption)      incomgrp(Per capita Income)

Frequency        |
Expected         |
Cell Chi-Square  |Low Inco|Middle I|High Inc|  Total
                 |me      |ncome   |ome     |
-----------------+--------+--------+--------+
Low Fuel Use     |    1   |    3   |   12   |   16
                 | 4.3333 |    6   | 5.6667 |
                 | 2.5641 |   1.5  | 7.0784 |
-----------------+--------+--------+--------+
Medium Fuel Use  |    8   |    7   |    2   |   17
                 | 4.6042 |  6.375 | 6.0208 |
                 | 2.5046 | 0.0613 | 2.6852 |
-----------------+--------+--------+--------+
High Fuel Use    |    4   |    8   |    3   |   15
                 | 4.0625 |  5.625 | 5.3125 |
                 |  0.001 | 1.0028 | 1.0066 |
-----------------+--------+--------+--------+
Total                13       18       17       48


           Statistics for Table of fuelgrp by incomgrp

Statistic                       DF       Value      Prob
----------------------------------------------------------
Chi-Square                       4      18.4040    0.0010
Likelihood Ratio Chi-Square      4      18.7393    0.0009
Mantel-Haenszel Chi-Square       1       7.2622    0.0070
Phi Coefficient                          0.6192
Contingency Coefficient                  0.5265
Cramer's V                               0.4378

WARNING: 33% of the cells have expected counts less
         than 5. Chi-Square may not be a valid test.

                    Sample Size = 48
```

Fig. 2.33. SAS Example B9: Fuel group × Income group cross-tabulation

If it is concluded that there is association between the two variables, several statistics are available for evaluating the strength of such association. The output resulting from including the measures option in a tables statement is shown in Fig. 2.36. For the interpretation of Gamma and τ_b, consider licgrp as the independent variable that is used to predict the dependent variable fuelgrp and that both variables are ordinal. The value of 0.5641 for Gamma indicates a positive association between the two variables. This implies that the ordering of the ranks of states for these two variables are positively correlated and that if the orders of ranks of percentage of licenses is used to predict the orders of ranks of fuel use, the proportional reduction in error compared to randomly assigning ranks of fuel use for pairs of states is 56%. The ordi-

```
                    Output from Proc Freq                          3

                        The FREQ Procedure

                    Table of fuelgrp by licgrp

    fuelgrp(Fuel Consumption)    licgrp(% Driving Licenses)

        Frequency    |
        Expected     |below 54|54 to 58|above 58|  Total
                     |   %    |   %    |   %    |
        -------------+--------+--------+--------+
        Low Fuel Use |    6   |    9   |    1   |   16
                     | 4.6667 |    6   | 5.3333 |
        -------------+--------+--------+--------+
        Medium Fuel Use|  7   |    5   |    5   |   17
                     | 4.9583 | 6.375  | 5.6667 |
        -------------+--------+--------+--------+
        High Fuel Use|    1   |    4   |   10   |   15
                     | 4.375  | 5.625  |    5   |
        -------------+--------+--------+--------+
        Total            14       18       16       48


        Statistics for Table of fuelgrp by licgrp

    Statistic                    DF       Value      Prob
    ------------------------------------------------------
    Chi-Square                    4      14.6905    0.0054
    Likelihood Ratio Chi-Square   4      16.2966    0.0026
    Mantel-Haenszel Chi-Square    1       9.9989    0.0016
    Phi Coefficient                       0.5532
    Contingency Coefficient               0.4841
    Cramer's V                            0.3912

    WARNING: 33% of the cells have expected counts less
             than 5. Chi-Square may not be a valid test.
```

Fig. 2.34. SAS Example B9: Fuel group × Licenses group cross-tabulation

nal measures of association Kendall's τ_b and Somers' D $R|C$ both have a value of 0.40 again confirming the positive association between these two variables. Rather than being concerned with the ranking of pairs of observations on the two variables (concordancy or discordancy) as discussed previously, Spearman's ρ is measures the strength of the relationship between the overall ranks of each observation (or subject) on the two variables.

```
                     Fisher's Exact Test
         ------------------------------------
         Table Probability (P)    5.544E-06
         Pr <= P                     0.0046

                    Sample Size = 48
```

Fig. 2.35. SAS Example B9: Fisher's exact test for Fuel × Licenses

```
                    Output from Proc Freq                          4

                         The FREQ Procedure

               Statistics for Table of fuelgrp by licgrp

      Statistic                              Value        ASE
      ------------------------------------------------------------
      Gamma                                  0.5641       0.1272
      Kendall's Tau-b                        0.4024       0.0983
      Stuart's Tau-c                         0.4010       0.0988

      Somers' D C|R                          0.4016       0.0979
      Somers' D R|C                          0.4031       0.0989

      Pearson Correlation                    0.4612       0.1049
      Spearman Correlation                   0.4640       0.1090

      Lambda Asymmetric C|R                  0.2667       0.1456
      Lambda Asymmetric R|C                  0.2903       0.1463
      Lambda Symmetric                       0.2787       0.1335

      Uncertainty Coefficient C|R            0.1553       0.0656
      Uncertainty Coefficient R|C            0.1547       0.0654
      Uncertainty Coefficient Symmetric      0.1550       0.0655

                         Sample Size = 48
```

Fig. 2.36. SAS Example B9: Measures of association: Fuel and Population

Similar to Pearson's correlation coefficient, the value of Spearman's ρ lies in the range -1 and $+1$, with these values indicating perfect negative or positive association, respectively. For example, if the ranks of one variable agrees perfectly with the ranks of the other variable, $\rho = +1$. Just as with Pearson's correlation coefficient, it is possible to conduct tests based on the t-statistic

$$t = \hat{\rho}\sqrt{\frac{n-2}{1-\hat{\rho}^2}}$$

where $\hat{\rho}$ is the sample rank-correlation coefficient, for testing hypotheses about the population rank-correlation coefficient ρ for sample sizes larger than 10. If the percentage of licenses is used to predict the fuel use category of a state, the nominal measure asymmetric lambda, $\lambda(R|C)$, can be used to obtain the proportional reduction in error. Here the value is 0.2903, so that that proportional reduction in error is 29% compared to prediction not based on licensing information.

Including cl along with measures as tables statement options will lead to the computation of asymptotic confidence intervals for the measures of association discussed earlier. The default confidence coefficient is .05, which may be changed by including an optional alpha= option with the required value. For some of the measures, adding a test statement will produce an asymptotic test of whether the measure is equal to zero as well as an asymptotic confidence interval. These association measures are Gamma, Kendall's τ_b, Stuart's

```
            Gamma                              Spearman Correlation Coefficient
--------------------------------              --------------------------------
Gamma                    0.5641               Correlation              0.4640
ASE                      0.1272               ASE                      0.1090
95% Lower Conf Limit     0.3148               95% Lower Conf Limit     0.2503
95% Upper Conf Limit     0.8134               95% Upper Conf Limit     0.6776

      Test of H0: Gamma = 0                       Test of H0: Correlation = 0

ASE under H0             0.1390               ASE under H0             0.1098
Z                        4.0587               Z                        4.2268
One-sided Pr >  Z        <.0001               One-sided Pr >  Z        <.0001
Two-sided Pr > |Z|       <.0001               Two-sided Pr > |Z|       <.0001

                                                        Sample Size = 48
```

Fig. 2.37. Result of TEST statement in PROC FREQ

τ_c, Somers' D, and Pearson's and Spearman's ρ. Fig. 2.37 shows the output resulting from including the statement `test gamma scorr;`. In both cases, the null hypotheses are rejected at reasonable α values. Finally, Fig. 2.38 illustrates how the `list` option may be used to format a cross-tabulation as a one-way table.

```
                                                   Cumulative    Cumulative
 taxgrp        fuelgrp      Frequency    Percent    Frequency      Percent
-------------------------------------------------------------------------
 High     Low Fuel Use          10        20.83        10          20.83
 High     Medium Fuel Use       10        20.83        20          41.67
 High     High Fuel Use          2         4.17        22          45.83
 Low      Low Fuel Use           6        12.50        28          58.33
 Low      Medium Fuel Use        7        14.58        35          72.92
 Low      High Fuel Use         13        27.08        48         100.00
```

Fig. 2.38. SAS Example B9: Tax group × Fuel group cross-tabulation as a list

2.3 Some Useful Base SAS Procedures

There are many Base SAS procedures that calculate a variety of statistics as well as others that perform utility functions. Some of these, such as `proc print`, `proc means`, `proc sort`, and `proc format`, were previously discussed or used in SAS Example programs. Some others such as `proc rank`, and `proc corr` will be used in examples to follow in later chapters. In this section, four useful Base SAS procedures will be briefly introduced and their use illustrated through SAS Example programs.

In Section 2.2, the `plots` option used in `proc univariate` produced a histogram (or a stem-and-leaf plot), a box plot, and a normal probability plot in low-resolution (or line-printer) graphics. See the SAS program displayed in Fig. 2.19 for an example of use of this option. Although high-resolution graphics created by SAS/GRAPH programs are preferable for use in presentations or publications, low-resolution graphics produced by some SAS procedures also play a role, for example, in routine exploratory data analysis or as diagnostic tools. In this section, two procedures that provide statements and options for constructing simple low-resolution scatter plots and charts are introduced.

2.3.1 The PLOT procedure

The general structure of a `proc plot` step is

```
PROC PLOT < options > ;
  BY variable(s) ;
  PLOT plot-request(s) </ option(s)>;
```

A number of `proc` statement options are available for `proc plot`. Although some of these are standard options such as the `data=` option for naming the data set to be analyzed, the `nomiss` for excluding observations with missing values, or the `nolegend` option that suppresses the legend that appears on top of the plot by default, a few are more specialized. Some of these special `proc` options are summarized below.

    $\boxed{\text{uniform}}$ option specifies that the same scale for the two axes be used for multiple plots produced by the use of the BY statement.

    $\boxed{\text{formchar} < (\text{position(s)}) > = \text{'formatting-character(s)'}}$ specifies printable characters to be used for drawing the borders (i.e., outlines) and their intersections on a plot. The `position` parameters identify elements in the border to be plotted: 1 and 2 identify vertical and horizontal lines, 7 identifies the intersection of two of those lines, and 3, 5, 9, and 11 identify the corners, respectively. *formatting-character(s)* define a list of characters assigned to each of these positions. By default, the characters $|, -, +, -, -, -,$ and $-$ are assigned to each of the above seven positions, respectively.

    $\boxed{\text{hpercent=}}$ option is used to list percentage(s) of the available horizontal space to be used for each plot (in case of multiple plots that are not overlaid).

    $\boxed{\text{vpercent=}}$ option is used to list percentage(s) of the available vertical space to be used for each plot.

The `plot` statements are called the *action* statements in a `proc plot` step. The operand of a `plot` statement consists of *plot-request(s)* followed by options if necessary.

2.3 Some Useful Base SAS Procedures

Plot-requests

The plot request(s) in `plot` statements specify the variables to be plotted on the vertical and horizontal axes, respectively and, optionally, the plotting symbol to be used. For example, the statement

```
plot height*weight;
```

requests a plot of the variables `height` by `weight`; that is, `height` appears on the vertical axis and `weight` appears on the horizontal axis. After `proc plot` determines the scales for the two axes, pairs of values of the two variables (as they occur in each observation) are used to determine the coordinates at which the points are plotted using plot symbols. When a plot symbol is not specified, as in the above example, `proc plot` uses the default scheme of sequentially using uppercase roman letters as plot symbols: 'A' to represent a single observation at a point, B to represent two observations at a point etc. Several more general formats for plot requests are available:

```
vertical*horizontal <$ label-variable>

vertical*horizontal='character' <$ label-variable>

vertical*horizontal=variable <$ label-variable>
```

All of these forms optionally allow the specification of a *label variable* that is preceded by a dollar sign ($). This is the name of a variable that contains strings of characters as values to be used for *labeling* the points plotted. For example, the action statement

```
plot height*weight='*';
```

produces a scatter plot consisting of points indicated by '*' symbols; the statement

```
plot height*weight='*' $ name;
```

will identify those points by 'names', the values of the variable `name`, corresponding to each observation in the data set. On the other hand, the statement

```
plot height*weight=sex;
```

results in the symbols 'M' or 'F', the values of the variable `sex`, to be used as the plot symbols. Plot statement options such as `outward=`, `penalties=`, and `placement=` may be used to control the actual placement of the point labels as described below. When plot request(s) contain multiple requests consisting of several variables, SAS provides syntax to abbreviate the requests. For example, the request (y1-y2)*(x1-x2) expands to y1*x1 y1*x2 y2*x1 y2*x2 and the request (y1-y2):(x1-x2) expands to y1*x1 y2*x2. Other forms are possible; for example, y*(x1-x3) implies y*x1 y*x2 y*x3 and the request (x1-x3) by itself is equivalent to x1*x2 x1*x3 x2*x3.

SAS Example B10

As in several previous SAS programs, the SAS program shown in Fig. 2.39 accesses the SAS data set named `fueldat` from a library (a folder, in this case) using the two-level name `mylib.fueldat`. The `proc plot` step contains three plot statements that illustrate forms of plot requests discussed earlier. The three resulting graphs are shown in Figs. 2.40, 2.41, and 2.42, respectively.

```
options ls=60 ps=40 nodate pageno=1;
options  formchar="|----|+|---+=|-/\<>*";
libname mylib 'C:\Documents and Settings\...\stat479\';

proc plot data=mylib.fueldat;
  plot fuel*roads;
  plot fuel*roads='+';
  plot fuel*roads='*' $ st;
  title 'Output from Sample Plot Statements';
run;
```

Fig. 2.39. SAS Example B10: Program

Some PLOT Statement Options

Many options are available with the `plot` statement to control the graphics that are output by `proc plot`. Whereas some of these are simple options for controlling the axes, plotting reference lines, and producing multiple plots, others are more complicated options that control the placement of labels at plot points. An abbreviated description of these options is as follows:

haxis= and vaxis= options specify the tick-mark values for the horizontal and vertical axes, respectively. For numeric variables, the axis specification is a list of values such as `haxis=5 10 15 20 25 30 35`. An alternate specification of this option is `haxis=5 to 35 by 5`. If the user wants `proc plot` to determine the minimum and maximum values for the tick marks, just `haxis=by 5` will work. For character variables, the possible tick-mark values are specified as character strings (for e.g., `haxis='Kansas'` 'Missouri' 'Iowa' 'Illinois' 'Nebraska'). For other types of data values, such as *date* values, special syntax is available. For example, `haxis='01MAY07'd to '01DEC07'd by month` specifies a sequence of eight dates as tick-mark values. For dates, the increment may be one of the keywords `day`, `week`, `month`, `qtr`, or `year`. Similar syntax is available for *datetime* and *time* data values.

hpos= and vpos= options control the number of print positions to be used for the horizontal and vertical axes. By default, the `linesize=` (or equivalently, `ls=`) and the `pagesize=` (or equivalently, `ps=`) system options are

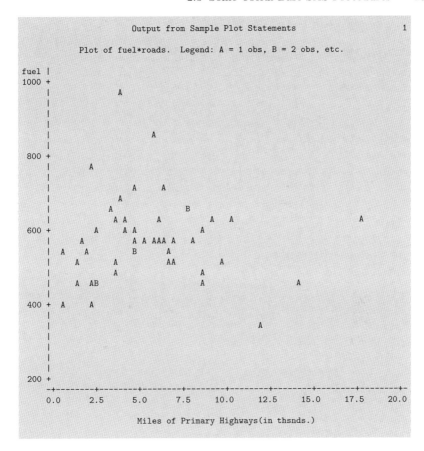

Fig. 2.40. Graph produced by plot fuel*roads;

used to determine the horizontal and vertical axis lengths, respectively. Users may change default values set under specific operating environments and displays for these options in an options statement.

href= and vref= options specify the positions on the horizontal and the vertical axis, respectively, at which lines will be drawn on the plot perpendicular to the respective axes. By default, the characters | and − respectively, will be used to draw the reference lines. The options hrefchar= and vrefchar= may be used to specify alternate choices (e.g. hrefchar='.').

box draws a solid line border around the plot.

contour <=*number-of-levels*> draws a contour plot using plotting symbols to generate degrees of shading where *number-of-levels* is the number of levels for dividing the range of the 'third' variable. The plot request must be of the form `vertical*horizontal=variable` where `variable` contains the "elevation" or "depth" values if the plot is topographical or function values if the plot shows contours of a two-dimensional surface corresponding to a function of two variables.

`overlay` overlays all plots that are specified in the `plot` statement on one set of axes. The variable names, or variable labels if they exist, from the first plot are used to label the axes. Overlaying is not meaningful unless the sets of variables are comparable, such as the same two variables measured for males and females, or under several different conditions etc. The axes are scaled to fit all sets of variable values unless they are scaled using `haxis=` and/or `vaxis=` options by the user.

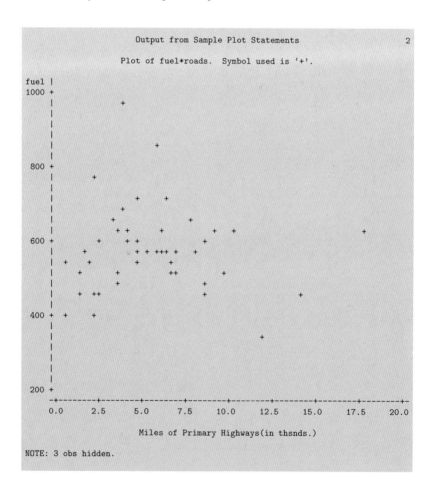

Fig. 2.41. Graph produced by `plot fuel*roads='+';`

2.3 Some Useful Base SAS Procedures

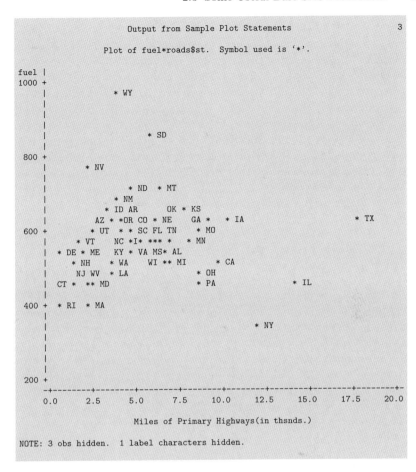

Fig. 2.42. Graph produced by plot fuel*roads='*' $ st;

split='*split-character*' may be used to indicate where to split the labels when the l= in a placement= option specifies that the label is to be split into more than one line.

placement=(expression(s)) is a facility for the user to control the placement of labels around the coordinates of the points plotted. If this option is omitted, proc plot uses a set of default expressions. See separate paragraph titled *Placement Expressions* below for some details and some examples.

| penalties< (*index-list*) >=*penalty-list* | `proc plot` determines the best placement positions for labels using a penalization scheme. If every label is placed with zero penalty, then no labels collide and all labels are near their plot symbols. When this is not possible, `proc plot` determines alternate placements that minimizes the total penalty. To do this, `proc plot` begins with a `default penalty table`. For example, a penalty of type 1 (i.e., *index-value=1*, a default value) is incurred when a nonblank character in the plot collides with an embedded blank in a label or there is not a blank or a plot boundary before or after each (line of the) label. The default penalty value assigned for that event is 1. The option `penalties(1)=2` would change this penalty to 2, a higher penalty. Generally, several penalty values may changed for a plot to obtain more preferable placements (e.g., `penalties(15 to 19)=2 3 4 10 15 25`).

Placement Expressions

The syntax for placement expressions is complicated because it provides a system for generating a large number of expressions by expanding expressions using the symbols * and : (in the same way as expanding plot requests). A single expression involves setting values for options `h=`, `l=`, `s=`, and `v=`. Each expression is called a *placement state*, which is basically a description of a position where the label can be placed. `proc plot` gives priority to the placement states in the order they occur in a list of placement states. However, it goes through several cycles of refining placement states to find a combination of placements for all labels that minimizes the *total penalty* as well, so the placement states preferred by the user may not always be given priority.

When defining a placement state, the options `h=` and `v=` specify the number of horizontal and vertical spaces to shift the label relative to the starting position, respectively. Both positive and negative integers are allowed, with positive integers signaling shifts to the right or upwards and negative integers indicating shifts to the left or downwards. The option `s=` specifies the starting position, with values `center`, `left`, or `right` indicating whether the string is centered, left-aligned, or right-aligned relative to the plotting symbol location. The option `l=` specifies the number of lines into which the label may be split. If a value is not set for any of these parameters, default values (`h=0`, `v=0`, `s=center`, `l=1`) are assigned.

Multiple placement states can be specified using expressions. For example, specifying `placement=(h=0 1)` results in defining two placement states: `(s=center l=1 h=0 v=0)` and `(s=center l=1 h=1 v=0)`. The expression `h=0 1 -1*v=1 -1` results in six different placement states. Other more complex examples are provided in the SAS documentation.

As an example, the following plot statement inserted into the proc step in the SAS Example B10 program, displayed Fig. 2.39, produces the graph displayed in Fig. 2.43:

2.3 Some Useful Base SAS Procedures   111

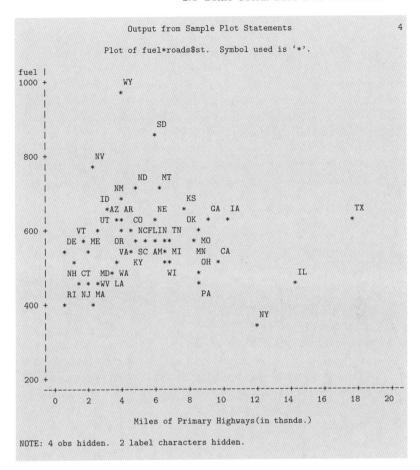

Fig. 2.43. Graph produced by plot fuel*roads='*' $ st; and placement= option

```
plot fuel*roads='*' $ st/
        haxis=0 to 20 by 2
        placement=((s=right left:h=1 -1)(v=1 -1*h=1 -1));
```

It is easy to observe that the placement of labels in this graph is different from that in Fig. 2.42. The main difference is that the placement states generated by the above placement specification allows for more positions for the labels to be placed around the plotting symbol. Usually, some experimentation is needed to determine a satisfactory set of placements, especially if the scatter plot is crowded.

SAS Example B11

This example illustrates the use of the `contour` option in `plot` statements. In addition, it also illustrates the creation of a SAS data set by generating observations entirely with programming statements without reading external data. When input data are not read, a data step goes through just a single iteration. When generating data for low-resolution plots, it is recommended that fewer observations than the number of positions available on the horizontal axis be generated for avoiding overplotting. This rule is not required to be followed when plotting contours, as the objective is to generate a shading using characters that are plotted contiguously; thus overplotting is not a problem.

The program shown in Fig. 2.44 calculates the density function of a bivariate normal distribution $\phi(x, y)$ given by the equation

$$\phi(x,y) = \frac{1}{2\pi\sigma_1\sigma_2\sqrt{1-\rho^2}}$$
$$\times \exp\left[\frac{1}{2(1-\rho^2)}\left\{\left(\frac{x-\mu_1}{\sigma_1}\right)^2 - \frac{2\rho(x-\mu_1)(y-\mu_2)}{\sigma_1\sigma_2} + \left(\frac{y-\mu_2}{\sigma_2}\right)^2\right\}\right]$$

where random variables X and Y have normal distributions with means μ_1 and μ_2 and variances σ_1^2 and σ_2^2, respectively, and ρ is the correlation between X and Y. A double `do loop` in the data step calculates the values of the density function $\phi(x, y)$ for values assigned to the variables x and y in the role of

```
data binorm;
mu1=4;sigma1=2;
mu2=10;sigma2=3;
rho=0.4;
   do x=-1 to 9 by .1;
      do y=1 to 18 by .1;
         z1=(x-mu1)/sigma1;
         z2=(y-mu2)/sigma2;
         v=z1**2-2*rho*z1*z2+z2**2;
         const=2*3.14159265*sigma1*sigma2*sqrt(1-rho**2);
         phi=(1/const)*exp(-v/(2*(1-rho**2)));
         output;
      end;
   end;
   drop z1 z2 v const;
run;

proc plot data=binorm nolegend;
   plot y*x=phi / contour=10
                  hpos=60 vpos=35
                  haxis=-1 to 9 by 2
                  vaxis=1 to 18 by 2  ;
   title 'Contour Plot of Bivariate Normal Density';
run;
```

Fig. 2.44. SAS Example B11: Program

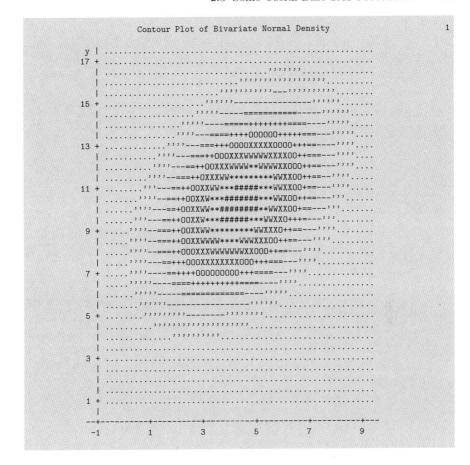

Fig. 2.45. Graph produced by `proc plot` using the `contour=` option

index variables for the two `do` statements. The ranges of values for x and y were determined by considering the possible values for random variables with the specified univariate normal distributions. In each iteration of the double `do` loop, for a pair of values of x and y a value *phi* is calculated and an observation output to the data set. The intermediate variables created were dropped from the data set. A plot of y versus x with *phi* as the third variable and a `contour=` option with 10 levels produced the contour plot shown in Fig. 2.45.

2.3.2 The CHART procedure

The procedure `chart` is typically used to draw low-resolution vertical or horizontal bar charts, block charts, or pie charts. Since the procedure statements in `proc chart` are similar to those available in the SAS/GRAPH procedure gchart discussed in Chapter 3, the discussion in this subsection will be limited to a brief discussion of the statements and some options and providing an example illustrating their use. The general structure of a proc chart step (which includes five of the procedure information statements to be illustrated) is

```
PROC CHART < option(s) >;
  BLOCK variable(s) < / option(s)>;
  BY variables;
  HBAR variable(s) < / option(s)>;
  PIE  variable(s) < / option(s)>;
  STAR variable(s) < / option(s)>;
  VBAR variable(s) < / option(s)>;
```

The proc statement options in proc chart are the data= option for naming the data set to be analyzed and the

```
formchar $<($position(s)$)>$ = 'formatting-character(s)'}
```

option for specifying printable characters to be used for various elements of a chart. The specifications are different from those used with proc plot, but an extensive discussion is not given here. For example, the position parameters identify elements in charts: 1 and 2 identify vertical and horizontal axes in bar charts, 7 identifies the tick marks in bar charts, and 9 identifies intersection of axes in bar charts, respectively. *formatting-character(s)* define a list characters assigned to each of these positions. By default, the characters $|, -, +,$ and $-$ are assigned to each of the above four positions, respectively.

The variable(s) that appear in any of the statements hbar, vbar, block, pie, or star statements above specifies the variable(s) for which these charts are produced. The options that may be specified following the slash in each of these statements enable one to customize the appearance of the charts. By default, if the variable specified is a numeric variable with continuous values, then, by default, the midpoints consist of the midpoints of the class intervals as determined by proc chart. For example, in vertical bar charts produced by the vbar statement, the midpoint values specify the midpoints of the intervals represented by the bar and these values are plotted at the bottom of each bar. By default, the statistic used to determine the sizes of bars, blocks, or pie sections is the class frequency. For example, in vertical bar charts, class frequency is the variable represented by the vertical axis. In that case the chart produced is a standard histogram. The user may use the option type= to specify different choices. For example, in bar charts, type=percent requests plotting the percent frequencies.

When the variable specified is discrete-valued such as a character category or a nominal variable or discrete-valued numeric variable such as a level or an ordinal variable, the actual values of the variable determine the placement of the bars, blocks, etc. The option discrete must be used to indicate whether the variable is a discrete-valued numeric variable. Otherwise, proc

2.3 Some Useful Base SAS Procedures 115

```
libname mylib 'C:\Documents and Settings\...\stat479\';

proc format;
   value ing 1 = 'Low'
             2 = 'Middle'
             3 = 'High';
run;

proc chart data=mylib.fueldat;
   vbar fuel;
   vbar fuel/midpoints =300 to 1000 by 100 type=percent;
   vbar incomgrp/discrete sumvar=fuel;
   vbar incomgrp/discrete sumvar=fuel type=mean subgroup=taxgrp;
   vbar incomgrp/discrete sumvar=fuel type=mean group=taxgrp;
   format incomgrp ing.;
   title 'Illustrating HBAR statement in PROC CHART';
run;
```

Fig. 2.46. SAS Example B12: Program

chart determines class intervals for the discrete values and incorrectly draws a histogram.

The value represented by each bar, block, etc. are then specified using sumvar= and the type= options. For example, the vbar statement vbar region/sumvar=sales type=mean; will produce bars representing sales averages from outlets in several regions. Operations needed to produce the bar chart such as the scaling of the vertical axis, determining the bar widths,

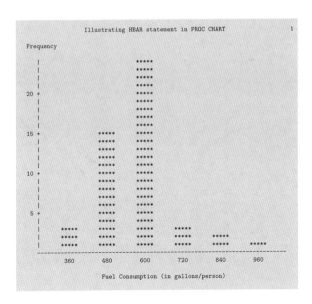

Fig. 2.47. Graph produced by vbar fuel;

and choosing the spacing between the bars are determined by `proc chart`. However, the user may use options to determine the class intervals and number of bars in a chart, produce side-by-side bars, or subdivide the bars. In some instances, `proc chart` may not accommodate user requests in order to produce a chart. For example, if the number of characters per line available (controlled by the `linesize=` option) is not sufficient to display all of the bars in a request for a vertical bar chart, `proc chart` substitutes a horizontal bar chart.

SAS Example B12

The SAS data set created in SAS Example B6 on fuel consumption data is analyzed in the SAS Example B12 program (see Fig. 2.46) to illustrate the use of `proc chart` for producing low-resolution charts. In this example, the `vbar` statement is selected and the effects of using several options discussed in this section are compared. The two-level name `mylib.fueldat` is used to access the previously saved SAS data set named `fueldat`, as was done in previous examples. The statement `vbar fuel;` produces the histogram shown in Fig. 2.47. As observed, `proc chart` has selected to draw six bars and placed them at the midpoints 360, 480, 600, 720, 840, and 960. In contrast, as shown in Fig. 2.48, the user opted to use eight bars centered at the midpoints 300, 400, 500, etc. and plot the percent frequency instead. The options needed to achieve this are `midpoints =300 to 1000 by 100` and `type=percent`.

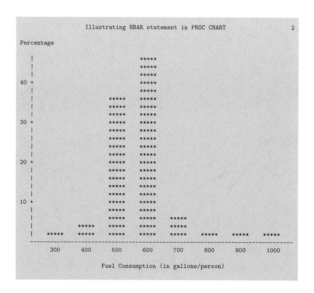

Fig. 2.48. Graph produced by `vbar fuel/midpoints =300 to 1000 by 100 type=percent;`

2.3 Some Useful Base SAS Procedures 117

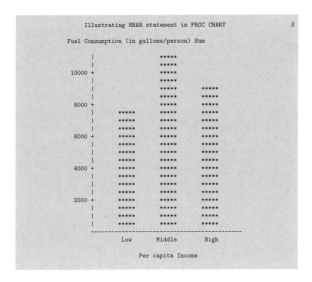

Fig. 2.49. Graph produced by vbar incomgrp/discrete sumvar=fuel;

The next vbar statement vbar incomgrp/discrete sumvar=fuel; uses the category variable incomgrp to determine the placement of the vertical bars. This variable is a grouping variable that divides the observations in the data set (i.e., states) into low-, middle-, and high-income groups and has corresponding numeric values of 1, 2, and 3. Thus, the option discrete is needed in order to force proc chart to regard this variable as a discrete-valued variable. As shown in Fig. 2.49, the variable plotted on the vertical axis is the variable fuel as specified in the option sumvar=fuel and the size of the bars is determined by default to be the total fuel consumption per person of states (denoted as 'Sum' in the plot) in each income group.

The next vbar statement also centers the bars at each income group but the option type=mean requests that each bar represent the mean fuel consumption for states in each income group rather than the sum, as shown in Fig. 2.50. The option subgroup=taxgrp subdivided each bar by the values of the category variable taxgrp. Recall that this variable has two values 'Low ' or 'High' depending on whether fuel tax is below 8 cents or not, respectively, for each state. In Fig. 2.50, for example, it can be observed that for middle-income states, the average fuel consumption for low-fuel-tax states is twice as large as those with higher fuel taxes.

The final vbar statement produces the plot shown in Fig. 2.51. The group=taxgrp option produces side-by-side bar charts for each value of the category variable taxgrp. Thus, it produces an analysis of mean fuel consumption for states in each income group within each tax group.

118   2 More on SAS Programming and Some Applications

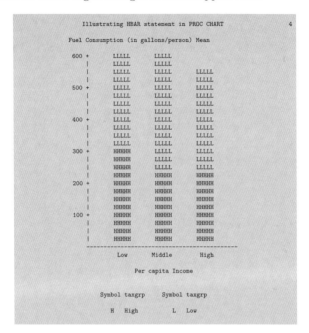

Fig. 2.50. Graph produced by vbar incomgrp/discrete sumvar=fuel type=mean subgroup=taxgrp;

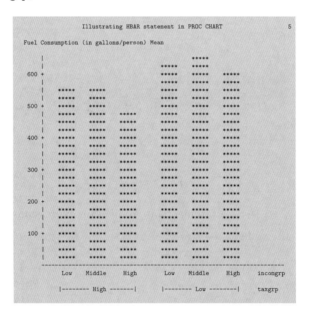

Fig. 2.51. Graph produced by vbar incomgrp/discrete sumvar=fuel type=mean group=taxgrp;

2.3.3 The TABULATE procedure

The tabulate procedure is an extremely versatile procedure for producing display-quality tables containing descriptive statistics. Using an extremely simple and flexible system of syntax, the user is able to customize the appearance of the tables incorporating labeling and formatting as well as to generate tabular reports that contain many of the same statistics that are computed by other descriptive statistical procedures. More recent innovations incorporated allow style elements (e.g., colors) to be specified to enhance the appearance of tables output in HTML and RTF formats using ODS graphics output. The statements available in `proc tabulate` and options that can be specified are too numerous to be described in detail in this text. A brief description useful for understanding the example presented below follows. The general structure of a `proc tabulate` step (that excludes several important statements) is

```
PROC TABULATE <options>;
  CLASS    variable(s) ;
  VAR      variable(s) ;
  BY       variable(s) ;
  TABLE    expression, expression, ... < / options >;
  KEYLABEL keyword='text' ... ;
```

Options available for the proc statement are `data=`, `out=`, `missing`, `order=`, `formchar<` (position(s)) `>='formatting-character(s)'`, `noseps`, `alpha=`, `vardef=`, `pctldef=`, `format=` , and `style=`. Since the reader is familiar with many of these options, only those that are relevant to this procedure are described. The default for `formchar=` is '|————|+|———' for positions 1 through 11. A value specified for the `format=` option is any valid SAS or user-defined format for printing each cell value in the table, the default being `best12.2`. The `style=` specifies the style element (or style elements) (for the Output Delivery System) to use for each cell of the table. For example, `style=[background=gray]` specifies that the background color for data cells be of gray color. Style elements can be specified in a *dimension expression* (described below) to control the appearance of analysis variable name headings, class variable name headings, class variable level value headings, data cells, keyword headings, and page dimension text.

The `table` statement is the primary statement in `proc tabulate`. The main components of table statements are dimension expressions. A simplified form of the table statement is

```
table   expression-1, expression-2, expression-3 </ options> ;
```

where *expression-1* defines the appearance of pages, *expression-2* defines the appearance of rows, and *expression-3* defines the appearance of the columns of the table. A *dimension expression* consists of combinations of the following elements separated by asterisks or blanks:

- Classification variable(s) (variables in the `class` statement)

- Analysis variables (variables in the `var` statement)
- Statistics (`n, mean, std, min, max`, etc.)
- Format specifications (e.g., f=7.2)
- Nested dimension expressions

If the statements

>   class region popgrp taxgrp;
>   var fuel income;

are present in a proc tabulate step, some valid examples of expressions are `region*popgrp`, `region*popgrp*taxgrp`, `region popgrp*taxgrp`, `(region popgrp)* taxgrp`, and `region*mean*fuel`. The expression `region*popgrp` defines a simple two-way table where the row variable is the nominal variable `region` and the column variable is the category variable `popgrp`. Note that the page dimension specification is omitted.

```
libname mylib 'C:\Documents and Settings\...\stat479\';

data fueldat3;
set mylib.fueldat;

if percent=<54 then licgrp=1;
else if 54<percent=<58 then licgrp=2;
else licgrp=3;
run;

proc format;
    value ing 1 = 'Low Income'
              2 = 'Middle Income'
              3 = 'High Income';
    value lg 1='below 54%'
             2='54 to 58%'
             3='above 58%' ;
run;

proc tabulate data=fueldat3;
    var fuel;
    class incomgrp taxgrp licgrp;
    format incomgrp ing. licgrp lg.;
    table incomgrp*taxgrp,fuel*(n mean stderr);
    table taxgrp*licgrp='Percent Driver Licenses',fuel*(n='Sample Size'*f=4.0
        (mean='Sample Mean' stderr='Standard Error of the Mean')*f=8.1);
    title 'Illustrating PROC TABULATE';
run;
```

Fig. 2.52. SAS Example B13: Program

The expression `popgrp*region*taxgrp` produces the same two-way tables for each level of `popgrp` on separate pages. The expression `(region popgrp)* taxgrp`, however, defines a two-way table where the rows are the levels of each of the variables `region` and `popgrp` (not combinations of the levels) and the columns are levels of `taxgrp`. The expression `region*mean*fuel` constructs a two-way table with the levels of region variable to appear on the rows. The column specification of `mean*fuel` causes the mean of the fuel consumption variable to be computed and appear as a single column. SAS documentation on `proc tabulate` provides ample examples of various combinations used to define dimension expressions; an extensive discussion is omitted here.

SAS Example B13

The SAS data set created in SAS Example B6 on fuel consumption data is used again in the SAS Example B13 program (see Fig. 2.52) to illustrate the use of `proc tabulate` for producing tables of statistics. Recall that `incomgrp` and `taxgrp` are two category variables created previously and included in the data set. In addition, another grouping variable `licgrp` is also created using the values of the variable `percent`, which contains the percentage with driving licenses in the population. From among several analysis variables present in the data set, `fuel` (per capita fuel consumption) is selected for the computation of statistics to be tabulated.

```
                       Illustrating PROC TABULATE                       1

                           | Fuel Consumption (in gallons/person) |
                           |--------------------------------------|
                           |    N    |    Mean    |    StdErr     |
        -------------------+---------+------------+---------------|
        |Per    |Fuel Tax |         |            |               |
        |capita |         |         |            |               |
        |Income |         |         |            |               |
        |-------+---------+---------+------------+---------------|
        |Low    |High     |   7.00  |   561.51   |    18.72      |
        |Income |---------+---------+------------+---------------|
        |       |Low      |   6.00  |   626.14   |    27.27      |
        |-------+---------+---------+------------+---------------|
        |Middle |High     |   7.00  |   547.93   |    26.69      |
        |Income |---------+---------+------------+---------------|
        |       |Low      |  11.00  |   649.03   |    35.21      |
        |-------+---------+---------+------------+---------------|
        |High   |High     |   8.00  |   463.14   |    20.02      |
        |Income |---------+---------+------------+---------------|
        |       |Low      |   9.00  |   590.71   |    49.73      |
```

Fig. 2.53. Output from PROC TABULATE: Page 1

In the SAS program, the `class` statement lists the three category variables and the `var` statement lists the analysis variable. A simple `table` statement is first used to produce a two-way table that tabulates the statistics sample size (n), the mean, and the standard error of the mean (stderr) for the variable `fuel`. The first dimension expression `incomgrp*taxgrp` defines the rows of the table, with the rows representing combinations of the levels of `incomgrp` and

```
                    Illustrating PROC TABULATE                        2
    ------------------------------------------------------------
    |                       | Fuel Consumption (in |
    |                       |    gallons/person)   |
    |                       |----------------------|
    |                       |Sam- |        |Standard|
    |                       |ple  | Sample |Error of|
    |                       |Size | Mean   |the Mean|
    |-----------------------+-----+--------+--------|
    |Fuel Tax|Percent       |     |        |        |
    |--------|Driver        |     |        |        |
    |High    |Licenses      |     |        |        |
    |        |--------------|     |        |        |
    |        |below 54%|  8 | 495.0 | 26.2 |
    |        |---------+----+-------+------|
    |        |54 to 58%|  10| 512.2 | 19.3 |
    |        |---------+----+-------+------|
    |        |above 58%|  4 | 597.1 | 27.5 |
    |--------+---------+----+-------+------|
    |Low     |below 54%|  6 | 552.6 | 42.3 |
    |        |---------+----+-------+------|
    |        |54 to 58%|  8 | 591.4 | 25.7 |
    |        |---------+----+-------+------|
    |        |above 58%| 12 | 680.5 | 37.5 |
    ------------------------------------------------------------
```

Fig. 2.54. Output from PROC TABULATE: Page 2

taxgrp. The second dimension fuel*(n mean stderr) defines the columns to be the sample size, the mean, and the standard error of the mean computed for the variable fuel. As can be observed, previously defined labels and formats are used for the variables and their levels. The default format of best12.2 is used for printing all cell values. The default formchar specification is used to select characters to draw the table borders, separators, and their intersections if the output is produced for low-resolution printing. This table is shown in Fig. 2.53.

The second table statement also produces a two-way table with taxgrp* licgrp defining the row and the same statistics computed for the fuel variable defining the columns. However, format specifications are added to control the formatting of the cell values in the table (e.g., n*f=4.0) Also, label parameters are added to assign more elaborate labels for class variable names (e.g., licgrp='Percent Driver Licenses') and statistic keyword headings (e.g., mean='Sample Mean'). These produce the table shown in Fig. 2.54.

2.4 Exercises

2.1 Write a SAS program containing an infile statement to access the data set used in SAS Example B5 (see Fig. 2.13) from a text file. Add proc step(s) to obtain the following low-resolution plots. Use labels in your data step for the variables and title your plots appropriately.

a. A scatter plot of weight against height using the data set used in SAS Example B5 (see Fig. 2.13). Use '*' as plot symbols and identify whether each point on your plot represents a male or a female student.
b. A vertical bar chart of enrollment in the biology class classified by gender.
c. A vertical bar chart of the average height of students classified by gender.
d. A horizontal bar chart of the average weight of students classified by year in school within each gender (i.e., side-by-side bar charts for each gender).
e. A vertical bar chart (i.e., a histogram) with six bars of the weight of students choosing your own midpoints. Use frequency as the statistic and subdivide bars by gender.

2.2 Ott and Longnecker (2001) present an example in which the number of cell clumps per algae species were fitted to a Poisson distribution. A lake sample was analyzed to determine the number of clumps of cells per microscope field. The data are summarized below for 150 fields examined. Here, x_i denotes the number of cell clumps per field and n_i, denotes the frequency of occurrence of fields of each cell clump count.

x_i	0	1	2	3	4	5	6	≤ 7
n_i	6	23	29	31	27	13	8	13

Write a SAS program to perform a chi-square goodness-of-fit at $\alpha = .05$ to test the hypothesis that the observed counts were drawn from a Poisson probability distribution.

2.3 Devore (1982) discussed an example in which it is examined whether the phenotypes produced from a dihybrid cross of tall cut-leaf tomatoes with dwarf, potato-leaf tomatoes obey the Mendelian laws of inheritance. There are four categories corresponding to the four possible phenotypes: tall cut-leaf, tall potato-leaf, dwarf cut-leaf, and dwarf-potato leaf, with respective expected probabilities p_1, p_2, p_3, and p_4. The null hypothesis of interest is

$$H_0: p_1 = \frac{9}{16}, \ p_2 = \frac{3}{16}, \ p_3 = \frac{3}{16}, \ p_4 = \frac{1}{16}$$

Write a SAS program to perform a chi-square goodness-of-fit at $\alpha = .05$ of this hypothesis given that the observed counts in each category in a sample of size 1611 are 926, 288, 293, and 104, respectively.

2.4 The following data, taken from Rice (1988), represent the incidence of tuberculosis in relation to blood groups in a sample of Eskimos. Is there any association of the disease and blood group?

Severity	O	A	AB	B
Moderate-Advanced	7	5	3	13
Minimal	27	32	8	18
Not Present	55	50	7	24

Write a SAS program with a `proc freq` step for performing a chi-square test using $\alpha = .05$ to answer the above question.

2.5 Ott et al. (1987) cited a study of migrants to determine whether "degree of kinship participation" in the extended family is independent of a family's socioeconomic status. Use the data below to compute a chi-square test of independence using `proc freq`. Use nominal measures of association, the contingency coefficient C and Cramer's V, to comment on the strength of association if present.

Socioeconomic Status	Degree of Kinship Participation		
	Low	Medium	High
Low	75	98	75
Minimal	182	211	123
Not Present	116	122	41

2.5 An example in Schlotzhauer and Littell (1997) presented data from an experiment conducted by an epidemiologist who classified the disease severity of dairy cows (none, low, high) by analyzing blood samples for the presence of a bacterial disease. The size of the herd that each cow belonged to was classified as large, medium, or small. One of the aims of this study was to determine if disease severity was affected by herd size.

Severity	Herd Size		
	Large	Medium	Small
None	11	88	136
Low	18	4	19
High	9	5	9

Use a SAS program to analyze these data using `proc freq`. Is there any association between two variables? Use the various measures to interpret any association present considering that the two variables are ordinal and that the experimenter is planning to use herd size for predicting disease severity. Obtain confidence intervals for the measures that you discuss.

2.6 Write a SAS program containing an `infile` statement to access the data set used in SAS Example B5 (see Fig. 2.13) from a text file. Use `proc tabulate` to obtain a table laid out as follows. The rows of the table consist of the combinations of year in school and gender, with the levels of gender appearing within each level of year. The column must present mean and standard deviation of the two variables height and weight. Use formats and labels to enhance your table. How can you add a *single* column containing the sample size formatted to print as a four-digit integer?

Exercises 2.7-2.15 concern the demographic data set on countries obtained from Ott et al. (1987) (see Table B.4 of Appendix B). To access this data set from a text file, include an appropriate `infile` statement in your SAS program. A `filename` statement may also be included if you prefer. Use the

following input statement to read these data:
`input country $20. birthrat deathrat inf_mort life_exp popurban`
                      `perc_gnp lev_tech civillib;`

2.7 Write a SAS program to create a SAS data set named `world` using the data in Table B.4. Label variables as appropriate. Create category variables as described below:

Variable	Groupings	Category Variable
Infant mortality	$< 24 = 1$ (low) $24 - 73 = 2$ (moderate) $\geq 74 = 3$ (high)	`infgrp`
Level of technology	$< 24 = 1$ (low) $\geq 24 = 2$ (high)	`techgrp`
Degree of civil liberties	$1, 2 = 1$ (low degree of denial) $3, 4, 5 = 2$ (moderate degree of denial) $6, 7 = 3$ (high degree of denial)	`civilgrp`

Use a `libname` statement and a two-level name to save the SAS data set in a folder in your computer.

2.8 In a SAS program, use `proc univariate` to compute 33.3 and 66.7 percentiles of the variables `birthrat`, `deathrat`, and `popurban` available in the SAS data set `world`. Access the SAS data set saved previously in Exercise 2.7. Use the `output` statement to save these statistics in a temporary SAS data set named, say, `stats`. Obtain a listing of this data set.

2.9 Use the printed output from the analysis performed in Exercise 2.8 to determine good cutoff values for creating additional category variables `birthgrp`, `deathgrp`, and `popgrp` (each with three categories) corresponding to these variables. In a data step of a new SAS program, access the SAS data set `world` saved previously. Add statements to the data step to create the category variables described in Exercice 2.7, name the resulting SAS data set `world2`, and save the new data set in the same folder.

2.10 In a new SAS program, use the SAS data set `world2` saved in Exercise 2.9 in a `proc univariate` step to compute descriptive statistics, extreme values, percentiles, and low-resolution plots for the variables `life_exp`, `perc_gnp`, and `popurban`. Use an appropriate option to calculate t-tests for the hypotheses that population means for each of these variables exceed 70 years, below $3000, and above 60%, respectively. Also, include an option for calculating 95% confidence intervals for these parameters. Interpret the results of the t-tests using the p-values printed. Comment on the shape of the distribution of each of these variables using the printed output produced. Do the Shapiro-Wilk tests for normality conducted for each variable above support your conclusions?

2.11 Add a `proc means` step containing appropriate `class` and `output` statements, to the same SAS program used in Exercise 2.10, to create a SAS data set named `stats1`. The data set `stats1` must contain sample means, standard errors of the means, and maximum and minimum values of the variables `birthrat`, `deathrat`, and `inf_mort` calculated separately for each of the nine groups defined by combinations of levels of the category variables `birthgrp` and `popgrp`. Use the `types` statement to ensure that statistics are calculated only for *combinations of levels* of `birthgrp` and `popgrp`. Suppress printed output in `proc means`. Also, add a `proc format` step to your SAS program to define user formats for use when printing all category variables. Obtain a listing of the data set `stats1`. Label all new variables on output.

2.12 Add a `proc plot` step to the SAS program used in Exercise 2.11, to use the SAS data set of sample statistics `stats1` created in the `proc means` step to obtain a low-resolution scatter plot of mean infant mortality versus `popgrp`. The plot symbols must identify the `birthgrp` using formatted values. Recall that a `proc format` was used to define user formats for the `popgrp` and `birthgrp` variables in this program.

2.13 In a new SAS program, use the SAS data set `world2` saved in Exercise 2.9 in a `proc freq` step to do the following:
 a. Obtain two-way frequency tables (in the cross-tabulation format) for the variable `techgrp` with `infgrp`, `techgrp` with `civilgrp`, and `popgrp` with `infgrp`. Compute chi-square statistics, cell χ^2, and cell expected values but no column, row, or cell percentages,
 b. Obtain a two-way frequency table (in the list format) for `infgrp` and `civilgrp`.
 c. Use the chi-square statistic to test hypotheses of independence between pairs of variables considered in part (a) and state your results.
 d. Use the contingency coefficient C to comment on the strength of association for the pair of variables `techgrp` and `civilgrp`.
 e. Use the values of Gamma, Kendall's Tau-b, and Spearman correlation coefficient to comment on the association between `techgrp` and `infgrp`.
 f. Use appropriate measures to evaluate the strength of association between `popgrp` and `infgrp`. Considering that `popgrp` is an independent variable useful for predicting `infgrp` for each country, interpret the appropriate measures of association. Explain.

2.14 Write a new SAS program to use the data set (named `world2`) saved in Exercise 2.9 in a `proc chart` step to construct the following low-resolution plots:
 i. A horizontal bar chart of mean life expectancy for each `infgrp` within each `techgrp` (i.e., side-by-side bar charts for each `techgrp`).

ii. A vertical bar chart (i.e., a histogram) of per capita GNP, choosing your own midpoints. Use frequency as the statistic and subdivide the bars by `civilgrp`.

2.15 Write a SAS program to use the data set (named `world2`) saved in Exercise 2.9, in a `proc tabulate` step to print a tabulation giving the sample size, sample mean, sample standard deviation, and the standard error of the mean of the variables `popurban` for subgroups of observations defined by the combination of values of `techgrp` and `deathgrp`. The row analysis consists of combinations of `techgrp` and `deathgrp` and the statistics for `popurban` must appear on the columns. Print the sample size without decimals, the sample mean with four decimal places, and the other two statistics with two decimal places each. Also, use appropriate text strings to label all statistics keyword headings (e.g., print Standard Deviation for `std`).

3
Statistical Graphics Using SAS/GRAPH

3.1 Introduction

Graphical methods have become a fundamental tool for statistical data analysis in recent times. In many situations, an analysis is incomplete without some graphical presentations, and in others, graphical methods are indespensible for supporting the arguments presented using classical statistical inference methods. This point cannot be more true than in the analysis of multivariate data. Whereas the traditional use of graphics is for presenting data summaries (say, for studying the sample distribution of a variable), the emphasis today is tilted more toward using graphics for examining and understanding the underlying structure of a data set. Sometimes some of the types of presentation graphics, such as histograms or pie charts, that are generally used for summarizing data also may be employed to convey a novel concept or a new idea using data or an interpretation not easily recognized just by looking at the numerical summaries. In this chapter, a selection of statistical graphics primarily useful for data analysis are presented and how these can be produced using SAS/GRAPH procedures are discussed. The emphasis is on the use of SAS/GRAPH statements to illustrate what capabilities are available and how they can be employed to obtain presentation quality graphics.

3.2 An Introduction to SAS/GRAPH

SAS/GRAPH software is a part of the SAS system that provides SAS procedures (procs) for producing high-resolution graphics. A general description of the structure of SAS/GRAPH programs, their operation, and detailed explanation of most common SAS/GRAPH statements is contained in Appendix A. The first subsection in this section provides details of statements available in two basic procedures: `proc gplot` and `proc gchart`. The second subsection provides examples of using some of these statements in SAS programs that also illustrate the use of simple SAS/GRAPH global statements.

3.2.1 Useful SAS/GRAPH procedures

In this subsection, statements available in two of the most useful SAS/GRAPH procedures are described in detail. In later sections in this chapter, the use of these statements will be illustrated by applying the procedures to obtain a variety of statistical graphics displays. For each procedure, statements and options that are most applicable for generating the types of graphs considered are summarized in a table format followed by a brief discussion of each option. For some options when the uses are not self-explantory, more elaborate information will be provided.

GPLOT procedure

The `gplot` procedure is one of the most useful SAS/GRAPH procedures. It produces two-dimensional graphs of bivariate data points and connects these points by regression lines and a variety of other interpolation methods provided. Selected statements and options available with `proc gplot` are summarized in Table 3.1. The primary *action* statement in `proc gplot` is the `plot` statement. The `plot` statement must have at least one *plot-request* but may contain multiple plot-requests, followed by *plot statement options* preceded by a slash. A plot-request can be any of the following forms:

$$y * x$$

$$y1 * x1 \quad y2 * x2 \ldots$$

$$y * x = z$$

$$y * x = n$$

where x, $x1$, etc. denote the variables plotted on the horizontal axis and y, $y1$, etc. denote those plotted on the vertical axes. z is a third variable, usually a classification variable associated with the subjects (or cases) on which the variables x and y have been measured. The number n denotes sequence number of the `symbol definition` statement associated with the symbol used to plot y against x. The use of n is illustrated in the SAS Example C1 program (see Fig. 3.2).

The global statements `axis`, `symbol`, `pattern`, `title`, and `footnote` and the `note` statement can be used for enhancing plots produced by the `gplot` procedure as described in Section A.2 of Appendix A. Some of the more important *plot statement options*, tabulated in the lower half of Table 3.1, are briefly described below:

vaxis=, haxis=, vminor=, hminor=

The `vaxis=` and `haxis=` allow the user to specify a list of major tick-mark values for the vertical and the horizontal axis, respectively. Various forms of

```
PROC GPLOT   < data=>   < gout=>   < annotate=>   <uniform> ;
  PLOT   plot-request(s) ... /   <options>
  AXIS statement         <options>
  SYMBOL statement       <options>
  PATTERN statement      <options>
  TITLE statement        <options>
  FOOTNOTE statement     <options>
  NOTE statement         <options>
  BY statement

PLOT statement options:
  vaxis=value-list   vminor=n      haxis=value-list   hminor=n
                     vm=n                             hm=n
  vaxis=axis-definition            haxis=axis-definition
  overlay            nolegend      noaxes
  legend=legend-definition         areas=n
  frame              cframe=color  caxis=color        ctext=color
  fr                 cfr=color     ca=color           ct=color
  annotate=data-set-name
  anno=data-set-name
  href=value-list    chref=color   lhref=line-type
                     ch=color      lh=line-type
  vref=value-list    cvref=color   lvref=line-type
                     cv=color      lv=line-type
```

Table 3.1. Summary of `GPLOT` procedure options

this list are discussed in the description of options for the `axis` statement in Subsection A.2.2. The `vminor=` and `hminor=` (or the shortened versions, `vm=` and `hm=`) options specify the number of minor tick marks drawn between each pair of major tick marks on the two axes, respectively. These options are useful when the user is content to accept the default settings for properties of these graphic elements such as color, size, font, etc. and thus does not need to provide global axis definitions. Otherwise, the list of major tick-mark values can be included in an `axis` definition along with the user's settings for their properties. These global axis definitions are then available to be specified as settings for the options `vaxis=` or `haxis=`, as illustrated in SAS Example C1 (see Fig. 3.2).

overlay

This option enables several plots created by plot-requests in a single `plot` statement to be placed in the same graph (i.e., to be plotted using the same set of axes). The axes are scaled to accommodate the range of values of all variables plotted on each axis but labeled only with the names or labels of the first pair of variables plotted. The primary application of this option is to overlay different graphical representations of a set of data on the same graph, and this is illustrated in the SAS Example C1 program (see Fig. 3.2). In the graph produced by the second `proc gplot` step, the data points and a regression line fitted to them appear in the same graph (see Fig. 3.4).

The `overlay` option can also be used to plot different sets of values for one variable (say, the variable plotted on the vertical axis) against the same set of values of the other variable. This will be the case when several time series or longitudinal data sets are plotted over the same time period or when several dependent variables are plotted against the same independent variable. The plots are meaningful only if the variables plotted on the vertical axis have been measured in the same units or in the same scale. In such applications, different symbols, colors, and/or line types may be used to identify each plot.

legend=

The `legend=` option allows the user to assign legend definitions if the legends created by default are not adequate. Legends are useful only when a third variable, z, is used in a plot request. They identify the symbols, colors, or line types that are used for plotting graphic elements for each value of the third variable z.

areas=

The `areas=` option allows the user to fill the areas below the plot line specified with different patterns. The data points must first be interpolated with one of the options such as `i=join` or `i=rl` so that these areas are generated. The `pattern` definitions are automatically assigned to the areas under each plot line in the order they are drawn. The first area is the area between the horizontal axis and the plot line that is drawn first, the second area is the area below the second plot line and the first, and so on. Care must be taken to make sure that the lines are drawn in order such that lines that are lower are drawn first to avoid overlapping of the fill patterns. The `pattern` definitions allow users to specify their own choices for the fill patterns as well as colors (see Table A.8 and related text).

3.2 An Introduction to SAS/GRAPH

> annotate=

If the user is interested in producing graphics using statements available in SAS/GRAPH procedures alone, knowledge of how to use annotate data sets is not essential. However, since the *annotate facility* is an integral part of SAS/GRAPH useful for adding user-drawn graphics elements for enhancing a graph produced by SAS/GRAPH procedures, some knowledge of this method is useful. As the value for this option in a `plot` statement, the user must provide the name of a SAS data set that contains special graphics commands in the form of observations in the data set. This data set, called the *annotate data set*, can be constructed using a SAS data step. The use of the *annotate facility* is discussed in Section A.3.

> href=, chref=, lhref=, vref=, cvref=, lvref=

The options `href=`, `chref=`, and `lhref=` allow the user to specify locations on the vertical axis at which reference lines are to be drawn parallel to the horizontal axis, their color, and the line-type, respectively. Similarly, the options `vref=`, `cvref=`, and `lvref=` specify properties of vertical reference lines.

GCHART procedure

SAS/GRAPH procedure `gchart` is typically used to draw vertical or horizontal bar charts, block charts, pie charts, or star charts. A *chart variable* and a *summary variable* are needed to produce these charts. The values of the chart variable are used to determine the *midpoints* that are plotted on the *midpoint axis* in the chart. The body of the chart itself displays various statistics (e.g., the mean) computed on the summary variable and drawn as bars, blocks, slices etc. and located at each of the midpoints. In many charts, a *response axis* is also drawn in the scale of the chart statistic displayed.

If the chart variable is a numeric variable with continuous values, then, by default, the midpoints consist of the midpoints of the class intervals (or bins) and the statistic plotted is the class frequency. These and other terms used for drawing bar charts with the `gchart` procedure are illustrated in Fig. 3.1 reproduced from SAS/GRAPH (SAS Institute Inc. ,2004). General form of the action statements available for use with `proc gchart`, and options available with these statements are summarized in Table 3.2. As an example, some of the options that may be used with the `hbar` statement are described briefly below.

> midpoints=

If the *chart variable* is a numeric variable, `proc gchart` will use an algorithm to derive *midpoints* for drawing bars, unless the mid points are specified using

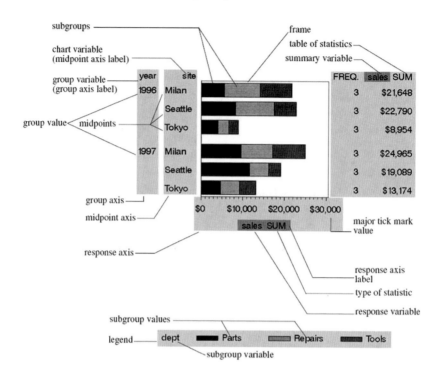

Fig. 3.1. Terms used with bar charts

a `midpoints=` option. The bars drawn will represent the frequencies of values falling in the class intervals (bins) defined by the midpoints, unless a `type=` option is used to request that cumulative frequencies, percent frequencies, or cumulative percentages are to be plotted instead. The possible values for the `type=` option are `freq`, `cfreq`, `pct`, and `cpct`. A `levels=` option may be used to specify the number of bars if the `midpoints=` option is not used and `proc gchart` will derive appropriate *midpoints*.

discrete

If the *chart variable* is a character variable, bars will be drawn at the different discrete values of the variable. If the *chart variable* is a discrete-valued but numeric variable (such as a category or a grouping variable) `discrete` option is needed to ensure that it is processed as a character variable.

sumvar=

When the `sumvar=var` option is used to specify a *summary variable* named *var*, the bars drawn represent the *sum* of the values of this variable for the observations corresponding to each value of the chart variable. The user

```
PROC GCHART  <data=>  <gout=>  <annotate=>  <uniform>;
  HBAR     chart variable(s) /   <options>
  VBAR     chart variable(s) /   <options>
  BLOCK    chart variable(s) /   <options>
  PIE      chart variable(s) /   <options>
  STAR     chart variable(s)/    <options>
  AXIS statement                 <options>
  PATTERN statement              <options>
  TITLE statement                <options>
  FOOTNOTE statement             <options>
  NOTE statement                 <options>
  BY statement

Options for HBAR include:

  ascending  descending  noheading  nolegend
  freq cfreq percent cpercent mean sum nostats
  axis=value-list        raxis=value-list    maxis=axis-definition
  axis=axis-definition   raxis=axis-definition

  group=variable         subgroup=variable
  sumvar=variable        type=statistic

  discrete               levels=number       midpoints=value-list

  frame                  caxis=color         ctext=color
  cframe=color           coutline=color

  patternid=method       legend=legend-definition

  name=quoted-string
```

Table 3.2. Summary of GCHART procedure options

may specify the statistic to be computed using the type= option. When the sumvar= option is used, possible values of this option are sum and mean.

subgroup=

The subgroup=*var* option causes each bar to be subdivided into a number of segments according to the values of the variable *var*. Thus, *var* is necessarily a discrete-valued variable. The effect is that the observations represented in each vertical bar will be subdivided into groups corresponding to each value of *var*. The statistic for each of these subgroups is computed and represented as a segment of the bar. The statistic could be a frequency count of a chart variable or a mean of a summary variable.

group=

The option `group=`*var*, on the other hand, groups the observations by the values of *var* before bars are drawn. This variable is also required to be a discrete-valued variable. The process of drawing bars for each value of the chart variable is repeated for each value of *var*. Thus, a separate set of bars is reproduced for each value of *var*. For example, the statistic of a summary variable is now calculated using only those observations corresponding to a given value of *var*.

maxis= , raxis=

The `maxis=` option specifies an axis definition to be used to draw the *midpoint axis* and `raxis=` is used to specify an axis defintion for the *response axis*.

3.2.2 Writing SAS/GRAPH programs

Generally, several Base SAS procedures such as `plot` or `chart` as well as SAS/STAT procedures such as `reg` produce low-resolution graphics (so-called line-printer graphics). Base SAS procedure `univariate` and SAS/STAT procedures such as `reg` or `boxplot` also provide mechanisms for producing high-resolution graphics in addition to line-printer graphics, using statements comparable to those available in SAS/GRAPH software, to control various attributes of the graphs.

SAS Example C1 and SAS Example C2 presented in this section illustrate SAS/GRAPH programs that contain simple `proc gplot` steps that employ some of the SAS/GRAPH statements and options discussed in Section 3.2.1. SAS Example C4 illustrates a SAS/GRAPH program that contain a simple `proc gchart` step.

SAS Example C1

The SAS Example C1 program shown in Fig. 3.2 illustrates features of a simple SAS/GRAPH program. The program contains examples of statements that are needed to produce some of the graphical tools for performing a statistical analysis. The data taken from Weisberg (1985) consist of atmospheric pressure (inches of mercury) and the boiling point of water (in degrees Fahrenheit) measured at different altitudes above sea level.

3.2 An Introduction to SAS/GRAPH

A simple linear regression line is to be fitted with the dependent variable y, taken to be 100 times the logarithm of pressure, `logpres`, with boiling point, `bpoint`, being the independent variable, x. The use of the SAS/GRAPH procedure `gplot` to obtain a scatter plot of `logpres` against `bpoint` is illustrated in the first `proc step` in Fig. 3.2. The only procedure statement needed in this proc step is `plot logpres*bpoint;`, which creates the graph shown in Fig. 3.3. The `plot` statement is a example of an `action` statement, that along with the `run` statement causes a graph to be drawn.

The graph shown in Fig. 3.4 improves on several features of the previous graph by using SAS/GRAPH *global* statements. SAS global statements such as `symbol` and `axis` are used not only to change the appearance of the various graphics elements such as symbols but also for some enhancements

```
goptions rotate=landscape targetdevice=pscolor   1
                         hsize=8 in vsize=6 in;

data forbes;
   input bpoint pressure @@;
   label bpoint='Boiling Point (deg F)'
         pressure= 'Barometric Pressure(in. Hg)';
   logpres=100*log(pressure);
datalines;
194.5 20.79 194.3 20.79 197.9 22.40 198.4 22.67 199.4 23.15
199.9 23.35 200.9 23.89 201.1 23.99 201.4 24.02 201.3 24.01
203.6 25.14 204.6 26.57 209.5 28.49 208.6 27.76 210.7 29.04
211.9 29.88 212.2 30.06
;
run;

proc gplot data = forbes;
   plot   logpres*bpoint;
run;

title1 j=c h=2.0 f=none c=darkviolet
                       'Analysis of Forbes(1857) data';
title2 j=c h=1.5 f=none c=mediumblue lspace=1.5   2
                       'on boiling point of water';

footnote1  h=1.5 f=italic c=blueviolet move=(8,+0)   3
                       'Source: S. Weisberg';
footnote2  h=1.3 f=centbi c=darkred move=(16,+0)
                       'Applied Linear Regression, 2005';

symbol1   c=red v=star i=none h=1.5;   4
symbol2   ci=darkcyan  i=rl v=none;

axis1 label=(c=magenta a=90 h=1   5
                       '100x log(pressure(in Hg))') value=(c=blue);
axis2 order=190 to 215 by 5 label=(c=magenta h=1) value=(c=blue );   6

proc gplot data = forbes;
   plot   logpres*bpoint=1
          logpres*bpoint=2/vaxis=axis1 haxis=axis2 overlay;   7
run;
```

Fig. 3.2. SAS Example C1: Program

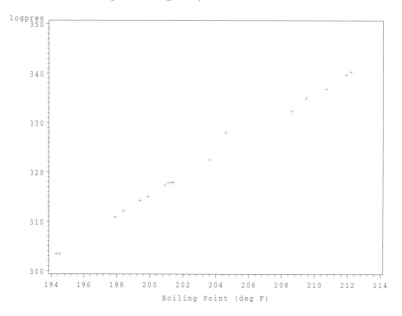

Fig. 3.3. SAS Example C1: Graph 1

such as adding titles, improving axis labels, and superimposing a regression line. The principal *global* statements used are the `title`, `footnote`, `symbol`, and the `axis` statements. Options available in each of these statements will be detailed in Section A.2. The choice of values specified for various options in each *global* statement in this example and their effects on the appearance of the graph are described in the paragraphs below. Note that the settings `rotate=landscape` and `hsize=8 in vsize=6 in` specified in the `goptions` changes the dimensions of the graphical output area while the graphical unit size remains unchanged. The terms *graphical output area* and *graphical unit size* are explained in Appendix A.

Title Statements in SAS Example C1

The `title` statements define the lines to appear in the upper part of the graphics output area, the number following the word `title` indicating the position it will appear in the title area (see Fig. A.7 of Appendix A). The settings `j=c, h=2, f=none,` and `c=darkviolet` for the specified options in the first title definition causes the first line of the title to be plotted center-justified, with text characters of height 2 (in default units) using the *default hardware font*, and in the predefined SAS color `darkviolet`. If `f=none` is omitted from the title statements, the software font named `complex` will be used for the first title line, instead of the hardware font.

Because of the goptions statement setting of `targetdevice=pscolor` [1] (see Fig. 3.2), SAS determines values for various device parameters for pro-

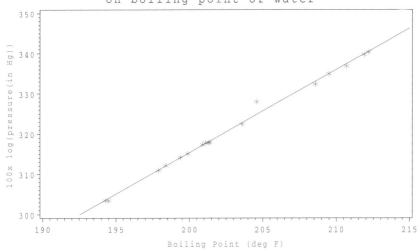

Fig. 3.4. SAS Example C1: Graph 2

ducing graphics output from the device entry for the `pscolor` device driver, unless they are overridden in SAS/GRAPH statements in the program. For example, the `Parameters` window in Fig A.2 shows that the value of Chartype is set to 1. Thus, the *default hardware font* will be the entry in the `Chartype` window that corresponds to a Chartype of 1 which, from Fig. A.4, is seen to be the `Courier` font.

The `lspace=1.5` ❷ used in the `title2` statement specifies the line spacing, in default units, between the two title lines, which, by default, is 1, but here is increased by 50%.

Footnote Statements in SAS Example C1

The `footnote` statements define the lines to appear in the lower part of the graphics output area (see Fig. A.7). The values for options specified in the first footnote statement, `h=1.5`, `f=italic`, `c=blueviolet`, and `move=(8,+0)` ❸, result in the first text line to be plotted beginning at the coordinate position (8, +0) in the predefined SAS color `blueviolet` using an italic font and text characters of height 1.5 in default units. The font named `italic` is a SAS/GRAPH software font for text, which the program will locate in the SAS/GRAPH font catalog.

Symbol Statements in SAS Example C1

The `symbol` statements define the appearance of the plot symbols of the points plotted and how they are interpolated. The keyword `symbol` followed by a number begins a symbol definition. The number can be used in a procedure action statement, such as the `plot` statement, to indicate that a particular symbol definition be applied in plotting points. For example, the statement `plot logpres*bpoint=1 logpres*bpoint=2/ ...` specifies that the plot `logpres` vs.`bpoint` be repeated, first using symbol definition 1 (`symbol1`) and again using symbol definition 2 (`symbol2`), respectively, for determining the appearance of the plot symbols and how the points are to be interpolated. The settings for options in symbol definition 1, `c=red v=star i=none h=1.5` **4** specify that the symbols plotted are red-colored stars of size 1.5 in default units with no interpolation (i.e., the symbols are not connected with lines or line segments). The settings in symbol definition 2, `ci=darkcyan i=rl v=none` on the other hand, specify that the points are to be interpolated by a linear regression line, plotted in the predefined SAS color `darkcyan`. The `v=none` option ensures that the points are not plotted a second time. Thus, the two symbol definitions result in the *overlaying* of a linear regression line fitted to the pairs of data values that were previously plotted as a scatter plot. The plot statement keyword option `overlay` **7** causes these two plots, the scatter plot of points and the regression line, to appear on the same graph (i.e., use the same pair of axes).

Axis Statements in SAS Example C1

The `axis` statements define the appearance of an axis. The keyword `axis` appended with a number followed by optional parameter specifications constitute an axis definition. Axis definitions can be used in procedure action statement(s) to designate that a particular axis definition be used for plotting the specified axis. In this example, the plot statement options `vaxis=axis1 haxis=axis2` specify that the vertical and horizontal axes be drawn using axis definition 1 (`axis1`) and axis definition 2 (`axis2`), respectively.

In axis definition 1, the `label=(c=magenta a=90 h=1 '100x log(pressure(in Hg))')` **5** option specifies that the given text string be rotated 90 degrees anticlockwise and printed using the color magenta and a character size of 1 default units. `c=    a=    h=   'text-string'` are *parameters* for the label option. The `value=(c=blue)` specifies that the values appearing at the major tick marks be of color blue. `c=` is one of the parameters for the `value` option. Thus, in Fig. 3.4, although the major tick-mark values for the vertical axis will remain those calculated by `proc gplot`, the plot color of those values will be changed to blue.

In Fig. 3.3, values calculated by `proc gplot` by default are used to determine the number of major tick marks and the values appearing at those tick marks in plotting both the horizontal and the vertical axes. The keyword specification `order=190 to 215 by 5` **6** in axis definition 2, however, overrides these default values. The consequence of this modification is seen in Fig. 3.4, as both the number of tick marks and their values differ from those in Fig. 3.3. This serves as an example of customizing the appearance of axes using the `axis` statement. The parameter values set for the keyword option `label=`, `(c=magenta h=1)` specifies the color and size of the text characters used to print the horizontal axis label, the actual text string printed being determined by `proc gplot` by default, to be the label for the variable `bpoint`, as specified in the label statement in the data step: `label bpoint='Boiling Point (deg F)'` ⋯.

```
data mileage;
input additive $ percent @@;
label additive='Gasoline Additive' percent='% Increase in Mileage';
datalines;
1 4.2 1  2.9 1 0.2 1 25.7 1  6.3 1  7.2 1  2.3 1 9.9 1 5.3 1  6.5
2 0.2 2 11.3 2 0.3 2 17.1 2 51.0 2 10.1 2  0.3 2 0.6 2 7.9 2  7.2
3 7.2 3  6.4 3 9.9 3  3.5 3 10.6 3 10.8 3 10.6 3 8.4 3 6.0 3 11.9
;
run;

title1 c=blue f=centb h=2 'Study Comparing % Mileage Increase';
title2 c=mediumblue f=swissi h=1.5 'from three different gasoline additives';

axis1 label=(c=darkviolet a=90 h=1.5) value=(c=blue);
axis2 offset=(.5 in) label=(c=darkviolet h=1.5) value=(c=blue );

symbol1 c=red v=dot i=none h=1;   ■1
symbol2 i=hiloj co=green cv=crimson h=1 v=dot w=2 l=2;
symbol3 i=stdmptj ci=crimson co=maroon v=dot h=1;
symbol4 i=boxt ci=maroon co=darkred v=star cv=vib bwidth=3;

proc gplot data=mileage;
plot percent*additive=1
     percent*additive=2   ■2
     percent*additive=3
     percent*additive=4 /vaxis=axis1 haxis=axis2;
run;
```

Fig. 3.5. SAS Example C2: Program

SAS Example C2

The SAS Example C2 program displayed in Fig. 3.5 illustrates several interpolation methods possible through the use of `symbol` statement options discussed previously. The data used for this example are the percentage increase in gas mileage (in mpg) obtained by 10 randomly assigned cars to each of three different gasoline additives, labeled 1, 2, and 3. Four different graphs are obtained using a single plot statement with four plot-requests each using a different symbol definition. First, a scatter plot of the data, obtained using `proc gplot` with `percent*additive=1`, is displayed in Fig. 3.6. Thus, `proc gplot` uses the definition `symbol1` ■1, where the options provided are `c=red`

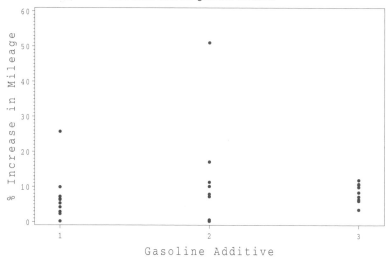

Fig. 3.6. SAS Example C2: Plot using `symbol1` definition without interpolation

`v=dot i=none h=1`, producing plot symbols that are red-colored dots of size 1 in cell units.

Fig. 3.7 displays the results of the plot-request `percent*additive=2`. The options `i=hiloj co=green cv=crimson h=1 v=dot w=2 l=2` used in the def-

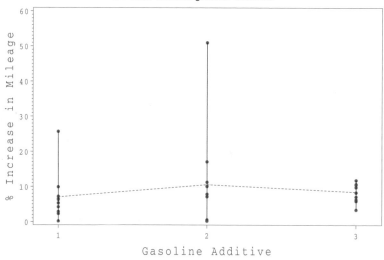

Fig. 3.7. SAS Example C2: plot using `symbol2` definition: `interpol=hilo`

inition `symbol2` ■ specifies that the data be interpolated by a solid line of width 2 connecting the smallest value to the largest for each additive, as shown in Fig. 3.7. A tick mark on each line indicates the *mean percentage mileage increase* for each additive.

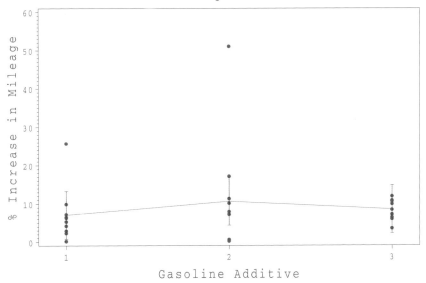

Fig. 3.8. SAS Example C2: Plot using `symbol3` definition with `interpol=std`

These means are connected by a line of type 2 (i.e., medium-length dashes) as a result of the use of the "j" suboption in `i=hiloj` and the `l=2` option. In addition to the solid line spanning the data values, the data points themselves are also plotted, using the the "dot" symbol. The color for the solid line (a bar) is specified by `co=green` and the color for the symbols is specified by `cv=crimson`. In the absence of a `ci=` specification specify the color of the plotted line connecting the means, by default, SAS/GRAPH uses the color specified using the `cv=` option.

In the definition `symbol3` ■, the option `i=stdmptj` is used to interpolate the data points for each additive with *error bars*. Error bars are solid lines drawn joining the two points defined by the sample mean ± $1s$, $2s$, or $3s$, where s is the standard error of the mean. By default, the pooled sample variance is used to calculate s, but that can be changed using a setting for the `var=` suboption. The options `v=dot h=1` define the plot symbol as before. The graph resulting from the plot-request `percent*additive=3` is displayed in Fig. 3.8. Again, the data points are also plotted, with the color of the

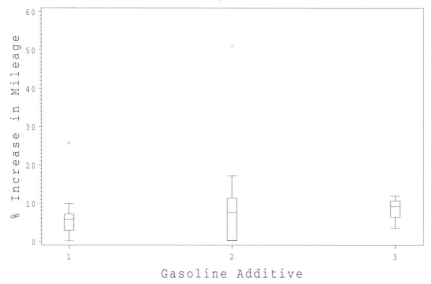

Fig. 3.9. SAS Example C2: Plot using `symbol4` definition with `interpol=box`

interpolated solid lines determined by the option `co=maroon`. The means for each additive are also joined by a solid line (by default) of color determined by the option `ci=crimson`. In the absence of a `cv=` option to specify the color of the symbols, SAS/GRAPH uses the color specified for the `ci=` option, by default.

The options `i=boxt ci=maroon co=darkred v=star cv=vib bwidth=3` used in the `symbol4` **1** statement specifies that the data be interpolated by *schematic box plots*, the `t` option adding tick marks to the ends of the whiskers. The plot-request `percent*additive=4` results in the graph displayed in Fig. 3.9. The outside values are plotted with the star symbol in the color specified in `cv=vib`. The color specified in `co=darkred` determines the color of the boxes and whiskers and the width of the box is changed to three times the default value by the use of `bwidth=3`.

SAS Example C3

The SAS Example C3 program shown in Fig. 3.10 illustrates the features of a simple SAS/GRAPH program for obtaining a histogram. The data are from an experiment, described in Ott and Longnecker (2001), to investigate whether the addition of an antibiotic to the diet of chicks promotes growth over a standard diet. An animal scientist raises and feeds 100 chicks in the same environment, with individual feeders for each chick. The weight gains for the 100 chicks are recorded after an 8-week period. The construction of a frequency table suggests class intervals of widths of 0.1 units beginning with a midpoint at the data value of 3.7. The histogram is displayed in Fig. 3.11 and the SAS program for drawing the histogram is shown in Fig. 3.10.

```
data chicks;
input wtgain @@;
label wtgain ='Weight gain (in gms) after 8-weeks';
datalines;
3.7  4.2  4.4  4.4  4.3  4.2  4.4  4.8  4.9  4.4
4.2  3.8  4.2  4.4  4.6  3.9  4.1  4.5  4.8  3.9
4.7  4.2  4.2  4.8  4.5  3.6  4.1  4.3  3.9  4.2
4.0  4.2  4.0  4.5  4.4  4.1  4.0  4.0  3.8  4.6
4.9  3.8  4.3  4.3  3.9  3.8  4.7  3.9  4.0  4.2
4.3  4.7  4.1  4.0  4.6  4.4  4.6  4.4  4.9  4.4
4.0  3.9  4.5  4.3  3.8  4.1  4.3  4.2  4.5  4.4
4.2  4.7  3.8  4.5  4.0  4.2  4.1  4.0  4.7  4.1
4.7  4.1  4.8  4.1  4.3  4.7  4.2  4.1  4.4  4.8
4.1  4.9  4.3  4.4  4.4  4.3  4.6  4.5  4.6  4.0
;
run;
  title1 c=blue 'Weight Gain of Chicks';
  title2 c=blue 'Histogram with 13 intervals';
  axis1 label=(c=steelblue h=1 ' Weight Gain ');
  axis2 order=0 to 15  by 1 label=(c=steelblue h=1 a=90 'Frequency');  1
  pattern1 c=magenta v=l3;  2

proc gchart data=chicks;
  vbar wtgain/midpoints=3.7 to 4.9 by .1   3
              maxis=axis1
              raxis=axis2 ;
run;
```

Fig. 3.10. SAS Example C3: Program

Two `title` statements, two `axis` statements, and a `pattern` statement comprise the set of global statements defined for use in the `proc gchart` step. The action statement is the following `vbar` 3 statement, with `wtgain` being the *chart variable*:

`vbar wtgain/midpoints=3.7 to 4.9 by .1 maxis=axis1 raxis=axis2;`

Since `wtgain` is a variable of type *numeric*, one of the options `midpoints=` or `levels=` must be specified for `proc gchart` to determine points on the horizontal axis (called the *midpoint axis*) at which the vertical bars are to be positioned. In this example, the option `midpoints=` specifies the midpoints of the class intervals (or *bins*) used to enumerate the class frequency counts. These frequencies determine the heights of the bars displayed because `freq` is the default *chart statistic*. The `midpoints=` option as shown in the current example results in a simple histogram that uses the class intervals determined by the midpoint values specified. If, for example, the option `levels=12` is used instead, thus specifying the number of midpoints (therefore the number of bars), a different histogram will result.

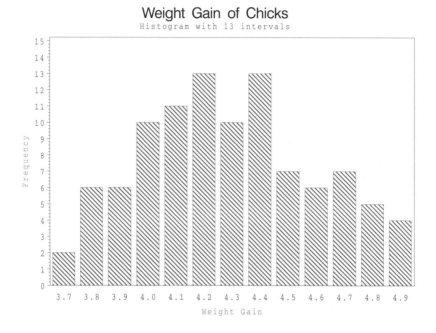

Fig. 3.11. SAS Example C3: Histogram

The axis definitions `axis1` and `axis2`, respectively, define the *midpoint axis* and *response* or *the vertical axis*. Note carefully that in the axis definition statement `axis2` ❶, the option `order=0 to 15 by 1` is used to specify the number of major tick marks and their values on the response axis. Also, the `pattern` ❷ definition supplied will be used to fill in the vertical bars using the color *magenta* and the fill pattern that corresponds to the setting `13`.

3.3 Quantile Plots

The methods discussed in this section are confined to the analysis of univariate data (i.e., observed values for a single variable) for a sample of experimental subjects.

Quantiles and Percentiles

Sample quantiles and percentiles are a primary tool for studying the distribution of a variable. The .8 quantile (or the 80th percentile) of a *theoretical distribution* is a value such that 80% of the probability mass of the distribution lies to the left and 20% lies to the right of the value of the quantile. For a given sample of data, due to discreteness, this definition cannot be directly applied to calculate sample quantiles. A variety of conventions can be used to compute a specified sample quantile or a percentile of a given set of data. SAS provides five definitions for computing specified sample quantiles in the univariate procedure through the procedure option pctldef=. The following general definition of sample quantiles closely follows that found in Chambers et al. (1983).

A number in the domain of the observations for a variable is defined as the pth quantile ($0 < p < 1$) of the variable if a fraction p of the observations fall below it and a fraction $(1-p)$ fall above it. Suppose that $x_{(1)}, x_{(2)}, \ldots, x_{(n)}$ denote the observations arranged in the increasing order of magnitude of a random sample of observations $x_1, x_2, \ldots, x_n$. Note that the the subscripts $(1), (2), \ldots, (n)$ represent the *ranks* of the data values, (1) being the smallest and (n) the largest. The *sample quantile function* $Q(u)$ given the ordered sample $x_{(1)}, x_{(2)}, \ldots, x_{(n)}$ is defined as

$$Q(u) = x_{(i)}, \quad \frac{i-1}{n} < u \leq \frac{i}{n}, \quad \text{for } i = 1, \ldots, n$$

The sample quantile function (also known as the *empirical distribution function*) is a step function and provides a method of computing sample quantiles, one of a number of nonparametric estimates of quantiles available. For a fixed p in $(0,1)$, the pth *sample quantile* $Q(p)$ is defined as $Q(p) = x_{(i)}$ when np is an integer or $Q(p) = x_{(i+1)}$ when np is not an integer, where $i = \lfloor np \rfloor$. This corresponds to the estimates computed by SAS procedure univariate using pctldef value equal to 3. An alternative weighted average estimator is

$$Q(p) = (1-g)x_{(i)} + gx_{(i+1)}$$

where $i = \lfloor (n+1)p \rfloor$ and $(n+1)p = i + g$. Thus, g is the fractional part of the number $(n+1)p$. This corresponds to the estimates computed by SAS using pctldef value equal to 4. This definition results in, for example, the .5 sample quantile or the 50th percentile to be computed as

$$Q(.5) = x_{((n+1)/2)} \text{ if n is odd}$$
$$= (x_{(n/2)} + x_{(n/2+1)})/2 \text{ if n is even}$$

which is the traditional formula for computing the sample median. The default value of 5 used by SAS for pctldef also leads to the sample median being computed this way. However, in general, the pctldef value equal to 5 gives an average estimator $Q(p) = (x_{(i)} + x_{(i+1)})/2$ when np is an integer and $Q(p) = x_{(i+1)}$ when np is not an integer, where $i = \lfloor np \rfloor$.

For the purpose of graphing quantile-related plots by hand, $Q(p)$ is defined to be $x_{(i)}$ whenever p is one of the fractions of $p_i = (i - .5)/n$ for $i = 1, \ldots, n$;

that is, the $(i-.5/n)$th quantile is the order statistic $x_{(i)}$. The following example illustrates this definition.

Example 3.3.1

For $n = 5$, if the ordered sample is $1.23, 2.64, 2.87, 3.41, 5.83$, the quantile function $Q(p)$ for the data set is

i	1	2	3	4	5
p_i	.1	.3	.5	.7	.9
$Q(p_i)$	1.23	2.64	2.87	3.41	5.83

Note that the i in the above table represent the rank of the data values. Thus, for a data set of five observations, the .5 quantile is the third largest observation. Any intermediate quantiles such as the .25 quantile or the 33rd percentile are approximated by linear interpolation. Note that the use of $p_i = (i-.5)/n$ is not unique as $p_i = i/(n+1)$ has also been used. However, for the purpose of descriptive comparisons, the convention used is immaterial. A more complex definition, $p_i = \frac{(i-\frac{3}{8})}{n+\frac{1}{4}}$, has been proposed and this definition is used in SAS procedures for constructing normal probability plots.

The graph of $Q(p_i)$ against p_i is called the *quantile plot* or the empirical cumulative probability plot or, simply, the probability plot. The quantile plot reveals many important characteristics of the distribution of the data.

SAS Example C4

Consider the ozone concentrations for Stamford given in Table B.2 of Appendix B. The `quantile function` or the ecdf for this data set is computed in Table 3.3, using the formula $p_i = (i-.5)/n$. SAS Example C4 illustrates

i	$Q(p_i)$	p_i
1	14	.0036765
2	14	.011029
3	23	.018382
4	24	.02574
5	24	.03309
6	24	.04044
⋮	⋮	⋮
132	206	.96691
133	212	.97426
134	215	.98162
135	230	.98897
136	240	.99632

Table 3.3. Quantiles of Stamford ozone data

the features of a SAS/GRAPH program, shown in Fig. 3.12, used to graph a quantile plot of the Stamford ozone data. Note that the data are entered in the same format they are tabulated (i.e., both Stamford and Yonkers ozone data continued in two columns in date order), with periods entered to indicate missing data values.

```
goptions rotate=landscape targetdevice=pscolor
                          hsize=8 in vsize=5 in;
data ozone;
input x1-x10;
drop x1-x10;
array var(10) x1-x10;  1
do i=1 to 10 by 2;
   stamford=var(i);           2
   yonkers =var(i+1);
   output;
end;
datalines;
66 47  61  36 152  76  80 66 113 66
52 37  47  24 201 108  68 82  38 18
 . 27   .  52 134  85  24 47  38 25
 . 37 196  88 206  96  24 28  28 14
 . 38 131 111  92  48  82 44  52 27
 .  . 173 117 101  60 100 55  14  9
49 45  37  31 119  54  55 34  38 16
 .  .   .   .   .   .
 .  .   .   .   .   .
 .  .   .   .   .   .
40 50 114  67  27  13 212 117 38 21
47 31  32  20   . 25   80  43 24 14
51 37  23  35  73 46   24  27 61 32
31 19  71  30  59 62   80  77 108 51
47 33  38  31 119 80  169  75  38 15
14 22 136  81  64 39  174  87  28 21
 . 67 169 119   . 70  141  47   . 18
71 45   .   . 111 74  202 114   .  .
run;
proc rank fraction data=ozone;  3
  var stamford;
  ranks p;
run;

title1 c=blue 'Quantile plot of Stamford Ozone data';
axis1 label=(c=darkviolet f=centb h=1.5 a=90 ' Quantile q(p)');
axis2 label=(c=darkviolet f=centb h=1.5 ' Fraction p=i/n');
symbol1 c=coral v=star i=none h=1;
proc gplot;  4
  plot stamford*p/vaxis=axis1 vm=9
                  haxis=axis2 hm=3
                  href=.25 .50 .75 chref=aqua lhref=2
                  vref=60 100 cvref=crimson lvref=20
                  frame ;
run;
```

Fig. 3.12. SAS Example C4: Program

The 10 columns form 10 variables x1,...,x10, which are declared in an array named var of length 10 **1**. The array elements var1, var3, var5, ...reference the variables $x_1, x_3, x_5,...$, respectively, and contain ozone measurements for Stamford, and the array elements var2, var4, var6,... reference the variables $x_2, x_4, x_6,...$, respectively, contain those for Yonkers. A do loop (with a counting variable i) and an output statement are used to create the two variables named stamford and yonkers containing the respective data values. **2**

The variables named x1,...,x10 are excluded from the data set ozone being created. For a simpler example of the use of array statement, see the SAS Example A6 program shown in Fig. 1.16. If the sample size is large, the graph of $Q(p_i)$ against $p_i = (i - .5)/n$ is identical to that against $p_i = i/n$, as the difference $.5/n$ causes no visible shift in the points plotted. The SAS procedure rank is first used to calculate the fractional ranks $p_i = i/n$ for values of the variable named stamford. This is easily accomplished by including the option fraction in the proc rank **3** statement and the procedure information statement ranks p;. This results in the creation of a new variable named p that contains the fractional ranks $p_i = i/n$, of the data values contained in the variable stamford. Note that these two variables will be written to a temporary SAS data set that may be given a name by the user with the out= option in the proc rank statement, if desired. This data set serves as the input SAS data set for the next procedure step using proc gplot.

The quantile plot is constructed in the proc gplot step **4** with the aid of some SAS/GRAPH global statements appearing before the step. These are a title statement, a symbol definition, and two axis definitions. The axis definitions are used to change a few appearance parameters; the default scales calculated by SAS/GRAPH for the two axes are considered adequate. The vm= and hm= options in the plot statement changes the number of minor tick marks on the vertical and the horizontal axis, respectively. The vref= and href= options specify values on the vertical and the horizontal axis, respectively, where reference lines are to be drawn. These lines are drawn, parallel to one axis, at the specified points on the other axis. For example, vertical lines will be drawn at values .25, .5, and .75 ,respectively, on the horizontal axis representing the 25th, 50th, and the 75th percentiles. The horizontal reference lines are drawn at the values 60 and 100 on the vertical axis. However, these lines were added to the graph subsequently, because values where the horizontal lines are to be drawn must be determined after first examining the quantile plot. Thus, it was possible to include the vref= option as shown here only on a subsequent execution of the SAS program. The options chref=, lhref=, cvref=, and lvref= define appearance characteristics (color and line type) of each of these reference lines.

The quantile plot of the ozone data for Stamford, the graphics output from the SAS Example C4 program, is displayed in Fig. 3.13. It can be used to ascertain, using the reference lines, that the median ozone concentration is about 80 ppb that the 75th percentile is roughly 120 ppb and that the IQR is

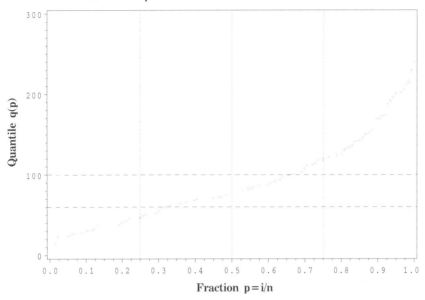

Fig. 3.13. SAS Example C4: Graph

about 70 ppb. Further, note that the local slope gives an indication of the local density of the data; that is, the flatter the curve (i.e., the smaller the local slope of the curve), the greater the concentration of points between the two data values that enclose that segment of the curve. In the quantile plot shown in Fig. 3.13 it is estimated by inspection that the curve can be approximated by a straight line that has the smallest slope between the ozone values 60 and 100 ppb. Thus, it could be concluded that the ozone concentration was inside of these values on at least 30% of the days in the period under study.

3.4 Empirical Quantile-Quantile Plots

A reliable comparison of the *shapes* of two distributions (empirical or theoretical) can be made using a quantile-quantile plot (Q-Q Plot). A Q-Q plot is made by plotting ordered pairs $\{Q_1(p), Q_2(p)\}$ in a scatter plot, where Q_1 and Q_2 are the respective quantile functions for the two distributions under consideration and where p runs over some suitable discrete set of values.

Suppose that Q_1 represents the quantile function for a sample $x_1, x_2, \ldots, x_n$ and Q_2 represents the quantile function for a data set $y_1, y_2, \ldots, y_m$. Adopting the convention of choosing the smaller of n or m (say, n), the values $p_i = \frac{i-.5}{n}$, $i = 1, 2, \ldots, n$, are used to calculate the sample quantiles of both samples. The sample quantiles $Q_1(p)$ that correspond to the above values of

p_i are the ordered values $x_{(1)}, x_{(2)}, \ldots, x_{(n)}$. The sample quantiles $Q_2(p_i)$ that correspond to the the same values of p_i are interpolated from the sample quantiles $y_1, y_2, \ldots, y_m$ that actually correspond to $p_i^* = \frac{i-.5}{m}$ $i = 1, 2, \ldots, m$. Note that if $n = m$, the sample quantiles $Q_1(p_i)$ and $Q_2(p_i)$ correspond to the the same values of p_i. Thus, when $n = m$, plotting $Q_1(p)$ against $Q_2(p)$ is equivalent to plotting the ordered pairs $(x_{(i)}, y_{(i)})$ in a scatter plot.

If the *shapes* of two distributions being compared are identical, then the resulting Q-Q plot is expected to be linear. Examining departures from linearity in a Q-Q plot is useful for making comparisons of the relative shapes of the two distributions. For example, if the y-distribution has "heavier tails" than the x-distribution, the plot may exhibit a pattern resulting from the y-quantiles being larger than the corresponding x-quantiles in the tails of the distribution; that is, the points will appear to move away from a straight line at both extremes. This effect will be examined further when theoretical Q-Q plots are discussed later.

SAS Example C5

Consider the ozone concentrations for Stamford and Yonkers, given in Table B.2, that were used in SAS Example C4. The observations for which a missing value is present at either Stamford or Yonkers were omitted, so that the two variables have equal sample sizes. The SAS Example C5 program, shown in Fig. 3.14, graphs the Q-Q plot of the ozone concentrations in Stamford and Yonkers. The first data step in the SAS Example C5 program is a modified version of the first data step in the SAS Example C4 program. Here, two data sets are formed, the first named `city1` and containing ozone concentrations of Stamford and the second named `city2` and containing those of Yonkers. For simplicity, only those observations that do not have a missing value for either city are retained. Thus the two data sets have the same sample size. The two samples are separately sorted in the increasing order of magnitude using `proc sort` ❶ and then combined into a single SAS data set named `ozone` using the `merge` statement ❷ in a separate data step to effect a one-to-one merge.

The Q-Q plot is then constructed executing a `proc gplot` step using the new data set `ozone` ❸, with several `title`, `axis`, and `symbol` SAS/GRAPH global statements used to enhance the plot. The two axes are defined to have the same scale so that valid comparisons of the two distributions can be made directly. A $y = x$ line is plotted as a reference. Executing the SAS Example C5 program creates the graph shown in Fig. 3.15, which displays an empirical Q-Q plot for Yonkers and Stanford ozone concentrations. The solid line is the $y = x$ line and has a slope of 1. All of the plotted points lie above this line, indicating that every Stamford quantile is larger than the corresponding Yonkers quantile. The points can be approximated by a straight line indicating that the distributions of ozone concentrations in the two cities have approximately the same shape. By locating the midpoint of this straight line,

3.4 Empirical Quantile-Quantile Plots   153

```
goptions rotate=landscape targetdevice=pscolor
                         hsize=8 in vsize=6 in;
data city1(keep=stamford) city2(keep=yonkers);
input x1-x10;
drop x1-x10 i;
array ozone(10) x1-x10;
do i=1 to 10 by 2;
   stamford=ozone(i);
   yonkers=ozone(i+1);
   if stamford=.|yonkers=. then delete;
   output city1;
   output city2;
end;
datalines;
66 47   61  36 152   76  80 66 113 66
52 37   47  24 201 108   68 82  38 18
 . 27    .  52 134   85  24 47  38 25
 . 37  196  88 206   96  24 28  28 14
 . 38  131 111  92   48  82 44  52 27
 .  .  173 117 101   60 100 55  14  9
49 45   37  31 119   54  55 34  38 16
  .  .   .   .   .    .   .  .
  .  .   .   .   .    .   .  .
  .  .   .   .   .    .   .  .
 . 67  169 119   .   70 141 47   . 18
71 45    .   . 111   74 202 114  .  .
run;

proc sort data=city1; 1
by stamford;
run;

proc sort data=city2; 1
by yonkers;
run;

data ozone;
merge city1 city2; 2
run;

   title1 h=3 f=triplex c=magenta 'Q-Q Plot of Ozone Concentrations';
   axis1 order= 0 to 250 by 50 c=black
         label=(c=blue h=1.5 a=90 ' Stamford Ozone ');
   axis2 order= 0 to 250 by 50 c=black
         label=(c=blue h=1.5 ' Yonkers Ozone');
   symbol1 v=star  h=1.5 cv=red;
   symbol2 i=join v=none c=cyan;

proc gplot data=ozone; 3
   plot stamford*yonkers=1 stamford*stamford=2 /vaxis=axis1 vm=1
                     haxis=axis2 hm=1
                     frame overlay ;
run;
```

Fig. 3.14. SAS Example C5: Program

it can be calculated that the medians of the two samples are approximately 75 and 125, respectively, thus giving an indication of the difference in the locations of the two distributions. The straight line has a slope of approximately 1.6, which implies that Stamford ozone concentrations are generally about

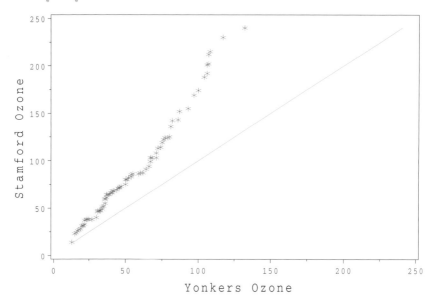

Fig. 3.15. SAS Example C5: Q-Q plot of Stamford and Yonkers ozone concentrations

60% larger in magnitude than those from Yonkers. Rather than implying that Stamford ozone concentration is 60% higher than in Yonkers on any given day, this indicates that the *variance* of the Stamford ozone concentrations is about $1.6^2 = 2.56$ (i.e., about 2.5 times that of the Yonkers ozone concentrations).

3.5 Theoretical Quantile-Quantile Plots or Probability Plots

An important special case of Q-Q plots called `probability plots` occurs when Q_2 represents quantiles from a theoretical distribution and Q_1 represents quantiles from a sample of data values. If n is the sample size, to construct this plot by hand, quantiles of the theoretical distribution that correspond to the values $p_i = \frac{i-.5}{n}$, $i = 1, 2, \ldots, n$, are plotted against the data values ordered in the increasing order of magnitude. If computations are performed on a computer, various other choices are avilable for p_i. These were discussed in Section 3.3. A probability plot enables one to visually assess how close the data appear to be a random sample from the theoretical distribution. In particular, when the standard normal quantiles are used for Q_2, a `normal probability plot` is obtained. If the data are a random sample from a normal population, the points should lie approximately in a straight line. Departures

3.5 Theoretical Quantile-Quantile Plots or Probability Plots 155

from a straight line indicates that the population distribution is different from a normal distribution.

From a practical point of view, normal probability plots are best used to judge whether a departure from normality is exhibited rather than for confirming that the sample is obtained from a normal distribution. How the distribution of a sample differs from a normal distribution can be identified by specific patterns of departures from a straight line in a normal probability plot. If a specific pattern is not observed, then it can be concluded that no evidence is provided by the plot to suspect the plausibility of the normality assumption. Figure 3.16 contains normal probability plots of computer-

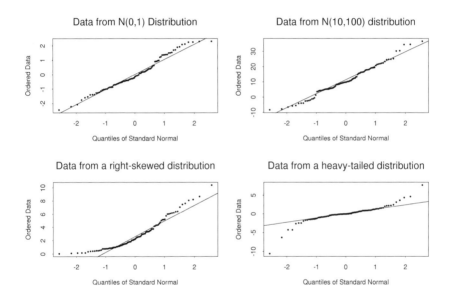

Fig. 3.16. Diagnostic use of normal probability plots

generated data from different populations. The plots in the top two panels show data from normal populations with different mean and variance parameters. The only effect of changing the mean (or location) or the variance (or scale) of the data is to change the slope or the intercept, respectively, of the straight line that approximates the points. The lower two panels show data drawn from populations that deviate from the normal distribution in ways described in each case.

Notice that the lower two plots show distinct patterns (the left graph shows a bowl-shape pattern and the other shows an inverted S-shaped pattern). These are markedly different from the deviations from a straight line seen in the top two graphs, which can be attributed as entirely due to random variation. Analogous to the patterns seen in the lower two graphs, a mound-shaped pattern will indicate a left-skewed distribution and an S-shaped pattern will indicate a short-tailed distribution, relative to the shape of a normal distribution. In addition to these four possibilities, if most point appear to conform to a straight-line pattern except for one or two points at either of the two extremes, it may be an indication of *extreme* or *outlying* values in a data set relative to data generated from a normal distribution.

SAS Example C6

Consider the rainfall data from seeded clouds shown in Table B.3. In practice, for computing a normal probability plot by hand, the standard normal tables could be used to approximate the normal quantiles. One would simply look up the z values corresponding to the p_i's for $i = 1, \ldots, n$ and then plot the ordered data against the corresponding z values. Table 3.4 displays the quantiles for the rainfall data from seeded clouds as well as the corresponding normal quantiles approximated from a standard normal table.

i	$Q(p_i)$	p_i	Z_i
1	4.1	.01923	−2.070
2	7.7	.076923	−1.425
3	17.5	.096154	−1.304
4	31.4	.13462	−1.108
5	32.7	.17308	−0.943
⋮	⋮	⋮	⋮
22	703.4	.82692	0.943
23	978.0	.86538	1.108
24	1656.0	.90385	1.304
25	1697.8	.94231	1.425
26	2745.6	.98077	2.070

Table 3.4. Quantiles of rainfall for seeded clouds

The SAS Example C6 program shown in Fig. 3.17 draws a normal probability plot of the seeded rainfall data using normal quantiles computed using `proc rank`. Note that both the data from control clouds as well as the seeded clouds are input to form the data set named `rainfall`, although the control data are not used in the present analysis. The `blom` specification for the `normal=` option ❶ used in the proc statement results in the computation of the p_i values using the approximation $p_i = \frac{(i-\frac{3}{8})}{n+\frac{1}{4}}$. The normal quantiles corresponding to these p_i values are then obtained using the inverse function of the standard normal cumulative distribution function (cdf) available in SAS.

3.5 Theoretical Quantile-Quantile Plots or Probability Plots

```
goptions rotate=landscape targetdevice=pscolor
                          hsize=8 in vsize=6 in;
data rainfall;
input control seeded @@;
logseed= log10(seeded);
datalines;
1202.6 2745.6 830.1 1697.8 372.4 1656.0 345.5 978.0 321.2 703.4
 244.3  489.1 163.0  430.0 147.8  334.1  95.0 302.8  87.0 274.7
  81.2  274.7  68.5  255.0  47.3  242.5  41.1 200.7  36.6 198.6
  29.0  129.6  28.6  119.0  26.3  118.3  26.1 115.3  24.4  92.4
  21.7   40.6  17.3   32.7  11.5   31.4   4.9  17.5   4.9   7.7
   1.0    4.1
;
run;
  title1 c=magenta f=centb 'Normal Probability Plots';
  title2 c=magenta f=centb h=1.5 ' of Seeded and Log Seeded Rainfall
                                                              Data';
  symbol1 c=steelblue v=dot i=none;
  symbol2 c=maroon v=diamond i=none;

  axis1 color=tomato label=(c=blue h=1.5 a=90 'Seeded Rainfall');
  axis2 color=tomato label=(c=violet h=1.5
                                  'Standard Normal Quantiles');

  axis3 color=crimson label=(c=navy h=1.5 a=-90 'Log Seeded Rainfall');

  footnote c=steelblue f=special h=1 'J J J' f=swissb h=1 ' Seeded'
           c=maroon f=special h=1
                       '       D D D' f=swissb h=1 ' Log Seeded';

proc rank normal=blom  1  data=rainfall out=quantiles;
  var seeded;
  ranks nscore1;  2
run;

proc gplot data=quantiles;
  plot  seeded*nscore1=1 /vaxis=axis1 vm=9 haxis=axis2 hm=4 frame;
  plot2 logseed*nscore1=2/vaxis=axis3 vm=9;  3
run;
```

Fig. 3.17. SAS Example C6: Program

The variable named `seeded` contains the seeded rainfall values in the data set named `rainfall`. The `ranks nscore1;` statement 2 causes the standard normal quantiles corresponding to the seeded rainfall data values to be computed and saved as values of a SAS variable named `nscore1`. This variable as well as the variable `seeded` are added to the SAS data set `quantiles` created in the `proc rank` step. This data set is then used in the `proc gplot` step to create a scatter plot of the ordered data values plotted against normal quantiles in a straightforward use of the `plot` statement. The SAS program, shown in Fig. 3.17, includes SAS/GRAPH global statements added to enhance the graph by adding titles, axis labels, symbols, etc.

The graphic output produced by executing the program for SAS Example C6, shown in Fig. 3.18, displays the **normal probability plot** of the seeded cloud data. A bowl-shaped pattern clearly can be observed, indicating that the data most likely came from a right-skewed distribution. This is to be expected when larger data values have much larger deviations from the center of the data distribution than smaller values in a data set. In such cases, a *logarithmic transformation* might be attempted in order to *normalize* the data.

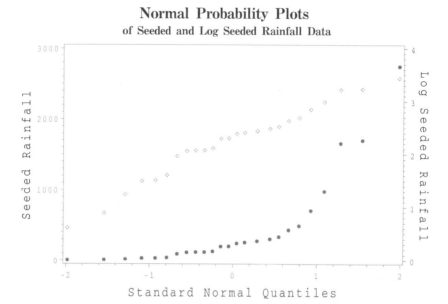

Fig. 3.18. SAS Example C6: Normal probability plots of rainfall data

To examine whether a log transformation will transform the data to near normality, a normal probability plot of `log seeded rainfall` is also plotted on the same graph using the `plot2` statement ❸. The vertical axis that corresponds to the quantiles of the transformed data is shown on the right-hand side of the graph. This is possible since the normal quantiles plotted on the horizonal axis is common to both plots. Also, note that since a logarithmic transformation is monotonic, the normal quantiles corresponding to the transformed data values are identical to those of the original data. Thus, they are not recomputed for the transformed data.

Typically, when different symbols, colors, and/or line types are used to distinguish among different plots, the values of a third variable are used to identify the plots. This is illustrated in the next example, SAS Example C7 (see Fig. 3.19), for which the value of the variable `drug` determines the `symbol` statement that defines the characteristics of each of the plots. In addition, the use of the third variable causes *legends* relating the line and/or symbol characteristics, with the value of the third variable to be produced automatically.

The user may modify the appearance, position, and content of these legends using `legend` statements to create alternative legend definitions. In the current example, although two plots with different characteristics are drawn, no legends were produced since a third variable is not present. Thus, a `footnote` statement is included to create descriptive text strings that serves the same purpose as a legend.

3.6 Profile Plots of Means or Interaction Plots

In this plot, the levels of one experimental factor (say, A) are plotted on the horizontal axis. The vertical axis represents the sample means of responses resulting from the application of combinations of the levels of A and the levels of a second factor (say, B) to experimental units, independently. This plot is sometimes called the *profile plot of means*, because the sample means corresponding to the same levels of factor B are joined by line segments, thus showing the pattern of variation of the average response across the levels of factor A at each level of B.

This plot is also known as the *interaction plot* because it is useful for interpreting significant interaction that may be present between the two factors A and B and may help in explaining the basis for such interaction. This type of a plot is also useful in *profile analysis* of independent samples in multivariate data analysis, where means of several variables, such as responses to test scores measured on different subjects, are compared across independent groups of subjects such as classrooms.

SAS Example C7

Consider the survival times of groups of four mice randomly allocated to each combination of three poisons and four drugs shown in Table 3.5. The experiment, described in Box et al. (1978) was an investigation to compare several antitoxins to combat the effects of certain toxic agents. The SAS Example C7 program shown in Fig. 3.19 draws an interaction plot of this data.

The data are entered so that each line of data includes responses to the four levels of `drug` for each level of `poison`. Four data lines represent the replicates for each level of `poison` so that there are 12 data lines in the input stream. The `trailing @` symbol in the `input` statements and an `output` in a `do loop` **1** enable a SAS data set to be prepared in the form required to be analyzed by procedures such as `means` or `glm`. As shown in Table 3.6, the data set named `survival` will have 48 observations, each corresponding to a response to a combination a level of `drug` and a level of `poison`.

The `means` **2** and `gplot` procedures are used in sequence to obtain a scatter plot of the cell means. Although the choice of the levels of the factor that is plotted on the horizontal axis is somewhat arbitrary, sometimes a prudent choice will make the profile of means more meaningful. In this example, by

| | \multicolumn{4}{c}{Drug} |
Poison	A	B	C	D
I	0.31	0.82	0.43	0.45
	0.45	1.10	0.45	0.71
	0.46	0.88	0.63	0.66
	0.43	0.72	0.76	0.62
II	0.36	0.92	0.44	0.56
	0.29	0.61	0.35	1.02
	0.40	0.49	0.31	0.71
	0.23	1.24	0.40	0.38
III	0.22	0.30	0.23	0.30
	0.21	0.37	0.25	0.36
	0.18	0.38	0.24	0.31
	0.23	0.29	0.22	0.33

Table 3.5. Mice survival time data (Box et al., 1978)

selecting the levels of `poison` to appear in the horizonal axis, the set of cell means that correspond to each level of `drug` will show the profile of the mean response of each `drug` to the three poisons.

Four different `symbol` statements are used to produce different symbols, colors, and line types for each level of `drug`. This is achieved by using the `drug` variable with values 1, 2, 3, and 4 as a third variable **3**. Two globally defined axis statements control the two axes and the tick marks on the horizontal axis

Poison	Drug	Time
1	1	0.31
1	2	0.82
1	3	0.43
1	4	0.45
1	1	0.45
1	2	1.10
⋮	⋮	⋮
3	3	0.24
3	4	0.31
3	1	0.23
3	2	0.29
3	3	0.22
3	4	0.33

Table 3.6. Data arranged for input to `proc means`

3.6 Profile Plots of Means or Interaction Plots

```
goptions rotate=landscape targetdevice=pscolor
                        hsize=8 in vsize=6 in;

data survival;
input   poison 1. @;
    do drug=1 to 4;  ❶
        input time 3.2 @;
        output;
    end;
datalines;
1 31 82 43 45
1 45110 45 71
1 46 88 63 66
1 43 72 76 62
2 36 92 44 56
2 29 61 35102
2 40 49 31 71
2 23124 40 38
3 22 30 23 30
3 21 37 25 36
3 18 38 24 31
3 23 29 22 33
;
run;

proc means noprint mean nway;  ❷
  class poison drug;
  output out=meandata mean=cellmean;
run;

proc print data=meandata;
run;

title1 c=magenta h=2  'Analysis of Survival Times Data';
title2 c=salmon h=1.5 'Profile Plot of Cell Means';

symbol1 c=red i=join v=square h=1.5 l=2;
symbol2 c=green i=join v=diamond h=1.5 l=3;
symbol3 c=blue i=join v=triangle h=1.5 l=4;
symbol4 c=cyan i=join v=star h=1.5 l=5;

axis1 label=(c=steelblue h=1.5 a=90 'Cell Means (seconds)')
                                          value=(c=blue);
axis2 offset=(.2 in) label=(c=steelblue h=1.5 'Levels of Poison')
                                          value=(c=blue);

proc gplot data=meandata;
  plot cellmean*poison=drug/vaxis=axis1 haxis=axis2 hm=0;  ❸
run;
```

Fig. 3.19. SAS Example C7: Program

are suppressed by setting the option hm= to zero. The graphical output from the SAS Example C7 program is reproduced in Fig. 3.20.

As alluded to earlier, *legends* are produced automatically in this graph due to the use of a third variable. The legend associates values of the third variable with graphic elements that are parts of a graph, such as symbols, lines, and areas, using characteristics of these elements as shape, color, line type, and the pattern of shading. Sometimes users may desire to change the placement and

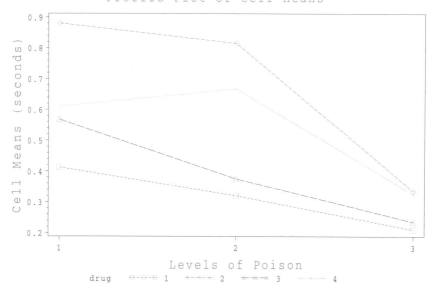

Fig. 3.20. SAS Example C7: Interaction plot of survival times data

appearance of legends using global `legend` definition statements. An extensive discussion of the `legend` statement is omitted in this book. However, even a simple `legend` statement may be sufficient to vastly improve the default appearance of the legends. Note that in SAS Example C7, the legends are placed at the bottom of the graph by default, and all four *legend entries* appear in a single row. This may be replaced by a more elaborate legend by including the legend definition

```
legend1 origin=(6, 4.5) in mode=share across=1 frame;
```

and modifying the plot statement as follows:

```
plot cellmean*poison=drug/vaxis=axis1 haxis=axis2 hm=0
                      legend=legend1;
```

The result is that the legend, enclosed in a box, is placed at a position with absolute coordinates (6, 4.5) in inches, with the legend entries appearing one to a row. The new version of the graph is shown in Fig. 3.21. The `mode=share` option enables the legends to be overlayed in the graphics output area. Note that the coordinates for use in the `origin=` are calculated using the same approach used for obtaining the coordinates for placement of `notes` or `footnotes` using the `move=` option (see Section A.2.2). In the present example note that `goptions` statement defines the graphics output area using the the options `hsize=8 in vsize=6 in`.

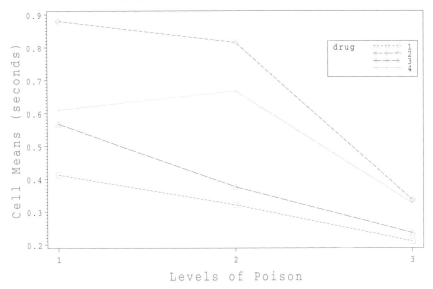

Fig. 3.21. SAS Example C7: Interaction plot with modified legends

3.7 Two-Dimensional Scatter Plots and Scatter Plot Matrices

3.7.1 Two-Dimensional Scatter Plots

The scatter plot is a statistical tool for analyzing relationship between two variables (say, x and y). Observations (x_i, y_i) for $i = 1, \ldots, n$ measured on the two variables that are paired in some way, such as those measured on the same objects or under the same experimental conditions, are plotted as points using a two-dimensional graph with the vertical axis representing variable y and the horizontal axis representing variable x. In addition to displaying the empirical bivariate distribution, many other useful observations regarding the joint distribution can be made from such plots. The graphical output resulting from executing the SAS Example C1 program (see Fig. 3.2) is displayed in Fig. 3.3. This is an example of a two-dimensional scatter plot. The data consisted of pairs of values of the logarithm of atmospheric pressure and the boiling point of water, measured at different altitudes above sea level. The graph is used to study the dependence of the boiling point of water on the logarithm of atmospheric pressure using linear regression and demonstrates a particular application of a scatter plot.

```
goptions rotate = landscape targetdevice = pscolor
                                 hsize = 8in vsize = 6in;
data ozone;
input x1-x10;
drop x1-x10;
array xvars(10) x1-x10;
do i=1 to 10 by 2;
    stamford = xvars(i);
    yonkers  = xvars(i+1);
    if stamford =. | yonkers =. then delete;
    output;
    end;
datalines;
66 47  61  36 152  76  80 66 113 66
52 37  47  24 201 108  68 82  38 18
 . 27   .  52 134  85  24 47  38 25
 . 37 196  88 206  96  24 28  28 14
 . 38 131 111  92  48  82 44  52 27
 .  .  173 117 101  60 100 55  14  9
49 45  37  31 119  54  55 34  38 16
 .  .   .   .   .   .   .  .   .  .
 .  .   .   .   .   .   .  .   .  .
 .  .   .   .   .   .   .  .   .  .
14 22 136  81  64  39 174 87  28 21
 . 67 169 119   .  70 141 47   . 18
71 45   .   . 111  74 202 114   .  .
;
run;

%include "C:\Documents and Settings\...\graphmacro.lib"; 1

%boxanno(data=ozone,xvar=yonkers,yvar=stamford,out=boxplots); 2

  title1 h=3 f=triplex c = blue 'Scatter plot of Ozone Data';
  axis1 order= 0 to 250 by 50 c = magenta
        label=(c=blue h=1.5 a=90  ' Stamford Ozone ');
  axis2 order= 0 to 250 by 50 c = magenta
        label=(c=blue h=1.5 'Yonkers Ozone ');
  symbol1 v=star color=red h=1.5 ;
proc  gplot  data = ozone;
   plot stamford*yonkers/vaxis = axis1 vm = 1
                        haxis = axis2 hm = 1
                        frame
                        annotate = boxplots; 3
run;
```

Fig. 3.22. SAS Example C8: Program

SAS Example C8

Another use of scatter plots is simply to study the relationship between two variables. This application is illustrated by the scatter plot of Stamford ozone data plotted against the Yonkers ozone data given in Table B.2, where cases with any missing value are omitted. The SAS Example C8 program shown in Fig. 3.22 draws a scatter plot of the ozone data using **proc gplot**. The only difference between this graph and the one drawn in SAS Example C1 is that the scatter plot is enhanced with univariate box plots drawn on the margins opposite the two major axes.

A SAS/Macro program named `boxanno` described in Friendly (1991) is used to create the *annotate data set* used for this purpose. Although a discussion of how to write SAS macros is omitted from this book, the use of macros written for various purposes is encouraged. The book by Friendly (1991) includes several macros for creating useful statistical graphics and is strongly recommended to advanced SAS users. A simple form of the `boxanno` macro ❷ call is

%boxanno(data=, xvar=, yvar=, out=);

where the values for the macro parameters are respectively the names of the data set to be plotted, the horizontal and vertical variables, and the output annotate data set. The `%include` statement ❶ refers to a file containing the code for the particular macro required or, ideally, as in SAS Example C8, a macro library that contains a collection of macros of interest to the user. The second option enables the user to avoid having to keep track of the other macro calls that may be made by the macro currently in use if all subordinate macros are included in the library.

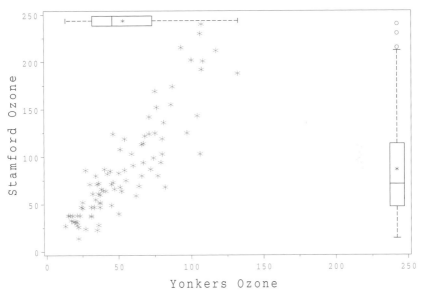

Fig. 3.23. SAS Example C8: A Scatter plot annotated with marginal box plots

The annotate data set created by the macro `boxanno` contains graphics commands required for drawing the marginal box plots. These commands are stored in the form of lines in the SAS data set named `boxplots`. A look at the text of the `boxanno` macro shows that the coordinates required for drawing the box plots have been calculated using `proc univariate` and a simple SAS data step program. The `annotate` = option names this data set **3** to be used for drawing the box plots. The resulting graph is shown in Fig. 3.23.

In this plot note the following:

a. The box plots displayed on the upper and right-hand-side axes show the marginal empirical distributions of the Stamford and Yonkers ozone data, respectively.
b. The scales on both axes are the same, enabling direct comparison to be made; for example, the range of ozone concentrations in Stamford is higher than the range of ozone concentrations in Yonkers.
c. There is a strong positive relationship between the two variables; that is on days in which ozone concentration is high in Yonkers, it is also high in Stamford.
d. Points lie mostly above the $y = x$ line, indicating that on the majority of the days the Stamford concentration is higher than at Yonkers.

3.7.2 Scatter Plot Matrices

When more than two variables are measured on observational units, the relationships among several variables may need to be analyzed simultaneously. One possible approach is to obtain bivariate scatter plots of all pairs of variables and display them arranged in a two-dimensional array of plots. For example, if three variables are present, six pairwise scatter plots are possible, since a pair of variables can be graphed in two possible ways (choosing one as the x-variable and the other as the y-variable, and vice versa). These six plots can be displayed as a 3×3 matrix of plots, each row-column combination of the matrix representing the position for placement of a plot. Although displaying only the lower (or upper) triangle of the matrix appears sufficient, most software programs display the entire matrix since it enables the user to observe patterns of relationships that may exist among the variables in a more straightforward manner. An additional numeric variable (usually an ordinal variable, category variable, or a grouping variable created from the values of another variable) may be represented in these scatter plots either by using different symbols or colors (or both) to identify different subsets of the observations.

In a multiple regression situation, scatter plot matrices are especially useful for establishing types of relationship that individual independent variables (regressors, explanatory variables) may have with the dependent variable (response). This will help in the model building stage, by suggesting the form the variables may enter the model (e.g., by suggesting possible transformations, etc. that linearize relationships). The scatter plot matrix will also help in studying pairwise collinearities that may exist among the independent variables.

3.7 Two-Dimensional Scatter Plots and Scatter Plot Matrices

```
data world;
infile  "C:\Documents and Settings\...\demogr.data";
input   country $20. birthrat deathrat inf_mort life_exp popurban
            perc_gnp lev_tech civillib;
label   birthrat ='Crude Birth Rate'
        deathrat ='Crude Death Rate'
        inf_mort ='Infant Mortality Rate'
        life_exp ='Life Expectancy(in yrs.)'
        popurban ='Percent of Urban Population'
        perc_gnp ='Per Capita GNP in U.S. Dollars'
        lev_tech ='Level of Technology'
        Civillib ='Degree of Civil Liberties';
;
run;
%include "C:\Documents and Settings\...\graphmacro.lib";

goptions device=win colors=(red,blue,magenta);  1

title1 'Demographic Data for 70 Countries';
%scatmat(data = world, var = popurban perc_gnp inf_mort life_exp,  2
         symbols = %str(x dot +), hsym=1.5,
         colors=BLUE MAGENTA VIOLET);
```

Fig. 3.24. SAS Example C9: Program

A SAS/Macro program named `scatmat` described by Friendly (1991) is used in the SAS Example C9 program displayed in Fig. 3.24 to obtain a scatter plot matrix of four variables in the demographic data set shown in Table B.4. The macro `scatmat` uses `proc gplot` to create the two-dimensional scatter plots for every pair of variables in a list provided by the user, stores them in a temporary graphics catalog, and uses `proc greplay` to display them in the final scatter plot matrix format. A primary use of SAS/GRAPH procedure `greplay` is to reproduce graphics output stored as SAS catalog entries, in separate panels of a template so that several graphs can be displayed in rectangular segments of a single window (or page). The template itself is also created in advance using `proc greplay`. A simple invocation of the `scatmat` macro is

%scatmat(data = ,var = ,group=);

where the values for the macro parameters, respectively, identify the name of the data set to be plotted, the list of the numeric variables separated by blanks, and the name of the grouping or category variable, if any.

Suppose that a scatter plot matrix of the four variables named `popurban`, `perc_gnp`, `inf_mort`, and `life_exp` in the SAS data set `world` is to be obtained. This SAS data set is created using data read from a file instead of data entered instream (see Fig. 3.24) (the data are displayed in Table B.4). Although there are many useful options available with the `infile` statement, for the purpose of reading a text file all that is needed is to supply the path name of the data file as a quoted string:

infile   "C:\Documents and Settings\...\demogr.data" ;

The `goptions` option `device=win` ◼1 specifies the display device to be the monitor (see Fig. 3.24). If this option is omitted, a device name will be requested via a pop-up dialog at run time. In addition to the required parameters `data=`, `var=`, the default values of the parameters `symbols=`, `hsym=`, and `colors=` are replaced by the respective values shown in the program ◼2. Note that the `group=` option is not needed, as the graphs are not grouped by the values of a category variable in this example.

The scatter plot matrix is reproduced in Fig. 3.25. From this plot it can be observed that, as one might expect, there is a strongly negative linear relationship between Infant Mortality Rate and Life Expectancy. There also appears to be linear relationships between each of these variables with Percent of Urban Population that, in each case, is not particularly strong. These two variables also appear to have differing nonlinear relationships with Per Capita GNP.

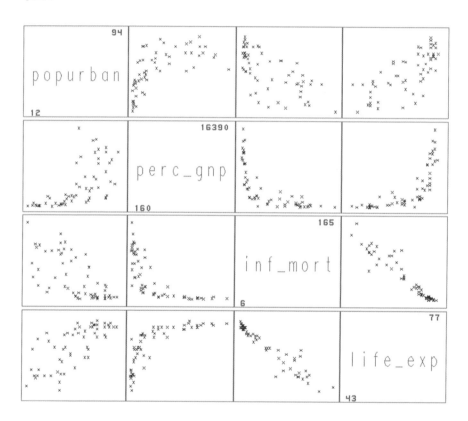

Fig. 3.25. SAS Example C9: Scatter plot matrix of four variables

3.8 Histograms, Bar Charts, and Pie Charts

The SAS/GRAPH procedure `gchart` is useful for drawing vertical and horizontal bar charts, block charts, pie charts, and star charts. A *chart variable* and a *summary variable* are needed to produce these charts. The values of the chart variable appear in the chart at a set of *midpoints*. The body of the chart itself displays various statistics (for e.g., the mean) computed on the summary variable, represented as bars, blocks, etc. located at the midpoint values. The terminology used for drawing bar charts with the `gchart` procedure was previously illustrated in Fig. 3.1. Although `proc gplot` can be manipulated to draw histograms by first computing a frequency table using a SAS base procedure such as `summary`, it is preferable to take advantage of bar-chart-drawing facilities provided in `proc gchart`. Unusual situations may require the use of `proc gplot` for plotting bar charts such as when a histogram may require nonstandard features (e.g., constructing frequency tables with bin intervals of varying widths).

SAS Example C10

In the SAS Example C10 program, four simple charts are produced using the demographic data set on countries (see Table B.4) used earlier in SAS Example C9. This example illustrates how SAS graphs, saved as *graphics catalog entries* are accessed subsequently to be displayed as panels in a composite graph using *templates* produced by `proc greplay`. This example is thus in two parts: SAS Example C10 (Part A) to prepare the four charts and SAS Example C10 (Part B) to display them. Note that the `proc format` step shown in Fig. 3.26 was executed prior to executing the program for SAS Example C10 (Part A), so that the predefined formats are available for use in the current program. These user-defined formats are used to print values of the category variables `techgrp`, `infgrp`, and `civilgrp`, respectively **1**. (The use of user-defined formats is discussed in Chapter 2.)

```
proc format;
    value bf 1='Low'
             2='Moderate'
             3='High' ;
    value gf 1=' Low'
             2='Medium'
             3='High';
    value cf 1='High'
             2='Moderate'
             3='Low' ;
run;
```

Fig. 3.26. Illustration of use of `proc format`

In the SAS Example C10 program (Part A) displayed in Fig. 3.27, the countries data set is again read from a file using an infile statement. In this instance, however, several new variables and variable attributes are included in the SAS data set named `world` created. New grouping or category variables (named `infgrp`, `techgrp`, and, `civilgrp`) are generated using cut off values determined in advance, so that the sample of data is divided into subgroups. Labels and user-defined formats are assigned to the category variables as well as to other variables used in the analysis. The statement

```
goptions device=win nodisplay;
```

suppresses the four graphs drawn from being displayed on any device. However, the device-dependent graphics output produced will be saved as *catalog entries* (by default, in a temporary catalog named `work.gseg`). For the purposes of this example, storing these graphs in a temporary catalog works, as they are accessed in the program for SAS Example C10 (Part B) during the same SAS session and, hence, need not be retained permanently. If necessary, they may be saved in a user library using the method for saving SAS data sets used in SAS Example B2 in Chapter 2. If the above `goptions` statement is omitted (or commented out by entering an asterisk as the first character in the statement), the four graphs will be displayed as usual in a SAS graphics windows.

The `proc gchart` action statements `hbar`, `block`, `vbar`, and `pie` are used to draw the four graphs in SAS Example C10 (Part A). Some of the options that can be used with these statements were described in detail in Section 3.2.1. In the present program, different forms of the statements are used and are discussed as examples of usage of these options.

Horizontal Bar Chart

```
hbar infgrp/
    discrete sumvar=life_exp type=mean subgroup=civilgrp
                         group=techgrp nostats name='myplot1';
```

In the above, the values of the *chart variable* `infgrp` are to be used as midpoints for the purpose drawing a bar chart. Thus, although `infgrp` is a numeric-valued variable, it is required to be considered an ordinal variable instead of one with continuous values. The `discrete` option indicates this fact to `proc chart`. Thus, the horizontal bars will be drawn with midpoints situated at each of the discrete values of `infgrp`, the lengths of these bars determined by the magnitude of a statistic computed on the values of a *summary variable*. Here, the `sumvar=` option specifies the *summary variable* to be `life_exp`, and `type=mean` specifies the statistic to be the sample mean. The `subgroup=civilgrp` **2** results in each bar being divided into three segments because the variable `civilgrp` has three possible values: 1, 2, and 3. Thus, a segment of a horizontal bar represents the mean of the variable `life_exp` for the subset of observations corresponding to a value of `civilgrp` within the subset corresponding to a value of infgrp. The global pattern definitions provided will be used to fill-in each of these segments using the specified color and the fill pattern. The three definitions are applied sequentially to fill in the three segments of each horizontal bar.

The option group=techgrp ❷ first groups the observations by the values of techgrp, which is an ordinal variable as required. The process of drawing bars for each value of infgrp as described earlier is now repeated for each value of techgrp. Thus, a separate set of horizontal bars is produced for each value of techgrp for which the means of the variable life_exp are calculated using the subset of observations corresponding to each value of techgrp. The *formatted* values of the levels of techgrp, Low, Moderate, and High, are plotted on the vertical axis (group axis) because a format for techgrp is included in the data set world. The resulting horizontal bar chart is shown in the lower left corner of Fig. 3.28.

By default, the values of the statistic computed and the number of observations (i.e., the frequency) used to calculate the statistic will be printed alongside each bar. The option nostats suppresses these from appearing in the plot. The name= option allows the user to specify the name of the catalog entry for the chart saved.

Block Chart

```
block lev_tech/
    midpoints= 10 25 60 sumvar=life_exp type=mean
                    group=civilgrp noheading name='myplot2';
```

Here, the *chart variable* is lev_tech, which is a numeric continuous variable, with values in the range 0 to 100. The midpoints= 10 25 60 option specifies that three class intervals with the given midpoints be set up. The options sumvar=life_exp and type=mean will result in the means for the variable life_exp to be computed for the observations falling in each of these classes. The set of means thus computed will be used for determining the block heights, with the blocks placed at each of the midpoints.

However, the presence of option group=civilgrp causes the observations falling in each of the intervals of lev_tech also to be separated into subsets defined by the three levels of civilgrp prior to the means being computed. The blocks are then plotted in rows that correspond to the levels of civilgrp (i.e., 1, 2, and 3 (in increasing order)). Each row contains blocks plotted at the midpoints 10, 25, and 60 of the three lev_tech classes. The *formatted* values of the levels of civilgrp are High, Moderate, and Low so they will appear on the group axis instead of the values 1, 2, and 3, because a format for the variable civilgrp is present in the data set world. The resulting block chart is shown in the upper left corner of Fig. 3.28.

Vertical Bar Chart

```
vbar inf_mort/
    midpoints= 10 to 150 by 20 type=percent
                    subgroup=techgrp name='myplot3';
```

```
data world;
infile  "C:\Documents and Settings\...\demogr.data";
input country $20. birthrat deathrat inf_mort life_exp popurban
              perc_gnp lev_tech civillib;

if inf_mort<25 then infgrp=1;
else if 25<= inf_mort<80 then infgrp=2;
else if inf_mort>= 80 then infgrp=3;

if lev_tech<15 then techgrp=1;
else if 15<=lev_tech<35 then techgrp=2;
else if lev_tech>= 35then techgrp=3;

if civillib<3 then civilgrp=1;
else if 3<= civillib<6 then civilgrp=2;
else if civillib>=6 then civilgrp=3;

label inf_mort='Infant Mortality Rate'
      life_exp='Life Expectancy in yrs.'
      lev_tech='Level of Technology'
       techgrp='Technology'
         infgrp='Infant Mortality'
         civilgrp='Civil Liberties';

format techgrp bf. infgrp bf. civilgrp cf.;  1
run;

goptions device=win nodisplay;
 pattern1 c=magenta v=s;
 pattern2 c=cyan v=s;
 pattern3 c=blue v=s;

proc gchart data=world;
  title 'Horizontal Bar Chart of Mean Life Expectancy';
  hbar infgrp/
       discrete sumvar=life_exp type=mean subgroup=civilgrp
                      group=techgrp nostats name='myplot1';  2
run;
proc gchart data=world;
  title 'Block Chart of Mean Life Expectancy';
  block lev_tech/
        midpoints= 10 25 60 sumvar=life_exp type=mean group=civilgrp
                                  noheading name='myplot2';
run;
proc gchart data=world;
  title 'Vertical Bar Chart of Infant Mortality';
  vbar inf_mort/
        midpoints= 10 to 150 by 20 type=percent
                          subgroup=techgrp name='myplot3';
run;
proc gchart data=world;
  title1 'Pie Chart of Mean Infant Mortality';
  title2 h=1.5 f=swissb 'by Level of Technology';
  pie  techgrp/discrete  3
       sumvar=inf_mort type=mean percent=inside value=inside
                          slice=outside noheading name='myplot4';
run;
```

Fig. 3.27. SAS Example C10 (Part A): Program

3.8 Histograms, Bar Charts, and Pie Charts   173

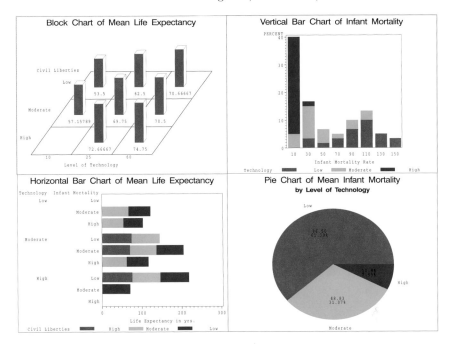

Fig. 3.28. SAS Example C10: Replay of four charts as panels

In this instance, also the *chart variable* inf_mort is a numeric continuous-valued variable, and the option midpoints= 10 to 150 by 20 will result in eight class intervals with respective midpoints $10, 30, 50, \ldots, 150$. The type=percent will cause the percent frequency counts in each class to be plotted as vertical bars placed at each of the midpoints.

The option subgroup=techgrp results in each of the bars being subdivided into three segments, corresponding to percent frequency counts for the three lev_tech groups. The global pattern definitions supplied are again used to fill in these segments using the color and the fill pattern specified in each. The three definitions are applied sequentially to fill in the three segments of each vertical bar. The *formatted* values of the levels of civilgrp, High, Moderate, and Low, will appear on the group axis instead of the values 1, 2, and 3, because a format for the variable civilgrp is present in the data set world. The resulting block chart is shown in the upper right corner of Fig. 3.28.

Pie Chart

```
pie   techgrp/discrete
      sumvar=inf_mort type=mean percent=inside value=inside
                   slice=outside noheading name='myplot4';
```

```
goptions reset=all;
proc greplay tc=c10b nofs;
tdef matrix4 des='Four Squares of Equal Size'

1/  llx=0    lly=0
    ulx=0    uly=50
    urx=50   ury=50
    lrx=50   lry=0
    color=magenta
2/  llx=0    lly=50
    ulx=0    uly=100
    urx=50   ury=100
    lrx=50   lry=50
    color=green
3/  llx=50   lly=50
    ulx=50   uly=100
    urx=100  ury=100
    lrx=100  lry=50
    color=blue
4/  llx=50   lly=0
    ulx=50   uly=50
    urx=100  ury=50
    lrx=100  lry=0
    color=red;

template matrix4;
list template;

goptions display ;
proc greplay nofs igout=work.gseg tc=c10b template=matrix4;
    treplay 1:myplot1
            2:myplot2
            3:myplot3
            4:myplot4;
run;

quit;
```

Fig. 3.29. SAS Example C10 (Part B): Program

Again, the *chart variable* `techgrp` will be considered a discrete-valued variable because of the `discrete` option ❸. Thus, the number of slices will correspond to the three values of `techgrp`, the relative areas of the slices being determined by the mean of the *summary variable* `inf_mort`, as specified in `sumvar=inf_mort` and `type=mean`. The options `percent=inside value=inside slice=outside` control how the slices are to be labeled. The percentage represented by each slice is printed inside, the chart statistic value (here, the mean of `inf_mort`) for each slice is also printed inside, and the midpoint value of each slice (here, the formatted values of `techgrp`) is printed outside. The resulting pie chart is shown in the lower right corner of Fig. 3.28.

In the SAS Example C10 (Part B) program shown in Fig. 3.29, the SAS/GRAPH procedure `proc greplay` is used to display the four graphs produced and temporarily saved in SAS Example C10 (Part A). The top part of the program defines a *template* (named `matrix4`) containing four square areas of equal size arranged in a 2×2 matrix structure. The bottom part of the program accesses the graphs from `work.gseg`, inserts them in the template in the sequence specified, and displays them as a single graph.

3.9 Other SAS Procedures for High-Resolution Graphics

There are several SAS procedures that create high-resolution graphics provided that the SAS/GRAPH module is available as part of the local SAS software installation. In this section, three examples are provided to illustrate their use. In particular, the base SAS procedure `univariate` is used to construct a histogram and a normal probability plot, and the SAS/STAT procedure `boxplot` is used to construct side-by-side box plots, all in high-resolution graphics. These examples make use of global SAS/GRAPH statements or equivalent action statement options to enhance the graphics produced.

SAS Example C11

```
data chicks;
input wtgain @@;
label wtgain ='Weight gain (in gms) after 8-weeks';
datalines;
3.7  4.2  4.4  4.4  4.3  4.2  4.4  4.8  4.9  4.4
4.2  3.8  4.2  4.4  4.6  3.9  4.1  4.5  4.8  3.9
4.7  4.2  4.2  4.8  4.5  3.6  4.1  4.3  3.9  4.2
4.0  4.2  4.0  4.5  4.4  4.1  4.0  4.0  3.8  4.6
4.9  3.8  4.3  4.3  3.9  3.8  4.7  3.9  4.0  4.2
4.3  4.7  4.1  4.0  4.6  4.4  4.6  4.4  4.9  4.4
4.0  3.9  4.5  4.3  3.8  4.1  4.3  4.2  4.5  4.4
4.2  4.7  3.8  4.5  4.0  4.2  4.1  4.0  4.7  4.1
4.7  4.1  4.8  4.1  4.3  4.7  4.2  4.1  4.4  4.8
4.1  4.9  4.3  4.4  4.4  4.3  4.6  4.5  4.6  4.0
;
run;

proc univariate plot;
 var wtgain;
 histogram wtgain/midpoints = 3.6 to 4.9 by 0.1
                 cfill=darkcyan pfill = x4 ctext=deepskyblue
                 caxes=darkgray normal(color=magenta mu=4.3 sigma=0.3);
 title c= firebrick  'Histogram of Chick Data using Proc Univariate';
run;
```

Fig. 3.30. SAS Example C11: Program

In an experiment to investigate whether the addition of an antibiotic to the diet of chicks promotes growth over a standard diet described in Ott and Longnecker (2001), an animal scientist rears and feeds 100 chicks in the same environment, with individual feeders for each chick. The weight gain for the 100 chicks are recorded after an 8-week period. The construction of a frequency table suggests class intervals of width 0.1 units beginning with a midpoint at the smallest data value of 3.6. The SAS Example C11 program, shown in Fig. 3.30, uses the `histogram` action statement in the SAS/BASE procedure `proc univariate` to construct a histogram with these midpoints specified using the option `midpoints = 3.6 to 4.9 by 0.1`, with the options `cfill=` and `pfill=` added to specify a fill color and a fill pattern for the bars. The `normal` option specifies that a normal density curve be superimposed on the histogram. The `mu=` and `sigma=` specify values for the parameters μ and σ. By default, the procedure fits a normal density to the data using the sample mean and sample standard deviation. The output is shown in Fig. 3.31.

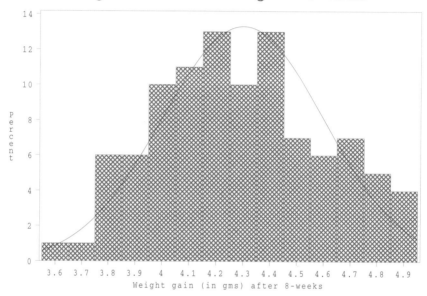

Fig. 3.31. SAS Example C11: Output

SAS Example C12

The data used in SAS Example C6 is used again to obtain a high-resolution graph of the normal probability plot of the logarithm of the seeded rainfall values (see Table B.3) using the `probplot` action statement ❶ in the SAS/BASE procedure `proc univariate`. The SAS program, shown in shown in Fig. 3.32, also contains a `symbol` statement to define a plot symbol and an `inset` ❷ statement to draw an *inset* containing summary statistics. An inset is typically used to display additional information to aid in the interpretation of the contents of a graph. The location of the inset can be controlled using the

3.9 Other SAS Procedures for High-Resolution Graphics

```
goptions reset = all;
goptions rotate = landscape targetdevice = pscolor
                              hsize = 8in vsize = 6in;
data rainfall;
input control seeded @@;
logseed = log10(seeded);
datalines;
1202.6 2745.6 830.1 1697.8 372.4 1656.0 345.5 978.0 321.2 703.4
 244.3  489.1 163.0  430.0 147.8  334.1  95.0 302.8  87.0 274.7
  81.2  274.7  68.5  255.0  47.3  242.5  41.1 200.7  36.6 198.6
  29.0  129.6  28.6  119.0  26.3  118.3  26.1 115.3  24.4  92.4
  21.7   40.6  17.3   32.7  11.5   31.4   4.9  17.5   4.9   7.7
   1.0    4.1
;
run;

title1 c = magenta f = triplex  h = 1.5
              ' Normal Probability Plot  of Log Seeded Rainfall';
symbol1 c = darkgreen v = dot  i = none;

proc univariate data=rainfall;
   probplot logseed/normal(mu=est sigma=est) caxis=blue
                ctext= red height= 2 vm=9
                vaxislabel= 'Log Seeded RainFall' pctlminor;
   inset mean std /format=5.3 header='Reference Line Stats'
                   refpoint=bl position=(70,10)
                   cfill=ywh ctext=maroon font=centb;
run;
```

Fig. 3.32. SAS Example C12: Program

position= option (pos= for short). Here, the location is specified as a pair of coordinates (x, y) in axis percent units. This produced the inset placed in the position shown in Fig. 3.33. The coordinates may be specified directly using axis units as well; for example, in position=('07JUL94'd, 3950) data, the coordinates refer to values plotted on the axes. In the above two cases, the refpoint= option may be used to specify the corner of the inset, the bottom left (bl), top left (tl), bottom right (br), or the top right (tr), respectively, is to be placed at the coordinates given in the position= parameter. The default reference point is bl.

More simply, the inset position may be specified as one of the eight compass points n, ne, e, se, s, sw, w, or nw. In the current example, a value of se for position would produce an inset placed at the rightmost lower corner of the frame. The inset may also be placed in one of the four margins surrounding the plot area using the margin postions lm, rm, tm, or bm to indicate the centers of the left, right, top, or bottom margins, respectively. Options such as cfill= ,cframe=, cheader=, and ctext= are available to control the color of such elements of the inset as the background, the frame, header text, and text color.

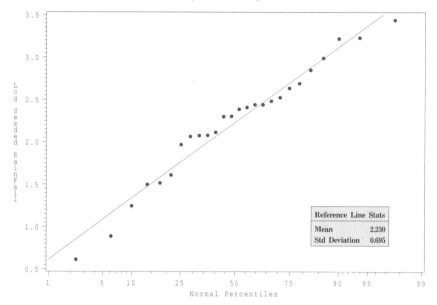

Fig. 3.33. SAS Example C12: Output

SAS Example C13

The data in Table B.5 taken from Koopmans (1987) give hydrocarbon (HC) emissions at idling speed, in parts per million (ppm), for automobiles of various years of manufacture. The data were extracted via random sampling from that of an extensive study of the pollution control existing in automobiles in current service in Albuquerque, New Mexico.

The SAS Example C13 program, shown in Fig. 3.34, uses the SAS/STAT procedure `proc boxplot` to construct side-by-side box plots for the five periods. Note that since the number of data values for each period is different, a new technique is employed to input the data. An artificially created data value for the variable `hc` of -1 is first inserted at the end of the set of data values for each period (note that this need not be at the end of every data line). A `do until` loop is used to read the data values for `hc` until a -1 is encountered, holding the data line after reading each value and using an `output;` statement to write an observation. The loop is restarted after the processing of all hc values for a period is completed. A new value for the variable `period` is read after each completion of the `do until` loop.

The output shown in Fig. 3.35 is used in the following to compare and comment on features such as shape, location, dispersion (spread), and ouliers, if any, of the distributions of HC emission in the five periods. Most of the parameter values are self-explantory. The value `schematicid` ❶ for the `boxstyle=`

3.9 Other SAS Procedures for High-Resolution Graphics

```
data emissions;
input period @;
hc=1;
do until (hc<0);
      input hc @;
      if (hc<0) then return;
      output;
      end;
label period = 'Year of Manufacture'
      hc = 'Hydrocarbon Emissions (ppm)';
datalines;
1 2351 1293 541 1058 411 570 800 630 905 347 -1
2 620 940 350 700 1150 2000 823 1058 423 270 900 405 780 -1
3 1088 388 111 558 294 211 460 470 353 71 241 2999 199 188
    353 117 -1
4 141 359 247 940 882 494 306 200 100 300 190 140 880 200
    223 188 435 940 241 223 -1
5 140 160 20 20 223 60 20 95 360 70 220 400 217 58 235 1880
    200 175 85 -1
;
run;

proc format;
 value pp 1='Pre-1963'
          2='1963-1967'
          3='1968-1969'
          4='1970-1971'
          5='1972-1974';
run;

proc sort data=emissions;
by period;
run;

axis1 minor = (n=9) label = (c=maroon a = 90 r = 0 h=1.5);
title c=darkblue j=c 'Box Plots of Hydrocarbon Emissions by Period';

proc boxplot data=emissions;
    plot hc * period /boxstyle = schematicid
                      cboxfill = white
                                 cboxes= red
                      idsymbol = star
                      height = 3
                      vaxis = axis1
                                 caxis=blue
                                 ctext= darkcyan
                      nohlabel;
      format period pp.;
        id hc;
run
```

Fig. 3.34. SAS Example C13: Program

option specifies a *schematic* box plot to be constructed in which an id variable value is used to label the symbols plotted identifying observations outside the upper and lower fences. In Fig. 3.34, the statement id hc; specifies that the values of the variable hc are to be used for this purpose. A schematic box plot is the standard style of box plots where the whiskers are drawn to the lower and upper adjacent values from the edges of the box. Lower and upper adjacent values are respectively the smallest observed value inside the lower fence and the largest observed value inside the upper fence. Options to specify other styles of box plots and various modifications are available. A format statement **2** specifies a user-defined format for labeling the values of the variable `period`.

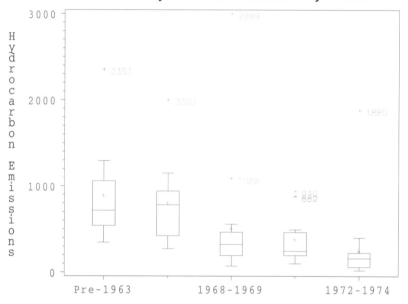

Fig. 3.35. SAS Example C13: Output

Side-by-side box plots is one of the most useful methods available for visually comparing features of sample distributions. They enable the comparison of the location, spread, and shape of the distributions by examining the relative positions of the median and the mean, the heights of the boxes which measure the interquartile ranges (IQRs), and the relative placing of the medians (and the means) between the ends of the boxes (i.e., the quartiles), the relative lengths of the whiskers, and the presence of outside values at either end of the whiskers. By observing the presence of trends in these characteristics, experimenters will be able to compare distributions across populations defined over time, location, treatments, or predefined experimental or observational groups. As an example, some observations regarding the empirical distributions of hydrocarbons over the five periods of study that may be made using Fig. 3.35 and useful conclusions that may be drawn from these observations are itemized as follows:

a. There is a decreasing trend in the magnitudes of location and spread of HC levels over years of manufacture except in the first two periods. The location, as measured by the median, and spread, as measured by the IQR and the lengths of the whiskers, appear to follow the same pattern. During the first two periods, there appears to be no significant shift of both location and the spread.

b. The observed shapes of the distribution of HC levels have some similarity over the five periods. For example, the sample mean is larger than the median in all five samples and the upper (or right) whisker is longer in all but one. There is also at least one outside value on the upper side. Thus, all of the evidences indicates right-skewed distributions for all five periods.

c. There is an apparent change in HC emission coincident with the establishment of federal emission control standards in 1967-1968 (period 3). This is clearly observed as both the location and spread of the data decrease significantly following period 2.

d. Statistical analysis of data (say, using the one-way anova model) may not be straightforward because assumptions such as normality and homogeneity of variances across periods may not hold. The graphical analysis shows that these two assumptions may not be plausible for this data; particulary, there is an obvious heterogeneity of variances as observed by the differences in the spread of the data as measured by the heights of the boxes.

3.10 Exercises

3.1 Edit the program from SAS Example C1 in the program editor and execute the program under the Windows system. Click on the graph window to make it active and view the graph. Note that the target device driver is `pscolor`, a SAS native driver, and hence the graph observed in the SAS graph window closely matches the characteristics of its device entry. Then do the following:

a. Use the `File` → `Print...` item from the top menu bar to print the graphs directly to a printer accessible to your computer.

b. Use `File` → `Export as Image...` item from the top menu bar to save the graphics output to a file named `c1.jpg` (or of any other graphics filetype available on the system).

c. Use any software available on your system (e.g., MS Photoshop or MS Word) to obtain a printed output from the file `c1.jpg`.

d. Replace the `goptions` statement with the following two statements:

```
filename gsasfile "[insert path to folder here]\c1.jpg";
goptions rotate=landscape gaccess=gsasfile device=jpeg
                gsfmode=append hsize=8 in vsize=6 in;
```

Using these statements is another way to save the graphics file created by a SAS/GRAPH program in a folder of your choice. Use any available software to view or print the file c1.jpg that you locate in your folder. Note that the filetype .jpg used above may be replaced with a filetype of your choice and the device, hsize, and vsize specifications amended appropriately. For example, if the filetype is changed to .ps, the settings device=ps600c, hsize=8 in vsize=6 in may be more appropriate.

3.2 Groundwater quality is affected by the geological formations in which it is found. A study of chlorine concentration was carried out to determine whether differences in quality existed for water tables on the east and west sides of the Rio Grande River in New Mexico. The data from wells on each side in (milliequivalents), reported in Koopmans (1987)), are

West Side: 0.58, 0.38, 0.32, 0.55, 0.56, 0.62, 0.61, 0.63, 0.52, 0.53, 0.49, 0.37, 0.40, 0.62, 0.44, 0.18, 0.24, 0.21

East Side: 0.34, 0.24, 1.03, 0.68, 0.29, 1.14, 0.34, 0.46, 0.53, 0.40, 0.33, 0.37, 0.40, 0.55, 0.76, 0.37, 0.40, 0.45, 0.30, 0.46, 0.12, 0.39, 0.65

Use proc gplot (as in SAS Example C2) to construct side-by-side box plots of chlorine concentrations for the two sides of the river. Use information provided by the box plots to describe and compare (location, dispersion, and shape of) chlorine concentration distributions on the two sides of the river. Compare these to normal distributions, commenting on how they may differ (shorter-tailed or longer-tailed, right-skewed or left-skewed, or the presence of extreme values).

3.3 Obtain a vertical bar chart of the nox variable from the pollution data set (see Tables B.6 and B.6). Examining the output from proc univariate for this variable, it is observed that nox values are highly dispersed. So it is necessary to use unevenly spaced intervals to obtain a useful bar chart for this variable. To do this, center the bars at the values, first from 0 through 30, incremented by 5, and then at 40, 60, and 100, respectively (i.e., 10 intervals in all). Subdivide the bars by the category variable density created in the data step so it has values 'Low', 'Medium', or 'High', depending on whether the value for the variable popn is less than 3000, larger than 3000 but less than 4000, or larger than 4000, respectively. Use different solid colors (e.g., cyan, magenta, etc.) to fill the different portions of the bars representing different values for the density variable.

3.4 In the following study reported Mason et al. (1989), two petroleum-product refineries that produce batches of fuel using the same refining process are compared. Random samples were selected from batches produced at each refinery and the cetane number measured on the fuel samples. Calculate descriptive statistics and graphics described below to compare the cetane measurements from the two refineries.

Refinery 1:

50.3580 47.5414 47.6311 47.6657 47.7793 47.2890 46.8472 48.2131 46.9531

48.4489 47.3514 48.3738 49.2652 47.2276 48.6901 47.5654 49.1038 49.8832 48.7042 47.9148

Refinery 2:

45.8270 46.8957 45.2980 45.4504 46.1336 46.6862 45.6281 46.1460 46.3159 45.2225 46.1988 47.4130 45.7478 45.5658

  a. Construct, by hand, back-to-back stem-and-leaf plots of cetane measurements from the two refineries. (Hint: Round the data to one decimal place and multiply by 10. Use leaves corresponding to the last digit with those in range 0–4 on one stem and 5–9 on the next stem).
  b. Use proc gplot (as in SAS Example C2) to construct side-by-side box plots of cetane measurements from the two refineries.
  c. Use the information provided by the above two plots to describe each distribution. Comment on whether each distribution is short-tailed, long-tailed, skewed, etc. with respect to a normal distribution.
  d. Use the above two plots to compare the distributions of cetane measurements from the two refineries with respect to their locations (central values), dispersions (spreads), and shapes (compared to a normal density).

3.5 Rice (1988) cites an experiment that was performed to determine whether two forms of iron (Fe^{2+} and Fe^{3+}) are retained differently. The investigators divided 108 mice randomly into 6 groups of 18 each; 3 groups were given Fe^{2+} in 3 different concentrations, 10.2, 1.2, and .3 millimolar, and 3 groups were given Fe^{3+} at the same three concentrations. The mice were given the iron orally and the percentage of iron retained was calculated by radioactively labeling the iron. The data for the six "treatments" of iron by each mouse are listed in the following table.

  a. Construct side-by-side box plots using proc boxplot as in SAS Example C13 for the six treatments. Place the plots on the x-axis in the order High, Medium, and Low doses for each of the two forms of iron, respectively. Compare and comment on features such as shape, location, dispersion and outliers of the six iron retention distributions.
  b. Comment on any observed trend in the median % iron retention over the levels and the forms of iron. What is the observed trend in dispersion (as measured by, say, IQR)? That is, compare the distributions of % iron retention across the six treatments.
  c. Statistical analysis of this data (e.g., using a one-way anova model) may be complicated by failure of assumptions such as homogeneity of variance and/or non-normal distributions. Do the box plots show evidence of these problems? Explain. If there is reason to believe that the assumptions fail based on the plots, a possible explanation is that each distribution is related to the median level of % iron retention in some way. Discuss whether there appears to be such a relationship and describe the relationship algebraically.

Fe^{3+}			Fe^{2+}		
10.2	1.2	0.3	10.2	1.2	0.3
0.71	2.20	2.25	2.20	4.04	2.71
1.66	2.93	3.93	2.69	4.16	5.43
2.01	3.08	5.08	3.54	4.42	6.38
2.16	3.49	5.82	3.75	4.93	6.38
2.42	4.11	5.84	3.83	5.49	8.32
2.42	4.95	6.89	4.08	5.77	9.04
2.56	5.16	8.50	4.27	5.86	9.56
2.60	5.54	8.56	4.53	6.28	10.01
3.31	5.68	9.44	5.32	6.97	10.08
3.64	6.25	10.52	6.18	7.06	10.62
3.74	7.25	13.46	6.22	7.78	13.80
3.74	7.90	13.57	6.33	9.23	15.99
4.39	8.85	14.76	6.97	9.34	17.90
4.50	11.96	16.41	6.97	9.91	18.25
5.07	15.54	16.96	7.52	13.46	19.32
5.26	15.89	17.56	8.36	18.4	19.87
8.15	18.3	22.82	11.65	23.89	21.60
8.24	18.59	29.13	12.45	26.39	22.25

3.6 Obtain a normal probability plot of the life_exp variable in the demographic data set on countries (see Table B.4) used in SAS Example C9. Use proc univariate as in the SAS Example C12 program. Use title and symbol statements to add a title and specify a plot symbol, choosing their colors. Use the normal option to add a reference line to the plot. Enhance the plot by adding options to specify colors for axis lines and text and adding appropriate minor tick marks. Use an inset statement to add statistics used for producing the reference line. Comment on the shape of the distribution of this variable compared to a normal distribution.

3.7 Write and execute a SAS program to obtain a quantile plot of the so2 variable of the pollution data set (see Table B.6 and B.6). Add minor tick marks to the axes to enable the reading of axis values more accurately. Estimate the median, the quartiles, and the IQR from this plot showing work. (You may draw vertical reference lines on the plot to indicate these quantile locations.) In the same SAS program, use proc univariate to check the accuracy of the values you estimated from the graph. By locating a region in the quantile plot where the local slope is smallest (i.e., flattest part of the curve), estimate the narrow range of values within which the sulfur dioxide levels lie for about 60% of the SMSAs in the sample. Find the worst 10% of sulphur dioxide levels in this sample.

3.8 Use proc gchart as in SAS Example C10 to obtain the following:
  a. A horizontal barchart of mean per capita GNP for each techgrp within each infgrp in the the demographic data set on countries (see Table B.4) (i.e., techgrp is the chart variable and infgrp is the group variable)

b. A vertical barchart (i.e., a histogram) of crude death rate, choosing your own midpoints. Use frequency as the chart statistic and subdivide bars by `civilgrp`

3.9 Insulin production from beta islets (insulin-producing cells) in the pancreas of obese rats, reported in Koopmans (1987), are reproduced below. In addition to measurements made at end of each of the first 3 weeks, data are also available for the first day of the experiment (labeled Week 0):

Week 0	Week 1	Week 2	Week 3
31.2	18.4	55.2	69.2
72.0	37.2	70.4	52.0
31.2	24.0	40.0	42.8
28.2	20.0	42.8	40.6
26.4	20.6	26.8	31.6
40.2	32.2	80.4	66.8
27.2	23.0	60.4	62.0
33.4	22.2	65.6	59.2
17.6	7.8	15.8	22.4

a. Construct side-by-side box plots using `proc boxplot`, as in SAS Example C13, for the four periods of study. Compare and comment on features such as shape, location, dispersion, and outliers of the distributions of insulin production for each week.

b. Comment on the observed trend in the median insulin production as time increases. What is the observed trend in dispersion (as measured by IQR)? That is, compare the distributions of insulin production across the period of study.

c. Statistical analysis of this data (e.g., using a one-way anova model) may be complicated by failure of assumptions such as homogeneity of variance, non-normal distributions, or presence of outliers. Do the box plots show evidence of any of these problems? Explain. If any of the above-mentioned problems exist, can you relate these problems to the median level of insulin production? Explain.

3.10 Obtain a normal probability plot of the `popurban` variable in the demographic data set on countries (see Table B.4) used in SAS Example C9. Use `proc univariate`, as was done in the SAS Example C12 program. Use `title` and `symbol` statements to add a title and specify a plot symbol, both in color. Use the `normal` option to add a reference line to the plot. Enhance the plot by adding options to specify colors for axis lines and text and adding appropriate minor tick marks. Use an `inset` statement to add statistics used for producing the reference line.

3.11 Write and execute a SAS program to obtain a quantile plot of the `infant mortality` variable in the demographic data set on countries (see Table B.4) used in SAS Example C9. Add minor tick marks to the axes to facilitate reading of the axis values more accurately. Estimate the median, the quartiles, and the IQR using this plot. Show work. In the same program, use `proc univariate` to check the accuracy of the values you estimated from the graph. Identify the narrowest range of values of infant mortality rate within which the largest number of countries fall by locating a part on the quantile plot where the local slope is "flattest". Likewise, identify the widest range of values of infant mortality rate within which the smallest number of countries fall by locating a part on the quantile plot where the local slope is "steepest".

3.12 Use the `boxanno` macro and `proc gplot` to obtain a scatter plot of variables with marginal box plots `life_exp` vs. `perc_gnp` (as in SAS Example C8) using the `demogr.data` data set used in the demographic data set on countries (see Table B.4) used in SAS Example C9. The sample correlation coefficient between these variables is calculated to be .74. Does this suggest a linear relationship between these two variables? Why or why not? Suggest a simple model (in words; no equation needed) that might fit the data better. By examining the ranges of the values of the two variables plotted, suggest a possible transformation of one of the variables so that the relationship may show more linearity. Repeat the plot to examine if your tranformation improves the linearity.

3.13 A preliminary study indicated that the Stamford ozone values had a distribution that is highly right-skewed relative to the normal distribution, suggesting a power transformation of the data with a power value less than 1. Further investigation also suggests that a power of 1/3, that is, a cube root transformation may be adequate. Use the technique shown in the SAS Example C6 program and the ozone data used in SAS Example C5 to obtain normal probability plots of Stamford ozone and data values obtained from transforming Stamford ozone values using a cube root transformation.

4
Statistical Analysis of Regression Models

4.1 An Introduction to Simple Linear Regression

In this section, a review of simple linear regression or straight-line regression is presented. Consider `bivariate data` consisting of ordered pairs of numerical values (x, y). Often such data arise by setting an X variable at certain fixed values and taking a random sample from the population of Y that, hypothetically, exists for each setting of X. The variable Y is called the dependent variable or the response variable and the variable X is called the independent variable or the predictor variable. The y-values observed at each x-value are assumed to be a random sample from a normal distribution with the mean $E(y) = \mu(x) = \beta_0 + \beta_1 x$; that is, the mean of the distribution is modeled as a linear function of x. The variance of the normal distributions at each x-value is assumed to have the same value σ^2. Thus, the y-values can be related to the x-values through the relationship

$$y = \beta_0 + \beta_1 x + \epsilon \tag{4.1}$$

where ϵ is a random variable (called `random error`) with mean zero (i.e, $E(\epsilon) = 0$), and variance σ^2. Equation 4.1 is called the *simple linear regression model* and β_0, β_1, and σ^2 are the *parameters* of the model. In the above representation, note that y is a random variable with mean $E(y) = \beta_0 + \beta_1 x$ and variance σ^2.

Estimation of Parameters

The first step in regression fitting is to obtain estimates of the above parameters of the model using the observed data. The *method of least squares* selects a model that minimizes the total error of prediction. The error or the *residual* is the difference between an observed value y for a given value of x and $\hat{y}$, the value of y predicted by the fitted model at that particular value of x. The method of least squares selects the line that produces the smallest value of the

M.G. Marasinghe, W.J. Kennedy, *SAS for Data Analysis*,
DOI: 10.1007/978-0-387-77372-8_4, © Springer Science+Business Media, LLC 2008

sum of squares of all residuals; that is, mathematically, it finds the estimates $\hat{\beta}_0$ and $\hat{\beta}_1$ that minimizes the quantity

$$\sum_i (y_i - \hat{y}_i)^2 = \sum_i (y_i - \hat{\beta}_0 - \hat{\beta}_1 x_i)^2$$

where (x_i, y_i), $i = 1, 2, \ldots, n$, are all pairs of observations available. The values $\hat{\beta}_0$ and $\hat{\beta}_1$ are called the least squares estimates of β_0 and β_1, respectively. The fitted regression equation $\hat{y} = \hat{\beta}_0 + \hat{\beta}_1 x$ is usually called the prediction equation and is used to calculate the predicted value $\hat{y}$ for a specified value of x. Since the mean of y is $E(y) = \beta_0 + \beta_1 x$, the slope $\hat{\beta}_1$ is the change in the mean of y for a unit change in the x-value, estimated by the fitted model.

Statistical Inference

Following the fitting of a least squares line to the data, it is of interest to measure how well the line estimates the population means. This is achieved by the computation of quantities necessary to perform statistical inference about the parameters of the model, such as hypothesis tests and confidence intervals. The first step is usually to construct an analysis of variance that partitions the total sum of squares, SSTot= $\sum (y - \bar{y})^2$, into two parts: the sum of squares due to regression, SSReg= $\sum (\hat{y} - \bar{y})^2$, and the sum of squares of residuals, SSE= $\sum (y - \hat{y})^2$. This algebraic partition is represented in the following anova table for regression:

Source	df	Sum of Squares	Mean Square	F
Regression	1	SSReg	MSReg=SSReg/1	MSReg/MSE
Error	$n-2$	SSE	MSE=SSE/$(n-2)$	
Total	$n-1$	SSTot		

This table provides several important statistics that are useful for assessing the fit of the model. First, the F-ratio, F = MSReg/MSE, is used to test the hypothesis that $H_0 : \beta_1 = 0$ vs. $H_a : \beta_1 \neq 0$. Second, the MSE from the above table provides the least squares estimate of σ^2. This is useful for calculating the standard errors of the estimates $\hat{\beta}_0$ and $\hat{\beta}_1$, commonly denoted by $s_{\hat{\beta}_0}$ and $s_{\hat{\beta}_1}$, respectively. Test statistics for performing t-tests and confidence intervals for the β_0 and β_1 coefficients are calculated using these standard errors. For example, a t-statistic for testing $H_0 : \beta_1 = 0$ vs. $H_a : \beta_1 \neq 0$ is given by

$$t = \frac{(\hat{\beta}_1 - 0)}{s_{\hat{\beta}_1}}$$

and a $(1-\alpha)100\%$ confidence interval for β_1 is given by

$$\hat{\beta}_1 \pm t_{\alpha/2,(n-2)} \times s_{\hat{\beta}_1}$$

Third, the ratio $R^2 =$ SSReg/SSTot, called the *coefficient of determination*, measures the proportion of variation in y explained by using $\hat{y}$ to predict y. A simple linear regression model using the predictor variable x with a larger R^2 does better in predicting y than one with a smaller R^2. Note that R^2 does not say how *accurate* the prediction is nor does it say that a straight line is the best function of x that could be used to model the variation in y.

4.1.1 Simple linear regression using PROC REG

Consider the following problem. An investigation of the relationship between traffic flow x (thousands of cars per 24 hours) and lead content y of bark on trees near the highway ($\mu g/g$ dry weight) yielded the data in Table 4.1 reported in Devore (1982).

Traffic Flow, x	8.3	8.3	12.1	12.1	17.0	17.0	17.0	24.3	24.3	24.3	33.6
Lead Content, y	227	312	362	521	640	539	728	945	738	759	1263

Table 4.1. Lead content in trees near highways (Devore, 1982)

SAS Example D1

In the SAS Example D1 program (see Fig. 4.1), a simple linear regression model is fitted to these data and statistics for making some of the statistical inferences about the model discussed earlier are computed using `proc reg`. This program produces printed output that helps to perform an elementary regression analysis of the data. It contains SAS statements necessary to compute an analysis of variance, estimated values of the parameters and their standard errors, associated t-statistics and confidence intervals, and the predicted values and residuals. It also produces, in high-resolution graphics, a scatter plot of the data overlaid with the fitted regression line by the use of a `plot` statement. Various other diagnostic statistics and plots necessary for examining the adequacy of the model and accompanying assumptions will be deferred to later examples.

The data are entered in a straightforward manner, where a pair of values for x and y separated by a blank comprise each line of data, so that the statement `input x y;` is all that is necessary to read the data and create the SAS data set (named `d1` in this example). The statement

```
label x='Traffic Flow' y='Lead Content';
```

```
data d1;
input x y;
label x='Traffic Flow' y='Lead Content';
datalines;
 8.3  227
 8.3  312
12.1  362
12.1  521
17.0  640
17.0  539
17.0  728
24.3  945
24.3  738
24.3  759
33.6 1263
;
run;

proc reg data=d1;
   model y=x/p clb;
   plot y*x/ nomodel nostat;
   title 'Simple Linear Regression of Lead Content Data';
run;
```

Fig. 4.1. SAS Example D1: Program

adds descriptive labels to the variables x and y as illustrated in SAS Example A10. In the procedure statement, the option `data=d1` specifies the SAS data set to be processed in the `proc reg` step. The use of such options is described in Section 1.8 of Chapter 1. The model statement `model y=x/p clb;` specifies model (4.1) as `y=x`. In this representation of the model, an intercept β_0 is presumed to be present and the error term is omitted. Various options may be included, following a slash (solidus) symbol, as keywords separated by at least one blank. For example, the option `noint` may be used to specify that a model be fitted omitting the intercept β_0. In the present example, the option `p` (for `predicted`) requests that predicted values be computed and printed along with the residuals **4**, as shown on page 2 of the SAS output (Fig. 4.3). The `clb` option specifies that $(1-\alpha)100\%$ confidence intervals for the β coefficients be computed. By default, $\alpha = .05$ is used; the `alpha=` option allows the user to specify an alternate confidence coefficient. For example,

$$\text{model y=x/p clb alpha=.1;}$$

would produce 90% confidence intervals for the β's. The anova table given below is constructed directly from the SAS output resulting from the model statement **1**, shown in page 1 of the SAS output (Fig. 4.2).

Since the p-value is smaller than a level of significance selected for this study (say, $\alpha = .05$), the null hypothesis of $H_0 : \beta_1 = 0$ is rejected and it is concluded that the lead content on tree bark can be modeled by a simple linear regression using traffic flow as a predictor variable. The R^2 value for this fit, also extracted from page 1 of the SAS output, is 0.9143, showing that 91.3% of the variation in the response is explained by the fitted model. This is supported by the graph shown in Fig. 4.4a. Additionally, the parameter estimates (estimates of the coefficients), their standard errors, and the t-statistics discussed in Section 4.1, are found on page 1 of the SAS output (Fig. 4.2) **2**.

Source	df	Sum of Squares	Mean Square	F	p-value
Regression	1	815,966	815,966	96.0	< .0001
Error	9	76,493	8,499		
Total	10	892,459			

The prediction equation is calculated to be $\hat{y} = -12.84 + 36.18\, x$; thus the fitted line has a positive slope. The value of the t-statistic is 9.80 with a p-value $< .0001$, which again shows that the hypothesis $H_0 : \beta_1 = 0$ will be rejected. The 95% confidence interval for β_1 is computed to be $(27.83, 44.54)$, as shown on page 1 **3**. Thus with 95% confidence, the increase in mean lead content on tree bark that results from an increase in traffic flow by 1000 cars in 24 hours lies in the above interval.

The above prediction equation must be used with caution. In practice, it is recommended that predictions be made only within the range of x-values observed. That is because extrapolation outside this range may cause problems, as the model may no longer be valid. For example, the intercept of the fitted line is $\hat{y} = -12.84$ (i.e., the predicted value given by the prediction equation at $x = 0$). Thus, when there is no traffic flow, the lead content of bark on trees near the highway is predicted to be -12.84 when the model is extrapolated to $x = 0$! This argument shows that in such situations, the routine test of $\beta_0 = 0$ (i.e., the y-intercept at $x = 0$) also will not make sense.

The SAS statement

```
plot y*x/nomodel nostat;
```

is an example of the use of the `plot` statement for creating high-resolution graphics within `proc reg`. It produces a scatter plot with the y-variable on the vertical axis and the x-variable on the horizontal axis. Unless the `lineprinter` option is specified in the `proc reg` statement, by default, plots in high-resolution graphic format are produced in a SAS graphics window. Other options such as `ctext=` and `caxis=` and SAS/GRAPH global statements such as `symbol, axis`, and `legend` statements may be used to enhance these plots. For example, options `ctext=vip, caxis=blue`, and the statement `symbol1 v=dot c=red;` were used to enhance the plots shown in Fig. 4.4.

More than one plot request can be included in the same plot statement to obtain multiple plots. For example, `plot y*x r.*x;` will produce two separate graphs. The variables can be any variable specified in the `var` or `model` statements or variables specified in the form `keyword.`, where `keyword` is an abbreviation for a regression diagnostic statistic. These keywords are listed as options available for the `output` statement to be discussed later. Two examples of variables of this form are `residual.` (or, in abbreviated form, `r.`)

and `predicted.` (or, `p.`), whose values are the residuals and the predicted values, respectively. Thus, the plot request `r.*x` will produce a plot of the residuals against the x-variable. A request of the form `r.*p.` will produce a plot of the residuals against the predicted values, a very useful diagnostic plot as will be discussed below.

The plot statement, `plot y*x/nomodel nostat;`, used in the SAS Example D1 program produces a scatter plot of y against x superimposed with a line plot of the fitted regression line. The options `nomodel` and `nostat` suppresses the prediction equation and statistics such as R^2 that appear by default on the margins of the plot. The resulting graph is shown in Fig. 4.4a.

The inclusion of the SAS statement

$$\text{plot r.*x r.*p. student.*nqq.;}$$

in the SAS Example D1 program would produce three high-resolution plots of the residuals against the x-variable, residuals against the predicted values, and a normal probability plot of the *studentized residuals*. Note that the keyword

```
              Simple Linear Regression of Lead Content Data                1

                              The REG Procedure
                              Model: MODEL1
                         Dependent Variable: y Lead Content

                             Analysis of Variance  1

                                    Sum of           Mean
    Source                  DF     Squares         Square    F Value    Pr > F

    Model                    1      815966         815966      96.01    <.0001
    Error                    9       76493      8499.17298
    Corrected Total         10      892459


              Root MSE              92.19096    R-Square     0.9143
              Dependent Mean       639.45455    Adj R-Sq     0.9048
              Coeff Var             14.41712

                            Parameter Estimates  2

                                 Parameter      Standard
    Variable   Label       DF     Estimate         Error    t Value    Pr > |t|

    Intercept  Intercept    1    -12.84155       72.14287     -0.18      0.8627
    x          Traffic Flow 1     36.18385        3.69290      9.80      <.0001

                            Parameter Estimates

    Variable    Label         DF     95% Confidence Limits  3

    Intercept   Intercept      1    -176.04007      150.35696
    x           Traffic Flow   1      27.82994       44.53776
```

Fig. 4.2. SAS Example D1: Output (page 1)

```
           Simple Linear Regression of Lead Content Data              2

                          The REG Procedure
                            Model: MODEL1
                    Dependent Variable: y Lead Content

                           Output Statistics

                   Dependent      Predicted
           Obs     Variable         Value       Residual

             1      227.0000      287.4844      -60.4844
             2      312.0000      287.4844       24.5156
             3      362.0000      424.9830      -62.9830
             4      521.0000      424.9830       96.0170
             5      640.0000      602.2839       37.7161
             6      539.0000      602.2839      -63.2839
             7      728.0000      602.2839      125.7161
             8      945.0000      866.4260       78.5740
             9      738.0000      866.4260     -128.4260
            10      759.0000      866.4260     -107.4260
            11          1263          1203       60.0643


           Sum of Residuals                              0
           Sum of Squared Residuals                  76493
           Predicted Residual SS (PRESS)            112492
```

Fig. 4.3. SAS Example D1: Output (page 2)

student. refers to the studentized residuals and the keyword nqq. refers to the corresponding standard normal quantiles used for constructing the *normal probability plot*.

Graphical tools may be used to help identify cases for which assumptions about the distribution of ϵ is not valid. Most of the plots used for this purpose involve some form of the residuals from fitting the model, $(y_i - \hat{y}_i)$, $i = 1, \ldots, n$. Graph of residuals versus x produced by plot r.*x is shown in Fig. 4.4b. If the model is correct, the residuals are expected to scatter evenly and randomly around the zero reference line as the value of x changes. If a nonlinear pattern is apparent, it usually indicates a need for higher-order terms in x or a model that is not linear in the parameters; that is, a systematic pattern of the residuals is an indication of model inadequacy. This plot may also show a departure from the homogeneity of variance assumption, as a marked decrease or increase of the spread of the residuals around zero may indicate a dependence of the variance of y on the actual values of x. A few points that stand out due to a comparatively larger or smaller residual than the others may also highlight outliers.

The graph of residuals versus predicted values produced by plot r.*p., is shown in Fig. 4.4c. This scatter plot should also show no systematic pattern and should indicate random scatter of residuals around the zero reference line if the straight-line model is adequate. If the homogeneity of variance

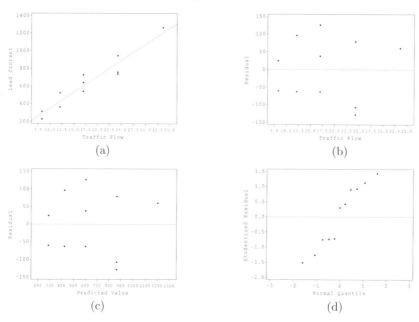

Fig. 4.4. (a) Graph of y versus x produced by `plot y*x/nomodel nostat;`, (b) graph of residuals versus x produced by `plot r.*x`, (c) graph of residuals versus predicted produced by `plot r.*p.`, (d) normal probability plot of the studentized residuals produced by `plot student.*nqq.`

assumption is not plausible, a pattern indicating a steady increase or decrease in spread of the residuals around the zero reference line as $\hat{y}_i$ increases will also be present. This pattern may show up along with a curved pattern in both this and the previous plot if nonlinearity is also present in addition to heterogeneous variance.

The graph produced by `plot student.*nqq.` appears in Fig. 4.4d. This displays a normal probability plot where *studentized residuals* are graphed against the corresponding percentiles of the standard normal distribution. *Internally* studentized residuals that are used here are a version of standardized residuals and are defined in Section 4.2.1. This plot will show an approximate straight-line pattern unless the normality assumption about the ϵ's is questionable. More discussion on how to correctly interpret a normal probability plot appears in Chapter 3 in Section 3.5.

The above three residual plots do not appear to show any inadequacy of the model or any serious violation of the model assumptions made concerning the straight-line model fitted to the lead content data. Although a few points may appear to have a larger residual in magnitude than the others, there is clearly no overall pattern showing a systematic variation of the residuals.

4.1.2 Lack of fit test using PROC ANOVA

Whenever regression data contain more than one response (y) value at one or more values of x, responses are said to be replicated. The data set shown in Table 4.1 contains replicated responses. In such cases, the sum of squares of residuals, SSE, can be partitioned into two parts: sum of squares representing pure experimental error, SSE_{Pure}, and sum of squares due to lack of fit, SS_{Lack}. Introducing a new notation, the responses within a subset i of observations with the same x-value are represented by y_{ij}, $j = 1, \ldots, n_i$, where n_i is the number of observations in the subset and $i = 1, \ldots, g$, where g is the number of such subsets. The partitioning of SSE is given by

$$SSE = SSE_{Pure} + SS_{Lack} \qquad (4.2)$$

$$\sum_i \sum_j (y_{ij} - \hat{y}_{ij})^2 = \sum_i \sum_j (y_{ij} - \bar{y}_{i.})^2 + \sum_i \sum_j (\bar{y}_{i.} - \hat{y}_{ij})^2 \qquad (4.3)$$

$$(n-2) = (n-g) + (g-2) \qquad (4.4)$$

where $n = \sum n_i$ is the total number of observations, and Equation 4.4 represents the *degrees of freedom* for each sum of squares. The hypotheses of interest are

$$H_0 : E(y) = \beta_0 + \beta_1 x$$
$$H_a : E(y) \neq \beta_0 + \beta_1 x$$

The test for lack of fit is an F-test. The F-statistic is the ratio of **mean squares** for lack of fit and pure experimental error. The above partitioning can be used to motivate the lack of fit using the following argument. Consider the sum of squares due to lack of fit $\sum_i \sum_j (\bar{y}_i - \hat{y}_{ij})^2$. If the true relationship among Y population means is indeed $E(y) = \beta_0 + \beta_1 x$, then one would expect this sum of squares to be small because both $\bar{y}_{i.}$ and $\hat{y}_{ij}$ are estimates of the mean of the Y population at a given x_i. If $E(y)$ deviates from $\beta_0 + \beta_1 x$, then one would expect the sum of squares due to lack of fit to be larger. The F-test is usually performed by supplementing the anova table as follows:

Source	df	Sum of Squares	Mean Square	F
Lack of Fit	$g-2$	SS_{Lack}	$MS_{Lack} = SS_{Lack}/(g-2)$	MS_{Lack}/MSE_{Pure}
Pure Error	$n-g$	SSE_{Pure}	$MSE_{Pure} = SSE_{Pure}/(n-g)$	
Total Error	$n-2$	SSE		

As an example, these calculations are first performed by hand for the data in Table 4.1 used previously in SAS Example D1. From the previous analysis, SSE=76,493 with 9 degrees of freedom. The following table simplifies the computation of SSE$_{\text{Pure}}$, noting that $g = 5$.

i	x_i	y_i			$(y_i - \bar{y})^2$			$\sum(y_i - \bar{y})^2$	$n_i - 1$
1	8.3	227	312		1,806.25	1,806.25		3,612.50	1
2	12.1	362	521		6,320.25	6,320.25		12,640.50	1
3	17.0	640	539	728	18.78	9,344.44	8,525.44	17,888.67	2
4	24.3	745	738	759	17,161.00	5,776.00	3,025.00	25,962.00	2
5	33.6	1,263			0			0	0
Total								60,103.67	6

Thus, SSE$_{\text{Pure}}$=60,103.67 with 6 degrees of freedom. The lack of fit F statistics is computed in the following anova table:

Source	df	Sum of Squares	Mean Square	F
Lack of Fit	3	16,389.33	5,463.11	0.55
Pure Error	6	60,103.67	10,017.28	
Total Error	9	76,493		

As expected, the test fails to reject $H_0 : E(y) = \beta_0 + \beta_1 x$; thus, the means of the populations at each value of X are modeled adequately by a simple linear regression model.

SAS Example D2

The SAS Example D2 program (see Fig. 4.5) illustrates how to obtain SSE$_{\text{Pure}}$ whenever there are replicated observations. In this example, the SAS procedure **proc anova** is employed to compute the pure error sum of squares simply by using the variable x as a *treatment* or a *classification* variable. This is done by including the statement **class x;** in the **proc anova** step. The model statement remains the same as in the SAS Example D1 program (omitting the options used earlier): **model y=x;**. The output from executing the SAS Example D2 program is reproduced in Fig. 4.6.

The information needed is in the line

```
Error       6    60103.6667    10017.2778  ❶
```

in Fig. 4.6. Thus, SSE$_{\text{Pure}}$=60,103.67 with 6 degrees of freedom, the same numbers obtained from hand calculation.

```
data d2;
input x y;
label x='Traffic Flow' y='Lead Content';
datalines;
 8.3  227
 8.3  312
12.1  362
12.1  521
17.0  640
17.0  539
17.0  728
24.3  945
24.3  738
24.3  759
33.6 1263
;
run;

proc anova data=d2;
  class x;
  model y=x;
  title 'Calculating Pure Error: Lead Content Data';
run;
```

Fig. 4.5. SAS Example D2: Program

4.1.3 Diagnostic use of case statistics

In addition to the residuals and studentized residuals, several other statistics related to each observation (or *case statistics*, as they are commonly called in modern regression literature) are computed by computer programs and are output on request. For example, the `proc reg` in SAS will output case statistics labeled as `Cook's D`, `RStudent`, and `Hat Diag` when certain options are specified in the `model` statement. These case statistics measure how well a specific data point fits the regression line. If the point is a large distance away from the center of the fitted line in the x-direction it is said to be a *high leverage point* and is called an *x-outlier*. A high leverage point will exhibit a comparatively large value for the `Hat Diag` statistic.

If the point is a large distance away from the fitted line in the y-direction, it will have a large residual or studentized residual. A statistical test procedure is available to determine whether a studentized residual is sufficiently large for the case to be declared a *y-outlier*. The Cook's D case statistic measures the *influence* a data point will have on the estimated parameters and/or overall fit statistics; that is, it measures whether the deletion of a data point will markedly change the estimates of the parameters β_0 and β_1, as well as MSE and R^2 values. If a large Cook's D value occurs, then that single data point is identified as an influential case.

A high leverage point that is also a y-outlier will most likely be a highly influential case and will have to be examined for validity by the experimenter. This is because including the specific case in the data set may have a substan-

```
                 Calculating Pure Error: Lead Content Data                    1
                            The ANOVA Procedure
                           Class Level Information

              Class           Levels      Values

                x               5         8.3 12.1 17 24.3 33.6


                         Number of observations     11
                 Calculating Pure Error: Lead Content Data                    2
                            The ANOVA Procedure

Dependent Variable: y    Lead Content

                                    Sum of
Source                     DF      Squares      Mean Square    F Value    Pr > F

Model                       4     832355.0606   208088.7652      20.77    0.0012

Error                       6      60103.6667    10017.2778  1

Corrected Total            10     892458.7273


            R-Square      Coeff Var       Root MSE         y Mean

            0.932654      15.65183        100.0864         639.4545


Source                     DF     Anova SS      Mean Square    F Value    Pr > F

x                           4     832355.0606   208088.7652      20.77    0.0012
```

Fig. 4.6. SAS Example D2: Output

tial effect on the predictions made using the fitted model. A more detailed discussion follows in Sections 4.2.1 and 4.2.2.

SAS Example D3

In this example, an artificial data set is used to illustrate the above concepts by examining, both numerically and graphically, the effects of changing a single case on a variety of statistical measures including case statistics. Four SAS programs were used to obtain the SAS output of case statistics displayed in Figs. 4.9 to 4.12 and the graphs shown in Figs. 4.8a to 4.8d. Four different values for case number 4 are used with the same 10 other observations, to create Artificial Data Sets 1 to 4, analyzed in these SAS programs. Artificial Data Set 1 is used in the program for SAS Example D3 shown in Fig. 4.7 with case number 4 set to (8.0 8.3). The other three SAS programs are all similar to SAS Example D3 program except that case number 4 is set respectively to the values (17 12.9), (8.0 5.8), and (17 8.5) in each.

4.1 An Introduction to Simple Linear Regression

```
data d1;
input x y;
datalines;
  5.0   7.0
 10.0   8.2
  7.0   8.0
  8.0   8.3
 11.0  10.0
  3.0   7.2
  1.0   4.3
  6.0   8.8
  4.0   5.8
  2.0   5.7
  9.0  10.1
;
run;
symbol1 cv=red v=dot i=none h=1;
proc reg data=d2;
  model y=x/p r influence;
  plot y*x/nomodel nostat haxis= 1 to 17 by 1
                    vaxis= 4 to 14 by 1 ctext=blue caxes=dap;
  title c=maroon h=2 'Artificial Data Set 2';
run;
```

Fig. 4.7. SAS Example D3: Program

The SAS Example D3 program illustrates the use of the `influence` option with the `model` statement. This option must be used in conjunction with the

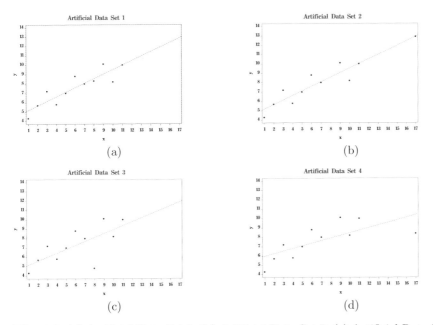

Fig. 4.8. (a) Artificial Data Set 1, (b) Artificial Data Set 2, (c) Artificial Data Set 3, (d) Artificial Data Set 4

r option and leads to the computation of additional diagnostic case statistics as seen in the output from this program (Fig. 4.9). The main case statistic of interest here is the column titled Hat Diag H **2**, which lists the h_{ii}'s discussed later in Section 4.2.1. The case statistic tabulated under the column titled RStudent **3** consists of the *externally* studentized residuals (t_i's) also discussed in Section 4.2.1. These can be used to test for outliers using the Bonferroni procedure described there.

```
                    Artificial Data Set 1                          2

                        The REG Procedure
                         Model: MODEL1
                      Dependent Variable: y

                         Output Statistics

       Dependent  Predicted    Std Error               Std Error   Student
 Obs   Variable     Value    Mean Predict   Residual   Residual   Residual 1

  1     7.0000      7.0982      0.2786      -0.0982     0.836      -0.117
  2     8.2000      9.5164      0.4284      -1.3164     0.770      -1.710
  3     8.0000      8.0655      0.2786      -0.0655     0.836      -0.0783
  4     8.3000      8.5491      0.3143      -0.2491     0.823      -0.303
  5    10.0000     10.0000      0.4970    -2.05E-15     0.728     -28E-16
  6     7.2000      6.1309      0.3662       1.0691     0.801       1.334
  7     4.3000      5.1636      0.4970      -0.8636     0.728      -1.187
  8     8.8000      7.5818      0.2657       1.2182     0.840       1.450
  9     5.8000      6.6145      0.3143      -0.8145     0.823      -0.990
 10     5.7000      5.6473      0.4284       0.0527     0.770       0.0685
 11    10.1000      9.0327      0.3662       1.0673     0.801       1.332

                         Output Statistics
                              4                    2
                           Cook's            Hat Diag
 Obs   -2-1 0 1 2            D    RStudent 3    H        Cov Ratio    DFFITS

  1   |        |    |      0.001   -0.1108    0.1000     1.4019     -0.0369
  2   |   ***| |    |      0.452   -1.9616    0.2364     0.7556     -1.0913
  3   |        |    |      0.000   -0.0739    0.1000     1.4043     -0.0246
  4   |        |    |      0.007   -0.2868    0.1273     1.4208     -0.1095
  5   |        |    |      0.000  -2.66E-15   0.3182     1.8563     -0.0000
  6   |     |**|    |      0.186    1.4042    0.1727     0.9847      0.6416
  7   |   **|  |    |      0.329   -1.2186    0.3182     1.3205     -0.8325
  8   |     |**|    |      0.105    1.5617    0.0909     0.8177      0.4938
  9   |    *|  |    |      0.071   -0.9883    0.1273     1.1518     -0.3774
 10   |        |    |      0.001    0.0646    0.2364     1.6556      0.0359
 11   |     |**|    |      0.185    1.4012    0.1727     0.9863      0.6403
```

Fig. 4.9. Diagnostics for the fit of Artificial Data Set 1

In addition, in the SAS Example D3 program the `plot` statement that is available in `proc reg` is used to produce a plot of the data superimposed with the fitted regression line. The `plot` statement in `proc reg` produces high-resolution graphics unless the `lineprinter` option is specified on the `proc` statement, in which case low-resolution plots are created instead. Global graphics statements such as goptions, symbol, axis, etc. may be used to enhance these plots. In this example, a symbol statement and options in the plot and title statements are employed for that purpose.

```
                       Artificial Data Set 2                               2

                           The REG Procedure
                            Model: MODEL1
                          Dependent Variable: y

                           Output Statistics

       Dependent  Predicted    Std Error              Std Error   Student
  Obs  Variable   Value     Mean Predict   Residual   Residual    Residual

   1    7.0000    7.1184       0.2854      -0.1184     0.829      -0.143
   2    8.2000    9.5428       0.3245      -1.3428     0.815      -1.649
   3    8.0000    8.0882       0.2646      -0.0882     0.836      -0.105
   4   12.9000   12.9371       0.6578      -0.0371     0.580      -0.0640
   5   10.0000   10.0277       0.3621      -0.0277     0.799      -0.0347
   6    7.2000    6.1486       0.3477       1.0514     0.805       1.306
   7    4.3000    5.1788       0.4340      -0.8788     0.762      -1.154
   8    8.8000    7.6033       0.2688       1.1967     0.835       1.434
   9    5.8000    6.6335       0.3126      -0.8335     0.819      -1.017
  10    5.7000    5.6637       0.3888       0.0363     0.786       0.0462
  11   10.1000    9.0579       0.2942       1.0421     0.826       1.262

                           Output Statistics

                         Cook's               Hat Diag    Cov
  Obs    -2-1 0 1 2        D     RStudent        H       Ratio    DFFITS

   1   |     |       |   0.001   -0.1348     0.1060    1.4092    -0.0464
   2   |  ***|       |   0.216   -1.8604     0.1370    0.7145    -0.7413
   3   |     |       |   0.001   -0.0995     0.0911    1.3890    -0.0315
   4   |     |       |   0.003   -0.0603     0.5629    2.8930    -0.0685
   5   |     |       |   0.000   -0.0327     0.1705    1.5254    -0.0148
   6   |     |**     |   0.159    1.3681     0.1573    0.9864     0.5910
   7   |   **|       |   0.216   -1.1781     0.2450    1.2173    -0.6712
   8   |     |**     |   0.107    1.5391     0.0940    0.8315     0.4956
   9   |   **|       |   0.075   -1.0197     0.1271    1.1355    -0.3890
  10   |     |       |   0.000    0.0436     0.1966    1.5746     0.0215
  11   |     |**     |   0.101    1.3110     0.1126    0.9663     0.4670
```

Fig. 4.10. Diagnostics for the fit of Artificial Data Set 2

Figure 4.9 displays the case statistics resulting from fitting a simple linear regression model to Artificial Data Set 1. It is observed that none of the leverages (Hat Diag H) is numerically larger than 0.36 (i.e., twice the average of all leverage values $2/n$, here $4/11 \approx 0.36$ is used as a cutoff to identify high-leverage points). Thus, there are no cases indicated as possible x-outliers. This is also apparent from inspecting Fig. 4.8a, where the x-values are evenly spread out between 0 and 12. The largest externally studentized residual (RStudent) value of -1.96 is not significant at .05 Recall that these statistics have t-distributions and critical values (Bonferroni adjusted for multiple testing)

202   4 Statistical Analysis of Regression Models

are available in Tables B.10 and B.11 of Appendix B. The absolute value of RStudent is used to perform a two-sided test of an hypothesis whether a case is an outlier. An $\alpha = .05$ critical value for $k = 1$ and $n = 11$ from Table B.10 is 3.90. Thus, the presence of y-outliers can also be ruled out. It has been suggested that, as a rule of thumb, cases with influence (Cook's D 4) values numerically larger than $4/n$ may be considered for further investigation. Here, only case number 2 exceeds that value. As discussed in Section 4.2.1, this is seen to be inflated due to relatively large values of the studentized residual and the leverage values for this case.

```
                          Artificial Data Set 3                            2

                              The REG Procedure
                              Model: MODEL1
                              Dependent Variable: y

                              Output Statistics

       Dependent  Predicted    Std Error                 Std Error    Student
 Obs   Variable     Value    Mean Predict   Residual     Residual    Residual

   1    7.0000     6.8436      0.4645        0.1564       1.394       0.112
   2    8.2000     8.9436      0.7142       -0.7436       1.284      -0.579
   3    8.0000     7.6836      0.4645        0.3164       1.394       0.227
   4    4.8000     8.1036      0.5241       -3.3036       1.372      -2.407
   5   10.0000     9.3636      0.8286        0.6364       1.213       0.525
   6    7.2000     6.0036      0.6105        1.1964       1.336       0.895
   7    4.3000     5.1636      0.8286       -0.8636       1.213      -0.712
   8    8.8000     7.2636      0.4429        1.5364       1.401       1.097
   9    5.8000     6.4236      0.5241       -0.6236       1.372      -0.454
  10    5.7000     5.5836      0.7142        0.1164       1.284       0.0906
  11   10.1000     8.5236      0.6105        1.5764       1.336       1.180

                              Output Statistics

                  Cook's                    Hat Diag        Cov
 Obs   -2-1 0 1 2    D       RStudent          H          Ratio       DFFITS

   1   |     |    |   0.001    0.1059        0.1000      1.4023       0.0353
   2   |   *|    |   0.052   -0.5566        0.2364      1.5361      -0.3097
   3   |     |    |   0.003    0.2146        0.1000      1.3902       0.0715
   4   |****|    |   0.423   -3.8034        0.1273      0.1839      -1.4525
   5   |    |*   |   0.064    0.5024        0.3182      1.7445       0.3432
   6   |    |*   |   0.084    0.8845        0.1727      1.2694       0.4042
   7   |   *|    |   0.118   -0.6910        0.3182      1.6530      -0.4721
   8   |    |**  |   0.060    1.1111        0.0909      1.0448       0.3514
   9   |    |    |   0.015   -0.4334        0.1273      1.3844      -0.1655
  10   |    |    |   0.001    0.0855        0.2364      1.6543       0.0476
  11   |    |**  |   0.145    1.2098        0.1727      1.0932       0.5528
```

Fig. 4.11. Diagnostics for the fit of Artificial Data Set 3

From inspecting the case statistics resulting from fitting a simple linear regression model to Artificial Data Set 2 displayed in Fig. 4.10, case number 4 is identified as an x-outlier since the leverage (Hat Diag H) is numerically larger than .36. As observed from Fig. 4.8b, the point corresponding to case number 4 lies away from the centroid of the other x-values, far to the right. An externally studentized residual (RStudent) value of -1.86 is again not significant clearly showing that there are no y-outliers. Influence (Cook's D) values are all small indicating that no observations are influential. Comparing the summary statistics of the fits of Artificial Data Sets 1 and 2 (see Fig. 4.13) shows that the fits are almost identical (only the MSE increased slightly), so the introduction of an x-outlier did not affect the fit appreciably, as it was not a y-outlier.

```
                        Artificial Data Set 4                          2

                            The REG Procedure
                              Model: MODEL1
                          Dependent Variable: y

                            Output Statistics

        Dependent  Predicted     Std Error                Std Error   Student
  Obs   Variable     Value     Mean Predict   Residual    Residual   Residual

   1     7.0000     7.0886       0.4354       -0.0886      1.265     -0.0700
   2     8.2000     8.4700       0.4951       -0.2700      1.243     -0.217
   3     8.0000     7.6411       0.4036        0.3589      1.275      0.281
   4     8.4000    10.4040       1.0036       -2.0040      0.884     -2.266
   5    10.0000     8.7463       0.5524        1.2537      1.218      1.029
   6     7.2000     6.5360       0.5305        0.6640      1.228      0.541
   7     4.3000     5.9834       0.6621       -1.6834      1.162     -1.448
   8     8.8000     7.3649       0.4100        1.4351      1.273      1.127
   9     5.8000     6.8123       0.4768       -1.0123      1.250     -0.810
  10     5.7000     6.2597       0.5931       -0.5597      1.199     -0.467
  11    10.1000     8.1937       0.4488        1.9063      1.260      1.513

                            Output Statistics

                             Cook's             Hat Diag      Cov
  Obs     -2-1 0 1 2           D     RStudent      H         Ratio     DFFITS

   1   |      |       |       0.000   -0.0660    0.1060     1.4141    -0.0227
   2   |      |       |       0.004   -0.2054    0.1370     1.4512    -0.0818
   3   |      |       |       0.004    0.2665    0.0911     1.3680     0.0843
   4   |  ****|       |       3.307   -3.2601    0.5629     0.5340    -3.6997
   5   |      |**     |       0.109    1.0329    0.1705     1.1879     0.4683
   6   |      |*      |       0.027    0.5183    0.1573     1.4058     0.2239
   7   |    **|       |       0.340   -1.5594    0.2450     0.9859    -0.8884
   8   |      |**     |       0.066    1.1467    0.0940     1.0303     0.3693
   9   |     *|       |       0.048   -0.7931    0.1271     1.2462    -0.3026
  10   |      |       |       0.027   -0.4456    0.1966     1.5000    -0.2204
  11   |      |***    |       0.145    1.6517    0.1126     0.7931     0.5883
```

Fig. 4.12. Diagnostics for the fit of Artificial Data Set 4

The case statistics resulting from fitting a simple linear regression model to Artificial Data Set 3 (see Fig. 4.11) show that, again, the leverages (Hat Diag H) are all smaller than .36. Thus, there are no x-outliers, as confirmed by inspecting Fig. 4.8c, where the x-values are, again, evenly spread out between 0 and 12. An externally studentized residual (RStudent) value of -3.8 probably has a p-value close to .05. Thus case number 4 is close to being a y-outlier. The influence (Cook's D) value for case number 4 is quite large; thus, this case may also be considered influential. As discussed in Section 4.2.1, this is inflated due to the fact that it is a y-outlier (large studentized residual **1**) although it is not an x-outlier. Comparing the summary statistics of the fits of Artificial Data Sets 1 and 3 (see Fig. 4.13) shows that both R^2 and MSE have changed substantially although the fitted line is almost identical. Thus, the presence of a y-outlier that is not an x-outlier may not substantially change the fitted line but may affect the estimate of the error variance, which, in turn, affects the hypothesis tests, confidence intervals, and prediction intervals.

In Fig. 4.12, the leverage (Hat Diag H) for case number 4 is again large indicating that case number 4 is an x-outlier. This is also verified from Fig. 4.8d, where this point lies far to the right of the rest of the data. An externally studentized residual (RStudent) value of -3.26 probably is not significant at .05 but is still quite large. The influence (Cook's D) value for case number 4 is extremely large and, thus, this case is highly influential. Again, this is inflated due to the fact that it is a y-outlier (large studentized residual) as well as an x-outlier (high leverage). Comparing the summary statistics of the fits of Artificial Data Sets 1 and 4 (see Fig. 4.13), shows that all summary statistics have changed substantially. Thus, the presence of a y-outlier that is influential drastically affects the fit of a model.

Model	$\hat{\beta}_0$	$\hat{\beta}_1$	MSE	R^2
1	4.69	0.48	0.78	0.78
2	4.69	0.48	0.77	0.88
3	4.74	0.42	2.16	0.50
4	5.70	0.28	1.79	0.50

Fig. 4.13. Summary of fit statistics for Artificial Data Sets 1-4

4.1.4 Prediction of new y values using regression

There are two possible interpretations of a y prediction at a specified value of x. Recall that the prediction equation for the lead content data is $\hat{y} = -12.84 + 36.18\,x$, where $x =$ traffic flow in thousands of cars per 24 hours and $y =$ lead content of bark on trees near the highway in $\mu g/g$ dry weight. If $x = 10$ is substituted in this equation, the value $\hat{y} = 348.96$ is obtained. This **predicted value** of y can be interpreted as either

- The estimate of the average or mean lead content of bark $E(y)$ near all highways with traffic flow of 10,000 cars per 24 hours is 348.96 $\mu g/g$ dry weight.,

or

- The lead content of bark y of a specific highway with a traffic flow of 10,000 cars per 24 hours is 348.96 $\mu g/g$ dry wt.

```
data d4;
input x y;
label x='Traffic Flow' y='Lead Content';
datalines;
 8.3   227
 8.3   312
12.1   362
12.1   521
17.0   640
17.0   539
17.0   728
24.3   945
24.3   738
24.3   759
33.6  1263
10.0    .    1
15.0    .
;
run;
legend1 position=(bottom right inside) across=2
        cborder=red offset=(0,0) shape=symbol(3,1)
        label=none value=(h=.8 f=swissxb);
symbol1 cv=blue v=dot i=none h=1;
symbol2 ci=red r=1;
symbol3 ci=magenta r=2;
symbol4 ci=green r=2;

proc reg data=d4;
   model y=x/clm cli;  2
   plot y*x/pred conf nomodel nostat legend=legend1 ctext=stb caxes=vip;
   title c=darkmagenta h=2 f=centx 'Prediction Intervals: Lead Content Data';
run;
```

Fig. 4.14. SAS Example D4: Program

The difference in the two predictions is that the standard error of predictions will be different. Since it is possible to more accurately predict a mean than an individual value, the first type of prediction will have less error than the the second type. Thus, the confidence interval calculated for the mean of y, $E(y)$, for a given x will be narrower than the prediction interval calculated for a new value y at a given x.

SAS Example D4

In `proc reg`, these intervals are calculated for the observations in the input data set and printed as part of the output of case statistics. However, if these intervals are required for observations with new x-values, then these observations must be included as cases in the input data set with missing value indicators (periods) as the corresponding y-values. In the SAS Example D4 program (see Fig. 4.14), which is a modified version of the SAS Example D1 program, the original lead content data have been supplemented by adding two cases each with values of 10.0 and 15.0 **1**, as values of x and missing values for y. `proc reg` fits the regression model using only the original lead content data and calculates the two types of prediction interval for all cases, including those with the new x-values. The case statistics are printed on page 2 of the output from the SAS Example D4 program as displayed in Fig. 4.15.

```
                    Prediction Intervals: Lead Content Data                 2

                              The REG Procedure
                               Model: MODEL1
                         Dependent Variable: y Lead Content

                              Output Statistics

             Dependent    Predicted      Std Error
    Obs      Variable       Value      Mean Predict          95% CL Mean  3

      1       227.0000     287.4844       45.4206        184.7359    390.2328
      2       312.0000     287.4844       45.4206        184.7359    390.2328
      3       362.0000     424.9830       35.3804        344.9470    505.0190
      4       521.0000     424.9830       35.3804        344.9470    505.0190
      5       640.0000     602.2839       28.0543        538.8206    665.7471
      6       539.0000     602.2839       28.0543        538.8206    665.7471
      7       728.0000     602.2839       28.0543        538.8206    665.7471
      8       945.0000     866.4260       36.1835        784.5731    948.2788
      9       738.0000     866.4260       36.1835        784.5731    948.2788
     10       759.0000     866.4260       36.1835        784.5731    948.2788
     11           1263         1203       63.8739            1058        1347
     12              .     348.9969       40.6376        257.0684    440.9255
     13              .     529.9162       29.9605        462.1408    597.6915

                              Output Statistics

                    Obs      95% CL Predict  4        Residual

                      1     54.9967    519.9721       -60.4844
                      2     54.9967    519.9721        24.5156
                      3    201.6021    648.3640       -62.9830
                      4    201.6021    648.3640        96.0170
                      5    384.2911    820.2767        37.7161
                      6    384.2911    820.2767       -63.2839
                      7    384.2911    820.2767       125.7161
                      8    642.3876        1090        78.5740
                      9    642.3876        1090      -128.4260
                     10    642.3876        1090      -107.4260
                     11    949.2204        1457        60.0643
                     12    121.0843    576.9095               .
                     13    310.6292    749.2032               .
```

Fig. 4.15. Prediction intervals: Lead content data

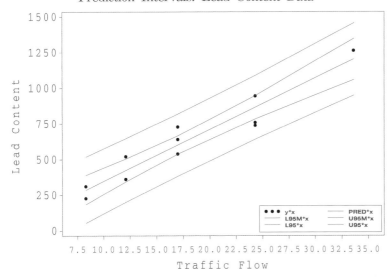

Fig. 4.16. Prediction Intervals: Lead Content Data

Assuming normally distributed data and the first type of prediction, `proc reg` calculates $(1-\alpha)100\%$ *confidence interval* ❸ for $E(y) = \beta_0 + \beta_1 x$, for each x-value in the data, including those cases with missing values specified for y. The second type of prediction is used to calculate a $100(1-\alpha)\%$ *prediction interval* ❹ for a new observation y at each x-value in the data, including the new x-values specified. For a default value of $\alpha = .05$, these are obtained by specifying `clm` and/or `cli` respectively, for the confidence intervals and the prediction intervals as model statement options:

```
model y=x/clm cli;
```

The `alpha=` option may be used to change the default value of the confidence coefficient, noting that this will affect the intervals calculated for the β coeficients using the `clb` option if it is used in the same model statement. These sets of intervals are tabulated under the headings 95% CL Mean ❸ and 95% CL Predict ❹, respectively, in Fig. 4.15.

The statement `plot y*x/pred conf nomodel nostat legend=legend1;` included in Fig. 4.14 produces line plots of confidence and prediction intervals to enhance the previous plot of the data with the regression line (see Fig. 4.4a) as displayed in Fig. 4.16. Several global SAS/GRAPH statements (`symbol` statements and a `legend` statement) and plot statement options (`ctext=`, `caxes=`) are used to enhance features of this graph.

4.2 An Introduction to Multiple Regression Analysis

In this section the simple linear regression model introduced in Section 4.1 is extended to the multiple linear regression case to handle situations in which the dependent variable y is modeled by a relationship that involves more than a single independent or explanatory variable, say $x_1, x_2, \ldots, x_k$. The study of multiple regression analysis may thus be viewed as the approximation of the functional relationship that may exist between a variable y and another set of variables $x_1, x_2, \ldots, x_k$. This relationship may sometimes model a postulated theoretical relationship between y and the x's, whereas at other times it may simply be a mathematical equation that approximates such a relationship. Such an equation is useful for prediction of a value for y when the values of the x's are known.

Multiple Regression Model

When this approximation is linear in the unknown parameters, it is called a multiple linear regression model and is expressed in the form

$$y = \beta_0 + \beta_1 x_1 + \cdots + \beta_k x_k + \epsilon$$

The variable y is the response or the dependent variable, which is a realization of a random variable Y observed for a fixed set of values of the explanatory variables $x_1, x_2, \ldots, x_k$. The coefficients $\beta_0, \beta_1, \ldots, \beta_k$ are unknown constants and ϵ is an unobservable random variable representing the error in observing y. Under this model, the y-values observed at each x-value are a random sample with the mean $E(y) = \mu(x) = \beta_0 + \beta_1 x_1 + \cdots + \beta_k x_k$, i.e., the mean is modeled as a function of the x values. This function is linear in $\beta_0, \beta_1, \ldots, \beta_k$, and is a hyperplane in the parameter space of the coefficients. Multiple regression data consist of n observations or **cases** of $k+1$ values denoted by

$$(y_i, x_{i1}, x_{i2}, \ldots, x_{ik}), \quad i = 1, 2, \ldots, n$$

Using this full notation, the model may be rewritten as

$$y_i = \beta_0 + \beta_1 x_{i1} + \beta_2 x_{i2} + \cdots + \beta_k x_{ik} + \epsilon_i \quad i = 1, 2, \ldots, n.$$

In addition to the fact that the random errors for the individual observations have mean zero, they are also assumed to be uncorrelated random variables with the same variance σ^2. The parameters $\beta_0, \beta_1, \ldots, \beta_k$ and σ^2 can be estimated using the least squares method. For the purpose of making valid statistical inference, it may be further assumed that the errors $\epsilon_1, \epsilon_2, \ldots, \epsilon_n$ are a random sample from a normal distribution.

Note that the definition of the regression model above admits models that may include squared terms, product terms, etc. of *quantitative* explanatory variables (x's). Thus, a model of the form

4.2 An Introduction to Multiple Regression Analysis

$$y = \beta_0 + \beta_1 x_1 + \beta_2 x_2 + \beta_3 x_1^2 + \beta_4 x_2^2 + \beta_5 x_1^3 + \beta_6 x_2^3 + \beta_7 x_1 x_2 + \epsilon$$

can also be expressed in the form of a multiple regression model

$$y = \beta_0 + \beta_1 x_1 + \beta_2 x_2 + \beta_3 x_3 + \beta_4 x_4 + \beta_5 x_5 + \beta_6 x_6 + \beta_7 x_7 + \epsilon$$

where the values for the variables x_3, x_4, x_5, x_6, and x_7 are obtained by substituting the values of x_1 and x_2 in the expressions $x_3 = x_1^2$, $x_4 = x_2^2$, $x_5 = x_1^3$, $x_6 = x_2^3$, and $x_7 = x_1 x_2$, respectively. The inclusion of a product term such as $x_1 x_2$ in a model allows the investigator to perform a statistical test of the absence (or presence) of *interaction* between two independent variables (say) x_1 and x_2. The concept of interaction will be discussed in a later section.

Estimation of Parameters

The *method of least squares* is used to obtain the estimates of the regression coefficients using the observed data. The least squares estimates of the β's denoted by $\hat{\beta}_0, \hat{\beta}_1, \ldots, \hat{\beta}_k$ are obtained by minimizing the sum of squares of residuals:

$$Q = \sum_{i=1}^{n} \{y_i - (\beta_0 + \beta_1 x_{1i} + \cdots + \beta_k x_{ki})\}^2$$

where $y_i, x_{1i}, x_{2i}, \ldots, x_{ki}$, $i = 1, \ldots, n$, denotes the n observations (or cases). The estimates are found by setting the partial derivatives of Q with respect to each of the β coefficients equal to zero. The resulting set of equations are linear in the β's and is called the *normal equations*. These are solved to yield the estimates of the parameters denoted by $\hat{\beta}$'s. The prediction equation or the fitted regression model is then

$$\hat{y} = \hat{\beta}_0 + \hat{\beta}_1 x_1 + \cdots + \hat{\beta}_k x_k$$

The predicted or fitted values of y_i, $i = 1, \ldots, n$, that correspond to the n observations (or cases) are calculated by substituting the n observed values of the explanatory variables $x_{1i}, x_{2i}, \ldots, x_{ki}$, $i = 1, \ldots, n$, in the prediction equation to obtain $\hat{y}_i$, $i = 1, \ldots, n$. The predicted or fitted value, y_{new} say, that corresponds to a new case $x_{1,new}, x_{2,new}, \ldots, x_{k,new}$ is calculated by substituting these values in the prediction equation.

Matrix Notation

In matrix notation, the linear regression model may be expressed in the form

$$\mathbf{y} = X\boldsymbol{\beta} + \boldsymbol{\epsilon}$$

where

4 Statistical Analysis of Regression Models

$$\mathbf{y} = \begin{bmatrix} y_1 \\ y_2 \\ \vdots \\ y_n \end{bmatrix}, \quad X = \begin{bmatrix} 1 & x_{11} & \cdots & x_{k1} \\ 1 & x_{12} & \cdots & x_{k2} \\ \vdots & \vdots & & \vdots \\ 1 & x_{1n} & & x_{kn} \end{bmatrix}, \quad \boldsymbol{\beta} = \begin{bmatrix} \beta_0 \\ \beta_1 \\ \vdots \\ \beta_k \end{bmatrix}, \quad \text{and } \boldsymbol{\epsilon} = \begin{bmatrix} \epsilon_1 \\ \epsilon_2 \\ \vdots \\ \epsilon_n \end{bmatrix}$$

In this notation, the sum of squares to be minimized is

$$Q = (\mathbf{y} - X\boldsymbol{\beta})'(\mathbf{y} - X\boldsymbol{\beta})$$

and the resulting normal equations are

$$X'X\boldsymbol{\beta} = X'\mathbf{y}.$$

The solution to the normal equations gives the least squares estimate $\hat{\boldsymbol{\beta}}$ of $\boldsymbol{\beta}$:

$$\hat{\boldsymbol{\beta}} = (X'X)^{-1}X'\mathbf{y}$$

where $(X'X)^{-1}$ is the inverse of the $X'X$ matrix assumed to be nonsingular. Note that $X'X$, a $(k+1) \times (k+1)$ matrix, is an important quantity associated with multiple regression computations. An analysis of variance table for testing the hypothesis

$$H_0: \beta_1 = \beta_2 = \cdots = \beta_k = 0 \quad \text{versus} \quad H_a: \text{at least one } \beta \neq 0.$$

where matrix expressions are used to define computational formulas for the various sums of square, is given as follows:

Source	df	SS	MS	F
Regression	k	$\hat{\boldsymbol{\beta}}'X'\mathbf{y} - n\bar{y}^2$	MSReg= SSReg/k	MSReg/MSE
Error	$n-k-1$	$\mathbf{y}'\mathbf{y} - \hat{\boldsymbol{\beta}}'X'\mathbf{y}$	MSE=SSE/$(n-k-1)$	
Total	$n-1$	$\mathbf{y}'\mathbf{y} - n\bar{y}^2$		

where $\bar{y} = \sum_{i=1}^{n} y_i/n$, $\mathbf{y}'\mathbf{y} = \sum_{i=1}^{n} y_i^2$, SSReg denotes the regression sum of squares given by $\hat{\boldsymbol{\beta}}'X'\mathbf{y} - n\bar{y}^2$, and SSE denotes the residual or error sum of squares given by $\mathbf{y}'\mathbf{y} - \hat{\boldsymbol{\beta}}'X'\mathbf{y}$. The null hypothesis is rejected if the above F-statistic exceeds the α level upper percentage point of the F-distribution with $(k, n-k-1)$ degrees of freedom. The mean square for error (MSE) denoted by $s^2 =$ SSE/$(n-k-1)$, is an unbiased estimate of σ^2, the variance of the random errors; that is, $\hat{\sigma}^2 = s^2$. Another quantity of statistical interest computed from the above analysis is the coefficient of determination or the multiple correlation coefficient R^2 given by

$$R^2 = \text{Regression SS/Total (Corrected) SS} = \text{SSReg/SSTot}$$

where SSTot $= \mathbf{y}'\mathbf{y} - n\bar{y}^2$. R^2 measures the proportion of the variability of y explained by the fitted regression model. Further, suppose that elements of the inverse of the $(X'X)^{-1}$ matrix are denoted by c_{ij}. Thus,

$$(X'X)^{-1} = \begin{bmatrix} c_{00} & c_{01} & c_{02} & \cdots & c_{0k} \\ c_{10} & c_{11} & c_{12} & \cdots & c_{1k} \\ \vdots & \vdots & & & \vdots \\ c_{k0} & c_{k1} & & \cdots & c_{kk} \end{bmatrix}$$

Then the standard error of the least squares estimate of the mth regression coefficients $\hat{\beta}_m$ is given by $s_{\hat{\beta}_m} = c_{mm}^{1/2}s$. Thus, a t-statistic for testing the hypothesis $H_0: \beta_m = 0$ versus $H_a: \beta_m \neq 0$ is

$$t = \hat{\beta}_m/s_{\hat{\beta}_m}$$

and a $(1-\alpha)100\%$ confidence interval for β_m is

$$\hat{\beta}_m \pm t_{\alpha/2,(n-k-1)} \times s_{\hat{\beta}_m}$$

where $t_{\alpha/2,(n-k-1)}$ is the upper $\alpha/2$ critical value of the t-distribution with $(n-k-1)$ degrees of freedom.

4.2.1 Multiple regression analysis using PROC REG

The data, given in Table B.7, are taken from Draper and Smith (1981). The explanatory variables x_1, x_2, x_3, and x_4 are percentages of the primary chemical components of clinkers from which cement is made and the response variable y is the heat evolved measured in calories per gram of cement. The objective is to model the heat produced as a linear function of the composition of clinkers. These data will be used here to illustrate procedures in SAS that are useful in the process of regression model building. The purpose of SAS Example D5 is fitting a multiple regression model to these data and producing statistics discussed in Section 4.2. Diagnostics and residual plots necessary for examining the adequacy of the model and accompanying assumptions will be calculated for the same regression model in later sections.

SAS Example D5

In the SAS Example D5 program (see Fig. 4.17), the multiple linear regression model

$$y = \beta_0 + \beta_1 x_1 + \beta_2 x_2 + \beta_3 x_3 + \beta_4 x_4 + \epsilon \tag{4.5}$$

is fitted to these data using `proc reg`. This program produces output containing statistics for making some of the statistical inferences about the model as discussed in the previous section. It contains SAS statements necessary to compute an analysis of variance, estimated values of the parameters, their

```
data cement;
input x1-x4 y;
datalines;
 7 26  6 60  78.5
 1 29 15 52  74.3
11 56  8 20 104.3
11 31  8 47  87.6
 7 52  6 33  95.9
11 55  9 22 109.2
 3 71 17  6 102.7
 1 31 22 44  72.5
 2 54 18 22  93.1
21 47  4 26 115.9
 1 40 23 34  83.8
11 66  9 12 113.3
10 68  8 12 109.4
;
run;

proc reg simple corr;
   model y = x1 x2 x3 x4/clb xpx i;
   title 'Regression Analysis of Hald Data';
run;
```

Fig. 4.17. SAS Example D5: Program

standard errors, associated t-statistics, and confidence intervals. Additionally, options used with the model statement requests that information about the normal equations be output.

The output pages are displayed in Figs. 4.18, 4.19, and 4.20. The keyword options simple and corr used with proc reg results in the output of descriptive statistics for each of the variables and the correlation matrix **1** displayed on page 1 (see Fig. 4.18). The specification of xpx and i results in the output of the $X'X$ matrix **2** shown on page 2 (see Fig. 4.18) and the inverse of the $X'X$ matrix **3** shown on page 3 (see Fig. 4.19).

The $X'X$ matrix and $X'\mathbf{y}$ vector displayed below are extracted from Fig. 4.18:

$$X'X = \begin{pmatrix} 13 & 97 & 626 & 153 & 390 \\ 97 & 1139 & 4922 & 769 & 2620 \\ 626 & 4922 & 33050 & 7201 & 15739 \\ 153 & 769 & 7201 & 2293 & 4628 \\ 390 & 2620 & 15739 & 4628 & 15062 \end{pmatrix}, \quad X'\mathbf{y} = \begin{pmatrix} 1240.5 \\ 10032 \\ 62027.8 \\ 13981.5 \\ 34733.3 \end{pmatrix}$$

The normal equations can thus be constructed as follows:

$$13\beta_0 + 97\beta_1 + 626\beta_2 + 153\beta_3 + 390\beta_4 = 1240.5$$
$$97\beta_0 + 1139\beta_1 + 4922\beta_2 + 769\beta_3 + 2620\beta_4 = 10032$$
$$626\beta_0 + 4922\beta_1 + 33050\beta_2 + 7201\beta_3 + 15739\beta_4 = 62027.8$$
$$153\beta_0 + 769\beta_1 + 7201\beta_2 + 2293\beta_3 + 4628\beta_4 = 13981.5$$
$$390\beta_0 + 2620\beta_1 + 15739\beta_2 + 4628\beta_3 + 15062\beta_4 = 34733.3$$

Note that the last element in the row (or the column) labeled y of the matrix output on page 2 is the total sum of squares $\mathbf{y}'\mathbf{y} = \sum_{i=1}^n y_i^2 = 121088.09$. The solutions to the above normal equations are the estimates of $\boldsymbol{\beta}$ and are obtained from page 3 of the output from `proc reg` (see the last column under y in Fig. 4.19) or from page 4 (see Fig. 4.20). These are reported below rounded

Regression Analysis of Hald Data

The REG Procedure

Descriptive Statistics

Variable	Sum	Mean	Uncorrected SS	Variance	Standard Deviation
Intercept	13.00000	1.00000	13.00000	0	0
x1	97.00000	7.46154	1139.00000	34.60256	5.88239
x2	626.00000	48.15385	33050	242.14103	15.56088
x3	153.00000	11.76923	2293.00000	41.02564	6.40513
x4	390.00000	30.00000	15062	280.16667	16.73818
y	1240.50000	95.42308	121088	226.31359	15.04372

Correlation

Variable	x1	x2	x3	x4	y
x1	1.0000	0.2286	-0.8241	-0.2454	0.7307
x2	0.2286	1.0000	-0.1392	-0.9730	0.8163
x3	-0.8241	-0.1392	1.0000	0.0295	-0.5347
x4	-0.2454	-0.9730	0.0295	1.0000	-0.8213
y	0.7307	0.8163	-0.5347	-0.8213	1.0000

Regression Analysis of Hald Data

The REG Procedure
Model: MODEL1

Model Crossproducts X'X X'Y Y'Y

Variable	Intercept	x1	x2
Intercept	13	97	626
x1	97	1139	4922
x2	626	4922	33050
x3	153	769	7201
x4	390	2620	15739
y	1240.5	10032	62027.8

Model Crossproducts X'X X'Y Y'Y

Variable	x3	x4	y
Intercept	153	390	1240.5
x1	769	2620	10032
x2	7201	15739	62027.8
x3	2293	4628	13981.5
x4	4628	15062	34733.3
y	13981.5	34733.3	121088.09

Fig. 4.18. SAS Example D5: Output (pages 1-2)

to five significant digits:

$$\hat{\boldsymbol{\beta}} = (62.405,\ 1.5511,\ 0.51017,\ 0.10191,\ -0.14406)'.$$

Further, the last element in the row (or the column) labeled $\mathbf{y}$ on page 3 is the SSE, computed using the formula $\mathbf{y}'\mathbf{y} - \hat{\boldsymbol{\beta}}'\mathbf{X}'\mathbf{y}$. Rounded to six digits, it is equal to 47.8636, giving $s^2 = 47.8636/(13 - 4 - 1) = 5.983$ ⟨4⟩. The following analysis of variance table for the regression is constructed from the output at the bottom of page 3:

Source	df	Sum of Squares	Mean Square	F	p-value
Regression	4	2667.90	666.97	111.5	< .0001
Error	8	47.86	5.98		
Total	12	2715.76			

Obviously, the null hypothesis $H_0 : \beta_1 = \beta_2 = \beta_3 = \beta_4 = 0$ is rejected at $\alpha = .01$, say, since the p-value is less than .01. The estimate s^2 of σ^2 is given by the MSE= 5.98. The R^2 value ⟨5⟩ is also reported on page 3, here equal to .9824, meaning that about 98% of the variation in the heat evolved is explained by the fitted multiple regression model involving all four explanatory variables, each giving the percentage of a chemical component of clinkers.

The inverse of the $X'X$ matrix is also available from page 3 and reproduced here with elements rounded to six digits:

$$(X'X)^{-1} = \begin{pmatrix} 820.655 & -8.44180 & -8.45778 & -8.63454 & -8.28974 \\ -8.44180 & 0.0927104 & 0.0856862 & 0.0926374 & 0.0844550 \\ -8.45778 & 0.0856862 & 0.0875603 & 0.0878666 & 0.0855981 \\ -8.63454 & 0.0926374 & 0.0878666 & 0.0952014 & 0.0863919 \\ -8.28974 & 0.0844550 & 0.0855981 & 0.0863919 & 0.0840312 \end{pmatrix}$$

To illustrate the use of the elements of the inverse of the $X'X$ matrix, the standard error of, say, $\hat{\beta}_1$ is computed as follows:

$$s_{\hat{\beta}_1} = \sqrt{c_{11}}\, s = \sqrt{0.0927104} \cdot \sqrt{5.983} = 0.74477$$

and, thus, a 95% confidence interval for β_1 is given by

$$\hat{\beta}_1 \pm t_{.025,8} \cdot s_{\hat{\beta}_1} \equiv 1.5511 \pm (2.306)(0.74477) \quad \text{giving} \quad (-0.16634, 3.2685).$$

These values can be verified from page 4 of the output (see Fig. 4.20), where the parameter estimates and their associated statistics are tabulated (see under **Parameter Estimate, Standard Error**, etc.). By inspecting these, it is observed that none of the p-values for the t-test for testing $\beta_m = 0$ for

$m = 1, 2, 3,$ and 4 in model (4.5) is less than .05; thus, none of these hypotheses can be rejected at $\alpha = .05$. This is also clearly reflected in the fact that the 95% confidence intervals for these coefficients all contain zero. Upon further examination, it is seen that the reason for both of these results is that the standard errors of the estimates of the coefficients are comparatively large (i.e., they are all near 0.7, larger than some of the estimates themselves!). Obviously, this indicates large sampling variability of the estimated regression coefficients.

It appears that the conclusions from these individual t-test, (or confidence intervals) contradict the result of the F-test of $H_0 : \beta_1 = \beta_2 = \beta_3 = \beta_4 = 0$. However, it is erroneous to infer from the results of the one-at-a-time t-tests

```
                     Regression Analysis of Hald Data              3
                           The REG Procedure
                             Model: MODEL1
                          Dependent Variable: y

                  X'X Inverse, Parameter Estimates, and SSE  3

Variable           Intercept              x1                  x2

Intercept         820.65457471       -8.441801862        -8.457779848
x1                 -8.441801862       0.0927104019        0.0856862094
x2                 -8.457779848       0.0856862094        0.0875602572
x3                 -8.634538775       0.0926373566        0.0878666397
x4                 -8.289743778       0.0844549553        0.0855980995
y                  62.4053693         1.5511026475        0.5101675797

                  X'X Inverse, Parameter Estimates, and SSE

Variable             x3                   x4                   y

Intercept         -8.634538775        -8.289743778        62.4053693
x1                 0.0926373566        0.0844549553        1.5511026475
x2                 0.0878666397        0.0855980995        0.5101675797
x3                 0.0952014097        0.0863919188        0.1019094036
x4                 0.0863919188        0.0840311912       -0.144061029
y                  0.1019094036       -0.144061029        47.86363935


                          Analysis of Variance

                                Sum of          Mean
Source                DF       Squares         Square    F Value    Pr > F

Model                  4      2667.89944      666.97486   111.48    <.0001
Error                  8        47.86364        5.98295  4
Corrected Total       12      2715.76308

              Root MSE                2.44601    R-Square     0.9824  5
              Dependent Mean         95.42308    Adj R-Sq     0.9736
              Coeff Var               2.56333
```

Fig. 4.19. SAS Example D5: Output (Page 3)

```
                Regression Analysis of Hald Data                    4

                          The REG Procedure
                          Model: MODEL1
                          Dependent Variable: y

                          Parameter Estimates

                        Parameter        Standard
        Variable    DF   Estimate           Error   t Value   Pr > |t|

        Intercept    1   62.40537        70.07096      0.89     0.3991
        x1           1    1.55110         0.74477      2.08     0.0708
        x2           1    0.51017         0.72379      0.70     0.5009
        x3           1    0.10191         0.75471      0.14     0.8959
        x4           1   -0.14406         0.70905     -0.20     0.8441

                          Parameter Estimates

          Variable      DF      95% Confidence Limits

          Intercept      1     -99.17855        223.98929
          x1             1      -0.16634          3.26855
          x2             1      -1.15889          2.17923
          x3             1      -1.63845          1.84227
          x4             1      -1.77914          1.49102
```

Fig. 4.20. SAS Example D5: Output (page 4)

that the coefficients in the model are all zero simultaneously. Rather, the contradictory nature of the results of the t-tests and the F-test must be taken as an indication that there are, possibly large, correlations among the explanatory variables. This condition, called *multicollinearity*, discussed in detail in Section 4.2.4, could lead to the situation that one or more explanatory variables may exhibit little or no effects on the response variable in the presence, in the same model, of other explanatory variables that are highly correlated with one or more of them. From the correlation matrix reported on page 1 (see Fig. 4.18), it is clear that pairs of variables ($x1$, $x3$) and ($x2$, $x4$) are highly (negatively) correlated with each other and, therefore, a model containing only one variable in each pair may turn out to be a model that exhibits less multicollinearity.

In general, this situation may indicate the need to select a subset of the explanatory variables to be included in a model for which multicollinearity does not have substantial effects on the sampling variability and therefore the accuracy of the estimated parameters. Procedures for *variable subset selection* are discussed in Section 4.4.

4.2.2 Case statistics and residual analysis

Many other statistical quantities are computed for use in residual analysis or in diagnostic plots. Some of these were introduced and discussed in the context of the simple linear regression.

These are necessary for checking the adequacy of the model or for assessing the plausibility of assumptions made in formulating a model. Others allow testing for presence or absence of outliers and assessing their effects on the fitted model if they are present. Collectively, these statistics are called *diagnostic statistics* or *case statistics*. The programs for SAS Examples D1, D2, and D3 illustrated the use of SAS statements in `proc reg` to produce these case statistics and also obtain residual plots. In SAS Example D3, in particular, an artificial data set was used to illustrate the concepts associated with several of these case statistics. In this subsection, several of these statistics are formally defined and their use demonstrated in the case of the multiple regression model.

Predicted or Fitted Values

The predicted values that correspond to the observed explanatory variables are calculated using the prediction equation

$$\hat{\mathbf{y}} = X\hat{\boldsymbol{\beta}}$$

where

$$\hat{y}_i = \hat{\beta}_0 + \hat{\beta}_1 x_{1i} + \cdots + \hat{\beta}_k x_{ki}, \quad i = 1, \ldots, n$$

Note that the regression sum of squares defined earlier as SSReg in the anova table, can also be expressed as

$$\sum_{i=1}^{n}(\hat{y}_i - \bar{y}_i)^2$$

Residuals

The residuals that correspond to the observed data are expressed in vector form as $\mathbf{e} = \mathbf{y} - \hat{\mathbf{y}}$, where the elements $\mathbf{e} = (e_1, e_2, \ldots, e_n)'$ are calculated as $e_i = y_i - \hat{y}_i$, $i = 1, \ldots, n$. Note that the residual sum of squares defined in the in the anova table ass SSE can also be expressed in the form

$$\text{SSE} = \sum_{i=1}^{n}(y_i - \hat{y}_i)^2 = \sum_{i=1}^{n} e_i^2 = \mathbf{e}'\mathbf{e}$$

Hat Matrix

The predicted values $\hat{\mathbf{y}} = X\hat{\boldsymbol{\beta}}$ may be expressed in the form

$$\hat{\mathbf{y}} = X(X'X)^{-1}X'\mathbf{y}$$
$$= H\mathbf{y}$$

where $H = X(X'X)^{-1}X'$ is an $n \times n$ symmetric matrix called the *hat matrix*. Let h_{ij} denote the ijth element of H. It can be shown that the ith diagonal element of H satisfies,

$$\frac{1}{n} \leq h_{ii} \leq \frac{1}{d}$$

where d is the number of times the ith observation is replicated. Thus, a specific value of h_{ii} is considered to be relatively small if it is near $\frac{1}{n}$ or relatively large if it is near $\frac{1}{d}$. Note that for a case that is not replicated, the upper bound is 1.

It is easier to visualize the relationship of the magnitude of h_{ii} to the position of a case in the space of explanatory variables in simple liner regression. In general, cases with relatively larger values of h_{ii} will correspond to those cases with x-values further away from the average of the x-values (i.e., center of the x-space).

The elements of the hat matrix are useful since the variance of the predicted values and the residuals among several other quantities can be expressed in terms of the elements of this matrix. For example, the variance of $\hat{y}_i$ is $\sigma^2 h_{ii}$, the standard deviation of $\hat{y}_i$ is $\sigma \sqrt{h_{ii}}$, and, hence, the standard error of $\hat{y}_i$ is $s \sqrt{h_{ii}}$, for $i = 1, 2, \ldots, n$. Noting that the vector of residuals may be expressed in the form $\mathbf{e} = \mathbf{y} - \hat{\mathbf{y}}$,

$$\mathbf{e} = \mathbf{y} - H\mathbf{y}$$
$$= (I - H)\mathbf{y}.$$

It is easily shown that the variance of e_i is $\sigma^2 (1 - h_{ii})$, the standard deviation of e_i is $\sigma \sqrt{(1 - h_{ii})}$, and, hence, the standard error of e_i is $s \sqrt{(1 - h_{ii})}$ for $i = 1, 2, \ldots, n$. It is clear that the magnitudes of the standard errors of both $\hat{y}_i$ and e_i for the ith case depend on the magnitude of h_{ii} as the value of s remains a fixed number for a fitted model. For example, standard errors for predicted values will be larger and the standard errors for the residuals will be smaller for cases for which the diagonal elements of the hat matrix are closer to 1 than it is for those cases for which they are small.

Confidence Interval for the Mean $E(y_i)$

A $(1-\alpha)100\%$ confidence interval for the mean of the ith observation $E(y_i) = \beta_0 + \beta_1 x_{1i} + \cdots + \beta_k x_{ki}$ is

$$\hat{y}_i \pm t_{\alpha/2, (n-k-1)} \times s \sqrt{h_{ii}}$$

Prediction Interval for y_i

A $100(1-\alpha)\%$ prediction interval for ith observation y_i is

$$\hat{y}_i \pm t_{\alpha/2,(n-k-1)} \times s \sqrt{1+h_{ii}}$$

Studentized Residuals

Studentized residuals, denoted by r_i, $i = 1, \ldots, n$, are a standardized version of the ordinary residuals, which are useful for the detection of outliers. It is common practice in statistics to use standardization when comparing statistics that are heterogeneous in variance. An *internally studentized* version of the residuals is obtained by directly dividing the residuals by their respective standard errors, as

$$r_i = e_i/(s\sqrt{1-h_{ii}})$$

for $i = 1, \ldots, n$. The maximum in absolute value of studentized residuals can be used as a basis of a test for the presence of a single *y-outlier* (i.e., an outlier in the y-direction), using the tables of percentage points reproduced in Tables B.8 and B.9. The null hypothesis H_0 : No Outliers is rejected in favor of H_a : A Single Outlier Present, if the computed value of $\max_i |r_i|$ exceeds the appropriate percentage point obtained from this table. It is common practice to use studentized residuals for normal probability plots.

Externally Studentized Residuals

Related statistics, denoted usually by t_i, $i = 1, \ldots, n$, are the ordinary residuals, standardized using $s_{(i)}^2$, the error mean square obtained from a regression model fitted with the ith case deleted, instead of s^2; that is, externally studentized residuals are defined as

$$t_i = e_i/(s_{(i)}\sqrt{1-h_{ii}})$$

for $i = 1, \ldots, n$. The advantage of this statistic is that since each t_i can be shown to have a t-distribution with $n - k - 2$ degrees of freedom, it can be used to construct a test for *y-outliers* (i.e., outliers in the y-direction). Since it is not known in advance which of the observations may be outliers, n t-tests must be performed using each of the n externally studentized residuals; that is, each of the n t_i values is compared to an appropriate critical value obtained from the t-distribution to test if the case is a y-outlier.

One approach is to use the Bonferroni method to obtain a conservative critical value for this multiple testing procedure. The critical value chosen is the $(\alpha/n) \times 100\%$ percentage point of the t-distribution with $n-k-2$ degrees of freedom. This test guarantees only that the Type I error will not exceed α (as opposed to being exactly equal to α); thus, it will only provide a conservative test. To use this method, special t-tables are needed because percentile points for small probability values are not available in ordinary t-tables. For example, for $n = 25, p = 3$, and $\alpha = .05$, the critical value needed corresponds to the $(.025/25) = .001$th upper percentile point of a t-distribution with 20 df.(Note that this will be a two-tailed test using $|t_i|$; hence, .025 must be used instead of .05.) Since the required percentage points are not available from ordinary t-table a special table was constructed (Weisberg, 1985). This table has been reconstructed by the authors and appear as Tables B.10 and B.11.

Leverage

Recall that
$$\hat{\mathbf{y}} = H\mathbf{y}$$
where H is the $n \times n$ hat matrix. Recall also that
$$\text{var}(\hat{y}_i) = h_{ii}\sigma^2$$
$$\text{var}(e_i) = (1 - h_{ii})\sigma^2$$
These imply that "larger" h_{ii} causes $\text{var}(\hat{y}_i)$ to be larger and $\text{var}(e_i)$ to be smaller. Hence, by examining diagonal elements of the hat matrix, it can be determined that an observed value is going to be predicted well or not by the regression on the x-values. Rewriting $\hat{\mathbf{y}} = H\mathbf{y}$ as
$$\hat{y}_i = h_{ii} y_i + \sum_{j \neq i} h_{ij} y_j$$
it is observed that when h_{ii} is closer to 1, the predicted value $\hat{y}_i$ will be closer to y_i and, therefore, e_i will be closer to zero. (Remember that $\frac{1}{n} \leq h_{ii} \leq \frac{1}{d}$.) For this reason, h_{ii} is called the **leverage** of the ith observation or case: It measures the effect that y_i will have on determining $\hat{y}_i$. However, note that the actual magnitude of $\hat{y}_i$, and therefore that of e_i, depends on **both** h_{ii} and y_i. As a rule of thumb, cases with leverages numerically larger than $2(k+1)/n$ (i.e., twice the average of all h_{ii}'s) may be marked for further investigation. Note also that h_{ii} actually measures how far away the case $\mathbf{x}_i = (x_{1i}, x_{2i}, \ldots, x_{pi})$ is from the other $\mathbf{x}$'s; that is, a large h_{ii} may indicate an x-outlier.

Influence

Cases whose deletion causes major changes in the fitted model are called **influential**. A diagnostic tool that measures the influence of the ith case on the fit of the model is known as the Cook's distance statistic and is defined by
$$D_i = \frac{1}{k'} \left\{ \frac{e_i}{s\sqrt{1-h_{ii}}} \right\}^2 \left(\frac{h_{ii}}{1-h_{ii}} \right)$$
$$= \frac{1}{k'} r_i^2 \left(\frac{h_{ii}}{1-h_{ii}} \right)$$

where $k' = k+1$. D_i measures the importance of the ith case on the fitted model. The fact that D_i may be partitioned as

$$D_i = \text{const} \times \text{studentized residual}^2 \times \text{monotone increasing function of } h_{ii}$$

indicates that a large D_i may be due to a large r_i, or a large h_{ii}, or both. So some cases with large leverages may actually not be influential because r_i is small, indicating that these cases actually fit the model well. Cases with relatively large values for both h_{ii} and r_i should be of more concern. As demonstrated in SAS Example D3, some cases that are not x-outliers but are significant y-outliers may cause inflation of the predictive variance and thus may be influential. As a rule of thumb, cases with Cook's D numerically larger than $4/n$ are flagged for further investigation.

SAS Example D5 (continued)

Although a residual analysis is usually deferred until an appropriate model is selected by identifying a subset of explanatory variables, some options available in `proc reg` are used with the model fitted in SAS Example D5 to illustrate the options needed to produce various case statistics discussed in this section.

In addition to the options specified on the `model` statement in the SAS Example D5 program, the option `p` (for `predicted`) requests that predicted values and the residuals be part of the case statistics output. If instead, or in addition, the option `r` (for `residual`) is specified, standard errors of the predicted values, residuals, standard errors of the residuals, studentized residuals, and Cook's D statistics (the column titled Cook's D) are calculated. The output resulting from including the `r` option includes every column shown in the output on page 5 of the SAS output (see Fig. 4.21) up to and including the column labeled Cook's D. If in addition, the option `influence` is added i.e., if the following model statement is used

```
model y = x1 x2 x3 x4/clb xpx i r influence;
```

the entire set of case statistics shown on page 5 is produced. The set of case statistics labeled DFFITS on page 5 and DFBETAS that usually appear on page 6 (not shown here), although important and useful, are omitted from discussion in this book.

Using the criteria described in this subsection, it is observed that none of the studentized residuals (or externally studentized residuals) is large enough for a case to be judged a y-outlier. The column labeled Hat Diag H also indicates that for none of the cases does h_{ii}, exceed the value $(10/13 \approx .8)$. The only case with a large enough Cook's D value $(> 4/13 \approx .308)$ is case number 8, but it is so only because of the relatively large values for h_{ii} and r_{ii}, with neither of those indicating any abnormality. In Section 4.2.3, residuals from the above fit are analyzed using several residual plots.

```
                    Regression Analysis of Hald Data                         5

                              The REG Procedure
                              Model: MODEL1
                           Dependent Variable: y

                             Output Statistics

       Dependent   Predicted     Std Error                 Std Error   Student
Obs    Variable     Value      Mean Predict   Residual     Residual   Residual

 1      78.5000    78.4952       1.8145       0.004760      1.640     0.00290
 2      74.3000    72.7888       1.4120       1.5112        1.997     0.757
 3     104.3000   105.9709       1.8579      -1.6709        1.591    -1.050
 4      87.6000    89.3271       1.3291      -1.7271        2.053    -0.841
 5      95.9000    95.6492       1.4627       0.2508        1.960     0.128
 6     109.2000   105.2746       0.8619       3.9254        2.289     1.715
 7     102.7000   104.1487       1.4820      -1.4487        1.946    -0.744
 8      72.5000    75.6750       1.5634      -3.1750        1.881    -1.688
 9      93.1000    91.7217       1.3270       1.3783        2.055     0.671
10     115.9000   115.6185       2.0471       0.2815        1.339     0.210
11      83.8000    81.8090       1.5956       1.9910        1.854     1.074
12     113.3000   112.3270       1.2544       0.9730        2.100     0.463
13     109.4000   111.6943       1.3480      -2.2943        2.041    -1.124

                             Output Statistics

                            Cook's                Hat Diag     Cov
Obs     -2-1 0 1 2            D     RStudent         H        Ratio    DFFITS

 1    |         |     |     0.000   0.002715      0.5503     4.3353    0.0030
 2    |        |*     |     0.057   0.7345        0.3332     2.0173    0.5193
 3    |      **|      |     0.301  -1.0581        0.5769     2.1948   -1.2356
 4    |       *|      |     0.059  -0.8240        0.2952     1.7413   -0.5333
 5    |         |     |     0.002   0.1198        0.3576     3.0041    0.0894
 6    |         |***  |     0.083   2.0170        0.1242     0.2252    0.7594
 7    |        *|     |     0.064  -0.7218        0.3671     2.1514   -0.5497
 8    |      ***|     |     0.394  -1.9675        0.4085     0.3649   -1.6352
 9    |         |*    |     0.038   0.6459        0.2943     2.0684    0.4171
10    |         |     |     0.021   0.1973        0.7004     6.3297    0.3016
11    |         |**   |     0.171   1.0859        0.4255     1.5583    0.9345
12    |         |     |     0.015   0.4394        0.2630     2.3089    0.2625
13    |      **|      |     0.110  -1.1459        0.3037     1.1854   -0.7568
```

Fig. 4.21. SAS Example D5: Output (Page 5)

4.2.3 Residual plots

The use of residual plots as a part of the analysis of residuals from regression has been common practice during the past 25 years. Many of these plots have been described in the regression texts by Draper and Smith (1981) and Weisberg (1985). The most useful of these plots are the scatter plots of residuals (or studentized residuals) versus each of the explanatory variables (i.e. e_i or r_i versus each of the x_i's) and the scatter plot of residuals (or studentized residuals) versus the predicted values (i.e., e_i or r_i versus the $\hat{y}_i$'s).

Any curvature pattern in the scatter plot of e_i versus x_i may suggest a need for inclusion of higher-order terms in that particular x in the model or a transformation either of the x variable or the response y. This plot is also useful for determining if any independent variables that were observed but not included in the fitted model will make any additional contribution if added to the model. Simply plot the e_i's from the fitted model against these variables one at a time to check whether these variables exhibit any systematic relationship (such as linearity) with the residuals.

The residuals versus the predicted values plot is a multipurpose diagnostic plot. Since the e_i's and the $\hat{y}_i$'s are nearly uncorrelated, this plot should show an even scatter of points around the zero line for e_i as the $\hat{y}_i$ values increase (or decrease), if the fitted model is the correct one. If the spread of the residuals around the zero line (which is a measure of the error variance σ^2) has a increasing or decreasing pattern as the value of $\hat{y}_i$ changes, then a dependence of the residual variance on the "mean" of the response variable may be suspected. For example, such would be the case if the response variable had a Poisson distribution. If any of the residuals versus x variable plots show such a pattern it would indicate a dependence of the residual variance on the particular x variable (i.e., that the residual variance is function of that x variable).

The residual plots generated from artificial data serve as a rough guide to interpreting such plots. In practice, one rarely encounters plots that clearly indicate a pattern as recognizable as those illustrated. If a plot is difficult to interpret it is perhaps best to declare it nonconclusive rather than drawing a possibly erroneous conclusion.

Sometimes marginal box plots are appended to the scatter plot of e_i versus $\hat{y}_i$ as an aid to the examination of the distribution of the residuals and the fitted values. Also, to aid the visual examination of the scatter of the residuals around zero, it is useful draw a horizontal reference line at $e = 0$. If the model assumptions about the errors (i.e., that ϵ_i's are uncorrelated and randomly distributed) are valid, one would expect the residuals to be evenly distributed around this line. Additionally, if the spread around the line remains approximately the same throughout the range of values of $\hat{y}$, it would suggest that the variance of e_i (and, therefore, the variance of y_i) is constant at all values of x_i.

```
insert data step to create the SAS dataset 'cement' here

title  c=darkpurple h=2 'Residual Plots: Regression Analysis of Hald Data';
symbol v=dot h=1 c=red;

proc reg noprint;
  model y = x1 x2 x3 x4/r;
  plot r.*p. r.*x1 r.*x2/nostat modellab='Full Model:' ctext=blue caxes=stbg;
  plot student.*nqq./noline nostat nomodel;
run;
```

Fig. 4.22. SAS Example D6: Program

SAS Example D6

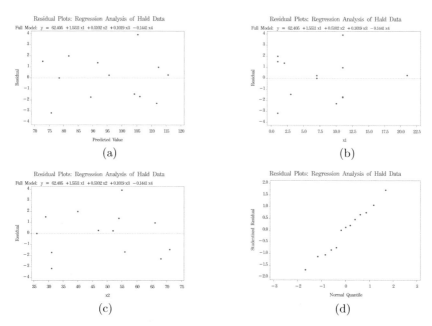

Fig. 4.23. (a) Graph of residuals versus predicted, (b) graph of residuals versus x_1, (c) graph of residuals versus x_2, (d) normal probabilty plot of the studentized residuals

In addition, the normal probability plot of the residuals (or, preferably, the studentized residuals) is useful for checking the model assumption of normality of the errors (ϵ_i's) and also for determining the presence of any outliers. As described in Chapter 3, Section 3.5, a better approach for assessing a normal probability plot is to examine the plot to check whether a specific pattern of the points is identifiable that conforms to a long-tailed, short-tailed, or a skewed distribution relative to a normal distribution rather than judge whether the points deviate from a straight line.

The plot statements in the proc step of the SAS Example D6 program (shown in Fig. 4.22) results in the the four graphs shown in Fig. 4.23. The global `symbol` statement is inserted to change the characteristics of the plot symbol. Options are also added to other statements to add color or resize character sizes. Notice that the three graphs constructed from the first plot statement contains a horizontal reference line drawn at a residual value of zero (i.e., $e = 0$). None of the residual plots shown in Fig. 4.23a, Fig. 4.23b, or Fig. 4.23c indicate any discernible pattern, the residuals being distributed evenly and in a band of equal width around the reference line. The points in the normal probability plot in Fig. 4.23d do not deviate sufficiently from a straight line to reject the assumption that the errors are normally distributed.

4.2.4 Examining relationships among regression variables

In simple linear regression, the relationship between response y and explanatory variable x can be visually examined easily using a scatter plot of y versus x. This plot gives a direct visual impression of the contribution of x on the regression as well as how well the data fits the regression. In the case of multiple regression, however, the scatter plot of y against any one of the x variables, called the *partial response plot*, may be less useful for this purpose. Although still a good indicator of the strength of the linear relationship between y and the particular x variable, this plot does not take into account the effects of the other explanatory variables and, therefore, may not help in determining the *contribution* of individual explanatory variables to the overall regression fit.

Recall that plotting residuals from a multiple regression against a variable that has been observed but not considered for the current model is a good way to determine if there is evidence for the variable to be included in the model. Thus, to check the effect of an explanatory variable, the residuals from fitting multiple regression on the rest of the x's may be plotted against the explanatory variable left out.

The *partial regression residual plot* (also called the *added-variable plot*) is an improvement on this idea. It is a graphical tool that allows the display of the relationship between y and a single x variable when both variables are adjusted for the effect of the other explanatory variables in the model. Suppose that the interest is in the relationship between variables y and x_m, after both have been adjusted for the effect of the other x's in the full model. The *partial regression residual plot* is obtained as follows:

- Compute the residuals from the regression fit of y on all x's except x_m. These, denoted by $y^*(x_m)$, represent the part of the variation of y remaining after removing the effect of the other x's.
- Compute the residuals from the regression fit of x_m on all x's except x_m. These, denoted by x_m^*, represent the part of the variation of x_m remaining after removing the effect of the other x's.
- Plot $y^*(x_m)$ against x_m^* to obtain the *partial regression residual plot* for the variable x_m.

The partial regression residual plot possesses several important properties that provide valuable insight about the multiple regression fit of y on the x's. The intercept of the regression of $y^*(x_m)$ on x_m^* is exactly zero and the slope of the fitted straight line is identical to the partial slope estimate of the full regression of y on the x's (i.e., $\hat{\beta}_m$). Additionally, the residuals from the simple linear regression of $y^*(x_m)$ on x_m^* are identical to those from the full regression. Thus, the standard error of $\hat{\beta}_m$ and the MSE s^2 would also agree if the degrees of freedom $n-2$ assigned to it is replaced by $n-k-1$. Thus, the partial regression residual plot summarizes the regression of y on x_m adjusted for the other x's. A strong linear relationship indicated in this plot corresponds to a strong contribution by x_m to the overall model even in the presence of the other x variables. Thus, the indication of a weak linear relationship in the partial regression residual plot of x_m is evidence that x_m may not contribute significant additional predictive information to the regression when the other x variables are present in the model. This would indicate that x_m may be strongly correlated to one or more of the other x variables in the model, a condition identified as *multicollinearity*.

One would immediately suspect multicollinearity problems if a large sampling variance is observed for the estimated coefficient of an x variable in the statistics computed from a fitted multiple regression. The statistic, called the *variance inflation factor* (hereinafter called VIF), computed for the each coefficient β_m is a direct measure of the effect of multicollinearity in the estimation of the parameter. Computationally,

$$\text{VIF}_m = \frac{1}{(1-R_m^2)}$$

where R_m^2 is the coefficient of determination (or the multiple correlation coefficient) of the regression of x_m on the other x variables (including the intercept). A high R_m^2 indicates that x_m is nearly a linear combination of the other x variables in the model.

The VIF_m value is the factor by which the variance is inflated over what it would be if the variable x_m were completely uncorrelated with all other x variables. A relatively large VIF for a coefficient (as a rule of thumb, a value in excess of 10) indicates that the multicollinearity that exists among the x variables is adversely affecting the estimation of that particular coefficient. If a variable is completely uncorrelated with all other x variables, the VIF value of its estimated coefficient will be exactly 1. On the other hand, if a predictor is highly collinear with one or more of the other predictors as indicated by a high R_m^2 value, it will produce a large VIF_m. The options `collin`, `collioint`, and `tol` in `proc reg` provide statistics called *collinearity diagnostics* for detecting dependencies among the regressor variables and also determine when these may begin to affect the regression estimates.

SAS Example D7

In the SAS Example D7 program displayed in Fig. 4.24, the full model used in earlier examples (Model 1) as well as a reduce model containing only the variables x_1 and x_2 (Model 2) are fitted to the Hald cement data using `proc reg`. The options `clb`, `vif`, and `partial` are specified requesting the 95% confidence intervals and the VIFs for the respective coefficients and partial regression residual plots for each variable in the respective model be output. The parts of the output giving the information on the parameter estimates and other fit statistics of interest appear on Pages 1 and 7, reproduced here in Figs. 4.25 and 4.26, respectively.

```
insert data step to create the SAS dataset 'cement' here

title 'Regression Analysis of Hald Data';

proc reg;
  model y = x1 x2 x3 x4/clb vif partial;
  model y = x1 x2/clb vif partial;
run;
```

Fig. 4.24. SAS Example D7: Program

```
            Regression Analysis of Hald Data                      1

                        The REG Procedure
                          Model: MODEL1
                       Dependent Variable: y

                       Analysis of Variance

                              Sum of         Mean
Source              DF       Squares       Square    F Value   Pr > F

Model                4    2667.89944    666.97486     111.48   <.0001
Error                8      47.86364      5.98295
Corrected Total     12    2715.76308


             Root MSE              2.44601    R-Square    0.9824
             Dependent Mean       95.42308    Adj R-Sq    0.9736
             Coeff Var             2.56333


                        Parameter Estimates

                    Parameter     Standard                       Variance
Variable    DF      Estimate        Error    t Value  Pr > |t|   Inflation

Intercept    1      62.40537     70.07096       0.89    0.3991          0
x1           1       1.55110      0.74477       2.08    0.0708   38.49621
x2           1       0.51017      0.72379       0.70    0.5009  254.42317
x3           1       0.10191      0.75471       0.14    0.8959   46.86839
x4           1      -0.14406      0.70905      -0.20    0.8441  282.51286

                        Parameter Estimates

               Variable    DF     95% Confidence Limits

               Intercept    1     -99.17855     223.98929
               x1           1      -0.16634       3.26855
               x2           1      -1.15889       2.17923
               x3           1      -1.63845       1.84227
               x4           1      -1.77914       1.49102
```

Fig. 4.25. SAS Example D7: Output (page 1)

From the fit statistics for Model 1 appearing on page 1, it is observed that none of the coefficients is significantly different from zero at $\alpha = .05$. This is also confirmed by the fact that the 95% confidence intervals for these coefficients all contain zero **1**. Of course, this is a consequence, as noted earlier, of the standard errors of all estimates being extremely large. The VIFs **2** for all four estimates are larger than 10, but extremely large for x_2 and x_4, indicating that although there is severe multicollinearity among all four variables, each of these variables is highly collinear with the others.

```
                   Regression Analysis of Hald Data                    7

                            The REG Procedure
                             Model: MODEL2
                          Dependent Variable: y

                          Analysis of Variance

                                Sum of           Mean
    Source            DF       Squares         Square    F Value   Pr > F

    Model              2    2657.85859     1328.92930     229.50   <.0001
    Error             10      57.90448        5.79045
    Corrected Total   12    2715.76308

               Root MSE              2.40634    R-Square    0.9787
               Dependent Mean       95.42308    Adj R-Sq    0.9744
               Coeff Var             2.52175

                          Parameter Estimates

                      Parameter      Standard                          Variance
    Variable    DF     Estimate         Error    t Value   Pr > |t|   Inflation

    Intercept    1     52.57735       2.28617      23.00    <.0001           0
    x1           1      1.46831       0.12130      12.10    <.0001     1.05513
    x2           1      0.66225       0.04585      14.44    <.0001     1.05513

                          Parameter Estimates

                   Variable    DF    95% Confidence Limits

                   Intercept    1     47.48344     57.67126
                   x1           1      1.19803      1.73858
                   x2           1      0.56008      0.76442
```

Fig. 4.26. SAS Example D7: Output (page 7)

The partial regression residual plots produced by the `partial` option are only available in low-resolution graphics (i.e., displayed in the Output window instead of SAS Graphics window) by default. As of SAS Version 9.1, ODS Statistical Graphics statements may be used to direct the output to destinations, such as a browser window, that support ODS Graphics. This facility is an experimental extension to the Output Delivery System (ODS) in SAS that will not be discussed in this book. The graphs as displayed in Figs. 4.27 and 4.28, were saved as PostScript files using ODS Statistical Graphics statements.

The partial regression residual plots for the full model are displayed in Fig. 4.27. It is clear that none of these plots shows strong linearity leading to the conclusion that, in the presence of the other variables in the model, none of these variables contributes significant additional information to the prediction of y, the heat evolved.

The second model analyzed in this example, Model 2, fits the model containing only the two variables x_1 and x_2, chosen because they showed only a small correlation ($\approx .23$) between them. As expected, this model displays no multicollinearity problems, as seen from examining the fit statistics for Model 2 appearing in page 7 (see Fig. 4.26). The VIFs for both variables are small and so are the standard errors of estimates of both coefficients. The p-values for the t-statistics are both extremely small and the confidence intervals also show that coefficients are both positive. The partial regression residual plots for the Model 2 displayed in Fig. 4.28 show clearly a significant linear rela-

Regression Analysis of Hald Data

The REG Procedure
Model: model1
Partial Regression Residual Plot

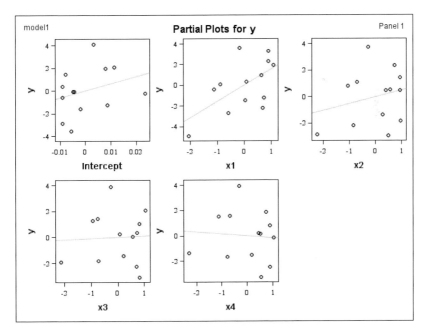

Fig. 4.27. Partial regression residual plots: Model 1

Regression Analysis of Hald Data

The REG Procedure
Model: model2
Partial Regression Residual Plot

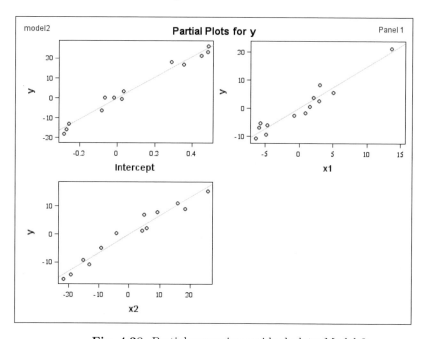

Fig. 4.28. Partial regression residual plots: Model 2

tionship of each variable with y in the presence of the other, showing that any collinearity present will not affect the estimation of the respective coefficients in a multiple regression model.

The examples used earlier are intended to illustrate diagnostic tools available for identifying multicollinearity and how these can be obtained using `proc reg`. The possible remedies available for multicollinearity require further study of this problem. The remedy suggested here of excluding variables may not be advisable in situations when the omitted variable may provide predictive information for data not used for building the regression model. Although the presence of multicollinearity may affect the estimation of some coefficients as seen earlier, the fitted model is still useful for estimating the mean responses or making predictions. One consequence of high variability of a coefficient estimate is that it may be infeasible to use the actual estimate to measure the effect of the variable on the expected response.

4.3 Types of Sums of Squares Computed in PROC REG and PROC GLM

4.3.1 Model comparison technique and extra sum of squares

In Section 4.4, F-statistics based on Type II sum of squares will be used in `proc reg` to obtain F-tests for individual parameters. In general, this approach for testing the significance of one or more parameters in a regression model is called the *model comparison technique* and the difference in the error sum of sum of squares of the two models thus compared is called the *extra sum of squares*.

A hypothesis about any one of the β's in the model, say $H_0 : \beta_2 = 0$, may be tested using a t-test. Sometimes a hypothesis may involve testing whether two or more of the β's in a model are zero against the alternative hypothesis that at least one of these β's is not zero. To introduce the model comparison technique, assume that k is the total number of explanatory variables (x's) in `Model1` and that ℓ is the number of explanatory variables considered to be in a smaller model, `Model2`, where, obviously, $\ell < k$. For convenience and ease of presentation, it is assumed here that the explanatory variables in the two models are ordered so that $x_{\ell+1}, x_{\ell+2}, \ldots, x_k$ is the subset of explanatory variables excluded from `Model1`. The two models are thus

Model1: $\quad y = \beta_0 + \beta_1 x_1 + \cdots + \beta_k x_k + \epsilon$

Model2: $\quad y = \beta_0 + \beta_1 x_1 + \cdots + \beta_\ell x_\ell + \epsilon$

Thus, $k - \ell$ represents the number of variables excluded from `Model1` to obtain `Model2`. It is of interest to test the hypothesis $H_0 : \beta_{\ell+1} = \beta_{\ell+2} = \cdots = \beta_k = 0$ against H_a : at least one of these is not zero, about the coefficients in `Model1`. To formulate a test of this hypothesis based on model comparison technique, first fit the two models and compute the respective residual sums of squares. The F-test statistic for testing H_0 is

$$F = \frac{[\text{SSE}(\texttt{Model2}) - \text{SSE}(\texttt{Model1}))]/(k - \ell)}{\text{SSE}(\texttt{Model1})/(n - k - 1)}$$

The numerator and denominator degrees of freedom for the F-distribution associated with F-statistic are $k - \ell$ and $n - k - 1$, respectively. When this computed F exceeds a critical value for a specified α level obtained from the F-tables, H_0 is rejected.

Reduction Notation

A system of notation for representing differences in residual sums of squares resulting from fitting two regression models was introduced by Searle (1971). Denoted by $R(\ )$ and called the reduction notation, this representation is useful for discussing types of sums of squares and for representing statistics useful in variable subset selection methods computed by computer programs. Consider, for example, the model

$$y = \beta_0 + \beta_1 x_1 + \beta_2 x_2 + \beta_3 x_3 + \beta_4 x_4 + \beta_5 x_5 + \epsilon .$$

To determine whether one of the explanatory variables $x_1, \ldots, x_5$ in this model does not contribute to the regression, a hypothesis of the form, say $H_0 : \beta_2 = 0$, may be tested. To put it in the framework of the model comparison technique, consider the above model to be `Model1`. The corresponding analysis of variance table (only the sources of variation, df, and the sums of squares are shown here) is

Source	df	SS
Regression	5	$\hat{\beta}' X' \underline{y} - n\bar{y}^2 = \text{SSR} \, (\beta_1, \beta_2, \beta_3, \beta_4, \beta_5)$
Residual	$n - 6$	$\sum y_i^2 - \hat{\beta}' X' \underline{y} = \text{SSE} \, (\beta_0, \beta_1, \beta_2, \beta_3, \beta_4, \beta_5)$
Total	$n - 1$	$\sum y_i^2 - n\bar{y}^2$

where n = number of observations in the data set. Suppose now that `Model2` consists of, say, the explanatory variables x_1, x_2, x_3, and x_4, so the model

$$y = \beta_0 + \beta_1 x_1 + \beta_2 x_2 + \beta_3 x_3 + \beta_4 x_4 + \epsilon$$

is fitted to the data. The corresponding analysis of variance table is

Source	df	SS
Regression	4	$\text{SSR}(\beta_1, \beta_2, \beta_3, \beta_4)$
Residual	$n - 5$	$\text{SSE}(\beta_0, \beta_1, \beta_2, \beta_3, \beta_4)$
Total	$n - 1$	$\sum y_i^2 - n\bar{y}^2$

Now, note that the residual sum of squares for the `Model1` is *always* smaller in magnitude than the corresponding sum of squares for `Model2`. The quantity representing the *reduction* in the residual sum of squares due to the addition of the variable x_5 to `Model2` is denoted by $R(\beta_5/\beta_0, \beta_1, \beta_2, \beta_3 \beta_4)$; that is

$$\begin{aligned} R(\beta_5/\beta_0, \beta_1, \beta_2, \beta_3, \beta_4) &= \text{SSE}(\beta_0, \beta_1, \beta_2, \beta_3, \beta_4) - \text{SSE}(\beta_0, \beta_1, \beta_2, \beta_3, \beta_4, \beta_5) \\ &= \text{SSE}(\texttt{Model2}) - \text{SSE}(\texttt{Model1}) \end{aligned}$$

$R(\beta_5/\beta_0, \beta_1, \beta_2, \beta_3, \beta_4)$ can now be used to formulate the F-statistic for testing $H_0 : \beta_5 = 0$ vs. $H_1 : \beta_5 \neq 0$ as

$$F = \frac{R(\beta_5/\beta_0, \beta_1, \beta_2, \beta_3, \beta_4)/(1)}{\text{SSE}(\beta_0, \beta_1, \beta_2, \beta_3, \beta_4, \beta_5)/(n-6)}$$

4.3 Types of Sums of Squares Computed in PROC REG and PROC GLM

The numerator and denominator degrees of freedom here for the F-statistic are 1 and $n - 6$, respectively. Note that when a single parameter is tested for zero, the numerator degrees of freedom of the F-test will always be equal to one and thus the F-test is equivalent to the t-test available for testing the same hypothesis. Note that this notation can be extended in an obvious way to the case when several variables are added or deleted.

4.3.2 Types of sums of squares in SAS

In the regression setting, proc glm in SAS computes two types of sums of squares associated with testing hypotheses about coefficients in the model. These are referred to as **Type I** and **Type II** sums of squares and are output by specifying ss1 or ss2, or both together, as options in the model statement.

Definition: Type I (or Sequential) Sums of Squares This is a partitioning of the regression sum of squares into component sums of squares each with one degree of freedom that represent the reduction in residual sum of squares (or, equivalently, the increase in regression sum of squares) as each variable is added to the model in the order specified in the model statement.

Definition: Type II (or Partial) Sum of Squares This is a partitioning of regression sum of squares into component sums of squares each with one degree of freedom that represent the reduction in residual sum of squares (or, equivalently, the increase in regression sum of squares) when each variable is added to a model containing the other variables specified in the model statement.

Type I and Type II Sums of Squares in Reduction Notation Consider the following model statement:

```
model y = x1 x2 x3 / ss1 ss2;
```

The resulting sums of squares computed by proc glm can be identified using the reduction notation developed above, as indicated below:

Effect	Type I SS	Type II SS
x1	$R(\beta_1/\beta_0)$	$R(\beta_1/\beta_0, \beta_2, \beta_3)$
x2	$R(\beta_2/\beta_0, \beta_1)$	$R(\beta_2/\beta_0, \beta_1, \beta_3)$
x3	$R(\beta_3/\beta_0, \beta_1, \beta_2)$	$R(\beta_3/\beta_0, \beta_1, \beta_2)$

These sums of squares are important and useful because they can be used to construct F-statistics to test certain hypotheses. For example, the Type I SS $R(\beta_2/\beta_0, \beta_1)$ may be used to test whether x_2 should be added to the model $y = \beta_0 + \beta_1 x_1 + \epsilon$, whereas the Type II SS $R(\beta_2/\beta_0, \beta_1, \beta_3)$ may be used to construct an F-statistic (or an equivalent t-statistic) to test the hypothesis $H_0 : \beta_2 = 0$ in the model $y = \beta_0 + \beta_1 x_1 + \beta_2 x_2 + \beta_3 x_3 + \epsilon$.

```
                        Types of SS: Hald Data                              2

                            The GLM Procedure

Dependent Variable: y

                                Sum of
Source                  DF      Squares      Mean Square   F Value   Pr > F

Model                    4     2667.899438    666.974859    111.48   <.0001

Error                    8       47.863639      5.982955

Corrected Total         12     2715.763077


          R-Square      Coeff Var      Root MSE        y Mean

          0.982376      2.563330       2.446008       95.42308


Source            DF      Type I SS     Mean Square   F Value   Pr > F

x1                 1     1450.076328    1450.076328    242.37   <.0001
x2                 1     1207.782266    1207.782266    201.87   <.0001
x3                 1        9.793869       9.793869      1.64   0.2366
x4                 1        0.246975       0.246975      0.04   0.8441


Source            DF     Type II SS     Mean Square   F Value   Pr > F

x1                 1      25.95091138    25.95091138    4.34    0.0708
x2                 1       2.97247824     2.97247824    0.50    0.5009
x3                 1       0.10909005     0.10909005    0.02    0.8959
x4                 1       0.24697472     0.24697472    0.04    0.8441


                                   Standard
      Parameter      Estimate        Error      t Value    Pr > |t|

      Intercept     62.40536930    70.07095921    0.89      0.3991
      x1             1.55110265     0.74476987    2.08      0.0708
      x2             0.51016758     0.72378800    0.70      0.5009
      x3             0.10190940     0.75470905    0.14      0.8959
      x4            -0.14406103     0.70905206   -0.20      0.8441
```

Fig. 4.29. SAS Example D8: Output

As will be discussed in Section 4.4, the Type II sums of squares computed in subset selection procedures in `proc reg` are used to construct both F-to-enter and F-to-delete statistics for adding variables to a model and removing variables from a model, respectively, as illustrated via an example in that section.

SAS Example D8

The sample output page in Fig. 4.29 shows the portion of the output if a `proc glm` step includes a model statement of the form

```
model y = x1 x2 x3 x4/ss1 ss2;
```

Note carefully that none of the variables in the above model statement must appear in a class statement as these are to be regarded as regression-type variables as opposed to classificatory variables for computational purposes. It is informative to observe from this output, for example, that the reduction in the residual sum of squares for adding variable x_2 to a model with only x_1 currently in the model (Type I SS: $R(\beta_2/\beta_0, \beta_1) = 1207.782266$ **1**) is considerably larger than the reduction in the residual sum of squares for adding variable x_2 to a model that contains x_1, x_3, and x_4 (Type II SS: $R(\beta_2/\beta_0, \beta_1, \beta_3, \beta_4) = 2.97247824$ **2**).

Also, note that the p-values corresponding to the Type II F-tests are identical to those computed for the t-tests for the individual parameters given in the lower table in Fig. 4.29, because both sets of statistics test the same hypotheses about the coefficients, as discussed earlier.

4.4 Subset Selection Methods in Multiple Regression

An experimenter usually attempts to choose a subset of variables from a large number of explanatory variables (x's) measured to construct a "good" regression model. Here, "good" may be taken to mean that the model is adequate for prediction and at the same time is economical to use. The model containing all explanatory variables measured is called the *full model*, whereas a model containing a subset of those is called a *subset model*.

For the purpose of selecting a good model, some criteria for selecting one model over another are needed. Usually, statistical test procedures available for this purpose are based on the residual sum of squares remaining after fitting each model. If the inclusion of additional independent variables does not "significantly" decrease the residual sum of squares, then the additional variables may be excluded to obtain a more parsimonious model. If one is dealing with a smaller number of explanatory variables, then it is perhaps best to look at all possible subset models. This can be done using an appropriate program that fits all combinations of the explanatory variables. Although such a procedure may be expensive if the number of explanatory variables that have to be considered is large, algorithms that reduce the amount of computations necessary to obtain the "best" model according to some criterion, containing different numbers of explanatory variables (*subset models of a specified size*), are available.

First, some classical methods for variable subset selection are summarized below.

Forward Selection Method

In the forward selection method, explanatory variables are entered into the model, one at a time, at each stage testing whether there is a significant decrease in the residual sum of squares. For example, suppose that k explanatory variables $x_1, x_2, \ldots, x_k$ are available. First, one independent variable is selected to obtain the best simple linear model. Suppose that each of the explanatory variables is used to fit simple linear regression models of the type

$$y_i = \beta_0 + \beta_j x_{ji} + e_i, \quad i = 1, 2, \ldots, n$$

for each $j = 1, 2, \ldots, k$ and the Type II F-statistic for each model determined. Each of these F-statistics is distributed as a $F(1, n-2)$. The variable corresponding to the largest F-statistic is chosen to enter the model, if it exceeds the critical value of an F-distribution at some preassigned α level called the *significance level for entry*. Since the largest F-statistic does not have an F-distribution, this procedure is not equivalent to testing the hypothesis $H_0 : \beta_j = 0$ in the model chosen. Nevertheless, the critical value thus chosen functions as a cutoff value that is a monotone function of the denominator degrees of freedom and the α level chosen.

Suppose for simplicity that x_1 is the variable that is chosen to enter the model first. Thus, after the first step, the model is

$$y_i = \beta_0 + \beta_1 x_{1i} + e_i \tag{4.6}$$

The residual sum of squares for this model is denoted by $\text{SSE}(\beta_0, \beta_1)$. Now, the next variable to enter the above model is determined by considering each of the explanatory variables other than x_1, one at a time. Thus, a two-variable model is of the form

$$y_i = \beta_0 + \beta_1 x_{1i} + \beta_j x_{ji} + e_i$$

for $j \neq 1$. Suppose that the residual sum of squares for this model is denoted by $\text{SSE}(\beta_0, \beta_1, \beta_j)$. Then an F-statistic for testing $H_0 : \beta_j = 0$ is

$$F = \frac{\text{SSE}(\beta_0, \beta_1) - \text{SSE}(\beta_0, \beta_1, \beta_j)}{\text{SSE}(\beta_0, \beta_1, \beta_j)/(n-3)} = \frac{R(\beta_j/\beta_0, \beta_1)}{\text{MSE}(\beta_0, \beta_1, \beta_j)}$$

which is distributed as $F(1, n-3)$. This statistic, called the "F-to-enter" statistic for variable x_j, effectively tests whether the reduction in the residual sum of squares by adding or "entering" the variable x_j to model (4.6) is significant. By fitting each of the two-variable models using the explanatory variables $x_2, \ldots, x_p$, the largest of the F-to-enter statistics can be determined. The variable next chosen to enter the model is that corresponding to the largest F-to-enter statistic. If this exceeds the critical value from the F-table that corresponds to the preassigned α level and degrees of freedom 1 and $n-3$, the variable is entered. Again, the largest F-statistic will not have the above F-distribution; however, the critical value obtained from the F-table is used as a cutoff value, as earlier. This process is continued until, at any stage of the process, the F-to-enter statistic corresponding to the variable most recently considered as a candidate to enter the model fails to exceed the cutoff value.

Backward Elimination Method

A method that is the direct opposite of forward selection is the so-called backward elimination method. In this method, as a first step, a regression model with all variables in the model is computed:

$$y = \beta_0 + \beta_1 x_1 + \cdots + \beta_k x_k + \epsilon$$

Suppose that one variable, say x_j, is removed from the above model and the residual sum of squares determined. From the residual sums of squares of these two models, an F-statistic called "F-to-remove" or "F-to-delete" can be computed. This F-statistic will have an $F(1, n-3)$ distribution. An "F-to-remove" statistic is computed for each variable x_j, $j = 1, \ldots, k$, contained in the original model. From these, the smallest F-critical value is determined. The variable corresponding to the smallest "F-to-remove" is deleted from the model if this F-critical value fails to exceed the critical value from the F-table corresponding to a preassigned α level. This α level is called *significance level for deletion*. Again, note that the smallest F-statistic will not have an F-distribution; thus, this does not correspond to a test of the hypothesis $H_0 : \beta_j = 0$ in the above model.

Once a variable is removed, a regression model with the variables not yet removed is computed and entire process is repeated. The process is stopped any time a variable fails to be removed from the current model because it is not significant at the α level for deletion.

Stepwise Method

Another more commonly used procedure in computerized regression model building is the stepwise method. This is a combination of both forward selection and backward elimination. Two preassigned α levels are selected: one for entry of variables and one for removal. The procedure is similar to forward selection except that after each new variable is entered in a forward selection step, a single backward elimination step is performed on the current model. Obviously, the α level for entry may not be greater than or equal to the α level for deletion, for, otherwise, the most recently entered variable will be a candidate for immediate deletion. Some computer programs allow this, but the procedure is terminated when a variable to be entered to the model is one just deleted from it. In any case, the choice of these α levels is quite arbitrary as they are not true significance levels but are used to calculate cutoff values for the entry or removal of variables. In practice, the p-values printed for the F-statistics can be compared to the α levels to check whether model selected is affected by changes in the set of levels chosen.

Other Stepwise Methods

Stepwise methods based on finding models that maximize the improvement in R^2 have been proposed. One such method begins by finding the one-variable model producing the highest R^2 and then adds another variable that yields the largest increase in R^2. Once the two-variable model is obtained, each of the variables in the model is swapped with the variables not yet in the model to find the swap that produces the largest increase in R^2. The process continues until the method finds that no switch could increase R^2 further at which time it is stopped and the "best" two-variable model is declared to be found. Another variable is then added to the model, and the swapping process is repeated to find the "best" three-variable model, and so forth.

All-Subsets Methods

A possible alternative to one-variable-at-a-time sequential methods of selecting a good subset of variables is considering all possible models of each "size"; that is, find the best two-variable model by fitting all two-variable models, and so on. Algorithms that require performing only a fraction of the regressions required to find the best subsets of each size by doing a complete search have been developed. In practice, implementations of such methods in computer software usually incorporate the ability for the user to specify the number of "best" models of each size to be selected according to some criterion, such as the R^2 value. The user may also request the values of other statistics such as the MSE and the C_p statistic (these statistics are described below) be included among the information output for each of these models. Thus, the user has the option of selecting a model based on several criteria that may not be the optimal one chosen by such algorithms.

The above methods are all based on finding a subset of variables such that the inclusion of further variables does not decrease the residual sum of squares significantly. However, since all these methods add and/or delete variables one at a time, it is clear that the order of entering or deleting variables may lead to different models being selected. Further, these methods may be affected to various degrees by multicollinearity, if present.

Each of the above methods has its own merits as well as deficiencies. Thus, although one can provide arguments in favor of one or more of these methods, in practice it is recommended that criteria other than those based on the minimum sum of squares of residuals alone be used in selecting a "best" subset model. Some proposed criteria available in computer packages are briefly discussed below. Associated computational algorithms that determine the best subset of a given size, in the sense of minimizing or maximizing the specified criterion without computing all possible regressions, are available. These perform well at a reasonable cost.

Coefficient of Multiple Correlation R^2 This statistic, introduced first in Section 4.1, can be expressed in the form

$$R^2 = \text{SSReg}/\text{SSTot} = 1 - \text{SSE}/\text{SSTot}.$$

It measures the proportion of variance explained by fitted model and is obviously maximized for the full model. The objective is to find subset models with comparable R^2 values so that the inclusion of any of the explanatory variables left out will not increase it to any appreciable degree. Since SSTot remains a constant for a given data set, comparing models based on R^2 is equivalent to comparing models based on SSE. An adjusted R^2 statistic that adjusts for the degrees of freedom of the SSE has been proposed and is given by

$$R^2_{adj} = 1 - \frac{(n-1)}{(n-p-1)} \text{SSE/SSTot}$$

It can be shown that comparing models based on R^2_{adj} is equivalent to comparing models based on MSE. Since it is possible to have subset models with a smaller MSE than the full model, using R^2_{adj} or, equivalently, the MSE has been suggested as an alternative criterion.

Mallows' C_p Statistic For a subset model with p explanatory variables, this statistic is defined as

$$C_p = (\text{SSE}_p/s^2) - (n - 2p)$$

where s^2 = MSE for the full model (i.e., is the model containing all k explanatory variables of interest). SSE_p is the residual sum of squares for the subset model containing p explanatory variables *counting the intercept* (i.e., the number of parameters in the subset model). Usually C_p is plotted against p for the collection of subset models of various sizes under consideration. Acceptable models in the sense of minimizing the total bias of the predicted values are those models for which C_p approaches the value p (i.e., those subset models that fall near the line $C_p = p$ in the above plot).

To understand what is meant by *unbiased predicted values*, consider the full model to be

$$\mathbf{y} = X\boldsymbol{\beta} + \boldsymbol{\epsilon}$$

and the subset model to be of the form

$$\mathbf{y} = X_1\boldsymbol{\beta}_1 + \boldsymbol{\epsilon}$$

where $X = (X_1, X_2)$ and $\boldsymbol{\beta}' = (\boldsymbol{\beta}'_1, \boldsymbol{\beta}'_2)$ are conformable partitions of X and $\boldsymbol{\beta}$ from the full model. Let the ith predicted value from the full model be $\hat{y}_i$ and that from the subset model be denoted by $\hat{y}_i^*$. The mean squared error of a fitted value for the full model is given by the expression:

$$\text{mse}(\hat{y}_i) = \text{var}(\hat{y}_i) + [E(\hat{y}_i) - E(y_i)]^2$$

where $[E(\hat{y}_i) - E(y_i)]$ is called the *bias* in predicting the observation y_i using $\hat{y}_i$. If it is assumed that the full model allows unbiased prediction, the bias term must be zero; that is,

$$E(\hat{y}_i) - E(y_i) = 0$$

The mean squared error of a fitted value for the subset model is given by the expression

$$\mathrm{mse}(\hat{y}_i^*) = \mathrm{var}(y_i) + [E(\hat{y}_i^*) - E(y_i)]^2$$

which gives the bias in predicting y_i using the subset model fitted value to be

$$E(\hat{y}_i^*) - E(y_i)$$

Under the assumption that the full model is "unbiased", this bias term thus reduces to

$$E(\hat{y}_i^*) - E(\hat{y}_i)$$

The statistic C_p, as defined above, is a measure of the total mean squared error of prediction (MSEP) of a subset model scaled by σ^2, given by

$$\frac{1}{\sigma^2} \sum_{i=1}^{n} \mathrm{mse}(\hat{y}_i^*).$$

C_p has been constructed so that if the subset model is unbiased (i.e., if $E(\hat{y}_i^*) - E(\hat{y}_i) = 0$), then it follows that

$$\begin{aligned} C_p &= \frac{\mathrm{SSE}_p}{s^2} - (n - 2p) \\ &\approx \frac{(n-p)\sigma^2}{\sigma^2} - (n - 2p) \\ &= p \end{aligned} \quad (4.7)$$

Recall that p here denotes the total number of parameters in the subset model (i.e., including the intercept). Thus only those subset models that have C_p values close to p must be considered if *unbiasedness* in the sense presented earlier is a desired criterion for selection of a subset model.

However, the construction of the C_p criterion is based on the assumption that s^2, the MSE from fitting the full model, is an unbiased estimate of σ^2. If the full model happens to contain a large number of parameters (β's) that are possibly *not* significantly different from zero, this estimate of σ^2 will be inflated (i.e., larger than the estimate of σ^2 obtained from a model in which more variables are significant). This is because the variables that are not contributing to significantly decreasing the SSE are still counted toward the degrees of freedom when computing the MSE in the full model. If this is the case, C_p will not be a suitable criterion to use for determining a good model. Thus, models chosen by other methods that have lower MSE values than the models selected based on the C_p must be considered competitive.

The AIC Criterion Akaike's information criterion also takes into account the number of variables in the model:

$$\mathrm{AIC} = n \ln \mathrm{SSE} - n \ln n + 2(p+1)$$

This is an unbiased estimate of the expectation of the log-likelihood function under a normally distributed errors assumption. The better models are those with smaller computed AIC values. Because of the form of the above formula, its values are negative, and as with statistics like R^2 and C_p, AIC tends toward a constant value as the number of variables in the model increases.

4.4.1 Subset selection using PROC REG

The cement data (see Table B.7), used previously in SAS Example D5 to demonstrate multiple regression, is used again to illustrate the use of subset selection procedures using `proc reg`. The use of `proc reg` for subset selection is straightforward. The only action statement required is a `model` statement with accompanying options. The options enable a user to specify any of several variable subset selection procedures available in SAS, to specify other parameters required by these methods, to specify lists of statistics output by these methods, to coerce selected variables to be included in the final model, and to fit a model without an intercept. Any number of `model` statements may appear in the same procedure step.

SAS Example D9

```
insert data step to create the SAS dataset 'cement' here

proc reg corr ;
  model y = x1-x4/selection=f sle=.05;
  model y = x1-x4/selection=b sls=.1;
  model y = x1-x4/selection=stepwise sle=.15 sls=.15;
  model y = x1-x4/selection=rsquare sse cp;
  model y = x1-x4/selection=rsquare start=1 stop=3 best=2
                  sse mse aic cp;
  title 'Regression : Variable Subset Selection Techniques';
run;
```

Fig. 4.30. SAS Example D9: Program

In the program for SAS Example D9, displayed in Fig. 4.30, three subset selection methods, `forward`, `backward` and `stepwise`, are specified in the first three model statements. With the `forward` selection (`selection=f`), the α level for entry of variables, `sle` (or `slentry`), used is .05, and with `backward` (`selection=b`), an α level for deletion, `sls` (or `slstay`), of .1 is used. The default settings of both `sle` and `sls` equal to .15 were used with the `stepwise` selection method.

The `rsquare` option is used in a different model statement to fit all possible regression models in a specified range of sizes (i.e., number of independent variables to be included in the models), using the R^2 criterion. Here, the

```
                Regression : Variable Subset Selection Techniques                    2

                          Forward Selection: Step 1

              Variable x4 Entered: R-Square = 0.6745 and C(p) = 138.7308

                               Analysis of Variance

                                    Sum of           Mean
        Source             DF       Squares          Square      F Value    Pr > F

        Model               1      1831.89616      1831.89616     22.80     0.0006
        Error              11       883.86692        80.35154
        Corrected Total    12      2715.76308

                         Parameter       Standard
        Variable         Estimate        Error      Type II SS    F Value    Pr > F

        Intercept        117.56793       5.26221        40108      499.16    <.0001
        x4                -0.73816       0.15460     1831.89616    22.80     0.0006

------------------------------------------------------------------------------------

                          Forward Selection: Step 2

              Variable x1 Entered: R-Square = 0.9725 and C(p) = 5.4959

                               Analysis of Variance

                                    Sum of           Mean
        Source             DF       Squares          Square      F Value    Pr > F

        Model               2      2641.00096      1320.50048     176.63    <.0001
        Error              10        74.76211         7.47621
        Corrected Total    12      2715.76308

                         Parameter       Standard
        Variable         Estimate        Error      Type II SS    F Value    Pr > F

        Intercept        103.09738       2.12398        17615     2356.10   <.0001
        x1                 1.43996       0.13842      809.10480    108.22   <.0001
        x4                -0.61395       0.04864     1190.92464    159.30   <.0001

------------------------------------------------------------------------------------

        No other variable met the 0.0500 significance level for entry into the model.

                          Summary of Forward Selection

                  Variable    Number     Partial       Model
        Step      Entered     Vars In    R-Square     R-Square    C(p)     F Value    Pr > F

          1         x4           1        0.6745       0.6745    138.731    22.80     0.0006
          2         x1           2        0.2979       0.9725      5.4959  108.22    <.0001
```

Fig. 4.31. SAS Example D9: Output (pages 2 and 3)

keyword options start=, stop=, and best= are used to specify all possible subset models that include one variable up to three variables, be fitted but only output information on the best two models of each size, as determined by the largest R^2 values. Other options, `sse`, `mse`, and `cp`, request that these quantities be output for the fitted models in addition to the R^2 values. The output from the program SAS Example D9 appear in Figs. 4.31 through 4.36, separated into six parts and edited (e.g., page titles and a few page numbers are removed and some spacing reduced) for compactness and clarity, but retaining the ordering in the original output. Since it is informative for the user to relate quantities in this output to the computational techniques discussed in Section 4.3, at various points in the discussion of this output some of the computations are reproduced in the text using the notation developed in that section.

The output for the `forward` selection method are on pages 2 and 3 shown in Fig. 4.31. Note that two statistics of interest computed in each step are R^2 and Mallows' C_p, which were discussed earlier. The first variable to enter the model in the `forward` method is x_4, whose F-to-enter value, 22.80, is the largest among the four independent variables and exceeds the F-critical value at .05. Here, note that the Type II SS for x_4 is $= R(\beta_4/\beta_0) = 1831.89616$.

With x_4 in the model, x_1 is found to have the largest F-to-enter value, 108.22, which also exceeds the F-critical value at .05. Thus at the end of Step 2, variables x_4 and x_1 are in the model. With these two variables in the model, x_2 is found to have the largest F-to-enter value, which is computed to be 5.03 (but not shown). This does not exceed the F-critical value at .05, and so x_2 fails to enter the model at this stage. Hence, the final model selected by the `forward` method is the two-variable model containing x_4 and x_1. Computationally, note that Type II SS for x_1 and x_4 are respectively equal to $R(\beta_1/\beta_0, \beta_4) = 809.10480$ and $R(\beta_4/\beta_0, \beta_1) = 1190.92464$ and that F-to-enter x_1 to the model in Step 1 is computed as

$$F\text{-to-enter}(x_1) = \frac{R(\beta_1/\beta_0, \beta_4)/1}{\text{SSE}(\beta_0, \beta_1, \beta_4)/10} = 809.10480/7.4762 = 108.22$$

Note that the quantity F-to-enter x_3 to the model in Step 2, for example, cannot be computed from the quantities that are available in the output on page 2 in Fig. 4.31. To obtain these values, the `details=all` option must be specified with the model statement, as will be illustrated in SAS Example D10. However, note that F-to-enter x_3 to the model in Step 2 here is identical to the value for F-to-delete x_3 computed for the model $y = \beta_0 + \beta_1 x_1 + \beta_2 x_2 + \beta_4 x_4 + \epsilon$ in the backward elimination method (see Step 1 in the output shown in Fig. 4.32).

In Step 0 of the output resulting from `backward` (pages 4, 5, and 6 shown in Fig. 4.32), a model containing all explanatory variables x_1, x_2, x_3, and x_4 is fitted to the data. The Type II sums of squares and F-statistics (the partial sums of squares and accompanying F-statistics) correspond to the F-to-delete values for each of these variables. The smallest of these is for variable x_3, which is 0.02 and does not exceed the F critical value at .1; thus, x_3 is deleted from the model. Type II SS for x_3 is $R(\beta_1/\beta_0, \beta_2, \beta_3, \beta_4) = 0.10909$ and thus the F-to-delete x_3 from the model in Step 0 is computed as

```
                Regression : Variable Subset Selection Techniques                4
                              Backward Elimination: Step 0
                    All Variables Entered: R-Square = 0.9824 and C(p) = 5.0000
                                   Analysis of Variance
                                      Sum of             Mean
    Source                    DF      Squares           Square      F Value    Pr > F

    Model                      4    2667.89944        666.97486     111.48     <.0001
    Error                      8      47.86364          5.98295
    Corrected Total           12    2715.76308

                          Parameter       Standard
             Variable     Estimate          Error     Type II SS   F Value   Pr > F

             Intercept     62.40537        70.07096     4.74552      0.79     0.3991
             x1             1.55110         0.74477    25.95091      4.34     0.0708
             x2             0.51017         0.72379     2.97248      0.50     0.5009
             x3             0.10191         0.75471     0.10909      0.02     0.8959
             x4            -0.14406         0.70905     0.24697      0.04     0.8441
    ----------------------------------------------------------------------------------
                              Backward Elimination: Step 1
                 Variable x3 Removed: R-Square = 0.9823 and C(p) = 3.0182
                                   Analysis of Variance
                                      Sum of             Mean
    Source                    DF      Squares           Square      F Value    Pr > F

    Model                      3    2667.79035        889.26345     166.83     <.0001
    Error                      9      47.97273          5.33030
    Corrected Total           12    2715.76308

                          Parameter       Standard
             Variable     Estimate          Error     Type II SS   F Value   Pr > F

             Intercept     71.64831        14.14239    136.81003    25.67     0.0007
             x1             1.45194         0.11700    820.90740   154.01     <.0001
             x2             0.41611         0.18561     26.78938     5.03     0.0517
             x4            -0.23654         0.17329      9.93175     1.86     0.2054
    ----------------------------------------------------------------------------------
                              Backward Elimination: Step 2
                 Variable x4 Removed: R-Square = 0.9787 and C(p) = 2.6782
                                   Analysis of Variance
                                      Sum of             Mean
    Source                    DF      Squares           Square      F Value    Pr > F

    Model                      2    2657.85859       1328.92930     229.50     <.0001
    Error                     10      57.90448          5.79045
    Corrected Total           12    2715.76308

                          Parameter       Standard
             Variable     Estimate          Error     Type II SS   F Value   Pr > F

             Intercept     52.57735         2.28617   3062.60416   528.91     <.0001
             x1             1.46831         0.12130    848.43186   146.52     <.0001
             x2             0.66225         0.04585   1207.78227   208.58     <.0001
    ----------------------------------------------------------------------------------
         All variables left in the model are significant at the 0.1000 level.
```

Fig. 4.32. SAS Example D9: Output (pages 4-6)

```
                 Regression : Variable Subset Selection Techniques              7

                          Stepwise Selection: Step 1

              Variable x4 Entered: R-Square = 0.6745 and C(p) = 138.7308

                              Analysis of Variance

                                 Sum of           Mean
   Source               DF       Squares         Square      F Value    Pr > F

   Model                 1      1831.89616     1831.89616     22.80     0.0006
   Error                11       883.86692       80.35154
   Corrected Total      12      2715.76308

                    Parameter     Standard
        Variable    Estimate       Error      Type II SS  F Value  Pr > F

        Intercept   117.56793     5.26221        40108    499.16   <.0001
        x4           -0.73816     0.15460     1831.89616   22.80   0.0006
   --------------------------------------------------------------------------
                          Stepwise Selection: Step 2

              Variable x1 Entered: R-Square = 0.9725 and C(p) = 5.4959

                              Analysis of Variance

                                 Sum of           Mean
   Source               DF       Squares         Square      F Value    Pr > F

   Model                 2      2641.00096     1320.50048    176.63     <.0001
   Error                10        74.76211        7.47621
   Corrected Total      12      2715.76308

                    Parameter     Standard
        Variable    Estimate       Error      Type II SS  F Value  Pr > F

        Intercept   103.09738     2.12398         17615   2356.10   <.0001
        x1            1.43996     0.13842       809.10480  108.22   <.0001
        x4           -0.61395     0.04864      1190.92464  159.30   <.0001
   --------------------------------------------------------------------------
                          Stepwise Selection: Step 3

              Variable x2 Entered: R-Square = 0.9823 and C(p) = 3.0182

                              Analysis of Variance

                                 Sum of           Mean
   Source               DF       Squares         Square      F Value    Pr > F

   Model                 3      2667.79035      889.26345    166.83     <.0001
   Error                 9        47.97273        5.33030
   Corrected Total      12      2715.76308

                    Parameter     Standard
        Variable    Estimate       Error      Type II SS  F Value  Pr > F

        Intercept    71.64831    14.14239      136.81003    25.67   0.0007
        x1            1.45194     0.11700      820.90740   154.01   <.0001
        x2            0.41611     0.18561       26.78938     5.03   0.0517
        x4           -0.23654     0.17329        9.93175     1.86   0.2054
   --------------------------------------------------------------------------
```

Fig. 4.33. SAS Example D9: Output (page 7-9)

$$F\text{-to-delete}(x_3) = \frac{R(\beta_3\beta_0, \beta_1, \beta_2, \beta_4)/1}{\text{SSE}(\beta_0, \beta_1, \beta_2, \beta_3, \beta_4)/8} = 0.10909/5.98295 = 0.02$$

In Step 1, x_4 is deleted, since the F-to-delete value of 1.86 for this variable is also not exceeded at .1. F-to-delete x_2 from the model in Step 1 is computed as

$$F\text{-to-delete}(x_2) = \frac{R(\beta_2\beta_0, \beta_1, \beta_4)/1}{\text{SSE}(\beta_0, \beta_1, \beta_2, \beta_4)/9} = 26.78938/5.33030 = 5.03$$

and this is identical to F-to-enter x_2 to the model containing only x_1 and x_4. In Step 2 no variable qualifies for deletion since the F-to-delete values for both x_1 and x_3 are quite large and thus the final model selected by the `backward` method contains x_1 and x_2. Note that this model is different from the one selected by the `forward` selection method.

```
                    Stepwise Selection: Step 4
          Variable x4 Removed: R-Square = 0.9787 and C(p) = 2.6782
                         Analysis of Variance

                                Sum of           Mean
Source              DF          Squares          Square      F Value   Pr > F

Model                2         2657.85859      1328.92930    229.50    <.0001
Error               10           57.90448         5.79045
Corrected Total     12         2715.76308

                  Parameter      Standard
Variable          Estimate       Error       Type II SS   F Value   Pr > F

Intercept         52.57735       2.28617     3062.60416   528.91    <.0001
x1                 1.46831       0.12130      848.43186   146.52    <.0001
x2                 0.66225       0.04585     1207.78227   208.58    <.0001

            Bounds on condition number: 1.0551, 4.2205
---------------------------------------------------------------------
        All variables left in the model are significant at the 0.1500 level.
   No other variable met the 0.1500 significance level for entry into the model.
```

Fig. 4.34. SAS Example D9: Output (pages 7-9)

In the output resulting from the `stepwise` option appearing on pages 7-9, variables are added based on F-to-enter statistics as in forward selection, but a backward elimination step is performed immediately after each forward step, based on F-to-delete statistics. Here, x_4 is entered in Step 1 since the F-to-enter value of 22.80 is the largest and exceeds the F-critical value at .15. x_4 is not considered for deletion since it is the variable just selected. In Step 2, variable x_1 is entered with a F-to-enter value of 108.22 which obviously exceeds the F-critical value. The F-to-delete statistics for variables x_1 and x_4, as before, are given by corresponding Type II F-statistics in Step 2. Since both of these exceed the F-critical values at .15, no variable qualifies for deletion at this stage.

In Step 3, variable x_2 is added to the model with an F-to-enter statistic of 5.03 since this is exceeds the F-critical value at .15. Here, $R(\beta_2/\beta_0, \beta_1, \beta_4) = 26.78938$ and thus F-to-enter x_2 to the model in Step 2 is computed as

$$F\text{-to-enter}(x_2) = \frac{R(\beta_2/\beta_0, \beta_1, \beta_4)/1}{\text{SSE}(\beta_0, \beta_1, \beta_2 \beta_4)/9} = 26.78938/5.3303 = 5.03.$$

The F-to-delete statistics for x_1, x_3, and x_4 respectively are 154.0, 5.03, and 1.86. The smallest value 1.86 fails to exceed the F-critical value at .15 and, thus, x_4 is a candidate for deletion. Type II SS for x_4 is $R(\beta_4/\beta_0, \beta_1, \beta_2) = 9.93175$ and the F-to-delete x_4 from the model in Step 3 is computed as

$$F\text{-to-delete}(x_4) = \frac{R(\beta_4/\beta_0, \beta_1, \beta_2)/1}{\text{SSE}(\beta_0, \beta_1, \beta_2, \beta_4)/9} = 9.93175/5.3303 = 1.86.$$

x_4 is deleted in Step 4 giving the final model since no other variable qualifies for entry at .15 significance when x_1 and x_2 are already in the model.

```
          Regression : Variable Subset Selection Techniques            10
                         R-Square Selection Method

                    Number of Observations Read         13
                    Number of Observations Used         13

Number in
  Model     R-Square         C(p)              SSE      Variables in Model

     1       0.6745        138.7308         883.86692   x4
     1       0.6663        142.4864         906.33634   x2
     1       0.5339        202.5488        1265.68675   x1
     1       0.2859        315.1543        1939.40047   x3
-----------------------------------------------------------------------
     2       0.9787          2.6782          57.90448   x1 x2
     2       0.9725          5.4959          74.76211   x1 x4
     2       0.9353         22.3731         175.73800   x3 x4
     2       0.8470         62.4377         415.44273   x2 x3
     2       0.6801        138.2259         868.88013   x2 x4
     2       0.5482        198.0947        1227.07206   x1 x3
-----------------------------------------------------------------------
     3       0.9823          3.0182          47.97273   x1 x2 x4
     3       0.9823          3.0413          48.11061   x1 x2 x3
     3       0.9813          3.4968          50.83612   x1 x3 x4
     3       0.9728          7.3375          73.81455   x2 x3 x4
-----------------------------------------------------------------------
     4       0.9824          5.0000          47.86364   x1 x2 x3 x4
```

Fig. 4.35. SAS Example D9: Output (page 10)

Pages 10 and 11, displayed in Figs. 4.35 and 4.36, respectively, contains the output from the two model statements with the `rsquare` selection criterion (see the program for SAS Example D9 in Fig. 4.30 for the model statements). Page 10 displays the default output showing all possible models of different sizes ordered by decreasing R^2 values. The statistics printed are those requested by the options `SSE` and `cp`.

```
              Regression : Variable Subset Selection Techniques                11
                            R-Square Selection Method

                       Number of Observations Read         13
                       Number of Observations Used         13

Number in
  Model     R-Square      C(p)           AIC          MSE          SSE      Variables in Model

    1        0.6745     138.7308       58.8516      80.35154     883.86692  x4
    1        0.6663     142.4864       59.1780      82.39421     906.33634  x2
    -----------------------------------------------------------------------------------------
    2        0.9787       2.6782       25.4200       5.79045      57.90448  x1 x2
    2        0.9725       5.4959       28.7417       7.47621      74.76211  x1 x4
    -----------------------------------------------------------------------------------------
    3        0.9823       3.0182       24.9739       5.33030      47.97273  x1 x2 x4
    3        0.9823       3.0413       25.0112       5.34562      48.11061  x1 x2 x3
```

Fig. 4.36. SAS Example D9: Output (page 11)

On page 11, information is displayed only for the "best" models of sizes 1,2, and 3 explanatory variables, determined as those with the largest R^2 values, along with the variables in each model. Thus the options `start=`, `stop=`, and `best=` are useful for controlling the number of models of interest when a large number of explanatory variables is involved. The MSE and AIC are additional statistics requested using appropriate options. Note that C_p must be compared to the number of independent variables plus one; thus, the number printed in the output as 'Number in Model' must be incremented by one before comparing it to C_p. By inserting the statement

```
plot cp.*np./cmallows=blue vaxis=0 to 8 by 1
                              ctext=blue caxes=darkred;
```

in the `proc reg` step of SAS Example D9 program (shown in Fig. 4.30), the high-resolution plot of C_p versus p shown in Fig. 4.37 is obtained. Note that the `vaxis=` option may be used to scale the vertical axis so that models with very high C_p values are eliminated from this plot, as illustrated in the above statement. Note also the use of `ctext=` and `caxes=` options to add color.

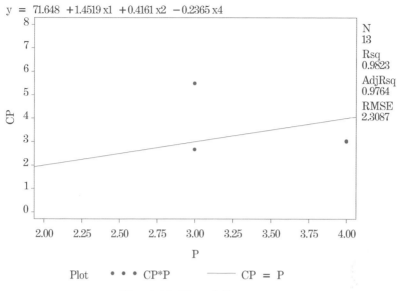

Fig. 4.37. Plot of C_p versus p

SAS Example D10

If the option details=all is included in the model statement when using a model selection procedure such as stepwise, a more detailed output is obtained. The sample output page in Fig. 4.38 shows the portion of the output in Step 3 of the stepwise procedure discussed previously, if the model statement was of the form

model y = x1-x4/selection=stepwise sle=.15 sls=.15 details=all;

This illustrates the additional information available if the user wants to keep track of the decision-making process. For example, from this output it can be seen immediately that the largest F-to-enter, to the model in Step 2 is 5.03 with a p-value $= .0517$. This is smaller than .15, the sle value; thus, the variable x_2 enters the model. The smallest F-to-delete from the model in Step 3 is 1.86 with a p-value=.2054. This is larger than the sls value .15 and thus the variable x_4 is removed from the model in Step 3.

4.4.2 Other options available in PROC REG for model selection

Several other model selection options are available in proc reg:

maxr Maximum R^2 Improvement Selection Method: The maxr method begins by finding the one-variable model producing the highest R^2. Then it adds

```
                Regression : Variable Subset Selection Techniques                    4
                            Stepwise Selection: Step 3
                              Statistics for Removal
                                   DF = 1,10

                         Partial        Model
            Variable     R-Square      R-Square     F Value      Pr > F

              x1          0.2979        0.6745      108.22       <.0001
              x4          0.4385        0.5339      159.30       <.0001

                               Statistics for Entry
                                    DF = 1,9

                                          Model
            Variable     Tolerance       R-Square     F Value     Pr > F

              x2         0.053247         0.9823       5.03       0.0517
              x3         0.289051         0.9813       4.24       0.0697

            Variable x2 Entered: R-Square = 0.9823 and C(p) = 3.0182

                               Analysis of Variance

                                 Sum of         Mean
   Source              DF       Squares        Square      F Value      Pr > F

   Model                3       2667.79035    889.26345     166.83      <.0001
   Error                9         47.97273      5.33030
   Corrected Total     12       2715.76308


                       Parameter       Standard
            Variable    Estimate        Error    Type II SS   F Value   Pr > F

            Intercept   71.64831       14.14239   136.81003    25.67    0.0007
            x1           1.45194        0.11700   820.90740   154.01    <.0001
            x2           0.41611        0.18561    26.78938     5.03    0.0517
            x4          -0.23654        0.17329     9.93175     1.86    0.2054
```

Fig. 4.38. SAS Example D10: Sample page of output obtained with `details=all`

another variable that produces the largest improvement in R^2. Then each variable in the model is replaced with a variable not in the model to find if such a swap will produce an increase in R^2. The swap that produces the largest increase in R^2 is output as the "best" two-variable model by `maxr`. Another variable is then added to the model, and the comparing-and-swapping process is repeated to locate the "best" three-variable model, and so forth. The `maxr` method is more expensive computationally than the `stepwise` method because it evaluates all combinations of variables before making a swap; whereas the `stepwise` method may remove the "worst" variable in a sequential fashion without considering the effects of replacing any of the other variables in the model by the remaining

4.5 Inclusion of Squared Terms and Product Terms in Regression Models

variables. Consequently, `maxr` typically takes much longer to run than `stepwise`.

minr Minimum R^2 Improvement Selection Method: The `minr` method closely resembles the `maxr` method, but the swap chosen at each stage is the one that produces the smallest increase in R^2.

adjrsq Adjusted R^2 Selection Method: This method is similar to the `rsquare` method, except that the adjusted R^2 statistic is used as the criterion for selecting models, and the method finds the models with the highest adjusted R^2 within the range of model sizes.

cp Mallows' C_p Selection Method: This method is similar to the `adjrsq` method, except that Mallows' C_p statistic is used as the criterion for model selection. Models are listed in ascending order of C_p.

groupnames Specifying Groups of Variables: Groups of variables can be specified to be treated as a single set during the selection process. For example,

```
model y=x1 x2 x3 /
          selection=stepwise groupnames='x1 x2' 'x3';
```

Another example is:

```
model mpg = cyl disp hp drat wt / selection=stepwise sle=.1
          sls=.2 groupnames='cyl' 'disp_hp_drat' 'wt';
```

4.5 Inclusion of Squared Terms and Product Terms in Regression Models

Although a complete discussion of regression modeling is beyond the scope of this book, what follows is a presentation of methods for obtaining analysis of such models in SAS. In general, models that contain terms of various powers (squared, cubed, etc.) and terms of crossed-products of of explanatory variables are called polynomial regression models. For example,

$$y = \beta_0 + \beta_1 x + \beta_2 x^2 + \beta_3 x^3 + \epsilon$$

is a *third-order model of one explanatory variable* and

$$y = \beta_0 + \beta_1 x_1 + \beta_2 x_2 + \beta_3 x_1^2 + \beta_4 x_1 x_2 + \beta_5 x_2^2 + \epsilon$$

is a *second-order model of two explanatory variables*. In practice, powers higher than the third or products that include more than three variables are rarely used since their interpretation becomes difficult. These are equations of surfaces and thus are primarily used to approximate an observed curvilinear relationship between the response variable and the explanatory variables. One might also consider inclusion of quadratic or cubic terms in a model if a first-order model is found to be inadequate following a residual analysis, usually by inspecting various residual-based diagnostic plots discussed in Section 4.1. For example, the plot in Fig. 4.39 shows a pattern of the residuals that indicates the need to include a squared term in x in the model.

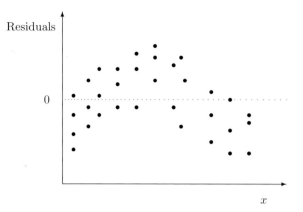

Fig. 4.39. Example of a plot of residuals versus an explanatory variable

4.5.1 Including interaction terms in the model

Two explanatory variables, x_1 and x_2, are said to interact if the expected change in the response y for unit change in x_1 depends on the value of x_2. To illustrate how interaction may be modeled by including a product term in the model, consider the first-order model in x_1 and x_2,

$$y = \beta_0 + \beta_1 x_1 + \beta_2 x_2 + \epsilon$$

(i.e., consider $E(y) = \beta_0 + \beta_1 x_1 + \beta_2 x_2$). This model does not contain a term to represent interaction and is called an *additive* model. The Figure 4.40a depicts the situation graphically when the variable x_2 is a ordinal or a qualitative variable. When the value of x_2 is kept fixed, $E(y)$ is a straight-line function

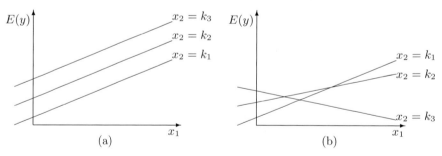

Fig. 4.40. (a) Model for no interaction and (b) model for interaction

of x_1. Different values of x_2 give parallel lines. Thus, the expected change in y for unit change in x_1 or the *partial slope* will remain β_1 regardless of the value of x_2. Now, suppose the model includes the product term $x_1 x_2$,

$$E(y) = \beta_0 + \beta_1 x_1 + \beta_2 x_2 + \beta_3 x_1 x_2$$

4.5 Inclusion of Squared Terms and Product Terms in Regression Models

$$= \beta_0 + (\beta_1 + \beta_3 x_2) x_1 + \beta_2 x_2$$

When the value of x_2 is kept fixed, $E(y)$ is still a straight-line function of x_1. However, the slope is $\beta_1 + \beta_3 x_2$, which obviously is a function of x_2. Therefore, the slope now depends on the value of x_2. So, in Fig. 4.40b, the lines are not parallel as the slopes are different for different values of x_2. So x_1 and x_2 are said to interact.

4.5.2 Comparing slopes of regression lines using interaction

When it is required to test whether the slopes of regression lines are the same under different conditions (e.g., levels of treatments), a quantitative explanatory variable (i.e., an x variable) of a special type called a dummy variable (also called an indicator variable) may be included in the model. This variable takes values 0, 1, ... to "indicate" the different conditions or different levels of a treatment, at which the slopes are being compared.

Example 4.5.1

The following data consist of blood coagulation time y measured on patients given 2 drugs (say, A and B) with 10 subjects assigned to each of 3 doses (10, 20, and 30 mg) of each drug.

Let y denote the response, blood clotting time (in minutes), and x_1 denote the drug dose, which is a quantitative variable. However, a qualitative variable (say, x_2) is required to represent the drug since it has values A and B. The value of x_2 is set to zero if a response is from a subject administered drug A and set to 1 if it is from drug B. Thus, the actual data has the following structure:

```
   y    x1    x2
---------------
  3.5   10    0
  4.2   10    0
   :     :    :
  4.8   30    0
   :     :    :
  5.6   10    1
  5.4   10    1
   :     :    :
  6.9   20    1
   :     :    :
  5.7   30    1
```

The interaction term $x_1 x_2$ is included in the model:

$$y = \beta_0 + \beta_1 x_1 + \beta_2 x_2 + \beta_3 x_1 x_2 + \epsilon$$

A test of whether the coefficient of the interaction term, β_3, is zero is equivalent to a test of whether the two regression lines corresponding to the two drugs have the same slope. The expected values (or mean) response $E(y)$ for each drug are (obtained by substituting $x_2 = 0$ and $x_2 = 1$, respectively) as follows:

$$\text{drug A:} \quad E(y) = \beta_0 + \beta_1 x_1$$

$$\text{drug B:} \quad E(y) = \beta_0 + \beta_1 x_1 + \beta_2 + \beta_3 x_1$$

$$= (\beta_0 + \beta_2) + (\beta_1 + \beta_3) x_1$$

These two equations are linear regression lines, each corresponding to the responses to the two drugs. If β_3 in the second is zero, then the two lines would have the same slope, but different y-intercepts. The situation is illustrated in Fig 4.41. If β_3 is positive, the slope for the line for drug B will be larger

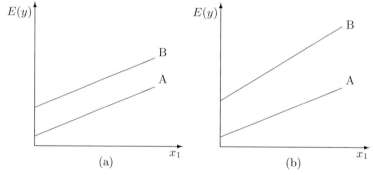

Fig. 4.41. (a) $\beta_3 = 0$: no interaction present and (b) $\beta_3 \neq 0$: interaction present

than that of the line for drug A. This situation is illustrated in the graphic Fig. 4.41b. This may be interpreted to indicate that *the mean blood clotting time* increases at a faster rate for drug B, as the drug dose level increases, compared to drug A. The test of same slope for the two lines thus may be directly performed using a SAS program to fit the model

$$y = \beta_0 + \beta_1 x_1 + \beta_2 x_2 + \beta_3 x_1 x_2 + \epsilon$$

and testing the hypothesis $H_0 : \beta_3 = 0$ vs. $H_a : \beta_3 \neq 0$ using the t-statistic (and the p-value) that corresponds to the $x_1 x_2$ term.

4.5.3 Analysis of models with higher-order terms with PROC REG

The data in Table 4.2 appears in Ott and Longnecker (2001) and consist of weight loss of a compound exposed to air at different humidity levels for different amounts of time. This is an experimental situation in which the

4.5 Inclusion of Squared Terms and Product Terms in Regression Models

values of the explanatory variables are levels of two experimental factors `time` and `humidity`. A multiple regression model

$$y = \beta_0 + \beta_1 x_1 + \beta_2 x_2 + \epsilon$$

was fitted to the data and various residual analyses were carried out. This

Weight Loss	Time	Humidity
4.3	4	0.2
5.6	5	0.2
6.4	6	0.2
8.0	7	0.2
4.0	4	0.3
5.2	5	0.3
6.6	6	0.3
7.5	4	0.4
4.0	5	0.4
5.6	6	0.4
6.4	7	0.4

Table 4.2. Weight loss (in pounds) of a compound exposed to air at different humidity levels for different amounts of time (in hours).

model is a *first-order additive model* in the two variables `time` and `humidity`.

SAS Example D11

The SAS Example D11 program shown in Fig. 4.42 is a SAS program to analyze these data. Fig. 4.43 displays the four high-resolution graphs produced by the SAS program for the analysis of the residuals from fitting the above model. Figure 4.44 shows the first part of the output from the program. Figure 4.43a is a plot of the data values against the levels of the `humidity` variable. Different colors, line types, and symbols identify the values of the other variable `time`. This plot indicates two features of the data. First, straight lines may not fit the points well for each level of time indicating possible **curvature** in the humidity variable. Second, the response patterns (or curvatures) appear to be different for each level of time, indicating possible **interaction** between time and humidity. The above model represents the situation that the expected response is a set of parallel straight lines at each value of one explanatory variable. Thus, a different situation must be represented by an alternative model that includes terms to model the observed deviations.

Further, by examining each of the residual plots, it is possible to reach the following conclusions:

- Fig. 4.43b shows no obvious pattern, so the assumption of constant variance is supported.

```
data evap;
input wtloss time humid;
label humid='Humidity' time='Time';
datalines;
4.3 4 .2
5.6 5 .2
6.4 6 .2
8.0 7 .2
4.0 4 .3
5.2 5 .3
6.6 6 .3
7.5 7 .3
3.2 4 .4
4.0 5 .4
5.6 6 .4
6.4 7 .4
;
run;

   symbol1 c=blue v=circle i=join ;
   symbol2 c=red v=x i=join;
   symbol3 c=cyan v=triangle i=join;
   symbol4 c=green v=square i=join;
   axis1 c=darkviolet label=(c=blue h=1.5 a=90   'Mean Weight Loss');
   axis2 c=darkviolet label=(c=blue h=1.5 'Humidity');

proc gplot;
plot wtloss*humid=time/vaxis=axis1 haxis=axis2;
title c=darkcyan h=2 'Plot of Weight Loss against Humidity';
run;

proc reg;
   model wtloss= time humid/r;
   plot r.*p./nostat ctext=blue caxis=darkviolet lline=20;
   plot r.*p. r.*humid r.*time/nostat;
   title c= darkcyan h=2 'Residual Analysis for Weight Loss';
run;
```

Fig. 4.42. SAS Example D11: Program

- Fig. 4.43c indicates some curvature in the humidity variable, so a quadratic term in humid is suggested.
- Fig. 4.43d does not suggest that a higher-order term in time is needed.

These suggest that the model may be modified by including a quadratic term in humid and an interaction term humid*time. However, when a cross-product term of a variable with a lower-order term is considered to form a second-order term, it is usual practice to also include a cross-product term (or an interaction term) with the higher-order term of the same variable, as well.

Thus, the model may be first written as, say

$$y = \beta_0 + \beta_1 x_1 + \beta_2 x_2 + \beta_3 x_2^2 + \epsilon$$

and then extended to

$$y = \beta_0 + \beta_1 x_1 + \beta_2 x_2 + \beta_3 x_2^2 + \beta_4 x_1 x_2 + \beta_5 x_1 x_2^2 + \epsilon$$

4.5 Inclusion of Squared Terms and Product Terms in Regression Models

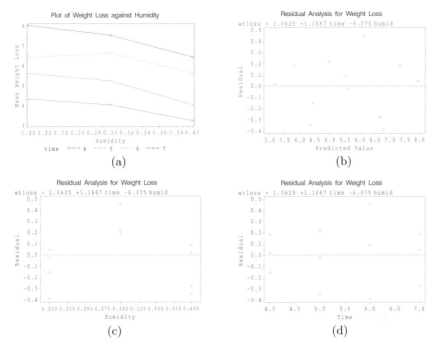

Fig. 4.43. (a) Graph of y versus x produced by PROC GPLOT; (b) graph of residuals versus predicted produced by plot r.*p.; (c) graph of residuals versus humidity produced by plot r.*humid; (d) graph of residuals versus time produced by plot r.*time

by considering the interaction to be between a function of x_1 and a function of (x_2, x_2^2). Instead of the data-driven model building approach used earlier, some investigators prefer to fit a complete second, or third, order model and use a model selection procedure to arrive at a suitable model. In either case, the basic approach is to approximate the true mean response by some polynomial function of the explanatory variables. There are two cautions to be aware of when using this type of modeling. One, as previously discussed, extrapolation of the fitted function beyond the range of the explanatory variables is to be avoided, as the true model is being approximated only in this range. Second, because of the presence of the higher-order transformations of variables already in the model, the multicollinearity problem is liable to arise. Thus, as discussed in SAS Example D13, centering of the independent variables is recommended when fitting models with higher-order terms.

SAS Example D12

SAS Example D12 shown in Fig. 4.45 displays the program used for the analysis of the larger of the above two models and Fig. 4.46 shows the first part of the output. Note that variables that correspond to the terms x_2^2, $x_1 x_2$,

```
                    Residual Analysis for Weight Loss                    1
                           Analysis of Variance

                                   Sum of            Mean
      Source             DF        Squares          Square     F Value    Pr > F

      Model               2        23.66792        11.83396    152.42     <.0001
      Error               9         0.69875         0.07764
      Corrected Total    11        24.36667

                     Root MSE              0.27864    R-Square    0.9713
                     Dependent Mean        5.56667    Adj R-Sq    0.9650
                     Coeff Var             5.00547

                           Parameter Estimates

                           Parameter      Standard
      Variable     DF      Estimate         Error     t Value    Pr > |t|

      Intercept     1       1.06250        0.50039      2.12      0.0627
      time          1       1.16667        0.07194     16.22      <.0001
      humid         1      -6.37500        0.98513     -6.47      0.0001
```

Fig. 4.44. SAS Example D11: Output (page 1)

and $x_1 x_2^2$ are created in the data step as the variables named humidsq, timehumid, and timehumidsq, respectively, using assignment statements with algebraic expressions on the right-hand side. This is required since polynomial

```
data evap;
input wtloss time humid;
humidsq=humid*humid;
timehumid=time*humid;
timehumidsq=time*humidsq;
datalines;
4.3 4 .2
5.6 5 .2
6.4 6 .2
8.0 7 .2
4.0 4 .3
5.2 5 .3
6.6 6 .3
7.5 7 .3
3.2 4 .4
4.0 5 .4
5.6 6 .4
6.4 7 .4
;
run;

proc reg;
   model wtloss= time humid humidsq timehumid timehumidsq/vif;
   plot r.*p. r.*humid r.*time/nostat;
   title 'Residual Analysis for Weight Loss';
run;
```

Fig. 4.45. SAS Example D12: Program

4.5 Inclusion of Squared Terms and Product Terms in Regression Models

effects such as humid*humid or time*humid cannot be specified directly in the model statement in proc reg.

```
                    Residual Analysis for Weight Loss                        1
                          Analysis of Variance

                                Sum of         Mean
    Source              DF      Squares        Square      F Value    Pr > F

    Model                5      24.08467       4.81693     102.49     <.0001
    Error                6       0.28200       0.04700
    Corrected Total     11      24.36667

              Root MSE              0.21679    R-Square    0.9884
              Dependent Mean        5.56667    Adj R-Sq    0.9788
              Coeff Var             3.89452

                          Parameter Estimates

                      Parameter      Standard                           Variance
    Variable    DF    Estimate       Error      t Value    Pr > |t|     Inflation

    Intercept    1     -1.14000       5.68110    -0.20      0.8476             0
    time         1      0.98000       1.01223     0.97      0.3704     327.00000
    humid        1      7.25000      40.17145     0.18      0.8627    2746.80000
    humidsq      1    -19.50000      66.64458    -0.29      0.7797    2746.80000
    timehumid    1      1.75000       7.15751     0.24      0.8150    4218.30000
    timehumidsq  1     -3.50000      11.87434    -0.29      0.7781    3167.30000
```

Fig. 4.46. SAS Example D12: Output (Page 1)

SAS Example D13

One of the main drawbacks of fitting higher-order polynomial models is clearly demonstrated in this output. It is observed that the VIF's for every term in the model are extremely large, indicating that the estimation of all coefficients is affected by multicollinearity. As can be inferred from the p-values of the t-tests, the effect of multicollinearity is that none of the coefficients turns out to be significantly different from zero.

Since the values of several variables are calculated from the values of other variables, high collinearity among these variables is to be expected. One of the methods available for reducing multicollinearity in polynomial models is to reexpress each variable as a deviation from its mean (i.e., as so-called *centered variables*). To obtain centered variables, the mean of the values for each variable is subtracted from the values of the variable. This transformation is performed in SAS Example D13.

In the SAS program shown in Fig. 4.47, the above two models are reanalyzed using the centered variables. The statistics from the fit of the larger polynomial model is shown in Fig. 4.48. Comparing with the fit of the same

```
data evap;
input wtloss time humid;
time=time-5.5;
humid=humid-.3;
humidsq=humid*humid;
timehumid=time*humid;
timehumidsq=time*humidsq;
datalines;
4.3 4 .2
5.6 5 .2
6.4 6 .2
8.0 7 .2
4.0 4 .3
5.2 5 .3
6.6 6 .3
7.5 7 .3
3.2 4 .4
4.0 5 .4
5.6 6 .4
6.4 7 .4
;
run;

proc reg;
  model wtloss= time humid humidsq timehumid timehumidsq/vif;
  model wtloss= time humid humidsq;
  title 'Regression Models with and without Interaction Terms';
run;
```

Fig. 4.47. SAS Example D13: Program

Residual Analysis for Weight Loss

Analysis of Variance

Source	DF	Sum of Squares	Mean Square	F Value	Pr > F
Model	5	24.08467	4.81693	102.49	<.0001
Error	6	0.28200	0.04700		
Corrected Total	11	24.36667			

Parameter Estimates

Variable	DF	Parameter Estimate	Standard Error	t Value	Pr > |t|	Variance Inflation
Intercept	1	5.82500	0.10840	53.74	<.0001	0
time	1	1.19000	0.09695	12.27	<.0001	3.00000
humid	1	-6.37500	0.76649	-8.32	0.0002	1.00000
humidsq	1	-38.75000	13.27592	-2.92	0.0267	1.00000
timehumid	1	-0.35000	0.68557	-0.51	0.6279	1.00000
timehumidsq	1	-3.50000	11.87434	-0.29	0.7781	3.00000

Fig. 4.48. SAS Example D13: Output (Edited Page 1)

model to the original data, the fit to the centered data shows much improvement. Ideal VIF values show that the multicollinearity effects have completely vanished. The *p*-values for time, humid and humidsq are all smaller than .05, showing that the corresponding coefficients are all different from zero at the .05 level. The individual *t*-tests for timehumid and timehumidsq are not significant, showing that the interaction of time with humidity and the square of humidity are nonexistent. This can also be verified by a test of the hy-

```
              Regression Models with and without Interaction Terms        2
                             Analysis of Variance

                                Sum of          Mean
   Source             DF       Squares         Square     F Value    Pr > F

   Model               3      24.06833        8.02278      215.14    <.0001
   Error               8       0.29833        0.03729
   Corrected Total    11      24.36667

                             Parameter Estimates

                           Parameter      Standard
   Variable      DF         Estimate         Error     t Value    Pr > |t|

   Intercept      1          5.82500       0.09656       60.33     <.0001
   time           1          1.16667       0.04986       23.40     <.0001
   humid          1         -6.37500       0.68275       -9.34     <.0001
   humidsq        1        -38.75000      11.82555       -3.28     0.0112
```

Fig. 4.49. SAS Example D13: Output (edited page 2)

pothesis $H_0 : \beta_4 = \beta_5 = 0$ against H_a : at least one of these is not zero, using the model comparison technique. The *F*-statistic needed can be constructed using the results of fitting the two models in Fig. 4.48 and Fig. 4.49:

$$F = \frac{[\text{SSE}(\text{Model1}) - \text{SSE}(\text{Model2})]/(5-3)}{\text{SSE}(\text{Model2})/(12-3-1)} = \frac{(0.29833 - 0.28200)/2}{0.28200/8} = 0.23$$

which is obviously not significant and thus H_0 is not rejected. It is to be noted that centering actually made the two variables humid and time exactly orthogonal. This is the reason for the VIFs to be small. Even when exact orthogonality does not occur, centering does reduce the collinearity among the explanatory variables.

4.6 Exercises

4.1 The following data are from an experiment that tested the performance of an industrial engine (Schlotzhauer and Littell, 1997). The experiment used a mixture of diesel fuel and gas from distilling organic materials. The horsepower (y) of the machine was measured at several speeds (x) measured in hundreds of revolutions per minute (rpm × 100).

x	22	20	18	16	14	12	15	17	19	21	22	20
y	64.03	62.47	54.94	48.84	43.73	37.48	46.85	51.17	58.00	63.21	64.03	59.63
x	18	16	14	12	10.5	13	15	17	19	21	23	24
y	52.90	48.84	42.74	36.63	32.05	39.68	45.79	51.17	56.65	62.61	65.31	63.89

Write and execute a SAS program to perform the computations to provide answers to the following questions. Extract material from the output and copy or attach them as the required answers.

a. Use the method of least squares to obtain estimates $\hat{\beta}_0$ and $\hat{\beta}_1$ of the parameters in the model $y = \beta_0 + \beta_1 x + \epsilon$.

b. Construct a plot that shows the scatter of (x, y) data points with y on the vertical axis.

c. Give the least squares prediction equation. Superimpose the least squares line on the scatter plot in part (b).

d. Compute a table of predicted values $\hat{y}$ and residuals $y - \hat{y}$ corresponding to the observed values y.

e. Identify the sums of squares $\sum(y-\bar{y})^2$, $\sum(\hat{y}-\bar{y})^2$, and $\sum(y-\hat{y})^2$ from your output. Verify that these give a decomposition of total variability in y into two parts and identify the parts by sources of variation.

f. Calculate the proportion of the total variability in the y-values that is accounted for by the linear regression model. Explain why this value is a measure of how well your model fits the data.

g. Give the point estimate s^2 of σ^2.

h. State the estimated standard errors of $\hat{\beta}_0$ and $\hat{\beta}_1$.

i. Construct 95% confidence intervals for β_0 and β_1.

j. Test the hypothesis $H_0 : \beta_1 = 0$ versus $H_a : \beta_1 \neq 0$ using a t-test. Give your conclusion using $\alpha = .05$ and the p-value.

k. Obtain plots of the residuals against x and the predicted values, respectively. Do these two plots suggest any inadequacies of this model? Explain why you reached your conclusion.

l. Obtain a normal probability plot of the *studentized residuals*. State the model assumption that you can verify using this plot. Is this a plausible assumption for this model?

4.2 The text McClave et al. (2000) discusses the following problem that relates the value of a home to its upkeep expenditure. Quality Home Improvement Center (QHIC) operates five stores in a large metropolitan area. The marketing department at QHIC wishes to study the relationship between x, home value (in thousands of dollars), and y, yearly expenditure on home upkeep (in dollars). A random sample of 40 homeowners is taken and asked to estimate their expenditures during the previous year on the types of home upkeep products and services offered by QHIC. Public records of the county auditor are used to obtain the previous year's assessed values of the homeowner's homes. The resulting x- and y-values are given in the following table. Use a SAS program to fit a simple linear regression model to this data. You must use SAS statements to produce output necessary to answer questions given below.

Home	Value of Home, x ($1000s)	Upkeep Expenditure, y (Dollars)	Home	Value of Home, x ($1000s)	Upkeep Expenditure, y (Dollars)
1	237.00	1,412.08	21	153.04	849.14
2	153.08	797.20	22	232.18	1,313.84
3	184.86	872.48	23	125.44	602.06
4	222.06	1,003.42	24	169.82	642.14
5	160.68	852.90	25	177.28	1,038.80
6	99.68	288.48	26	162.82	697.00
7	229.04	1,288.46	27	120.44	324.34
8	101.78	423.08	28	191.10	965.10
9	257.86	1,351.74	29	158.78	920.14
10	96.28	378.04	30	178.50	950.90
11	171.00	918.08	31	272.20	1,670.32
12	231.02	1,627.24	32	48.90	125.40
13	228.32	1,204.78	33	104.56	479.78
14	205.90	857.04	34	286.18	2,010.64
15	185.72	775.00	35	83.72	368.36
16	168.78	869.26	36	86.20	425.60
17	247.06	1,396.00	37	133.58	626.90
18	155.54	711.50	38	212.86	1,316.94
19	224.20	1,475.18	39	122.02	390.16
20	202.04	1,413.32	40	198.02	1,090.84

a. Obtain a scatter plot of the dependent variable y against the values of the independent variable x. Does this plot suggest that simple linear regression model might relate y to x?

b. Use SAS to fit the model

$$y = \beta_0 + \beta_1 x + \epsilon$$

to these data. From the output, report the following information on a separate sheet. You must extract this information from the output and write them in the form discussed in the text.

c. Give the least squares estimates of β_0 and β_1 and their standard errors, respectively. What is your prediction equation?

d. Use your prediction equation to estimate the expected increase in yearly upkeep expenditure for an additional $10,000 increase in home value. Show your work clearly.

e. What is the coefficient of determination for your regression equation? In you own words, explain what this means to you in terms of variability in yearly upkeep expenditure.

f. Construct a 95% confidence interval for β_1. State in words what this interval says about the expected increase in yearly upkeep expenditure.

g. Test the hypothesis $H_0: \beta_1 = 0$ against $H_a: \beta_1 \neq 0$ at $\alpha = .05$ using the p-value printed. State your decision.
h. Find the point estimate of the mean yearly upkeep expenditure of all homes worth \$220,000. Given that $SS_{xx} = 124,951.80$ for the QHIC data, calculate a 95% confidence interval for the mean yearly upkeep expenditure of all homes worth \$220,000.
i. Obtain a graph with plots of the 95% confidence interval and 95% prediction interval curves for the fitted regression line overlaid on a scatter plot of the original data.
j. Obtain a scatter plot of residuals against the x variable. Does the assumption of a constant error variance appear to be satisfied? Explain.
k. Obtain a normal probability plot of residuals and a plot of residuals against the predicted values variable. Do these plots indicate that any of the model assumptions are not plausible? In particular, is the assumption of normal errors reasonable? Explain.

4.3 To model the relationship between the dose level of a drug product and its potency, a pharmaceutical firm inoculated each of 15 test tubes were with a virus culture and incubated for 5 days at 30°C. Three test tubes were randomly assigned to each of the five dose levels to be investigated (2, 4, 8, 16, 32 mg). Each tube was injected with one dose level and a measure of the antiviral strength of the product was obtained. The experiment was described in Ott and Longnecker (2001) and the data are reproduced here:

x, Dose Level (mg)	y, Response
2	5, 7, 3
4	10, 12, 14
8	18, 15, 18
16	20, 21, 19
32	23, 24, 28

Write and execute a SAS program to perform the computations to provide answers to the following questions. You may use as many steps in your program as necessary.
a. Plot the data in a scatter plot. Does it appear that a simple linear regression model would be a good fit?
b. Compute an analysis of variance table for regression.
c. Compute a lack of fit test for this model and report the results in an anova table where SS_{Lack} and SSE_{Pure} are shown as a partition of SSE. What is your conclusion from this test? Use $\alpha = .05$.
d. Compute the predicted values and residuals. Plot the residuals against the dose level and the predicted values, respectively. Do these two plots suggest any inadequacies of this model? Explain why you reached your conclusion.

e. Many times a logarithmic transformation can be used on the dose levels to *linearize the response* with respect to the dose level. Plot the response against the logarithms (to the base 10) of the five dose levels and comment on this plot.

4.4 Experience with a certain type of plastic indicates that a relation exist between the hardness (measured in Brinell units) of items molded from the plastic (y) and the elapsed time since termination of the molding process (x). In a study to examine this relationship, 16 batches of the plastic were made, and from each batch, 1 test item was molded (Kutner et al., 2004; data slightly modified for ease of hand calculation without affecting results of the analysis). Each test item was randomly assigned to one of four predetermined time levels, and the hardness was measured after the assigned time had elapsed. The result are shown as follows:

x, Elapsed Time (hours)	y, Hardness
16	199, 205, 196, 200
24	218, 220, 215, 223
32	237, 234, 235, 230
40	250, 248, 253, 245

Answer the following questions. You must execute an appropriate SAS program and extract portions of the output to provide the answers.
  a. Plot the data in a scatter plot. Does it appear that a simple linear regression model would be a good fit?
  b. Use the least squares method to fit a simple linear regression model. What is your prediction equation? Add the straight line of your prediction equation to the plot in part a).
  c. Construct an analysis of variance table for the regression. Use the F-ratio to perform a test of $H_0 : \beta_1 = 0$ versus $H_a : \beta_1 \neq 0$ using $\alpha = .05$
  d. Compute pure experimental error sum of squares SSP_{exp} by hand and verify it using an additional SAS proc step.
  e. Perform a lack of fit test for this model. Report the results in an Anova table where SS_{Lack} and SSP_{exp} are shown as a partition of SSE as demonstrated in the text. What is your conclusion from this test? Use $\alpha = .05$.
  f. Compute the predicted values and residuals. Plot the residuals against `elapsed time` and the predicted values, respectively. Do these two plots suggest any inadequacies of this model? Explain how you reached your conclusion.

4.5 In an experiment to study the problem of predicting the tensile strength (y) of concrete beams from measurements of their specific gravity (x_1) and moisture content (x_2), a multiple regression was fitted to data for 10 specimens (Devore, 1982). The data are given as follows:

Obs.	Tensile Strength y	Specific Gravity x_1	Moisture Content x_2
1	11.24	0.499	11.1
2	12.64	0.558	8.9
3	12.93	0.604	8.8
4	11.32	0.441	8.9
5	11.68	0.510	8.8
6	11.90	0.528	9.9
7	10.73	0.418	10.1
8	11.70	0.480	10.5
9	11.12	0.406	10.5
10	11.41	0.467	10.7

Use a SAS program to fit the model $y = \beta_0 + \beta_1 x_1 + \beta_2 x_2 + \epsilon$. Extract and/or calculate answers for the following from the SAS output:

a. Give the analysis of variance including the F-ratio and the associated p-value. State the null and alternative hypotheses about the coefficients you will test using the F-ratio. Use the p-value to perform this test using $\alpha = .05$

b. Report the prediction equation and standard errors of $\hat{\beta}_1$ and $\hat{\beta}_2$. If the specific gravity is kept constant at 0.5, estimate the change in mean tensile strength of a concrete beam if its moisture content increases from 8.0 to 10.0 units. Does the mean tensile strength increase or decrease?

c. Using the estimate of β_1 and its standard error given in the SAS output, compute a 95% confidence interval for β_1. Test the hypothesis $H_0 : \beta_1 = 0$ versus $H_a : \beta_1 \neq 0$ using this interval, specifying the α level of the test.

d. The specifications stipulate that the mean tensile strength must increase by at least 1 unit for an increase in specific gravity of 0.1 if the moisture content is unchanged. Test this hypothesis using the above confidence interval.

e. Compute the predicted tensile strength $\hat{y}_{11}$ of a beam with $x_1 = 0.5$ and $x_2 = 9.0$. Extract the 95% confidence interval for $E(y_{11})$ from the SAS output. What does this interval say about the tensile strength of beams with the given specific gravity and moisture content values?

4.6 Ott and Longnecker (2001) discussed an experiment for comparing responses of rats to different doses of two drugs. Sixty rats of a particular strain were randomly allocated into two equal groups. The first group of rats received drug A , with 10 rats randomly assigned to each of 3 doses

(5, 10, and 20 mg). Similarly, the 30 rats in the second group was given drug B, with 10 rats randomly assigned to each of the 3 (5, 10, and 20 mg) doses. After each rat received its prescribed dose, an anxiety score was assigned to each on a 0 to 30 scale after a 30-minute observation period. The data are given in the following table.

Drug A			Drug B		
Drug Dose (mg)			Drug Dose (mg)		
5	10	20	5	10	20
15	18	20	16	19	24
16	17	19	17	21	25
18	18	21	18	22	23
13	19	18	17	23	25
19	20	19	15	20	25
16	16	17	15	18	23
15	15	18	15	20	24
16	19	21	18	21	22
17	18	20	17	22	26
15	16	17	16	19	24

The anxiety score for each drug is to be modeled as straight line function of the dosage level. Use a SAS program to fit the regression model $y = \beta_0 + \beta_1 x_1 + \beta_2 x_2 + \beta_3 x_1 x_2 + \epsilon$ to these data. Perform a test of the equality of slopes using the model comparison technique (use $\alpha = .05$). The t-statistic for interaction can also be used to perform the same test. Extract the value of the corresponding t-statistic and its p-value from the SAS output. What is your conclusion about the slopes of the two straight lines? Use the result to interpret the mean response to the two drugs at different levels of dosage.

4.7 The data in the following table (Kutner et al., 2005) are from a study relating the amount of body fat (y) to several possible predictor variables: triceps skinfold thickness (x_1), thigh circumference (x_2), and midarm circumference (x_3), measured on a random sample of healthy females 25-34 years old. The amount of body fat is obtained using a cumbersome and expensive procedure involving the immersion of a person in water, so it would be helpful if a reliable prediction equation based on easy-to-measure explanatory variables is available to estimate body fat. Use a SAS program to fit the full model

$$y = \beta_0 + \beta_1 x_1 + \beta_2 x_2 + \beta_3 x_3 + \epsilon$$

Write answers to the following questions extracting numbers from a SAS output. Hi-lite or circle values you use from the output and label them carefully.

Triceps Skinfold Thickness	Thigh Circumference	Midarm Circumference	Body Fat
19.5	43.1	29.1	11.9
24.7	49.8	28.2	22.8
30.7	51.9	37.0	18.7
29.8	54.3	31.1	20.1
19.1	42.2	30.9	12.9
25.6	53.9	23.7	21.7
31.4	58.5	27.6	27.1
27.9	52.1	30.6	25.4
22.1	49.9	23.2	21.3
25.5	53.5	24.8	19.3
31.1	56.6	30.0	25.4
30.4	56.7	28.3	27.2
18.7	46.5	23.0	11.7
19.7	44.2	28.6	17.8
14.6	42.7	21.3	12.8
29.5	54.4	30.1	23.9
27.7	55.3	25.7	22.6
30.2	58.6	24.6	25.4
22.7	48.2	27.1	14.8
25.2	51.0	27.5	21.1

a. Report $\hat{\beta}_0$, $\hat{\beta}_1$, $\hat{\beta}_2$, and $\hat{\beta}_3$. Explain what the estimate $\hat{\beta}_3$ tells you about the mean body fat.

b. Report s_ϵ, $s_{\hat{\beta}_0}$, $s_{\hat{\beta}_1}$, $s_{\hat{\beta}_2}$, and $s_{\hat{\beta}_3}$. Calculate by hand the t-statistic for testing $H_0 : \beta_3 = 0$ versus $H_a : \beta_3 \neq 0$, using $\hat{\beta}_3$ and its standard error. Compare it with the value in the output.

c. Construct an analysis of variance for the above regression. Report the coefficient of determination and interpret its value in the context of this problem.

d. Use the F-test statistic for testing $H_0 : \beta_1 = \beta_2 = \beta_3 = 0$ versus H_a : at least one β is not zero, and report the p-value for the test. State your decision using the p-value.

e. Use the t-test statistic for testing $H_0 : \beta_2 = 0$ versus $H_a : \beta_2 \neq 0$ and report the p-value for the test. State your decision based on the p-value. What does this say about the role of the variable thigh circumference in your model?

f. Construct a 95% confidence interval for β_2. Use this interval to test $H_0 : \beta_2 = 0$ versus $H_a : \beta_2 \neq 0$. What is the α level of this test.

g. Construct a 95% confidence interval for $E(y)$ at $x_1 = 20$, $x_2 = 50$, and $x_3 = 30$. Describe in words what this interval tells you about the population of individuals with these values for triceps skinfold thickness, thigh circumference, and midarm circumference.

h. Construct a 95% prediction interval for body fat of a new individual on whom the values $x_1 = 20$, $x_2 = 50$, and $x_3 = 30$ have been measured. Describe in words what this interval tells you.
i. Obtain plots of the residuals versus predicted values, x_1, x_2, and x_3, respectively. Does any pattern of the types discussed in class observed in the plots? Give your interpretation of each plot.
j. Obtain a normal probability plot of the *studentized residuals*. State the model assumption that you can verify using this plot. Is this a plausible assumption for this model?
k. Add a model statement to your SAS program to fit the model

$$y = \beta_0 + \beta_1 x_1 + \beta_3 x_3 + \epsilon$$

to the above data. Use the results of the t-tests and R^2 value to state why this model may be preferred compared to the first model.

4.8 A laundry detergent manufacturer wished to test a new product prior to market release. One area of concern was the relationship between the height of the detergent suds in a washing machine to the amount of detergent added and the degree of agitation in the wash cycle. For a standard-size washing machine tub filled to the full level, different agitation levels (measured in minutes) and amounts of detergent were assigned in random order, and sud heights measured. This problem appears in Ott and Longnecker (2001) and the data are reproduced in the table presented. Write and execute a SAS program to perform the computations needed to provide answers to the following questions. Add SAS statements and/or steps necessary to obtain all answers required. Hi-lite or circle values you use from the output and label them carefully.

a. By examining only the plot of height, y, against agitation, x_1, at each value of amount, x_2, suggest a multiple regression model to explain the variation in sud height. Be specific about whether you think a higher-order term x_1^2 or an interaction term $x_1 x_2$ is needed or not and give reasons for your choices.
b. The first order model

$$y = \beta_0 + \beta_1 x_1 + \beta_2 x_2 + \epsilon$$

was fitted to the data. Using the SAS output, construct an analysis of variance table and test the hypothesis of $H_0 : \beta_1 = \beta_2 = 0$ versus H_a : at least one of β_1 or $\beta_2 \neq 0$. Give the R^2 value. Is there a reason to look beyond the the first-order model because R^2 is extremely high? Explain.

Height (y)	Agitation (x_1)	Amount (x_2)
28.1	1	6
32.3	1	7
34.8	1	8
38.2	1	9
43.5	1	10
60.3	2	6
63.7	2	7
65.4	2	8
69.2	2	9
72.9	2	10
88.2	3	6
89.3	3	7
94.1	3	8
95.7	3	9
100.6	3	10

c. Use the residual plots to examine the plausibility of the model assumptions. By examining the residual plots, can you find reasons to suspect that the first-order model is not adequate? If so, what other terms would you consider adding to the model? Explain how you reached your conclusion.

d. Based on the above residual analysis, the model

$$y = \beta_0 + \beta_1 x_1 + \beta_2 x_2 + \beta_3 x_1^2 + \beta_4 x_2 x_1^2 + \epsilon$$

was fitted to the data. Using the sums of squares from the SAS output, construct an F-statistic to test the hypothesis $H_0 : \beta_3 = \beta_4 = 0$. Use a percentile from the F-table to test this hypothesis at $\alpha = .05$. What does the result of this test tell you?

e. Use residual plots from fitting the model given in part (d) to verify the model assumptions. What improvements do these plots show compared to the first-order model, if any?

f. Use a t-statistic and the corresponding p-value from the SAS output of the fit of the model in part (d) to test $H_0 : \beta_4 = 0$. What does the result of this test suggest?

Problems 4.9-4.14 concern the following example presented in Bowerman and O'Connell (2004):

A multiple regression analysis conducted to determine labor needs of U.S. Navy hospitals is reproduced from Bowerman and O'Connell (2004). The data set consists of five independent variables: average daily patient load (LOAD), monthly X-ray exposures (XRAY), monthly occupied bed days (a hospital has one occupied bed day if one bed is occupied for an entire day) (BEDDAYS), eligible population in the area (in 1000s) (POP), average length of patients' stay (in days) (LENGTH), and a dependent variable, monthly labor hours required (HOURS).

H	LOAD	XRAY	BEDDAYS	POP	LENGTH	HOURS
1	15.57	2463	472.92	18.0	4.45	566.52
2	44.02	2048	1,339.75	9.5	6.92	696.82
3	20.42	3940	620.25	12.8	4.28	1,033.15
4	18.74	6505	568.33	36.7	3.90	1,603.62
5	49.20	5723	1,497.60	35.7	5.50	1,611.37
6	44.92	1,1520	1365.83	24.0	4.60	1,613.27
7	55.48	5779	1,687.00	43.3	5.62	1,854.17
8	59.28	5969	1,639.92	46.7	5.15	2,160.55
9	94.39	8461	2,872.33	78.7	6.18	2,305.58
10	128.02	2,0106	3655.08	180.5	6.15	3,503.93
11	96.00	1,3313	2912.00	60.9	5.88	3,571.89
12	131.42	1,0771	3921.00	103.7	4.88	3,741.40
13	127.21	1,5543	3865.67	126.8	5.50	4,026.52
14	252.90	3,6194	7684.10	157.7	7.00	10,343.81
15	409.20	3,4703	12,446.33	169.4	10.78	11,732.17
16	463.70	3,9204	14,098.40	331.4	7.05	15,414.94
17	510.22	8,6533	15,524.00	371.6	6.35	18,854.45

Write and execute SAS programs to perform the computations necessary to provide answers to the following questions. Add SAS statements and/or steps needed to obtain all answers required.

4.9 Consider relating y (HOURS) to x_1 (XRAY), x_2 (BEDDAYS), and x_3 (LENGTH) by using the model

$$y = \beta_0 + \beta_1 x_1 + \beta_2 x_2 + \beta_3 x_3 + \epsilon$$

Plot y versus each of x_1, x_2, and x_3. Do the plots indicate that the above model might appropriately relate y to x_1, x_2, and x_3? Explain your answer. You may compute other statistics to support your answer.

4.10 The main objective of this regression analysis is to help the Navy evaluate the performance of its hospitals in terms of how many labor hours are used relative to how many labor hours are needed. The Navy selected hospitals 1 through 17 from hospitals that it thought were efficiently run and wishes to use a regression model based on efficiently run hospitals to evaluate the efficiency of questionable hospitals. For hospital 14, note that $x_1 = 36,194$, $x_2 = 7684.10$, and $x_2 = 7.00$. Using the SAS output, discuss the case diagnostics for hospital 14. Discuss statistical evidence to show that this case might not fit the above model very well.

4.11 Since the Navy wishes to use a regression model based on efficiently run hospitals, it follows that hospital 14 be removed from the data set if it is concluded that hospital 14 was inefficiently run. Using the results from fitting model defined in Exercise 4.9 to the data set modified by removing hospital 14, answer the following:

  a. Do all of the residuals on the output appear to have come from the same distribution? Provide evidence for supporting or rejecting your claim.
  b. Use the leverage values for cases 14, 15, and 16 (which are the original hospitals 15, 16, and 17) to determine if these hospitals are outliers with respect to their x-values? How does being an x-outlier affect other diagnostics of each of these cases?
  c. Use the Cook's distance measure for case 14 (the original hospital 15) to explain why removing hospital 14 from the data set has made the original hospital 15 noninfluential?
  d. Which hospital had the largest Cook's D when all 17 hospitals were used to perform the regression analysis? Does this hospital appear to be less influential after removing hospital 14 from the data set? If so, explain why.

4.12 For two hospitals not used in the above analysis whose efficiency the Navy questions, the values of XRAY, BEDDAYS, and LENGTH are 56,194, 14,077.88, and 6.89 and 6021, 1651.42, and 5.41, respectively. Use SAS to obtain predictions of the number of monthly labor hours used for these hospitals. Use the model in Exercise 4.9 fitted to data excluding hospital 14. Use the observed number of labor hours for these hospitals, $y = 17{,}207.3$ and $y = 1823.4$, respectively, to comment on the efficiency of these two hospitals.

4.13 Consider relating y (HOURS) to x_1 (LOAD), x_2 (XRAY), x_3 (BEDDAYS), x_4(POP), and x_5(LENGTH) by fitting the model

$$y = \beta_0 + \beta_1 x_1 + \beta_2 x_2 + \beta_3 x_3 + \beta_4 x_4 + \beta_5 x_5 + \epsilon$$

using all 17 hospitals.

  a. Use the scatter plot matrix and correlation matrices to carry out a preliminary assessment of the relationship between y and each of x_1, x_2, x_3, x_4, and x_5. Based on your assessment, which independent variables do you judge might be most strongly involved in multicollinearity?
  b. Do any least squares estimates of the regression coefficients have a sign (positive or negative) that is different from what you would intuitively expect (another consequence of multicollinearity)? Which two variables have the largest variance inflation factors? What is conspicuous about these variables?
  c. Obtain the partial regression residual plots for each of the variables in the model. Use these to comment on the effects of multicollinearity on the estimation of each coefficient in the fitted model.
  d. If the independent variables x_1 and x_4 are removed from the five-variable model above and use regression to relate y to x_2, x_3, and x_5

alone, the model fitted is identical to that in Exercise 4.9. Is the fit of that model less affected by multicollinearity than the five-variable model? Explain. Does x_3 seem to have additional importance in the smaller model than the larger one? Justify your answers.

4.14 Since previous analyses indicated that hospital 14 might have been inefficiently run, use SAS procedures to obtain the "best" model resulting from using possible combinations of the five explanatory variables after hospital 14 is removed from the data set.

   a. Use a SAS procedure to do all possible regressions containing no less than two and no more than four independent variables. Print statistics for only the four best models in each case. Construct a plot of the C_p statistic for all models with "reasonable" C_p values. Select "good" models each with two, three, and four independent variables, respectively, for the purpose of predicting monthly labor hours, *indicating your reasons for selection of each model*. There may be several possible choices; that is, there may be many "good" models but give arguments for each of your choices. Primarily, use s^2, R^2, and C_p in your arguments. Select one of these models as your final model and provide arguments supporting your choice.

   b. Use the following subset selection procedures:
      A. backward elimination, with significance level of .05 for deleting variables
      B. stepwise, with significance levels of .20 for entry and .10 for deletion of variables

   State the model(s) selected in each case and report estimates of parameters and the analysis of variance table for these models. Compare models selected from each procedure with final model selected in part (a). Comment on whether changing the cutoff levels up or down by small amount would change the models selected by these procedures. (You do not need to re-run programs; examine the p-values from outputs from the SAS programs used for procedures A and B.)

5
Analysis of Variance Models

5.1 Introduction

In Chapter 4, multiple regression models discussed involved quantitative variables as explanatory variables. As discussed in Section 4.2, the least squares method was used to obtain the estimates of the parameters of the model. Using the matrix form of the multiple regression model

$$\mathbf{y} = X\boldsymbol{\beta} + \boldsymbol{\epsilon}$$

this was done by solving the normal equations

$$X'X\boldsymbol{\beta} = X'\mathbf{y},$$

the solution to which gave the the least squares estimate $\hat{\boldsymbol{\beta}}$ of $\boldsymbol{\beta}$:

$$\hat{\boldsymbol{\beta}} = (X'X)^{-1}X'\mathbf{y}.$$

The *analysis of variance models* introduced in this chapter, although conceptually different, can also be represented in this framework, where the X matrix now represents the *design matrix*. The design matrix is constructed from the linear model describing the responses observed from a specific experiment. Linear models describing various experiments discussed in this chapter will be called analysis of variance models. Analysis of variance models used for describing responses from several experimental situations are discussed in detail in separate sections in this chapter. Here, a linear model is used as an example demonstrating how a design matrix is constructed from it and for discussing properties of the resulting normal equations. Consider the linear model

$$y_{ij} = \mu + \alpha_i + \tau_j + \epsilon_{ij}, \quad i = 1, 2, 3; \ j = 1, 2, 3, 4$$

This model, for example, may be used to describe the yield of corn, y_{ij}, observed from twelve 1-acre plots where combinations of 3 different levels of nitrogen ($i = 1, 2, 3$) and 4 levels of irrigation ($j = 1, 2, 3, 4$) were applied. In this model, α_i and τ_j represent the *effects* of nitrogen level i and irrigation level j, respectively, expressed as deviations from an overall mean μ. When formulating this model for this situation, it is assumed that the above effects and the random component ϵ_{ij} representing *experimental error* are additive. In this chapter, the $\epsilon_{11}, \epsilon_{12}, \ldots, \epsilon_{34}$ are assumed to be a random sample from a normal distribution with mean 0 and variance σ^2. The matrix form of the model is

$$\mathbf{y} = X\boldsymbol{\beta} + \boldsymbol{\epsilon}$$

where

$$\mathbf{y} = \begin{bmatrix} y_{11} \\ y_{12} \\ y_{13} \\ y_{14} \\ y_{21} \\ y_{22} \\ y_{23} \\ y_{24} \\ y_{31} \\ y_{32} \\ y_{33} \\ y_{34} \end{bmatrix}, \quad X = \begin{bmatrix} 1 & 1 & 0 & 0 & 1 & 0 & 0 & 0 \\ 1 & 1 & 0 & 0 & 0 & 1 & 0 & 0 \\ 1 & 1 & 0 & 0 & 0 & 0 & 1 & 0 \\ 1 & 1 & 0 & 0 & 0 & 0 & 0 & 1 \\ 1 & 0 & 1 & 0 & 1 & 0 & 0 & 0 \\ 1 & 0 & 1 & 0 & 0 & 1 & 0 & 0 \\ 1 & 0 & 1 & 0 & 0 & 0 & 1 & 0 \\ 1 & 0 & 1 & 0 & 0 & 0 & 0 & 1 \\ 1 & 0 & 0 & 1 & 1 & 0 & 0 & 0 \\ 1 & 0 & 0 & 1 & 0 & 1 & 0 & 0 \\ 1 & 0 & 0 & 1 & 0 & 0 & 1 & 0 \\ 1 & 0 & 0 & 1 & 0 & 0 & 0 & 1 \end{bmatrix}, \quad \boldsymbol{\beta} = \begin{bmatrix} \mu \\ \alpha_1 \\ \alpha_2 \\ \alpha_3 \\ \tau_1 \\ \tau_2 \\ \tau_3 \\ \tau_4 \end{bmatrix}, \quad \text{and } \boldsymbol{\epsilon} = \begin{bmatrix} \epsilon_{11} \\ \epsilon_{12} \\ \epsilon_{13} \\ \epsilon_{14} \\ \epsilon_{21} \\ \epsilon_{22} \\ \epsilon_{23} \\ \epsilon_{24} \\ \epsilon_{31} \\ \epsilon_{32} \\ \epsilon_{33} \\ \epsilon_{34} \end{bmatrix}$$

For example, the first line of this model is $y_{11} = \mu + \alpha_1 + \tau_1 + \epsilon_{11}$, the second line is $y_{12} = \mu + \alpha_1 + \tau_2 + \epsilon_{12}$, and so on. There are certain obvious differences from the matrix form of the multiple regression model of Section 4.2. The design matrix X, for example, consists entirely of 0's and 1's, their positions determined by the subscripts i and j. The 12 rows of X correspond to the 12 observations $y_{11}, y_{12}, \ldots, y_{34}$ in that order. Note that the ordering is determined by letting the first subscript i remain fixed and the second subscript j take values 1 through 4 representing the four different irrigation methods. The first subscript i takes the values 1 through 3 in that order and represents the three nitrogen levels. The eight columns of X represent the eight parameters $\mu, \alpha_1, \alpha_2, \alpha_3, \tau_1, \tau_2, \tau_3, \tau_4$, respectively. Thus, in any row of X, the presence of a 1 or 0 in any column indicates that the corresponding parameter, as determined by the column position, appears or not in the model for the observation represented by that row. For example, in row 7 of the above design matrix there is a 1 in columns 1, 3, and 7 and the rest of the columns are 0, indicating the presence of the parameters μ, α_2, and τ_3 in the model; thus, the model for y_{23}, the observation represented by row 7 of X, is $y_{23} = \mu + \alpha_2 + \tau_3 + \epsilon_{23}$.

The special structure of X results in a situation not usually encountered when attempting to solve the normal equations $X'X\boldsymbol{\beta} = X'\mathbf{y}$ to obtain the least squares estimates of the parameters in the multiple regression model.

5.1 Introduction

In that case, it is assumed that the matrix $X'X$ is nonsingular; that is, the inverse of the matrix can be calculated and therefore a unique solution to the normal equations exists that is given by $(X'X)^{-1}X'\mathbf{y}$. In analysis of variance models, the rank of the $X'X$ matrix is less than p, the number of parameters, and therefore there is an infinite number of solutions to the normal equations. Hence, these models are also called *less than full rank* models. The outcome of this is that the parameters μ, α_1, α_2, α_3, τ_1, τ_2, τ_3, and τ_4 cannot be uniquely estimated. In general, a nonunique solution to such a system of linear equations may be found by setting some of the unknown parameters to a constant (such as zero) and obtaining a solution for the rest of the parameters. This solution will be unique up to the constants chosen. Thus, when computer software produce *estimates* of parameters in a analysis of variance model, the numbers output are not unique and depend on the procedure adopted by the software to obtain a solution to the normal equations.

As an example, it can be easily shown that the $X'X$ matrix for the linear model given earlier is

$$X'X = \begin{bmatrix} 12 & 4 & 4 & 4 & 3 & 3 & 3 & 3 \\ 4 & 4 & 0 & 0 & 1 & 1 & 1 & 1 \\ 4 & 0 & 4 & 0 & 1 & 1 & 1 & 1 \\ 4 & 0 & 0 & 4 & 1 & 1 & 1 & 1 \\ 3 & 1 & 1 & 1 & 3 & 0 & 0 & 0 \\ 3 & 1 & 1 & 1 & 0 & 3 & 0 & 0 \\ 3 & 1 & 1 & 1 & 0 & 0 & 3 & 0 \\ 3 & 1 & 1 & 1 & 0 & 0 & 0 & 3 \end{bmatrix}$$

and that the normal equations required for obtaining the least squares estimates are given by

$$
\begin{aligned}
12\mu + 4\alpha_1 + 4\alpha_2 + 4\alpha_3 + 3\tau_1 + 3\tau_2 + 3\tau_3 + 3\tau_4 &= y_{..} \\
4\mu + 4\alpha_1 + \tau_1 + \tau_2 + \tau_3 + \tau_4 &= y_{1.} \\
4\mu + 4\alpha_2 + \tau_1 + \tau_2 + \tau_3 + \tau_4 &= y_{2.} \\
4\mu + 4\alpha_3 + \tau_1 + \tau_2 + \tau_3 + \tau_4 &= y_{3.} \\
3\mu + \alpha_1 + \alpha_2 + \alpha_3 + 3\tau_1 &= y_{.1} \\
3\mu + \alpha_1 + \alpha_2 + \alpha_3 + 3\tau_2 &= y_{.2} \\
3\mu + \alpha_1 + \alpha_2 + \alpha_3 + 3\tau_3 &= y_{.3} \\
3\mu + \alpha_1 + \alpha_2 + \alpha_3 + 3\tau_4 &= y_{.4}
\end{aligned}
$$

where $y_{..} = \sum_{i=1}^{3}\sum_{j=1}^{4} y_{ij}$, $y_{i.} = \sum_{j=1}^{4} y_{ij}$ for $i = 1, 2, 3$, and $y_{.j} = \sum_{i=1}^{3} y_{ij}$ for $j = 1, 2, 3, 4$. It can be shown that the rank of $X'X$ is 6, so only two of the parameters need to be set to constant to obtain a solution to the normal equations. Another way to obtain a solution to the normal equations in a *balanced design model* such as in the above example is to impose restrictions on parameters. For example, in the above example two such restrictions are needed. The so-called *sum-to-zero* restrictions are $\sum_{i=1}^{3} \alpha_i = 0$ and $\sum_{j=1}^{4} \tau_j = 0$ of this type. These can also be written as $\alpha_3 = -\alpha_1 - \alpha_2$ and $\tau_4 = -\tau_1 - \tau_2 - \tau_3$; thus, a solution for the other parameters can be obtained by eliminating α_3 and τ_4 from the equations. The solution to the normal equations under these restrictions is $\tilde{\mu} = \bar{y}_{..}$, $\tilde{\alpha}_i = \bar{y}_{i.} - \bar{y}_{..}$, for $i = 1, 2, 3$, and $\tilde{\tau}_j = \bar{y}_{.j} - \bar{y}_{..}$ for $j = 1, 2, 3, 4$. For the experiment described, most experimenters consider these quantities as the "estimates" of the respective parameters. Essentially, setting some parameters equal to a constant is also a restriction on the parameters. The method of computation used in `proc glm` in SAS produces estimates equivalent to those obtained by setting the last parameter for each effect equal to zero. Thus, in the above model, setting $\alpha_3 = 0$ and $\tau_4 = 0$ and solving the normal equations will result in the same estimates as those produced by SAS. These solutions are $\tilde{\mu} = \bar{y}_{..} + \bar{y}_{3.} + \bar{y}_{.4}$, $\tilde{\alpha}_i = \bar{y}_{i.} - \bar{y}_{3.}$ for $i = 1, 2$, and $\tilde{\tau}_j = \bar{y}_{.j} - \bar{y}_{.4}$ for $j = 1, 2, 3$.

Thus, it is evident that the solutions to the normal equations are not unique and, therefore, it is more useful to obtain estimates of "interesting" functions of the parameters that are unique. Linear functions of the parameters for which unique estimates exist are called *estimable functions*, and fortunately for the experimenter, some of the estimable functions of parameters of a given model are usefully interpreted. Estimates of these functions will be the same no matter which solution to the normal equations is used to compute them.

Analysis of variance models may be used to analyze data from

- Designed experiments
- Observational studies

The statistical analyses of these data based on analysis of variance models require some familiarity with the basic concepts associated with such studies. In the subsections that follow, a few of these ideas are briefly reviewed.

5.1.1 Treatment Structure

Treatments (or more generally, *factor levels*) are various settings of the conditions that are being compared in an experiment. Generally, treatments are applied to *experimental units* and a *response* (a value of the dependent variable) is measured from each experimental unit.

Example 1: Study effects of baking temperature on a commercial cake mixture
- Levels of temperature: 150°C, 170°C, 190°C, 210°C
- Response variable: Cross-section of cake
- Replications: The number of cakes baked at each temperature

Example 2: Same experiment as in Example 1 with an additional factor: a chemical additive
- Treatments: All combinations of 4 levels of temperature and 3 chemical additives forming 12 treatment combinations in a 4×3 *factorial treatment structure*

- Response Variable: Cross-section of cake
- Replications: It is needed to bake at least 24 cakes (i.e., 2 `Replications` per treatment combination in order to be able to estimate *experimental error variance*.

Example 3: Study the variation of number of traffic tickets issued in different precincts in a large U.S. city
- Levels of factor : Select 10 precincts at random.
- Response variable: Number of traffic tickets issued in a 6-month period
- Replications: From each precinct, select several police officers at *random* for each of whom a response is measured.
- Note: In this experiment, the interest is in measuring the variability of the number of traffic tickets issued from precinct to precinct within the city, rather than estimating the mean number of traffic tickets issued in a particular precinct.

Factors can be categorized into two basic types:

Fixed Factors
- Experimenter selects the levels of each of the factors to be included in the experiment.
- Interest is in the estimation and comparison of differences among these selected levels.

Random Factors
- Experimenter randomly samples the population of levels of each of the random factors.
- Interest here is in measuring the variability of the response over the population.

Example 1: Cake Baking Experiment
Both `Temperature` and `Additive` are fixed factors because the experimenter selected these levels to be studied in the experiment. The effects of the levels of Temperature and Additive will be compared in the analysis of the data resulting from this experiment.

Example 2: Study of Traffic Tickets
Both officers and precincts are random factors because the levels of these factors were selected from the available population of levels. The experimenter is not interested in the effects of a particular officer or a particular precinct but the variation of the number of traffic tickets issued.

5.1.2 Experimental Designs

Experimental designs describe how treatments (treatment combinations) are assigned to the experimental units for application to the experimental units and in the order in which observations are taken.

Example 1: Completely Randomized Design

Compare four fertilizers (A, B, C, D) on corn yield in a field experiment. Suppose 20 plots are available in a field as experimental units. If a *completely randomized design* (CRD) is used, the following design could be used to allocate the treatments to the plots:

A	B	C	D	D
B	D	A	A	C
C	A	B	D	C
B	A	B	C	D

Here, the four fertilizers are assigned at random to the 20 plots so that each fertilizer is applied to five plots, giving five replications of each treatment.

Example 2: Randomized Blocks Design

If, on the other hand, a *randomized complete block design* (RCBD) is used, the following allocation of the fertilizers to the plots may result:

		Blocks		
1	2	3	4	5
A	B	D	A	C
C	C	C	B	D
B	D	A	D	A
D	A	B	C	B

Here, the 20 plots are first grouped into 5 blocks (numbered from 1 to 5 in the diagram) of four plots each. The 4 fertilizers are assigned at random to the four plots in each block. Note that in the actual field layout, the plots within blocks may not be aligned across the blocks as shown above.

5.1.3 Linear Models

Data arising from designed experiments is represented by a linear model for the purpose of statistical analysis of such data. One advantage accrued from using such a model is that all effects to be estimated and hypotheses that need to be tested to answer the research questions related to the experiment may be formulated in terms of the *estimable functions* of the parameters of the model. In various linear models introduced in the chapter, parameters are estimated by the *least squares method*; that is, the solutions to the normal equations minimize the sum of squared deviations of the observations from their expected values. These estimates are equivalent to *maximum likelihood estimators* obtained by maximizing the likelihood function under the assumption that the observations are normally distributed. In either case, the estimators of the parameters (i.e., estimable functions of the parameters) are called "best" since they have the properties of being unbiased and having minimum variance.

Example 1: Cake Baking Experiment

In the cake baking experiment, for example, a model that may be used to describe the observations y_{ijk} is of the form

$$y_{ijk} = \underbrace{\mu + \alpha_i + \beta_j + \gamma_{ij}}_{\mu_{ij}} + \epsilon_{ijk}, \ i = 1, 2, 3, 4; \ j = 1, 2, 3; \ k = 1, 2.$$

where y_{ijk} is the area of cross-section of the kth replication of the ijth temperature-additive combination and ϵ_{ijk} is a random error assumed to be normally distributed with zero mean and constant variance σ^2. μ_{ij} = Expected Mean Response of Treatment Combination ij; that is, this model says that $E(y_{ijk}) = \mu_{ij}$ for each k.

From this formulation, μ_{ij}, $\alpha_i - \alpha_{i'}$, $\beta_j - \beta_{j'}$, σ^2, etc. may be estimated. For example, $\bar{y}_{ij\cdot}$ is the "best" estimate of μ_{ij}. Further, hypotheses about model parameters such as $H_0 : \gamma_{ij} = 0$ for all i, j vs. $H_a : \text{not} H_0$, or $H_0 : \alpha_i = \alpha_{i'}$ vs. $H_a : \alpha_i \neq \alpha_{i'}$, etc. may be tested.

Example 2: Study of Traffic Tickets. In the second example concerning the number of traffic tickets, the model may be represented by

$$y_{ij} = \mu + A_i + \epsilon_{ij}, \ i = 1, 2, \ldots, I; \ j = 1, 2, \ldots, n_i$$

where A_i is the effect of the ith precinct with $A_i \sim$ iid $N(0, \sigma_A^2)$, ϵ_{ij} is the effect of the jth officer in the ith precinct with $\epsilon_{ij} \sim$ iid $N(0, \sigma^2)$, and y_{ij} is the number of traffic tickets issued by the jth officer in the ith precinct.

Since the "I" precincts were chosen at random, "precinct" is considered to be a random factor and it is assumed that the effects of the precincts, A_i, are independently distributed as $N(0, \sigma_A^2)$ random variables. The officers were selected randomly within each precinct; hence "officer within precinct" denoted by officer/precinct or officer(precinct) is independently distributed as a random factor and it is assumed that ϵ_{ij} are $N(0, \sigma^2)$ random variables. The above is a one-way random model and the factor B (officer) is nested within factor A (precinct); that is, levels of B are nested within levels of A.

Using this model the *variance components* σ_A^2 and σ^2 may be estimated and hypothesis such as $H_0: \sigma_A^2 = 0$ vs. $H_a: \sigma_A^2 > 0$ about them may be tested.

5.2 One-Way Classification

Data generated from a study of several levels of a single factor in a completely randomized design are said to be in a one-way classification. In this situation, the levels of the factor are also sometimes called "treatments."

Model

A linear model appropriate for the response y_{ij} observed from the jth replication of the ith treatment is given by

$$y_{ij} = \mu + \alpha_i + \epsilon_{ij} \quad i = 1, 2, \ldots, t, \quad j = 1, \ldots, n_i$$

where α_i is the effect of the ith treatment expressed as the deviation of the *treatment mean* μ_i from an overall mean μ (i.e. $\alpha_i = \mu_i - \mu$) and ϵ_{ij} is the random experimental error associated with the ijth observation assumed to have normal distribution with zero mean and variance σ^2, usually expressed as $\epsilon_{ij} \sim \text{iid } N(0, \sigma^2)$. This model is equivalent to assuming that the observations for each treatment i, $y_{i1}, y_{i2}, \ldots, y_{in_i}$, is a random sample from the $N(\mu_i, \sigma^2)$ distribution where μ_i is the ith treatment mean. Thus, it is implied that μ_i is the mean of the population from which the sample corresponding to the ith treatment was drawn. Note that this also incorporates the assumption of *homogeneity of variance*; that is, populations corresponding to each treatment have a common variance σ^2. The above model may be reexpressed in terms of the treatment means as follows:

$$y_{ij} = \mu_i + \epsilon_{ij} \quad i = 1, 2, \ldots, t, \quad j = 1, \ldots, n_i$$

where $\mu_i = \mu + \alpha_i$. This model is is called the "means model" and the previous model the "effects model".

Estimation

The best estimates of μ_i and σ^2 are respectively

$$\hat{\mu}_i = \bar{y}_{i.} = (\sum_j y_{ij})/n_i, \quad i = 1, \ldots, t$$

$$\hat{\sigma}^2 = s^2 = \frac{\sum_i \sum_j (y_{ij} - \bar{y}_{i.})^2}{N - t}, \quad N = \sum_i n_i$$

and the best estimate of the difference between the effects of two treatments labeled p and q is

5.2 One-Way Classification

$$\widehat{\alpha_p - \alpha_q} = \bar{y}_{p\cdot} - \bar{y}_{q\cdot}, \quad p \neq q$$

with standard error given by

$$s_d = s\sqrt{\frac{1}{n_p} + \frac{1}{n_q}}$$

A $(1-\alpha)100\%$ confidence interval (C.I.) for $\alpha_p - \alpha_q$ (or $\mu_p - \mu_q$) is

$$(\bar{y}_{p\cdot} - \bar{y}_{q\cdot}) \pm t_{\alpha/2,\nu} \cdot s_d$$

where $t_{\alpha/2,\nu}$ = upper $\alpha/2$ percentage point of the t-distribution with ν df where $\nu = N - t$.

Testing Hypotheses

An analysis of variance (Anova) table corresponding to the above model is

SV	df	SS	MS	F	p-value
Trt	$t-1$	SS$_{\text{Trt}}$	MS$_{\text{Trt}}$	$F_c = $ MS$_{\text{Trt}}$/MSE	$\Pr(F > F_c)$
Error	$N-t$	SSE	MSE$(= s^2)$		
Total	$N-1$	SS$_{\text{Tot}}$			

The above F-statistic tests the hypothesis of equality of treatment means

$$H_0: \mu_1 = \mu_2 = \cdots, \mu_t \text{ versus } H_a: \text{ at least one inequality}$$

or, equivalently, the hypothesis of equality of treatment effects

$$H_0: \alpha_1 = \alpha_2 = \cdots = \alpha_t \text{ versus } H_a: \text{ at least one inequality}$$

To test the equality of means of two treatments labeled p and q (i.e., $H_0: \mu_p = \mu_q$ vs. $H_a: \mu_p \neq \mu_q$) or, equivalently, to test the equality of effects of two treatments labeled p and q (i.e., $H_0: \alpha_p = \alpha_q$ vs. $H_a: \alpha_p \neq \alpha_q$), use the following t-statistic:

$$t_c = \frac{|\bar{y}_{p\cdot} - \bar{y}_{q\cdot}|}{s_d}$$

The null hypothesis H_0 is rejected if $t_c > t_{\alpha/2,\nu}$, where $t_{\alpha/2,\nu}$ is the upper $\alpha/2$ percentile of the t-distribution with $\nu = N - t$ degrees of freedom.

Preplanned or a Priori Comparisons of Means

When the hypothesis $H_0: \mu_1 = \mu_2 = \cdots = \mu_t$ is rejected using the anova F-test, the inference is that at least one of the t population means differs from the rest. The next step is to identify the means that are different from each other. If the researcher had planned to compare treatment effects or means by suggesting specific questions about them based on the treatment structure, this would be a much simpler task. For example, "Is the average $(\mu_1 + \mu_2)/3$ different from $(\mu_3 + \mu_4)/3$?" would be a meaningful question if treatments 3 and 4 contained a component or an ingredient, say, that treatments 1 and 2 did not have.

Many times these questions may not result in a simple comparison of whether a difference like $\mu_2 - \mu_3$ is significant or not. It may be a question that requires a more complex comparison such as $\mu_1 - (\mu_2 + \mu_3)/2$ to be made. Not all questions can be formulated in the form of comparisons. To understand what kinds of question can be formulated as comparisons, define a **linear comparison** as a linear combination of the means $\mu_1, \mu_2, \ldots, \mu_t$ to be of the form:

$$\ell = a_1\mu_1 + a_2\mu_2 + \cdots + a_t\mu_t = \sum_{i=1}^{t} a_i\mu_i$$

for given numbers $a_1, a_2, \ldots, a_t$ with the restriction that the sum of these numbers is zero (i.e., $\sum_{i=1}^{t} a_i = 0$). Several examples of linear comparisons of means are provided to illustrate this definition.

Examples: Suppose the number of treatments is $t = 5$; that is, consider the population means $\mu_1, \mu_2, \mu_3, \mu_4,$ and μ_5.

- The linear combination $\ell = \mu_2 - \mu_3$ has a_i values

$$a_1 = 0,\ a_2 = 1,\ a_3 = -1,\ a_4 = 0,\ a_5 = 0$$

Note that $\sum a_i = 0$ as required; thus, ℓ is a linear comparison.

- The linear combination $\ell = (\mu_1 + \mu_2)/2 - (\mu_3 + \mu_4)/2$ has a_i values

$$a_1 = 1/2,\ a_2 = 1/2,\ a_3 = -1/2,\ a_4 = -1/2,\ a_5 = 0$$

Again, $\sum a_i = 0$; thus, ℓ is a linear comparison.

An estimate of a linear comparison is called a **linear contrast** of the means and is given by

$$\hat{\ell} = a_1\bar{y}_{1.} + a_2\bar{y}_{2.} + a_3\bar{y}_{3.} + \cdots + a_t\bar{y}_{t.}$$

where $\sum a_i = 0$. Recall that the sample treatment means $\bar{y}_{1.}, \bar{y}_{2.}, \ldots, \bar{y}_{t.}$ are the best estimates of the population means $\mu_1, \mu_2, \ldots, \mu_t$.

A linear comparison is a way to express a hypothesis among population means that is naturally suggested by the choice of levels of the factor under study. It is not surprising that an experimenter would like to make more than one such comparison for a given experiment. A comparison is *made* by the testing of the hypotheses,

$$H_0 : \ell = 0 \quad \text{versus} \quad H_a : \ell \neq 0$$

where ℓ is a linear combination of $\mu_1, \mu_2, \ldots, \mu_t$ at a specified level α.

To compute an F-statistic to test the above hypotheses, a "SS due to the contrast" needs to be computed. In the most general case, when the sample sizes for the treatments are different, say, $n_1, n_2, \ldots, n_t$, this SS is calculated using the formula

$$\text{SSC} = \frac{\hat{\ell}^2}{\sum_{i=1}^{t}(a_i^2/n_i)}$$

When the sample sizes are all equal to n, the above reduces to the form

$$\text{SSC} = \frac{n\hat{\ell}^2}{\sum_{i=1}^{t} a_i^2}$$

Example 5.2.1

The data shown in Table 5.1 are from an experiment in plant physiology described in Sokal and Rohlf (1995) and give the length (in coded units) of pea sections grown in tissue culture. The purpose of the experiment was to compare the effects of the addition of various sugars on growth as measured by this length. Four treatments representing three different sugars and one mixture of sugars, plus one control treatment with no sugar were used. Ten independent samples were obtained for each treatment in a completely randomized design. The model for the observations in terms of the means $\mu_1, \ldots, \mu_5$ of

		Sugars		
			1% glucose	
	2%	2%	+	2%
Control	glucose	fructose	1% fructose	sucrose
75	57	58	58	62
67	58	61	59	66
70	60	56	58	65
75	59	58	61	63
65	62	57	57	64
71	60	56	56	62
67	60	61	58	65
67	57	60	57	65
76	59	57	57	62
68	61	58	59	67

Table 5.1. Effect of sugars on growth of peas (Sokal and Rohlf, 1995)

populations that represent the five samples is

$$y_{ij} = \mu_i + \epsilon_{ij}, \quad i = 1, 2, \ldots, 5, \quad j = 1, \ldots, 10$$

and the analysis of variance (anova) table that is required for the analysis of this data is

SV	df	SS	MS	F	p-value
Sugars	4	1077.32	269.33	49.37	<0.0001
Error	45	245.50	5.46		
Total	49	1322.82			

Labeling the five treatments as Trt1, ..., Trt5, respectively, in the order they appear in the data table, the five treatment means are

Trt1	Trt2	Trt3	Trt4	Trt5
70.1	59.3	58.2	58.0	64.1

It would be more useful to test the preplanned (or a priori) comparisons given below rather than, say, test all pairwise differences among the five treatment means (or effects) to determine the means that are different. In this study, the experimenter could have planned to (i) compare the effect of the control with the average effect of the sugars, (ii) compare the effect of the mixed sugars with average effects of pure sugars, and (iii) compare the differences among the effects of the three pure sugars. The linear combinations

(i) $\mu_1 - \frac{1}{4}(\mu_2 + \mu_3 + \mu_4 + \mu_5)$
(ii) $\mu_4 - \frac{1}{3}(\mu_2 + \mu_3 + \mu_5)$
(iii) $\mu_2 - \mu_3$, $\mu_2 - \mu_5$, or $\mu_3 - \mu_5$

represent these comparisons, in terms of the five respective population means, $\mu_1, \ldots, \mu_5$. Consider testing

$$H_0: \mu_1 - \frac{1}{4}(\mu_2 + \mu_3 + \mu_4 + \mu_5) = 0 \quad \text{versus} \quad H_a: \mu_1 - \frac{1}{4}(\mu_2 + \mu_3 + \mu_4 + \mu_5) \neq 0$$

or, equivalently,

$$H_0: 4\mu_1 - \mu_2 - \mu_3 - \mu_4 - \mu_5 = 0 \quad \text{versus} \quad H_a: 4\mu_1 - \mu_2 - \mu_3 - \mu_4 - \mu_5 \neq 0$$

Here $\ell_1 = 4\mu_1 - \mu_2 - \mu_3 - \mu_4 - \mu_5$. The coefficients of the corresponding contrast are therefore

a_1	a_2	a_3	a_4	a_5
4	−1	−1	−1	−1

Thus,

$$\hat{\ell}_1 = 4\bar{y}_1 - \bar{y}_2 - \bar{y}_3 - \bar{y}_4 - \bar{y}_5$$
$$= 4 \times 70.1 - 59.3 - 58.2 - 58.0 - 64.1 = 40.8$$

where

$$\sum_{i=1}^{5} \frac{a_i^2}{n_i} = \frac{4^2}{10} + \frac{1^2}{10} + \frac{1^2}{10} + \frac{1^2}{10} + \frac{1^2}{10} = \frac{20}{10}$$

Thus,

$$\mathrm{SSC}_1 = \frac{\hat{\ell}_1^2}{\sum \frac{a_i^2}{n_i}} = \frac{(40.8)^2}{2} = 832.32$$

therefore
$$F_c = \frac{\mathrm{SSC}_1/1}{\mathrm{MSE}} = \frac{832.32}{5.456} = 152.56$$

Since $F_{.05,\,1,\,45} \approx 4.0$, we reject H_0 at $\alpha = .05$. Now, consider testing

$$H_0 : \mu_4 - \frac{1}{3}(\mu_2 + \mu_3 + \mu_5) = 0 \quad \text{versus} \quad H_a : \mu_4 - \frac{1}{3}(\mu_2 + \mu_3 + \mu_5) \neq 0$$

Since H_0 is equivalent to $\mu_2 + \mu_3 + \mu_5 - 3\mu_4 = 0$, the problem is equivalent to testing
$$H_0 : \ell_2 = 0 \quad \text{versus} \quad H_a : \ell_2 \neq 0$$

where $\ell_2 = \mu_2 + \mu_3 + \mu_5 - 3\mu_4$. Here, the contrast coefficients are

a_1	a_2	a_3	a_4	a_5
0	1	1	-3	1

giving $\hat{\ell}_2 = 59.3 + 58.2 + 64.1 - 3 \times 58.0 = 7.6$ and the divisor is

$$\frac{a_i^2}{n_i} = \frac{0^2}{10} + \frac{1^2}{10} + \frac{1^2}{10} + \frac{3^2}{10} + \frac{1^2}{10} = \frac{12}{10} = 1.2$$

Thus, similar to the above,
$$\mathrm{SSC}_2 = \frac{\hat{\ell}_2^2}{\sum \frac{a_i^2}{n_i}} = \frac{(7.6)^2}{1.2} = 48.1333$$

therefore
$$F_c = \frac{\mathrm{SSC}_2/1}{\mathrm{MSE}} = \frac{48.1333}{5.456} = 8.82$$

which leads us to reject H_0 at $\alpha = .05$, the same result as $F_{.05,\,1,\,45} \approx 4.0$. Computations of the F-tests for the other two comparisons follow in a similar fashion. The above F-tests performed for making comparisons of interest also may be carried out by equivalent t-tests. In the equal sample size case (i.e., $n_1 = n_2 = \cdots = n$), a test for a preplanned (or an a priori) comparison,

$$H_0 : \sum_i c_i \mu_i = 0 \quad \text{versus} \quad H_a : \sum_i c_i \mu_i \neq 0$$

is given by the t-statistic
$$t_c = \frac{|\sum_i c_i \bar{y}_{i\cdot}|}{s\sqrt{\sum c_i^2 / n}}$$

and we reject $H_0 :$ if $t_c > t_{\alpha/2, (N-t)}$ for a two-tailed test. For testing

$$H_0 : 4\mu_1 - \mu_2 - \mu_3 - \mu_4 - \mu_5 = 0 \quad \text{versus} \quad H_a : 4\mu_1 - \mu_2 - \mu_3 - \mu_4 - \mu_5 \neq 0$$

the t-statistic is
$$t_c = \frac{40.78}{\sqrt{5.456} \times \sqrt{2}} = 12.35$$
and H_0 is rejected as $t_{.025, 45} \approx 2.0$

Two contrasts $\hat{\ell}_1 = \sum_i a_i \bar{y}_{i\cdot}$ and $\hat{\ell}_2 = \sum_i b_i \bar{y}_{i\cdot}$ are said to be *orthogonal* whenever $\sum_i a_i b_i = 0$. This is defined only when $n_1 = n_2 = \cdots = n_t = n$. If all linear contrasts in a set
$$\hat{\ell}_1, \hat{\ell}_2, \ldots, \hat{\ell}_{t-1}$$
are pairwise orthogonal (i.e., every possible pair is orthogonal), then the set is said to be a *mutually orthogonal* set of linear contrasts. Given t means $\mu_1, \mu_2, \ldots, \mu_t$ and sample means $\bar{y}_{1\cdot}, \bar{y}_{2\cdot}, \ldots, \bar{y}_{t\cdot}$ (all based on the same number n of observations), it is the case that the *maximum number of mutually orthogonal contrasts that exist is* $(t-1)$. There are many $(t-1)$ sets of contrasts that are mutually orthogonal. In the previous example, the four comparisons made were mutually orthogonal.

Pairwise Comparisons of Means

One-at-a-time comparisons between pairs of mean μ_p and μ_q control *the per-comparison error rate*. These comparisons are carried out simply by

- doing a t-test of $H_0 : \mu_p = \mu_q$, or
- constructing a confidence interval for $\mu_p - \mu_q$, or
- equivalently, *when sample sizes are equal*, using the least significance difference (LSD) procedure. Note that the rejection region for the t-test for $H_0 : \mu_p - \mu_q = 0$ is equivalent to Reject H_0 if

$$|\bar{y}_{p\cdot} - \bar{y}_{q\cdot}| > \underbrace{t_{\alpha/2, (N-t)} \cdot s \cdot \sqrt{2/n}}_{\text{LSD}_\alpha}, \quad \text{where } n = \text{sample size}$$

The right member of this inequality is not a function of i or j. It is constant, and is called the *least significant difference* or LSD and is denoted here as LSD_α. In the LSD procedure, differences of pairs of the sample means are compared to the single computed value of LSD_α, to determine the pairs that are significantly different. To minimize the number of comparisons needed to be made, the $\bar{y}_{i\cdot}$'s are first arranged smallest to largest in value. Using the notation $\bar{y}_{(i)}$ for the ith smallest $\bar{y}$, the ordered means may be represented as

$$\bar{y}_{(1)} \leq \bar{y}_{(2)} \leq \bar{y}_{(3)} \leq \cdots \leq \bar{y}_{(t)}$$

Now, note that if the difference $\bar{y}_{(t)} - \bar{y}_{(1)}$, for example, does not exceed the LSD value, then all the differences $\bar{y}_{(t)} - \bar{y}_{(2)}, \bar{y}_{(t)} - \bar{y}_{(3)} \ldots, \bar{y}_{(t)} - \bar{y}_{(t-1)}$ will not exceed the LSD. It follows that computing all the above differences and comparing them to the LSD is avoided. The LSD procedure is based on this idea.

Often the findings are reported using the following scheme called the *underscoring procedure*:

- The ordered list of the computed values of the means are identified (in a separate line above them) by the treatment numbers (or the names identifying the corresponding treated populations) corresponding to the ordered means. For example, suppose the means are

$$\begin{array}{ccccc} \text{Trt5} & \text{Trt3} & \text{Trt1} & \text{Trt4} & \text{Trt2} \\ 9.5 & 10.5 & 11.6 & 12.2 & 13.5 \end{array}$$

- Connect the means by underscoring those pairs of means whose differences are less than the LSD_α in the following manner.
- Consider each mean in turn beginning from the smallest and moving right to the next largest mean and so on.
- On a separate line below the list starting from column 1, begin the underscore connecting means until a mean is found that is significantly different (i.e., difference is larger than the LSD_α) from the mean in column 1. Extend the line all the way and stop to the left to this mean.
- This line implies that those means that are connected with this line are not significantly different from the mean in column 1.
- Now, the procedure is restarted at column 2 and is repeated the same way as described above.
- For example, for an LSD_α value of 2.72, the underscoring procedure produces the following:

$$\begin{array}{ccccc} \text{Trt5} & \text{Trt3} & \text{Trt1} & \text{Trt4} & \text{Trt2} \\ 9.5 & 10.5 & 11.6 & 12.2 & 13.5 \end{array}$$

- Now, the populations whose identifying numbers or labels are joined by an underscore have means that are found to be not significantly different. To simplify the display, any line that is completely covered by (i.e., overlaps) another line and therefore is not needed can be deleted. Thus, the above reduces to

$$\begin{array}{ccccc} \text{Trt5} & \text{Trt3} & \text{Trt1} & \text{Trt4} & \text{Trt2} \\ 9.5 & 10.5 & 11.6 & 12.2 & 13.5 \end{array}$$

When an experimenter wants to make all possible pairwise comparisons, one of the multiple comparison procedures such as the Tukey procedure is recommended, because such procedures control the experimentwise error rate.

Multiple Comparisons of Pairs of Means

Following an anova F-test that rejects the hypothesis that the population means are equal to each other, one may conduct tests of equality of all pairs of the population means in order to ascertain which of these are actually different from each other. One-at-a-time comparisons (i.e., individual t-tests or confidence intervals for differences in pairs of means) may be used for this purpose. However, these tests are not adjusted for multiple inference; that is, the error rate controlled is the Type I error rate for each individual test. When pairwise comparisons of population means are made, *multiple comparison procedures* such as the Bonferroni method attempt to control the probability of making at least one Type I error. These procedures protect the experimenter against declaring too many pairs of means significantly different when they are actually not, when making all possible pairwise comparisons; that is, they ensure that the probability of making at least one Type I error is controlled and thus are more conservative than one-at-a-time comparisons. They are said to control the *experimentwise error rate*. Although SAS makes available several procedures through a variety of options, just three such procedures are discussed here.

- The Bonferroni method involves the use of a t-percentile corrected for the total number of pairwise comparisons. This procedure controls the probability of making at least one Type I error to be less than or equal to α.
- The Tukey procedure (also called Tukey-Kramer method when an adjustment for unequal sample sizes is incorporated) for all possible pairwise comparisons simultaneously, also called the HSD (honestly significant difference) procedure. This procedure controls the maximum experimentwise error rate.
- The Scheffé procedure is used for testing a set of comparisons (contrasts of the type $\sum c_i \mu_i$) simultaneously and controls the maximum experimentwise error rate for the set. The set of comparisons may include pairwise comparisons and any other contrast of interest, as well.

Instead of performing tests for all comparisons among t means, these procedures may also be carried out in a manner similar to the LSD procedure to minimize the number of comparisons. For example, LSD_α is replaced by

$$\text{HSD}_\alpha = q_{\alpha,t,\nu} \sqrt{\frac{s^2}{n}}$$

where $q_{\alpha,t,\nu}$ is the upper α percentage point of the Studentized range distribution with ν degrees of freedom for performing the Tukey procedure (for equal sample sizes). For the Bonferroni method, the t-percentile used in the LSD procedure $t_{\alpha/2,\nu}$ is replaced by $t_{\alpha/2m,\nu}$ where m is the total number of comparisons made. For all pairwise comparisons among t means, $m = t(t-1)/2$. For the Scheffé procedure, LSD_α is replaced by

$$S_\alpha = \sqrt{\widehat{\text{Var}}(\hat{\ell})} \sqrt{(t-1)F_{\alpha,df_1,df_2}}$$

where $df_1 = t-1$, $df_2 = \nu$, and $\widehat{\text{Var}}(\hat{\ell}) = s^2 \sum_i \frac{a_i^2}{n_i}$, and s^2 is the error mean square with ν degrees of freedom. The Tukey-Kramer method is more powerful than either the Bonferroni or the Scheffé methods when only pairwise comparisons are made.

Confidence intervals for all pairwise differences can also be adjusted for each of these methods by changing the percentile value used in their construction. For example, 95% Bonferroni-corrected confidence intervals for m pairwise differences are obtained by substituting $t_{0.025/m,\nu}$ in place of $t_{0.025,\nu}$.

In conclusion, a word of caution about drawing inferences from pairwise comparisons is warranted. The transitivity that one expects from logical relationships may not exist among the results of these hypothesis tests; that is, for example μ_A may be found be not significantly different from μ_B, and μ_B from μ_C, but it is possible that μ_A and μ_C are declared significantly different.

5.2.1 Using PROC ANOVA to analyze one-way classifications

Consider the data appearing in Table 5.2 (Box et al., 1978). These are the observed coagulation times (in seconds) of blood drawn from 24 animals randomly allocated to 4 different diets, labeled A, B, C, and D.

	Diet		
A	B	C	D
62	63	68	56
60	67	66	62
63	71	71	60
59	64	67	61
	65	68	63
	66	68	64
			63
			59

Table 5.2. Blood Coagulation Data (in seconds): Example E1

The one factor in this experiment is Diet with four levels, the levels being the four types of diet (may also be called four treatments). The experiment was conducted in a completely randomized design. SAS Example E1 illustrates the analysis of these data using proc anova.

SAS Example E1

The SAS Example E1 program shown in (see Fig. 5.1) is used to obtain the necessary analysis of the above data. The input statement with the trailing @@ is useful for inputting this type of data. This allows several observations to be continued on the same data line rather than using a new line for each observation. Notice that it is necessary to separate data values by at least one blank as in the case of list input. Although the sample sizes are unequal, proc anova may be used for the analysis of this data since it is a one-way classification. The class statement identifies the variables that appear in the model statement, which are classification variables. In this case, the variable diet is the classification variable with four classes (the four diets) and time is the dependent variable (y). Note that in specifying the model this way, a mean μ as well as an error term are implicitly assumed to be part of the model and thus are omitted from the statement.

The first page of the SAS output (see Fig. 5.2) is a result of the proc print statement and illustrates the appearance of the SAS data set produced by the data step. The class level information ■2 page resulting from the class statement and the analysis of variance (anova) table ■3 produced by proc anova as a result of the model statement follows (see Fig. 5.3). The experimenter may construct an analysis of variance table in the standard form given in statistics textbooks by extracting information from the above output. The analysis of variance (anova) table for the coagulation time data is

SV	df	SS	MS	F	p-value
Diet	3	228.0	76.0	13.57	.0001
Error	20	112.0	5.6		
Total	23	340.0			

Page 4 of the SAS output from proc anova results from the first means statement and contains the pairwise comparisons of means (see Fig. 5.4). The cldiff option on the means statement requests that the comparisons be given in the form of 95% confidence intervals ■4 on pairwise differences of means constructed using the t-percentage points (t or lsd option). For example, the 95%

```
data;
input diet time @@;
datalines;
1 62 1 60 1 63 1 59
2 63 2 67 2 71 2 64 2 65 2 66
3 68 3 66 3 71 3 67 3 68 3 68
4 56 4 62 4 60 4 61 4 63 4 64 4 63 4 59
;
run ;
proc print;
   title 'Blood Coagulation Data';
run;
proc anova;
   class diet ;
   model time=diet;
   means diet/t  cldiff;
   means diet/tukey alpha=.05;
   means diet/hovtest = bartlett; ■1
   means diet/hovtest ;
run;
```

Fig. 5.1. SAS Example E1: Program

Analysis of Blood Coagulation Data   1

Obs	diet	time
1	1	62
2	1	60
3	1	63
4	1	59
5	2	63
6	2	67
7	2	71
8	2	64
9	2	65
10	2	66
11	3	68
12	3	66
13	3	71
14	3	67
15	3	68
16	3	68
17	4	56
18	4	62
19	4	60
20	4	61
21	4	63
22	4	64
23	4	63
24	4	59

Fig. 5.2. SAS Example E1: Output (page 1)

confidence interval on $\mu_1 - \mu_2$ is $-5.0 \pm (2.086)(1.5275) = (-8.186, -1.814)$, where $t_{.025}(20) = 2.086$, $s\sqrt{1/n_1 + 1/n_2} = 1.5275$, and $s^2 = 5.6$.

Since the sample sizes n_i are unequal, `proc anova` would have produced confidence intervals being calculated instead of LSD pairwise comparisons, in any case. By specifying `alpha=p` as a `means` statement option, $(1-p)100\%$ confidence intervals may be obtained ($p = .05$ is the default when this option is omitted). The following is the set of 95% confidence intervals on the six pairwise differences of means extracted from the SAS output:

$$\mu_1 - \mu_2 : (-8.186, -1.814)$$
$$\mu_1 - \mu_3 : (-10.186, -3.814)$$
$$\mu_1 - \mu_4 : (-3.023, 3.023)$$
$$\mu_2 - \mu_3 : (-4.850, 0.850)$$
$$\mu_2 - \mu_4 : (2.334, 7.666)$$
$$\mu_3 - \mu_4 : (4.334, 9.666)$$

The intervals for $\mu_1 - \mu_4$ and $\mu_2 - \mu_3$ include zero, thus indicating that those pairs of means are not significantly different at an α level of .05. The main conclusion to be drawn is that mean coagulation times due to Diets B and C are similar but significantly larger than those due to Diets A and D, which are also similar. The LSD procedure may be replaced by one of several other

```
               Analysis of Blood Coagulation Data                    2

                        The ANOVA Procedure

                     Class Level Information  2

                Class          Levels      Values

                diet              4        1 2 3 4


           Number of Observations Read            24
           Number of Observations Used            24

               Analysis of Blood Coagulation Data                    3

                        The ANOVA Procedure

Dependent Variable: time               3

                              Sum of
Source                DF      Squares     Mean Square   F Value    Pr > F

Model                  3   228.0000000    76.0000000     13.57    <.0001

Error                 20   112.0000000     5.6000000

Corrected Total       23   340.0000000


          R-Square     Coeff Var      Root MSE     time Mean

          0.670588      3.697550      2.366432      64.00000


Source                DF     Anova SS     Mean Square   F Value    Pr > F

diet                   3   228.0000000    76.0000000     13.57    <.0001
```

Fig. 5.3. SAS Example E1: Output (pages 2 & 3)

more conservative procedures. bon, tukey, and scheffe are examples of options that may replace the t or lsd option for this purpose. The SAS output resulting from the statement means diet/tukey; is given on page 5 **4** (see Fig. 5.5). The following is the set of 95% confidence intervals on the six pairwise differences of means extracted from this output:

$$\mu_1 - \mu_2 : (-9.275, -0.725)$$
$$\mu_1 - \mu_3 : (-11.275, -2.725)$$
$$\mu_1 - \mu_4 : (-4.056, 4.056)$$
$$\mu_2 - \mu_3 : (-5.824, 1.824)$$
$$\mu_2 - \mu_4 : (1.423, 8.577)$$
$$\mu_3 - \mu_4 : (3.423, 10.577)$$

Notice that this option results in wider confidence intervals. This is because the Tukey procedure controls the *experimentwise error rate* 5 resulting in a more conservative procedure; that is, there is less of a chance of finding significant differences using this procedure. The confidence intervals based on the *t*-statistics (t or lsd option) controls the *per-comparison error rate* that guarantees only that the Type I error of each comparison will be below the specified alpha value. Procedures based on controlling experimentwise error rate are recommended for use when the fact that the experimenter will be making inferences using $t - 1$ comparisons among the means has to taken into account. Otherwise, the actual Type I error rate will be more than the specified significance level for an individual comparison. Which procedure is to be used depends on many factors: As a rule of thumb, it is recommended that one use Bonferroni or Tukey procedure when all pairwise comparisons are being made and use Scheffé procedure when in addition to all pairwise comparisons, other contrasts or comparisons among the means are also included in the inferences being made. Note that each of these three procedures

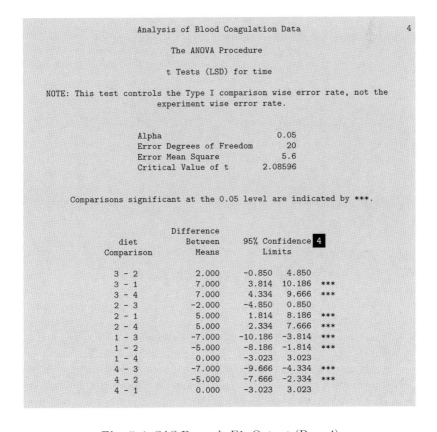

Fig. 5.4. SAS Example E1: Output (Page 4)

is progressively more conservative than the LSD procedure and, therefore, the corresponding confidence intervals will be progressively wider. Note, however, that in this example, the conclusions drawn from using the Tukey procedure are identical to those drawn using the LSD procedure.

```
                 Analysis of Blood Coagulation Data                    5

                          The ANOVA Procedure

              Tukey's Studentized Range (HSD) Test ■4■ for time

        NOTE: This test controls the Type I experiment wise error rate. ■5■


              Alpha                                   0.05
              Error Degrees of Freedom                  20
              Error Mean Square                        5.6
              Critical Value of Studentized Range  3.95829


         Comparisons significant at the 0.05 level are indicated by ***.


                         Difference       Simultaneous
              diet       Between          95% Confidence
           Comparison     Means              Limits

              3 - 2        2.000        -1.824    5.824
              3 - 1        7.000         2.725   11.275   ***
              3 - 4        7.000         3.423   10.577   ***
              2 - 3       -2.000        -5.824    1.824
              2 - 1        5.000         0.725    9.275   ***
              2 - 4        5.000         1.423    8.577   ***
              1 - 3       -7.000       -11.275   -2.725   ***
              1 - 2       -5.000        -9.275   -0.725   ***
              1 - 4        0.000        -4.056    4.056
              4 - 3       -7.000       -10.577   -3.423   ***
              4 - 2       -5.000        -8.577   -1.423   ***
              4 - 1        0.000        -4.056    4.056
```

Fig. 5.5. SAS Example E1: Output (page 5)

The option `hovtest=` ■1■ used in a `means` statements allows the user to specify that one of several tests for homogeneity of variance be calculated. The available selections are `bartlett, bf, levene,` and `obrien`. Although, traditionally, experimenters have used Bartlett's test for this purpose in practice, currently Levene's test is widely recognized to be the standard procedure for testing homogeneity of variance. Pages 6 and 8 (see Fig. 5.6) show the results of the Bartlett test and Levene test (produced by default), respectively, as a chi-square test and an F-test conducted in an anova table format. In this example, the p-values for these tests are large indicating that the null hypothesis of equal variances will not be rejected. One may examine this assumption visually by constructing a side-by-side box plot as illustrated in SAS Example C13 (see Fig. 3.35).

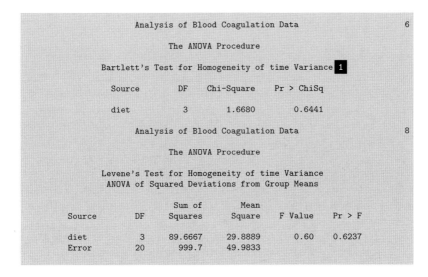

Fig. 5.6. SAS Example E1: Output (pages 6 and 8)

5.2.2 Making preplanned (or a priori) comparisons using PROC GLM

Consider the data (see Fig. 5.1) introduced in Section 5.2. These were used earlier in Section 5.2 to illustrate how to calculate F-statistics or t-statistics for testing hypotheses of the type

$$H_0 : \ell = 0 \quad \text{versus} \quad H_a : \ell \neq 0 .$$

where $\sum_{i=1}^{t} a_i \mu_i$ for given numbers $a_1, a_2, \ldots, a_t$, which satisfy $\sum_{i=1}^{t} a_i = 0$ at a specified level α.

The one factor in this experiment is labeled Sugar with five levels (treatments), four treatments representing three different sugars, one consisting of a mixture of sugars, and one a control treatment not containing any sugars. Random samples of size 10 were obtained for each treatment in a completely randomized design. SAS Example E2 illustrates the SAS program used for the analysis of these data.

SAS Example E2

The SAS Example E2 program (see Fig. 5.7) is used primarily to illustrate the use of contrast **1** and estimate **2** statements in proc glm to obtain F-test and t-tests, respectively, for making the comparisons suggested in Section 5.2. These comparisons were represented by the following linear combinations of the five respective population means, $\mu_1, \mu_2, \ldots, \mu_5$ as follows:

(i) $\mu_1 - \frac{1}{4}(\mu_2 + \mu_3 + \mu_4 + \mu_5)$

(ii) $\mu_4 - \frac{1}{3}(\mu_2 + \mu_3 + \mu_5)$
(iii) $\mu_2 - \mu_3$
(iv) $\mu_2 - \mu_5$

The input statements with trailing @ and a do loop were used for inputting the data in a straightforward way, with the *levels* for the classification variable sugar being identified by the numbers $1, 2, \ldots, 5$ in the data. The use of the trailing @ was described in Section 1.7.2 in Chapter 1. The data step first reads the level of sugar from a data line, holds the line, and then reads 10 numbers successively as values of the variable length, writing a pair of values for sugar and length each time through the loop as observations into the SAS data set named peas. The SAS output from proc print (not shown) may be examined to make sure that the data set has been created in the required format.

The analysis of variance (anova) tables produced by proc glm **3** as a result of the model statement appear on page 4 of the output (see Fig. 5.8). By default, proc glm computes two types of sums of squares (called Types I and III) **4** for each of the independent variables (i.e., variables appearing

```
data peas;
input sugar @;
do i=1 to 10;
 input length @;
 output;
end;
drop i;
datalines;
1 75 67 70 75 65 71 67 67 76 68
2 57 58 60 59 62 60 60 57 59 61
3 58 61 56 58 57 56 61 60 57 58
4 58 59 58 61 57 56 58 57 57 59
5 62 66 65 63 64 62 65 65 62 67
;
run;
proc print;
   title ' Effect of Sugars on the Growth of Peas';
run;
proc glm;
   class sugar ;
   model length =  sugar;
   means sugar/lsd alpha = .05;
   means sugar/tukey alpha = .05;

   contrast 'CONTROL VS. SUGARS' sugar 4 -1 -1 -1 -1; 1
   contrast 'SUGARS VS. MIXED  ' sugar 0  1  1 -3  1;
   contrast 'GLUCOSE=FRUCTOSE'   sugar 0  1 -1  0  0;
   contrast 'FRUCTOSE = SUCROSE' sugar 0  1  0  0 -1;

   estimate 'CONTROL VS. SUGARS' sugar 4 -1 -1 -1 -1; 2
   estimate 'SUGARS VS. MIXED  ' sugar 0  1  1 -3  1;
   estimate 'GLUCOSE=FRUCTOSE'   sugar 0  1 -1  0  0;
   estimate 'FRUCTOSE = SUCROSE' sugar 0  1  0  0 -1;
run;
```

Fig. 5.7. SAS Example E2: Program

```
                       Effect of Sugars on the Growth of Peas                    4

                                 The GLM Procedure

Dependent Variable: length
                                      Sum of
Source                        DF     Squares    Mean Square   F Value   Pr > F

Model                          4  1077.320000    269.330000     49.37   <.0001

Error                         45   245.500000      5.455556

Corrected Total               49  1322.820000


          R-Square      Coeff Var      Root MSE    length Mean

          0.814412       3.770928      2.335713       61.94000


Source                        DF     Type I SS    Mean Square   F Value   Pr > F

sugar                          4   1077.320000     269.330000     49.37   <.0001


Source                        DF   Type III SS    Mean Square   F Value   Pr > F

sugar                          4   1077.320000     269.330000     49.37   <.0001
```

Fig. 5.8. SAS Example E2: Output (page 4)

to the right of the equal sign) included in the model statement. In the case of one-way classification, these two sets of sums of squares are identical in magnitude, as is the case here. The experimenter may construct an analysis of variance table in the standard form by extracting information from the above output:

SV	df	SS	MS	F	p-value
Sugars	4	1077.32	269.33	49.37	<.0001
Error	45	245.50	5.46		
Total	49	1322.82			

Page 5 of the SAS output from `proc glm` (see Fig. 5.9) results from the first `means` statement and contains the pairwise comparisons of means using the LSD procedure. The means are arranged in decreasing order of magnitude down the page and the level of the corresponding treatment (sugar) in shown in the last column. The $LSD_{.05}$ is computed as 2.1039 and means that are not significantly different are grouped by the same letter in the first column of the output. This is comparable to the underscoring procedure described at the beginning of Section 5.2, which results in

Trt4	Trt3	Trt2	Trt5	Trt1
58.0	58.2	59.3	64.1	70.1

This can be interpreted to indicate that the mean lengths for treatment 1 (Control) and 5 (Sucrose) are significantly different from each other and from the other three treatments (Glucose, Fructose, and Mixed Sugars) but that there is no significant difference among those three. The output resulting from the second means statement `means diet/tukey;` is given on page 6 (see Fig. 5.10). In this case the differences are compared to $HSD_{.05}$ value calculated as 2.9681. This turns out to be larger than the $LSD_{.05}$ value as expected, but the outcome of the underscoring procedure is unchanged from that of the LSD procedure.

The four `contrast` statements result in the computation of F-statistics **1** for testing the four `single degree of freedom` comparisons of interest. The syntax of these statements is of the form

```
contrast 'label' effect_name contrast_coefficients
                                    < / options > ;
```

These results usually appear below the anova table in the SAS output but

```
                    Effect of Sugars on the Growth of Peas                    5

                              The GLM Procedure

                            t Tests (LSD) for length

    NOTE: This test controls the Type I comparison wise error rate, not the
                           experiment wise error rate.


                     Alpha                                 0.05
                     Error Degrees of Freedom                45
                     Error Mean Square                  5.455556
                     Critical Value of t                 2.01410
                     Least Significant Difference        2.1039


           Means with the same letter are not significantly different.


                  t Grouping         Mean       N    sugar

                            A       70.100     10      1

                            B       64.100     10      5

                            C       59.300     10      2
                            C
                            C       58.200     10      3
                            C
                            C       58.000     10      4
```

Fig. 5.9. SAS Example E2: Output from LSD procedure

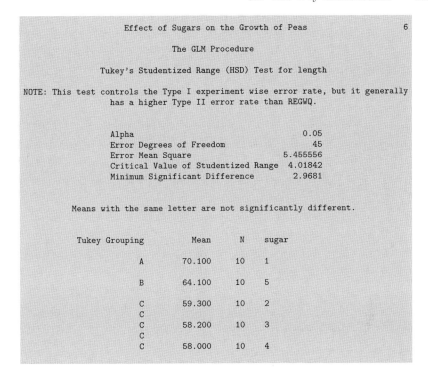

Fig. 5.10. SAS Example E2: Tukey procedure

on a separate page, page 7 in this output (see Fig. 5.11). The divisor mean square used for constructing the F-statistics is the same as Error MS from the anova table. The four `estimate` statements result in the computation of t-statistics ❷ for testing the same four comparisons. The F-tests above are equivalent to these t-tests as the numerator degrees of freedom is equal to 1 for each F-statistic. This is reflected by the observation that the p-values for corresponding tests are identical.

The p-values of the four F-tests and four t-tests, respectively, are identical, as they are testing the same hypotheses. the p-values indicate that the hypothesis comparing the effect of the control with the average effect of all sugars and the hypothesis comparing the effect of the mixed sugars with average effects of pure sugars are rejected at an α level of .05. This results in the finding that all sugars depress the mean pea lengths and the effect of the mixed sugars is less than the average effect of the pure sugars. The two hypotheses comparing the differences among the effects of the three pure sugars result in the findings that there is a significant difference between the Fructose and Sucrose means but no significant difference between the Fructose and Glucose means.

```
                    Effect of Sugars on the Growth of Peas                       7

                              The GLM Procedure

Dependent Variable: length

                                                                 ┌─┐
                                                                 │1│
                                                                 └─┘
Contrast                     DF    Contrast SS    Mean Square   F Value    Pr > F

CONTROL VS. SUGARS            1    832.3200000    832.3200000    152.56    <.0001
SUGARS VS. MIXED              1     48.1333333     48.1333333      8.82    0.0048
GLUCOSE=FRUCTOSE              1      6.0500000      6.0500000      1.11    0.2979
FRUCTOSE = SUCROSE            1    115.2000000    115.2000000     21.12    <.0001


                                                Standard        ┌─┐
Parameter                     Estimate             Error        │2│ t Value    Pr > |t|
                                                                └─┘
CONTROL VS. SUGARS           40.8000000         3.30319710        12.35    <.0001
SUGARS VS. MIXED              7.6000000         2.55864547         2.97    0.0048
GLUCOSE=FRUCTOSE              1.1000000         1.04456264         1.05    0.2979
FRUCTOSE = SUCROSE           -4.8000000         1.04456264        -4.60    <.0001
```

Fig. 5.11. SAS Example E2: Making preplanned comparisons

Finally, it is important to recognize, for example, that when an estimate statement such as

estimate 'CONTROL VS. SUGARS' sugar 4 -1 -1 -1 -1;

is used, the estimate output by **proc glm** is the estimate of $4\mu_1 - \mu_2 - \mu_3 - \mu_4 - \mu_5$) and not of $\mu_1 - \frac{1}{4}(\mu_2 + \mu_3 + \mu_4 + \mu_5)$, as one might mistakenly take the numerical value of the estimate (and its standard error) output to be. Although this will not affect the computed value of the t-statistic and the p-value, there may be instances when the actual estimate may be needed. One could use the **divisor=** option of the **estimate** statement to obtain the correct estimate without affecting the t-test by writing the two statements as follows:

estimate 'CONTROL VS. SUGARS' sugar 4 -1 -1 -1 -1/divisor=4;

estimate 'SUGARS VS. MIXED   ' sugar 0 1 1 -3 1/divisor=3;

These changes will result in estimates of 10.20 and 2.53 for the two comparisons with standard errors 0.8258 and 0.8529, respectively.

5.2.3 Testing orthogonal polynomials using contrasts

An experimenter may be interested in determining whether the observed mean response of a factor under study is related to the levels of the factor in some way. This relationship can be linear or curved. Orthogonal polynomials is a method of partitioning the sums of squares due to the factor into components that allows the experimenter to construct F-statistics with 1 degree of freedom each for the numerators. Each of these can then be used to test whether the relationship can be represented by a linear, quadratic, cubic, etc. function of the factor levels in a sequential fashion. To make the computations easier, this partitioning of the treatment sum of squares can be performed using appropriate orthogonal contrasts.

Car	Engine Size			
	300	350	400	450
1	16.6	14.4	12.4	11.5
2	16.9	14.9	12.7	12.8
3	15.8	14.2	13.3	12.1
4	15.5	14.1	13.6	12.0
Mean	16.1	14.4	13.0	12.1

Table 5.3. Effect of Engine Size on Gasoline Consumption

Example 5.2.2

It must be understood that the reparameterized linear model obtained using orthogonal polynomials in this fashion is not exactly equivalent to a regression model with the (centered) factor levels as regressor variables. Thus, care is needed about how the results of tests involving orthogonal polynomials are interpreted. A suggested procedure is to start with a test of a linear trend of the mean response on the factor levels and use a lack of fit test to check if the remaining treatment sum of squares is significant. If there is lack of fit, proceed by successively increasing the order of the polynomial and performing lack of fit tests.

To illustrate the technique consider the following example (Morrison, 1983). Suppose that a consumer research group wishes to study the gasoline consumption of large eight-cylinder passenger cars of a given model year. The cars of interest have been classified by their engine sizes of approximately 300, 350, 400, and 450 cubic inches. Four cars were drawn at random from each engine size, and each car driven over a standard urban route three times. These miles per gallon of fuel, recorded for the 16 cars, are shown in Table 5.3.

The analysis of variance computed for this data gave the following statistics:

SV	df	SS	MS	F
Engine size	3	38.25	12.7833	44.59
Error	12	3.44	0.2867	

According to this analysis, significant differences among the engine size means exist. It is now of interest to determine whether the decreasing gas mileage is a linear function of the engine volume or related in a more complex way to the engine volume.

The contrasts corresponding to the linear, quadratic, and cubic orthogonal polynomials have the coefficients given by

$$c_1' = [-3, -1, 1, 3]$$
$$c_2' = [1, -1, -1, 1]$$
$$c_3' = [-1, 3, -3, 1]$$

as can be obtained from a standard table of orthogonal polynomials (see Table B.12 of Appendix B). The sums of squares and F-statistics corresponding to the three single-degree of freedom contrasts are calculated to be:

Contrast	df	SS	F
Linear	1	37.538	131.00
Quadratic	1	0.810	2.83
Cubic	1	0.002	0.01
Total	3	38.350	-

Since the three contrasts are mutually orthogonal, the sum of squares correspond to a partitioning of the sum of squares for treatment (here Engine Size) with three degrees of freedom as evident from the analysis of variance table given above.

The linear trend contrast is significant with 1 and 12 degrees of freedom. The lack of fit F-statistic is not significant. The differences among the means seem to be explainable by a linear trend alone. These contrast sum of squares and corresponding F-tests are usually included in a complete anova table as follows:

SV	df	SS	MS	F
Engine Size	3	38.350	12.7833	44.59
Linear	1	37.538	37.5380	131.00
Lack of Fit	2	0.812	0.4060	1.42
Error	12	3.440	0.2867	

Since the lack of fit is not significant, a higher-order polynomial is not needed. The experimenter can conclude that there is a significant decreasing linear trend in the mean gas mileage as the engine size increases.

SAS Example E3

Part 1 of the SAS Example E3 program (see Fig. 5.12) illustrates how the sum of squares needed for testing linear trend can be obtained using a contrast statement in proc glm. In this example, the gas mileage data are read using a similar approach to that used in the program for SAS Example E2. The trailing @ symbol is used to hold the data line after engine size is input. Then successive input and output statements are executed in a do-end loop to

```
data mileage;
input size $ @;
    do   i=1 to 4;
    input  mpg @;
    output;
    end;
drop i;
datalines;
300 16.6 16.9 15.8 15.5
350 14.4 14.9 14.2 14.1
400 12.4 12.7 13.3 13.6
450 11.5 12.8 12.1 12.0
;
run;
proc glm data=mileage;
  class size;
  model mpg = size/p;
  means size/lsd;
  contrast 'Linear Trend' size -3 -1 1 3;
  estimate 'Linear Trend' size -3 -1 1 3;
  output out=stats1 p=fitted r=residual;
  title 'Analysis of Gas Mileage Data';
  run;

libname mylib 'C:\Documents and Settings\...\stat479';

data mylib.stats1;
set stats1;
run;
```

Fig. 5.12. SAS Example E3: Program (Part 1)

read each of the gas mileage values and write new observations into the SAS data set. Each observation output will have the current value of engine size and the gas mileage as they appear in the program data vector (PDV).

The model statement, similar to Example E1, codes the appropriate model for one-way classification. The option p used in the model statement results in producing the predicted **1** and residual values (see Fig. 5.14). The output statement causes these values to be written to a new SAS data set by the name of stats1 as values of new variables named fitted and residual, respectively, along with the variables in the original SAS data set (named mileage).

The contrast statement contains the coefficients taken from Table B.12 and correspond to values tabulated for *Number of levels*=4 and *Degree of polynomial*=1. These are $-3, -1, 1$, and 3, respectively, and each coefficient corresponds to a level of size. The contrast statement **2** is

```
contrast 'Linear Trend' size -3 -1 1 3;
```

The page 2 of the output is shown in Fig 5.13. Page 1, which is omitted here, shows the levels of the treatment factor (here size) must always be checked to verify that they are in the correct sequence. Page 2, as in SAS Example E2, provides the information necessary to construct the analysis of variance table.

```
                    Analysis of Gas Mileage Data                    2

                         The GLM Procedure

Dependent Variable: mpg

                              Sum of
Source              DF       Squares     Mean Square   F Value   Pr > F

Model                3    38.35000000    12.78333333     44.59   <.0001

Error               12     3.44000000     0.28666667

Corrected Total     15    41.79000000


          R-Square     Coeff Var      Root MSE     mpg Mean

          0.917684      3.844974      0.535413     13.92500

Source              DF     Type I SS    Mean Square   F Value   Pr > F

size                 3    38.35000000   12.78333333     44.59   <.0001

Source              DF    Type III SS   Mean Square   F Value   Pr > F

size                 3    38.35000000   12.78333333     44.59   <.0001
```

Fig. 5.13. SAS Example E3: Page 2

Results of both the contrast and estimate statements are shown in Fig. 5.15. The F-test has a p-value of $< .0001$; thus, the linear trend is significant. The

```
                    Analysis of Gas Mileage Data                    3

                         The GLM Procedure

      Observation       Observed        Predicted 1       Residual

             1         16.60000000      16.20000000      0.40000000
             2         16.90000000      16.20000000      0.70000000
             3         15.80000000      16.20000000     -0.40000000
             4         15.50000000      16.20000000     -0.70000000
             5         14.40000000      14.40000000      0.00000000
             6         14.90000000      14.40000000      0.50000000
             7         14.20000000      14.40000000     -0.20000000
             8         14.10000000      14.40000000     -0.30000000
             9         12.40000000      13.00000000     -0.60000000
            10         12.70000000      13.00000000     -0.30000000
            11         13.30000000      13.00000000      0.30000000
            12         13.60000000      13.00000000      0.60000000
            13         11.50000000      12.10000000     -0.60000000
            14         12.80000000      12.10000000      0.70000000
            15         12.10000000      12.10000000     -0.00000000
            16         12.00000000      12.10000000     -0.10000000
```

Fig. 5.14. SAS Example E3: Page 3

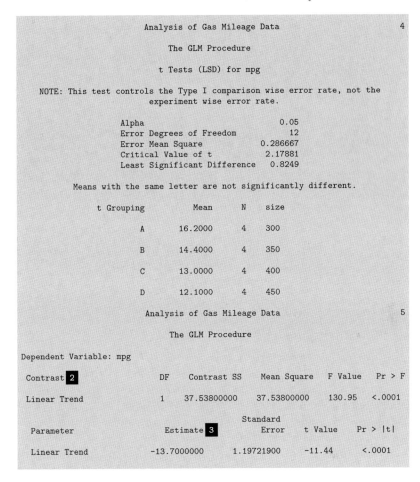

Fig. 5.15. SAS Example E3: Pages 4 and 5

output from the `estimate` statement ▮3 can be used to calculate the slope of the straight line fitted to the levels of engine size, using the formula

$$b = \frac{\sum c_i \bar{y}_{i.}}{\sum c_i^2}$$

where $\bar{y}_{i.}$ are the treatment means and $c_1, c_2, \ldots, c_t$ are the contrast coefficients. The estimate output is $\sum c_i \bar{y}_{i.} = -13.7$. Thus, the slope is calculated as $-13.7/(3^2 + 1^2 + 1^2 + 3^2) = -13.7/20 = -0.685$ per unit in the coded scale of the x variable (i.e., a decrease of 0.685 miles per gallon for every increase of 25 cubic inches in engine size). This estimate may be directly calculated in SAS by modifying the estimate statement to

```
estimate 'Linear Trend' size -3 -1 1 3/divisor=20;
```

```
libname mylib 'C:\Documents and Settings\mervyn\My Documents\Classwork\stat479';

proc rank normal=blom data=mylib.stats1 out=quantiles;
  var residual;
  ranks nscore;
run;

data stats2;
merge mylib.stats1 quantiles;
run;

  title1 ' ';
  axis1 c=vibg label=(c=darkviolet h=1.5 a=90 ' Residuals ');
  axis2 c=vibg label=(c=darkviolet h=1.5 ' Engine Size');
  axis3 c=vibg label=(c=darkviolet h=1.5 'Predicted Values');
  axis4 c=vibg label=(c=darkviolet h=1.5 a=90 'Predicted Values');
  axis5 c=vibg label=(c=darkviolet h=1.5 'Standard Normal Percentiles');
  symbol1 c=stp v=star i=none h=1.5;
  symbol2 c=dep v=dot  i=none h=1;
  symbol3 c=darkblue v=none i=join h=1;
  symbol4 c=bgr v=K f=special i=none h=1.5;
  symbol5 c=str v=M f=special i=none h=1;

proc gplot data=stats2;
  plot residual*size=1/vaxis=axis1 haxis=axis2 vref=0 cvref=red lvref=3;
  plot residual*fitted=2/vaxis=axis1 haxis=axis3 vref=0 cvref=red lvref=3;
  plot fitted*size=3 mpg*size=4/vaxis=axis4 haxis=axis2 overlay;
  plot residual*nscore=5/vaxis=axis1 vm=9 haxis=axis5 hm=4 frame;
run;
```

Fig. 5.16. SAS Example E3: Program (Part 2)

Thus, the slope is estimated as -0.685 with a standard error of 0.05986.

The new SAS data set stats1 is saved as a permanent file in a library (in this case, a folder under the Windows system) for use in part 2 of the SAS Example E3 program that will use the data to obtain standard plots described below. In this SAS program (see Fig. 5.16), proc gplot is used to obtain the standard residual plots of residuals against the levels of engine size and the residuals against the predicted values. Both these plots (see Figs. 5.17a and 5.17b) do not exhibit any outliers or trends in the dispersion of the points around zero (the reference line drawn at residual value equal to 0 using the option vref=0 is useful for ascertaining this) as values plotted on the x-axis change. Thus, there is evidence supporting the model assumption of homogeneity of variance and the adequacy of a first-order model in engine size to describe the variation in gas mileage.

The plot showing the observed values (see Fig. 5.17c) plotted against the levels of engine size also display the predicted values (as connected by line segments, so that they are easy to pick out). Note that the predicted value for all observations at each factor level is their sample mean. The data in stats1 are also used for producing the normal probability plot of the residuals (see Fig. 5.17d). The method used for obtaining this plot was described previously in relation to SAS Example C6 (see Section 3.6 in Chapter 3) and uses proc rank for calculating the standard normal quantiles (variable nscore) of the data needed for the plot.

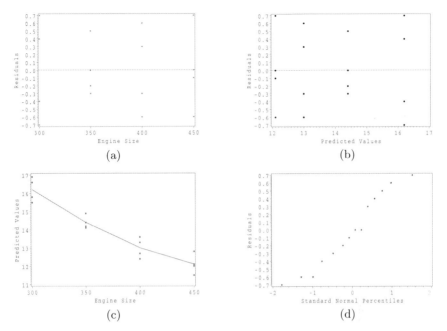

Fig. 5.17. SAS Example E3: Plots. (a) Residuals versus engine size; (b) residuals versus predicted values; (c) predicted values overlaid on observed values; (d) normal probability plot of the residuals

5.3 One-Way Analysis of Covariance

The technique of measuring an additional variable, say x, called a *covariate*, in addition to the response variable y on each experimental unit in designed experiments can be used to increase precision of an experiment. The analysis of data from such experiments involves adjusting the analysis of variance and estimation procedures to account for the regression variable x. The resulting model for y thus contains the measured variable x in addition to the usual effects for the treatment factor.

Model

A single-factor experiment in a completely randomized design where a single covariate is measured is considered below. Equal replication of sample size n is assumed for the purpose of discussion.

$$y_{ij} = \mu + \tau_i + \beta(x_{ij} - \bar{x}_{..}) + \epsilon_{ij} \quad i = 1, \ldots, t; \quad j = 1, \ldots, n$$

where τ_i is the ith treatment effect as in Section 5.2. It is also assumed that the random error ϵ_{ij} is distributed as iid $N(0, \sigma^2)$. The above model stipulates straight-line regression models for each treatment with the same slope β and different intercepts α_i for $i = 1, \ldots, t$, where $\alpha_i = \mu + \tau_i - \beta\bar{x}_{..}$, because the model expresses the relationship between y_{ij} and x_{ij} for each ith treatment as

$$\text{Treatment 1: } y_{1j} = \alpha_1 + \beta x_{1j} + \epsilon_{1j}, \, j = 1, \ldots, n$$
$$\text{Treatment 2: } y_{2j} = \alpha_2 + \beta x_{2j} + \epsilon_{2j}, \, j = 1, \ldots, n$$
$$\vdots$$
$$\text{Treatment } t: y_{tj} = \alpha_t + \beta x_{tj} + \epsilon_{tj}, \, j = 1, \ldots, n$$

Because of the stipulation that the straight lines have the same slope β, the above model is often called the *equal slopes* model.

Estimation

Note that since $E(y_{ij}) = \mu + \tau_i + \beta(x_{ij} - \bar{x}_{..})$, unlike the model in Section 5.2, the ith treatment mean now depends on different values of x_{ij}. For the equal slopes model, one treatment mean evaluated at $x_{ij} = \bar{x}_{..}$ and denoted by μ_i is of interest. This is usually called the "adjusted mean." Note that $\mu_i = \mu + \tau_i$. The best linear unbiased estimate of μ_i is

$$\hat{\mu}_i = \bar{y}_{i.}(\text{Adj.}) = \bar{y}_{i.} - b(\bar{x}_{i.} - \bar{x}_{..}) \text{ for } i = 1, \ldots, t$$

These estimates are called *adjusted treatment means* because the usual estimate $\bar{y}_{i.}$ is adjusted for regression on x_i using the estimated common slope b, where b is the usual least squares estimate of β given by

$$b = \frac{\sum_i \sum_j (x_{ij} - \bar{x}_{i.})(y_{ij} - \bar{y}_{i.})}{\sum_i \sum_j (x_{ij} - \bar{x}_{i.})^2}$$

where $\bar{y}_{i.} = (\sum_j y_{ij})/n$, $\bar{x}_{i.} = (\sum_j x_{ij})/n$, and $\bar{x}_{..} = (\sum_i \sum_j x_{ij})/tn$. Also, an unbiased estimate of the error variance σ^2 is given by $\hat{\sigma}^2 = s^2$, where s^2 is the Error MS from the analysis of covariance table below. It can be shown that the adjusted treatment means $\hat{\mu}_i$ are the predicted responses $\hat{y}_{ij}$ computed at the value of $x_{ij} = \bar{x}_{..}$ using the fitted regression models, for each $i = 1, \ldots, t$.

A $(1-\alpha)100\%$ confidence interval for $\mu_p - \mu_q$, the difference between the effects of two treatments labeled p and q, is

$$(\bar{y}_{p.}(\text{Adj.}) - \bar{y}_{q.}(\text{Adj.})) \pm t_{\alpha/2,\nu} s_d$$

where s_d, the standard error of the difference between two adjusted means $\bar{y}_{p.}(\text{Adj.}) - \bar{y}_{q.}(\text{Adj.})$, is given by

$$s_d = s \left\{ \frac{2}{n} + \frac{(\bar{x}_{p.} - \bar{x}_{q.})^2}{E_{xx}} \right\}^{1/2}$$

and $t_{\alpha/2,\nu}$ is the upper $\alpha/2$ percentile of the t-distribution with $\nu = t(n-1)-1$ degrees of freedom.

Testing Hypotheses

The presentation of the results of the analysis is somewhat complicated because the total sum of squares is partitioned in two ways: One partition shows the test of the main effects without covariate adjustment and one show it with covariate adjustment. It is convenient to present the results of both in a single compact analysis of variance table rather than two separate tables. In the following table, the analysis above the `Total SS` line shows the treatment sum of squares *unadjusted* for the covariate and the partition of the regression sum of squares from the `Error SS` to form the test of the regression parameter. The second analysis, shown below the `Total SS` line contains the treatment sum of squares *adjusted* for the covariate. This table is called the *analysis of covariance* table:

SV	df	SS	MS	F
Trt	$t-1$	SS_{Trt}	MS_{Trt}	MS_{Trt}/MSE_{Unadj}.
Error(Unadj.)	$t(n-1)$	SSE_{Unadj}.	MSE_{Unadj}.	
Regression	1	SS_{Reg}	MS_{Reg}	MS_{Reg}/MSE
Error(Adj.)	$t(n-1)-1$	SSE	$MSE(=s^2)$	
Total	$tn-1$	SS_{Tot}		
Trt(Adj.)	$t-1$	SS_{Trt}	MS_{Trt}	MS_{Trt}/MSE
Error(Adj.)	$t(n-1)-1$	SSE	$MSE(=s^2)$	

The F-statistic for `Trt` tests the hypothesis

$$H_0: \mu_1 = \mu_2 = \cdots = \mu_t \quad versus \quad H_a: \text{ at least one inequality}$$

when the covariate is not present in the model (i.e., without taking into account any adjustment due to the covariate). The divisor for computing the F-statistic is the MS for `Error(Unadj.)` with $t(n-1)$ df The F-statistic for `Regression` tests the hypothesis

$$H_0: \beta = 0 \quad versus \quad H_a: \beta \neq 0$$

and thus is a test of whether the covariate has an effect on the response as an explanatory variable in a linear regression.

The F-statistic for Trt(Adj.) tests the hypothesis that the adjusted treatment means are the same, or, equivalently, the treatment effects

$$H_0: \tau_1 = \tau_2 = \cdots = \tau_t \quad versus \quad H_a: \text{ at least one inequality}$$

when β is not zero (i.e., when the analysis of variance is adjusted for the covariate). This test is also equivalent to comparing the intercepts of the regression lines i.e.,

$$H_0: \alpha_1 = \alpha_2 = \cdots = \alpha_t \quad versus \quad H_a: \text{ at least one inequality}$$

If this hypothesis is rejected, then at least one pair of treatment effects (equivalently, adjusted treatment means) is different. One can proceed to make preplanned or pairwise comparisons of these means as in SAS Example E1 but using the adjusted treatment means. Note carefully that SS_{Trt} (and thus MS_{Trt} and the F-statistics) in the bottom table will be different in magnitude from those in the top table since those takes into account that a covariate is present in the model.

5.3.1 Using PROC GLM to perform one-way covariance analysis

The data displayed in Fig. 5.18 are results from an experiment on the use of two treatments, slow-release fertilizer (S) and a fast-release fertilizer (F), on the yield (grams) of peanut plants compared to a control (C), a standard fertilizer, described in Ott and Longnecker (2001). Ten replications of each treatment were grown in a greenhouse study.

		Fertilizer Treatments			
Control (C)		Slow Release (S)		Fast Release (F)	
Yield	Height	Yield	Height	Yield	Height
12.2	45	16.6	63	9.5	52
12.4	52	15.8	50	9.5	54
11.9	42	16.5	63	9.6	58
11.3	35	15.0	33	8.8	45
11.8	40	15.4	38	9.5	57
12.1	48	15.6	45	9.8	62
13.1	60	15.8	50	9.1	52
12.7	61	15.8	48	10.3	67
12.4	50	16.0	50	9.5	55
11.4	33	15.8	49	8.5	40

Fig. 5.18. Yield of peanut plants from three fertilizer treatments and their initial heights

Since the researcher recognized that the 30 peanut plants used were different in their development and health, the height (in centimeters) of each plant was recorded at the start of the experiment to be used as a covariate to adjust for this variation. This experiment is an example of a single-factor experiment in a completely randomized design in which a covariate is also measured on each experimental unit. Let y_{ij} and x_{ij} be the yield and the pretreatment measure of height of the jth plant treated with the ith fertilizer, respectively. The model is then

$$y_{ij} = \mu_i + \beta(x_{ij} - \bar{x}_{..}) + \epsilon_{ij}, \quad i = 1, 2, 3; \quad j = 1, \ldots, 10.$$

Thus, three regression lines (corresponding to the three fertilizers) with the same slope parameter β are stipulated by this model. The hypothesis of equality of the population means of yields due to the three fertilizers, $H_0 : \mu_1 = \mu_2 = \mu_3$ versus H_a : at least one inequality, is tested using the analysis of covariance discussed earlier.

SAS Example E4

The SAS Example E4 program (see Fig. 5.19) is used to obtain the necessary analysis of the above data. In the SAS program, once again the data are input in a straightforward format. The "effects" form of the model $E(y_{ij}) = \mu + \alpha_i + \beta x_{ij}$ is used to specify the model for analysis by `proc glm`. The `model` statement thus includes the term `fertilizer` to represent the treatment effect α_i and the term `height` to represent the covariate x_{ij}. Note that this variable does not appear in the `class` statement; thus, when it occurs on the right side of the model statement, it is recognized to be a regression-type variable and not a classificatory variable, by default. It is also important to note that the covariate appears after the treatment variable in the model statement. The design matrix

$$X = \begin{bmatrix} 1 & 1 & 0 & 0 & 45 \\ 1 & 1 & 0 & 0 & 52 \\ 1 & 1 & 0 & 0 & 42 \\ \vdots & \vdots & \vdots & \vdots & \vdots \\ 1 & 1 & 0 & 0 & 33 \\ 1 & 0 & 1 & 0 & 63 \\ 1 & 0 & 1 & 0 & 50 \\ 1 & 0 & 1 & 0 & 63 \\ \vdots & \vdots & \vdots & \vdots & \vdots \\ 1 & 0 & 1 & 0 & 49 \\ 1 & 0 & 0 & 1 & 52 \\ 1 & 0 & 0 & 1 & 54 \\ 1 & 0 & 0 & 1 & 58 \\ \vdots & \vdots & \vdots & \vdots & \vdots \\ 1 & 0 & 0 & 1 & 40 \end{bmatrix}$$

that results from the model statement is a 30×5 matrix as shown. The columns of X correspond to the parameters μ, α_1, α_2, α_3, and β, respectively.

The `lsmeans` statement is required for `proc glm` to generate the *adjusted treatment means* and their standard errors, instead of the ordinary sample means. Just as in the `means` statement, `lsmeans` statement lists effects that involve only classification variables. Note that the sample means produced by the `means` statement are not adjusted for the covariate. Page 2 of the SAS output (see Fig. 5.20; page 1 of the SAS output is not shown) shows that the ordering of the levels of factor `fertilizer` present in the input data is retained as requested by including the `order=data` as a `proc glm` option.

The information needed to construct the analysis of covariance table described earlier is found in the output resulting from the model statement.

```
data peanuts;
input fertilizer $ yield height;
label fertilizer='Fertilizer' yield='Yield' height='Height';
datalines;
C   12.2    45
C   12.4    52
.    .       .
.    .       .
.    .       .
C   12.4    50
C   11.4    33
S   16.6    63
S   15.8    50
.    .       .
.    .       .
.    .       .
S   16.0    50
S   15.8    49
F    9.5    52
F    9.5    54
.    .       .
.    .       .
.    .       .
F    9.5    55
F    8.5    40
;
run;

proc print data=peanuts;
run;

proc glm data=peanuts order=data;
  class fertilizer;
  model yield = fertilizer height;
  lsmeans fertilizer/stderr cl pdiff;
  contrast 'Modified vs. Standard' fertilizer 1 -.5 -.5;
  contrast 'Slow-release vs. Fast-release' fertilizer 0  1 -1;
  title 'Covariance Analysis of Peanut Fertilizer Data';
run;
  symbol1 v="C" cv=red f=centb h= 1.5 i=none;
  symbol2 v="F" cv=steelblue f=centb h= 1.5 i=none;
  symbol3 v="S" cv=magenta f=centb h= 1.5 i=none;
  axis1 c=dapk label=(c=blueviolet h=1.5 a=90 f=centb 'Yield');
  axis2 c=dapk label=(c=blueviolet h=1.5 f=centb  'Height');

proc gplot data=peanuts;
plot yield*height=fertilizer/vaxis=axis1 haxis=axis2;
run;
```

Fig. 5.19. SAS Example E4: Program

Specifically, the Total and Error(Adj.) degrees of freedom, sums of squares, and mean squares needed are extracted from those found in the corresponding columns for Corrected Total and Error **1**, respectively, given in the table shown on top part of page 3 of the SAS output (see Fig. 5.21). Values for fertilizer and height, respectively, from the Type I SS **2** part on the same SAS output page, provide the degrees of freedom, sums of squares, mean squares, and the corresponding F-statistics for both the Fertilizer(Unadj.) and Regression lines in the top portion of the analysis of covariance table.

```
                Covariance Analysis of Peanut Fertilizer Data                2

                           The GLM Procedure

                         Class Level Information

                    Class           Levels      Values

                    fertilizer         3        C S F

                 Number of Observations Read       30
                 Number of Observations Used       30
```

Fig. 5.20. SAS Example E4: Output (page 2)

The "unadjusted" Error SS is then obtained by summing the regression and adjusted error sums of squares. Thus, the top portion of the table is complete. To complete the `Fertilizer(Adj.)` line in the bottom portion of the table, the degrees of freedom, sums of squares, mean squares, and the corresponding F-statistics for `fertilizer` are obtained from the Type III SS part on the same page of SAS output. The completed table is

```
                Covariance Analysis of Peanut Fertilizer Data                3

                           The GLM Procedure

Dependent Variable: yield

                                    Sum of
Source                    DF       Squares    Mean Square   F Value   Pr > F

Model                      3   214.3759539    71.4586513    4447.85   <.0001

Error                     26     0.4177128     0.0160659      ■1

Corrected Total           29   214.7936667

                 R-Square      Coeff Var      Root MSE     yield Mean

                 0.998055       1.017537       0.126751      12.45667


Source                    DF     Type I SS    Mean Square   F Value   Pr > F

fertilizer                 2   207.6826667   103.8413333    6463.47   <.0001  ■2
height                     1     6.6932872     6.6932872     416.62   <.0001


Source                    DF    Type III SS   Mean Square   F Value   Pr > F

fertilizer                 2   213.9038045   106.9519022    6657.08   <.0001
height                     1     6.6932872     6.6932872     416.62   <.0001
```

Fig. 5.21. SAS Example E4: Output (page 3)

SV	DF	SS	MS	F	p-value
Fertilizer(Unadj.)	2	207.6827	103.8414	394.23	< .0001
Error	27	7.1110	0.2634		
Regression	1	6.6933	6.6933	416.62	<.0001
Error(Adj.)	26	0.4177	0.0161		
Total	29	214.7937			
Fertilizer(Adj.)	2	213.9038	106.9519	6657.08	<.0001
Error(Adj.)	26	0.4177	0.0161		

```
              Covariance Analysis of Peanut Fertilizer Data                4

                            The GLM Procedure
                           Least Squares Means

                                    Standard                      LSMEAN
         fertilizer   yield LSMEAN    Error      Pr > |t|         Number

              C        12.3141728    0.0410853   <.0001              1
              S        15.8858099    0.0401754   <.0001              2
              F         9.1700172    0.0417711   <.0001              3


                 Least Squares Means for effect fertilizer
                   Pr > |t| for H0: LSMean(i)=LSMean(j)  3

                          Dependent Variable: yield

                i/j            1            2            3

                 1                      <.0001       <.0001
                 2         <.0001                    <.0001
                 3         <.0001       <.0001


         fertilizer    yield LSMEAN  4      95% Confidence Limits

              C         12.314173       12.229721       12.398625
              S         15.885810       15.803228       15.968392
              F          9.170017        9.084155        9.255879


                 Least Squares Means for Effect fertilizer

                        Difference
                         Between        95% Confidence Limits for
           i    j         Means           LSMean(i)-LSMean(j)

           1    2       -3.571637       -3.688869       -3.454405
           1    3        3.144156        3.020055        3.268256
           2    3        6.715793        6.595528        6.836058


    NOTE: To ensure overall protection level, only probabilities associated with
          pre-planned comparisons should be used.
```

Fig. 5.22. SAS Example E4: Output (page 4)

First, the p-value for the F-test for `Regression` clearly shows that the hypothesis of $H_0 : \beta = 0$ is rejected, thus confirming that plant height is linearly related to seed yield. Second, from the p-value for the F-statistic for `Fertilizer(Adj.)`, the hypothesis of no difference in fertilizer effects is also rejected.

```
                  Covariance Analysis of Peanut Fertilizer Data              5

                             The GLM Procedure

Dependent Variable: yield

Contrast                       DF   Contrast SS    Mean Square   F Value   Pr > F

Modified vs. Standard           1    0.28305108    0.28305108      17.62   0.0003
Slow-release vs. Fast-release   1  211.6745206   211.6745206    13175.4   <.0001
```

Fig. 5.23. SAS Example E4: Output (page 5)

In the analysis of variance table shown earlier, the `Total` line is the result of ignoring the covariate. In this table, the F-statistic for testing no difference in fertilizer effects hypothesis (shown in the `Fertilizer(Unadj.)` line) actually is much smaller than the F-statistic for `Fertilizer(Adj.)`. It can be easily seen that this is due to the inflated error variance estimate given by the MSE (0.2634), because the mean squares for `Fertilizer(Adj.)` and `Fertilizer(Unadj.)` are similar in magnitude. This shows that, in other situations, it is a possible for differences that may exist among the treatments to go undetected if covariance adjustment is not taken into account if the effect of the adjustment is substantial.

The output shown on SAS output page 4 (see Fig. 5.22) is produced as a result of the `lsmeans` statement used in the program. By default, the statistics computed are identical to the adjusted treatment means $\bar{y}_{i.}(\text{Adj.})$ discussed previously in this section. These are displayed under LSMEAN on page 4 **4**. It is important to note that `proc glm` computes the `lsmeans` by setting the covariate values equal to their mean (i.e., $x_{ij} = \bar{x}_{..}$) as discussed previously. This implies that, implicitly, the option `at means` is in effect, as the default. This is appropriate, as the regression lines are parallel when the equal slopes model holds and, thus, the differences in `lsmeans` are the same at any value of x, and these have smallest standard errors. If the slopes were different, however, the `at` option would enable the user to request these to be computed at different covariate values considered interesting for comparison of the predicted responses at those values (e.g. by using an option like `at x=10`).

The `stderr` option on the `lsmeans` statement resulted in the standard errors of the adjusted treatment means to be also output. The `pdiff` option produced the second portion of this page of the SAS output, which gives p-

```
                Covariance Analysis of Peanut Fertilizer Data                    4

                            The GLM Procedure
                            Least Squares Means
                    Adjustment for Multiple Comparisons: Bonferroni

                                     Standard                           LSMEAN
        fertilizer    yield LSMEAN     Error       Pr > |t|             Number

        C              12.3141728    0.0410853      <.0001                 1
        S              15.8858099    0.0401754      <.0001                 2
        F               9.1700172    0.0417711      <.0001                 3


                    Least Squares Means for Effect fertilizer
                    t for H0: LSMean(i)=LSMean(j) / Pr > |t|

                           Dependent Variable: yield

              i/j              1               2               3

              1                            -62.6244         52.07806
                                             <.0001          <.0001
              2             62.62441                        114.7842
                             <.0001                          <.0001
              3            -52.0781        -114.784
                            <.0001          <.0001


         fertilizer    yield LSMEAN         95% Confidence Limits

         C              12.314173       12.229721        12.398625
         S              15.885810       15.803228        15.968392
         F               9.170017        9.084155         9.255879


                    Least Squares Means for Effect fertilizer

                             Difference        Simultaneous 95%
                              Between         Confidence Limits for
               i    j          Means          LSMean(i)-LSMean(j)

               1    2         -3.571637      -3.717580       -3.425694
               1    3          3.144156       2.989662        3.298649
               2    3          6.715793       6.566074        6.865511
```

Fig. 5.24. SAS Example E4: Use of `tdiff` and `adjust` options

values associated with testing pairwise differences in means (i.e., hypotheses of the form $H_0: \mu_i = \mu_j$ versus $H: \mu_i \neq \mu_j$ for all pairs (i,j)). The 95% confidence intervals for individual adjusted means and their pairwise differences were produced as a result of the `cl` option. The `alpha=` keyword option may be added to specify a confidence coefficient different from 95%.

It is possible to use the `contrast` statement to test single-degree of freedom comparisons of interest about treatment means. Here, the average effect of the fertilizers with the control is compared using the comparison $\tau_1 - (\tau_2 + \tau_3)/2$. Note that the sums of squares and the F-tests are also adjusted for the covariate. From the output on page 5 (see Fig. 5.23) this hypothesis is clearly re-

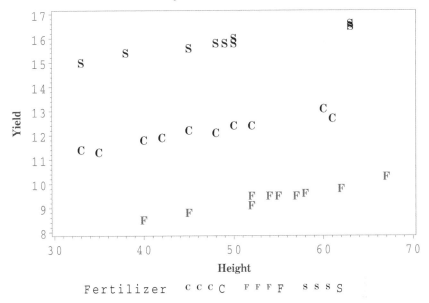

Fig. 5.25. SAS Example E4: Plot of yield versus height by fertilizer

jected, implying that the two, on the average, lower the mean yield compared to the control. In addition, by examining the tests and confidence intervals on page 4 (see Fig. 5.22), it is found that there is a significant difference between the two fertilizers.

The output on page 4 may be modified to include computed t-statistics for comparing the pairwise differences in the means (in addition to the p-values) using the option `tdiff` on the `lsmeans` statement as follows:

```
lsmeans fertilizer/stderr cl tdiff pdiff adjust=bon;
```

The new output is shown in Fig. 5.24. The pairwise confidence limits are different because of the option `adjust=bon` 5 option included in the modified `lsmeans` statement. This option causes a multiple comparison adjustment to be made to the p-values and confidence limits for the pairwise differences. Here, the adjustment requested is the Bonferroni adjustment. `tukey` and `scheffe` are two other adjustments available. The default is `adjust=t`, which really signifies no adjustment made for doing multiple comparisons.

The `proc gplot` step in the SAS program produced the graph shown in Fig. 5.25, which is a simple scatter plot of the yield of peanuts against the heights, identified by the fertilizer (slow-release, fast-release, or the control) received by individuals. It may be helpful if the regression lines fitted under the equal slope assumption are superimposed on this plot to check if such an

320   5 Analysis of Variance Models

```
                          The GLM Procedure

Dependent Variable: yield

                               Sum of
Source                DF      Squares     Mean Square   F Value   Pr > F

Model                  4    4869.432287    1217.358072   75772.9   <.0001

Error                 26       0.417713       0.016066

Uncorrected Total     30    4869.850000


              R-Square    Coeff Var    Root MSE    yield Mean

              0.998055    1.017537     0.126751    12.45667


Source                DF    Type I SS     Mean Square   F Value   Pr > F

fertilizer             3    4862.739000   1620.913000   100892    <.0001
height                 1       6.693287      6.693287    416.62   <.0001


Source                DF    Type III SS   Mean Square   F Value   Pr > F

fertilizer             3     378.7954363   126.2651454   7859.21  <.0001
height                 1       6.6932872     6.6932872   416.62   <.0001

                                  Standard
       Parameter     Estimate ■1     Error    t Value   Pr > |t|

       fertilizer C   9.52925636   0.13357349   71.34   <.0001
       fertilizer S  13.10089348   0.13958529   93.86   <.0001
       fertilizer F   6.38510075   0.15352310   41.59   <.0001
       height         0.05580995   0.00273429   20.41   <.0001
```

Fig. 5.26. SAS Example E4: Regression parameter estimates

assumption is supported by the data. To do this, modify the `model` statement in the `proc glm` step as follows:

    model yield = fertilizer height/noint solution;

This results in the output shown in Fig. 5.26.

SAS Example E5

In SAS Example E5, estimated values ■1 shown in Fig. 5.26 are used in a data step that calculates the end points ■2 of the fitted regression lines (in the second data step of the SAS program displayed in Fig. 5.27). The SAS data set created (named `lines`) is then appended to the data set `peanuts` containing the original data. To create a new graph with these regression lines superimposed on the graph in Fig. 5.25, three new SAS/GRAPH symbol statements are added to those in Fig. 5.19. The modified SAS program is shown in Fig. 5.27. It uses values printed in the previous output to generate a SAS data set named `lines` for drawing the fitted regression lines. This data

5.3 One-Way Analysis of Covariance

```
insert data step to create the SAS dataset 'peanuts' here

data lines; 2
fertilizer='CL';
height=30; yield= 9.52926+0.05581*height; output;
height=70; yield= 9.52926+0.05581*height; output;
fertilizer='SL';
height=30; yield=13.10089+0.05581*height; output;
height=70; yield=13.10089+0.05581*height; output;
fertilizer='FL';
height=30; yield= 6.3851 +0.05581*height; output;
height=70; yield= 6.3851 +0.05581*height; output;
run;

data appended; 3
set peanuts lines;
run;

proc print data=appended;
run;

symbol1 v="C" cv=red f=centb h= 1.5 i=none ;
symbol2 ci=red v=none i=join ;

symbol3 v="F" cv=magenta f=centb h= 1.5 i=none ;
symbol4 ci=magenta v=none i=join ;

symbol5 v="S" cv=steelblue f=centb h= 1.5 i=none ;
symbol6 ci=steelblue v=none i=join ;

axis1 c=dapk label=(c=blueviolet h=1.5 a=90 f=centb 'Yield');
axis2 c=dapk label=(c=blueviolet h=1.5 f=centb 'Height');

proc gplot data=appended; 4
plot yield*height=fertilizer/vaxis=axis1 haxis=axis2 nolegend;
run;
```

Fig. 5.27. SAS Example E5: Program

set is appended to the original SAS data set `peanuts` to form the SAS data set named `appended` 3. The new data set is then used as input to `proc gplot` 4 that uses the original data to plot the points that correspond to each fertilizer and overlays them with the estimated straight lines using the generated data.

The graph (see Fig. 5.28) verifies that it is feasible to model the yield as a linear function of height for each fertilizer and that the assumption of equal slopes is reasonable. The intercepts are clearly different, thus validating the result of the test of equal treatment means.

5.3.2 One-way covariance analysis: Testing for equal slopes

In the beginning of Section 5.3, the equal slopes model for a single-factor experiment in a completely randomized design in which a single covariate is measured was introduced as

$$y_{ij} = \alpha_i + \beta x_{ij} + \epsilon_{ij}, \quad i=1,\ldots,t; \ j=1,\ldots,n$$

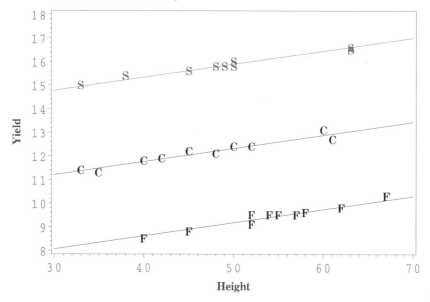

Fig. 5.28. SAS Example E5: Plot of yield versus height by fertilizer

The above model is not the most general model for the analysis of data from this experiment because of the often unrealistic assumption of equal slopes, although it is useful in certain situations. A more general model is the unequal slopes model

$$y_{ij} = \alpha_i + \beta_i x_{ij} + \epsilon_{ij}, \quad i = 1, \ldots, t; \; j = 1, \ldots, n$$

where different slopes β_i, $i = 1, \ldots, t$, are assumed for the t regression lines relating the responses y_{ij} to the covariates x_{ij}. The advantage of this model over the equal slopes model is that a hypothesis of whether the slopes are indeed the same (i.e. $H_0 : \beta_1 = \beta_2 = \cdots = \beta_t = \beta$ versus H_a : at least one inequality) may be tested as part of the inference from this model.

If this hypothesis is rejected, then a hypothesis of equal treatment means $\mu_{ij} = E(y_{ij}) = \alpha_i + \beta_i x_{ij}$ is of interest. In this case, however, such a test is not equivalent to the test of equality of the intercepts $H_0 : \alpha_1 = \alpha_2 = \cdots = \alpha_t$ as was the case in the equal slopes model. This is because the differences in means now depend on the value of β_i. Thus, the difference in intercepts will actually depend on the value x_{ij} at which the intercepts are compared. These comparisons are therefore dependent on the value of the covariate at which the comparisons are made. In practice, the treatment means are compared at several choices of the covariate values such as the the mean, the median, the minimum or the maximum, or at values of the covariate that are of special interest to the experimenter.

5.3 One-Way Analysis of Covariance

Since the above model may be expressed as

$$y_{ij} = \alpha_i + \bar{\beta}x_{ij} + (\beta_i - \bar{\beta})x_{ij} + \epsilon_{ij}, \quad i = 1,\ldots,t; j = 1,\ldots,n$$

this model is used in `proc glm` to obtain the sum of squares for testing the equal slopes hypothesis. This model is fitted to the cholesterol data taken from Milliken and Johnson (2001) in SAS Example E6. Thirty-two female subjects were assigned completely at random to one of four different diets. The response variable is the cholesterol level determined after being on the diet for 8 weeks. The cholesterol levels of the subjects measured before the experiment began was used as a covariate. The data appear in Fig 5.29.

Cholesterol Measurements							
Diet 1		Diet 2		Diet 3		Diet 4	
Post-	Pre-	Post-	Pre-	Post-	Pre-	Post-	Pre-
174	221	211	203	199	249	224	297
208	298	211	223	229	178	209	279
210	232	201	164	198	166	214	212
192	182	199	194	233	223	218	192
200	258	209	248	233	274	253	151
164	153	172	268	221	234	246	191
208	293	224	249	199	271	201	284
193	283	222	297	236	207	234	168

Fig. 5.29. Prediet and postdiet cholesterol levels by diet

SAS Example E6

To fit the unequal slopes model using `proc glm`, the `model` statement in the `proc glm` step in the SAS program shown in Fig. 5.19 is modified as shown in SAS Example E6 program (see Fig. 5.30). Recall that `diet` is a classificatory variable and `pre_chol` is a regression variable. The `pre_chol*diet` term in the above model is called a *discrete by continuous* interaction because of this reason. The resulting output is shown in Fig. 5.31.

The `proc glm` output in Fig. 5.31 can be used to obtain tests for three hypotheses of interest. First, the Type III SS for `pre_chol*diet` and the corresponding F-statistic provides a test of the hypothesis $H_0: \beta_1 = \beta_2 = \cdots = \beta_t = \beta$ versus H_a: at least one inequality (i.e., that the slopes are all the same). Second, the Type I SS for `pre_chol` and the corresponding F-statistic provides a test of the hypothesis $H_0: \beta = 0$ if the equal slopes model is used following the result of the previous test. This can be used to determine if the `post_chol` can be modeled as a linear function of the `pre_chol` at all.

The Type III SS for `diet` and the corresponding F-statistic provide a test of the hypothesis $H_0: \mu_{1j} = \mu_{2j} = \cdots = \mu_{tj}$ versus H_a: at least one inequality

```
data women;
input diet $ post_chol pre_chol;
datalines;
1 174 221
1 208 298
 .  .  .
 .  .  .
 .  .  .
1 208 293
1 193 283
2 211 203
2 211 223
 .  .  .
 .  .  .
 .  .  .
2 224 249
2 222 297
3 199 249
3 229 178
 .  .  .
 .  .  .
 .  .  .
3 199 271
3 236 207
4 224 297
4 209 279
 .  .  .
 .  .  .
 .  .  .
4 201 284
4 234 168
;
run;
proc glm data=women;
class diet;
model post_chol = diet pre_chol pre_chol*diet;
run;
```

Fig. 5.30. SAS Example E6: SAS Program

at the value $x_{ij} = 0$ for all i. Thus, this is equivalent to the test that the intercepts of the regression lines are the same (i.e., $H_0 : \alpha_1 = \alpha_2 = \cdots = \alpha_t$ versus H_a : at least one inequality at the value $x_{ij} = 0$ for all i); that is, the regression lines for the diets intersect at the pre_chol value of zero. Since the value of pre_chol can never be zero, this test is not a particularly useful.

Thus, the above test is omitted from the following adjusted analysis of covariance table. However, comparisons of these means (and therefore the intercepts), may be made at any selected value(s) using adjusted least squares means. To obtain adjusted least squares means modify the model statement in the proc glm step as follows:

```
model post_chol = diet  pre_chol*diet/noint solution;
```

and include the two lsmeans statements

```
lsmeans diet/cl tdiff adjust=bon at pre_chol=190;

lsmeans diet/cl tdiff adjust=bon at pre_chol=250;
```

SV	df	SS	MS	F	p-value
Diet(Unadj.)	3	4,593.84			
Regression	1	1.90	1.90	0.01	.9374
Error	27	8,187.72	303.25		
Regression	3	2,334.81	778.27	3.19	.0417
Error(Adj.)	24	5,852.91	243.87		
Total	31	12,783.47			

The results from the `solution` option in the modified model statement are shown in Fig. 5.32. The estimates of the coefficients available in this output are used for drawing the fitted lines shown in Fig. 5.33.

Options available for the `lsmeans` statement are used to perform multiple comparisons of the adjusted post_chol means by constructing Bonferroni adjusted confidence intervals at two different values of the prediet cholesterol levels using the `at` option. The values of 190 and 250 were selected because they are values just below the "desired" level of 200 and just above the "high-risk" level of 240, respectively. Extracts from the SAS output from the `lsmeans` statements are shown in Fig. 5.34.

```
                          The SAS System                              2

                         The GLM Procedure

Dependent Variable: post_chol

                                    Sum of
Source                    DF       Squares     Mean Square    F Value    Pr > F

Model                      7    6930.55635      990.07948       4.06    0.0045

Error                     24    5852.91240      243.87135

Corrected Total           31   12783.46875


            R-Square     Coeff Var      Root MSE    post_chol Mean

            0.542150      7.408809      15.61638         210.7813


Source                    DF     Type I SS     Mean Square    F Value    Pr > F

diet                       3    4593.843750    1531.281250       6.28    0.0027
pre_chol                   1       1.903672       1.903672       0.01    0.9303
pre_chol*diet              3    2334.808924     778.269641       3.19    0.0417


Source                    DF    Type III SS    Mean Square    F Value    Pr > F

diet                       3    3718.772130    1239.590710       5.08    0.0073
pre_chol                   1       1.824611       1.824611       0.01    0.9318
pre_chol*diet              3    2334.808924     778.269641       3.19    0.0417
```

Fig. 5.31. SAS Example E6: Unequal slopes model output (page 2)

All pairs of diet means are found to be not significantly different at the prediet cholesterol level of 250 (since the intervals for all differences included zero); however, at the prediet cholesterol level of 190, diet 1 post_chol mean is found to be significantly lower than both diets 3 and 4 means. Note that the default confidence coefficient was 95% for these intervals.

Parameter		Estimate	Standard Error	t Value	Pr > |t|
diet	1	137.6323082	27.26507534	5.05	<.0001
diet	2	195.7360754	32.03129844	6.11	<.0001
diet	3	223.7329893	33.70599704	6.64	<.0001
diet	4	276.6032780	23.66995633	11.69	<.0001
pre_chol*diet	1	0.2333029	0.11125081	2.10	0.0467
pre_chol*diet	2	0.0450224	0.13673614	0.33	0.7448
pre_chol*diet	3	-0.0232319	0.14761695	-0.16	0.8763
pre_chol*diet	4	-0.2332730	0.10379713	-2.25	0.0341

Fig. 5.32. SAS Example E6: Regression parameter estimates in the unequal slopes model

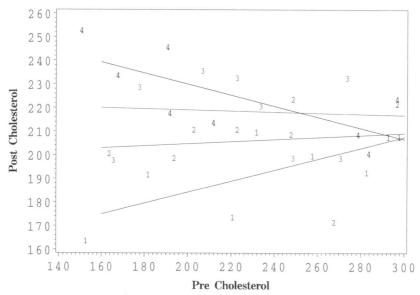

Fig. 5.33. SAS Example E6: Plot of postdiet cholesterol versus prediet cholesterol by diet

```
                        The GLM Procedure
                 Least Squares Means at pre_chol=190
              Adjustment for Multiple Comparisons: Bonferroni

                  post_chol        Standard                    LSMEAN
       diet         LSMEAN           Error      Pr > |t|       Number

        1         181.959856        7.837460     <.0001           1
        2         204.290336        7.844175     <.0001           2
        3         219.318925        7.586851     <.0001           3
        4         232.281416        6.429979     <.0001           4


                          Difference        Simultaneous 95%
                           Between         Confidence Limits for
             i     j         Means          LSMean(i)-LSMean(j)

             1     2       -22.330480      -54.211226     9.550266
             1     3       -37.359069      -68.720812    -5.997326
             1     4       -50.321560      -79.468042   -21.175079
             2     3       -15.028589      -46.404206    16.347028
             2     4       -27.991080      -57.152490     1.170330
             3     4       -12.962491      -41.555581    15.630598


                 Least Squares Means at pre_chol=250
              Adjustment for Multiple Comparisons: Bonferroni

                  post_chol        Standard                    LSMEAN
       diet         LSMEAN           Error      Pr > |t|       Number

        1         195.958029        5.632193     <.0001           1
        2         206.991682        6.116555     <.0001           2
        3         217.925010        6.620583     <.0001           3
        4         218.285039        6.251569     <.0001           4


                          Difference        Simultaneous 95%
                           Between         Confidence Limits for
             i     j         Means          LSMean(i)-LSMean(j)

             1     2       -11.033653      -34.939131    12.871825
             1     3       -21.966981      -46.957773     3.023811
             1     4       -22.327010      -46.519475     1.865455
             2     3       -10.933328      -36.848180    14.981524
             2     4       -11.293357      -36.439236    13.852521
             3     4        -0.360029      -26.539850    25.819792
```

Fig. 5.34. SAS Example E6: Comparison of means in the unequal slopes model

Following the method in SAS Example E5, a graph containing plots of the fitted models of the postdiet cholesterol values as straight lines of prediet cholesterol predictors superimposed on the scatter plots of the data values was constructed and is shown in Fig. 5.33. As suggested from the results of the multiple comparisons procedure of the adjusted diet means, it is observed from this graph that the differences of average postdiet cholesterol among the diets are larger for those individuals with higher prediet cholesterol than those with lower prediet levels.

5.4 A Two-Way Factorial in a Completely Randomized Design

A two-way factorial treatment structure consists of all combinations of levels of two factors under study in the experiment. The design employed is a completely randomized design (CRD) if these treatment combinations have been applied completely randomly to the experimental units. In the following discussion, it is assumed that all combinations of a two-way factorial with "a" levels of factor A and "b" levels of factor B are used and an equal number of replications per each treatment combination are obtained. The cake-baking experiment discussed earlier is an example of this set-up.

Model

The model is

$$y_{ijk} = \underbrace{\mu + \alpha_i + \beta_j + \gamma_{ij}}_{\mu_{ij}} + \epsilon_{ijk}, \quad i=1,\ldots,a; \quad j=1,\ldots,b; \quad k=1,\ldots,n$$

where $\mu_{ij} = E(y_{ijk})$ is the mean of an observation in the ijth cell of the two-way classification and are called the *cell means*. It is also assumed that the random error ϵ_{ijk} is distributed as iid $N(0, \sigma^2)$. A model expressed in terms of the cell means is called the "means model." If the cell means are partitioned into the sum of effects α_i of level i of A, β_j of level j of B, and an interaction effect γ_{ij} of the ith level of A and the jth level of B, an "effects model" is said to be in use.

The n observations corresponding to the ijth treatment combination y_{ijk}, $k=1,\ldots,n$, are assumed to be a random sample from the $N(\mu_{ij}, \sigma^2)$ distribution under this model. It is convenient and useful to express the hypotheses of interest to the experimenter in terms of "averaged" means or marginal means of the cell means that are defined as follows:

$$\text{Factor A Means:} \quad \bar{\mu}_{i.} = (\sum_j \mu_{ij})/b, \quad i=1,\ldots,a$$

$$\text{Factor B Means:} \quad \bar{\mu}_{.j} = (\sum_i \mu_{ij})/a, \quad j=1,\ldots,b$$

A table of cell means as shown in Fig. 5.35 is a visual illustration of the two-way classification model with equal sample sizes in each cell. The "averaged" means defined above for the two factors appear in the margins of this table.

Hypotheses Testing

The usual format of analysis of variance (anova) table for computing the required F-statistics for testing hypotheses of interest in a two-way classification is

5.4 A Two-Way Factorial in a Completely Randomized Design

		Levels of Factor B					
		1	2	...	j	...	b	
	1	μ_{11}	μ_{12}	...		...	μ_{1b}	$\bar{\mu}_{1.}$
	2	μ_{21}	μ_{22}				μ_{2b}	$\bar{\mu}_{2.}$
Levels of Factor A	i				μ_{ij}			$\bar{\mu}_{i.}$
	a	μ_{a1}	μ_{a2}	...		...	μ_{ab}	$\bar{\mu}_{a.}$
		$\bar{\mu}_{.1}$	$\bar{\mu}_{.2}$	...	$\bar{\mu}_{.j}$	...	$\bar{\mu}_{.b}$	

Fig. 5.35. Means model: Cell means and marginal means

SV	df	SS	MS	F	
Treatment	$ab-1$	SS_{Trt}			
A	$a-1$	SS_A	MS_A	MS_A/MSE	(1)
B	$b-1$	SS_B	MS_B	MS_B/MSE	(2)
A*B	$(a-1)(b-1)$	SS_{AB}	MS_{AB}	MS_{AB}/MSE	(3)
Error	$ab(n-1)$	SSE	MSE		
Total	$abn-1$	SS_{Tot}			

The F-statistics from the anova table are used to perform tests of the following hypotheses of interest:

(1) Use this F-statistic to test main effects of Factor A. Using the marginal means for Factor A, these are expressed as

$$H_0: \bar{\mu}_{1.} = \bar{\mu}_{2.} = \cdots = \bar{\mu}_{a.} \text{ vs. } H_a: \text{at least one inequality}$$

(2) Use this F-statistic to test main effects of Factor B. Using the marginal means for Factor B, these are expressed as

$$H_0: \bar{\mu}_{.1} = \bar{\mu}_{.2} = \cdots = \bar{\mu}_{.b} \text{ vs. } H_a: \text{at least one inequality}$$

(3) Use this F-statistic to test interaction of Factors A and B. Using the marginal means for both Factors A and B and the cell means, these are expressed as

$$H_0: (\mu_{ij} - \bar{\mu}_{i.} - \bar{\mu}_{.j} + \bar{\mu}_{..}) = 0 \text{ for all combinations of } (i,j) \text{ vs.}$$
$$H_a: \text{at least one } (\mu_{ij} - \bar{\mu}_{i.} - \bar{\mu}_{.j} + \bar{\mu}_{..}) \neq 0$$

(equivalent to H_0 : no interaction present   vs. H_a : interaction present)

An approach for using these in practical situations and how to proceed based on the result of each test are discussed below and in the several examples to follow. Although the interpretation of the results of the experiment appear to be simpler using the "means model" and associated means, in practice many experimenters resort to the "effects model" for such purposes. SAS procedures available for the analysis of data from designed experiments usually require that the effects models be used to describe the model equation to the program. An understanding of the theory of the linear model is necessary to clarify complications that result from the usage of the "effects model." An attempt will be made to illustrate some of these differences in the course of the discussions of the examples below.

Estimation

The best estimates of the cell means μ_{ij} and the marginal means $\bar{\mu}_{i.}$ and $\bar{\mu}_{.j}$ respectively are given by

$$\hat{\mu}_{ij} = \bar{y}_{ij.} = (\sum_k y_{ijk})/n$$

$$\hat{\bar{\mu}}_{i.} = \bar{y}_{i..} = (\sum_j \sum_k y_{ijk})/bn$$

$$\hat{\bar{\mu}}_{.j} = \bar{y}_{.j.} = (\sum_i \sum_k y_{ijk})/an$$

An estimate of the error variance σ^2 is $\hat{\sigma}^2 = s^2$, where s^2 is the MSE value obtained from the anova table. The standard error of the difference in the pair of Factor A means at levels i and i' is

$$\text{s.e.}(\bar{y}_{i..} - \bar{y}_{i'..}) = s\sqrt{2/bn}$$

and the standard error of the difference in the pair of Factor B means at levels j and j' is

$$\text{s.e.}(\bar{y}_{.j.} - \bar{y}_{.j'.}) = s\sqrt{2/an}$$

Thus, (1-α)100% confidence intervals for the differences in a pair of Factor A and B means are respectively given by

$$\bar{\mu}_{i.} - \bar{\mu}_{i'.} : \quad (\bar{y}_{i..} - \bar{y}_{i'..}) \pm t_{\alpha/2,\nu} \cdot s \cdot \sqrt{2/bn}$$

$$\bar{\mu}_{.j} - \bar{\mu}_{.j'} : \quad (\bar{y}_{.j.} - \bar{y}_{.j'.}) \pm t_{\alpha/2,\nu} \cdot s \cdot \sqrt{2/an}$$

where $t_{\alpha/2,\nu}$ is the upper $\alpha/2$ percentile of the t-distribution with ν df and ν is the degrees of freedom for MSE equal to $ab(n-1)$.

Differences in factor means ($\bar{\mu}_{i.}$ or $\bar{\mu}_{.j}$) may not measure actual differences in the *cell means* for Factor A or Factor B, respectively at levels of the other factor when interaction is present (i.e. when the model is nonadditive). Thus, the interpretation of main effects depends on whether interaction effect is found to be significant or not.

- The F-test for interaction in the anova table is a test whether the model is additive. It is recommended that this test be performed prior to making inferences from the main effects tests.
- If the interaction effect turns out to be not significant, then essentially the effects of the two factors A and B may be interpreted independently of each other. The F-tests for Factors A and B in the anova table are then used to test for main effects of A and B. If either or both of these main effect F-tests are significant, then the averaged marginal means may be compared (say, using preplanned comparisons or multiple comparison procedures) and significant comparisons interpreted as usual.
- If interaction F-test is significant, then the model is non-additive. This implies that care must be taken in interpreting main effect hypotheses of Factors A and B because there is significant interaction. The F-tests for main effects may still be performed, but the results may not be meaningful because differences in averaged means may not reflect the differences of the effects of one factor at each level of the other factor.
- An interaction plot may be useful for identifying whether differences in main effect means (marginal means) are affected significantly by interaction. If it is found that this is the case, comparisons of cell means of one factor over the levels of the other factor (e.g., $\mu_{12}-\mu_{13}$), may be necessary. If preplanned comparisons of the factor means are available, interesting interaction comparisons may be constructed that will aid in interpreting the significant interaction.

5.4.1 Analysis of a two-way factorial using PROC GLM

The data shown in Table 5.4 are survival times of groups of four animals randomly allocated to each of all combinations of three poisons and four drugs. The experiment was an investigation to combat the effects of certain toxic agents. This example is from Box et al. (1978) where a standard analysis as well as an analysis based on transforming the data using a variance stabilizing transformation are performed. It is assumed that the observations in each cell of the above classification are random samples of size 4 from normal distributions with means μ_{ij} and the same variance σ^2. The model is thus

$$y_{ijk} = \mu_{ij} + \epsilon_{ijk}, \quad i = 1, 2, 3; \quad j = 1, 2, 3, 4; \quad k = 1, 2, 3, 4$$

where $\epsilon_{ij} \sim$ iid $N(0, \sigma^2)$. A nonadditive model $\mu_{ij} = \mu + \alpha_i + \beta_j + \gamma_{ij}$ for the cell means is considered for partitioning the treatment sum of squares given the two-way factorial treatment structure.

SAS Example E7

In SAS Example E7, the `glm` procedure in SAS is used to obtain the appropriate analysis of variance table. The SAS program is shown in Fig. 5.36. The

Poison	Drug			
	A	B	C	D
I	0.31	0.82	0.43	0.45
	0.45	1.10	0.45	0.71
	0.46	0.88	0.63	0.66
	0.43	0.72	0.76	0.62
II	0.36	0.92	0.44	0.56
	0.29	0.61	0.35	1.02
	0.40	0.49	0.31	0.71
	0.23	1.24	0.40	0.38
III	0.22	0.30	0.23	0.30
	0.21	0.37	0.25	0.36
	0.18	0.38	0.24	0.31
	0.23	0.29	0.22	0.33

Table 5.4. Survival times data

class statement must precede the model statement and declare the classification variables, here the two factors poison and drug. Note that the "effects model" is used for formulating the model statement, by including a term for each effect (except a term for μ that is assumed to be in the model by default). The poison× drug interaction term is specified as poison*drug in the model statement. The means statement with the lsd option requests that all pairwise comparisons be made for both poison and drug means. In this case, by default, the lsd procedure is performed because the sample sizes are equal. On the other hand, if the sample sizes were unequal, confidence intervals would be constructed for all pairwise differences of the main effects. These could be specifically requested by including the option cldiff in the means statement. The confidence coefficient used by default is 95%; this could be changed by using the option alpha=.

In the course of analyzing two-way factorial data using SAS, the means and gplot procedures are used prior to the glm procedure to obtain a scatter plot of the cell means. This plot, used as an example of an interaction plot in Chapter 3 (see Section 3.6) and displayed in Fig. 3.20, shows a *profile* of the means across the levels of one factor at the same level of the other factor. In Fig. 3.20, levels of poison are plotted on the x-axis and the points corresponding to the same levels of drug are connected with line segments. Not only do the line segments allow the pattern of the mean response of each drug to the three poisons to be observed visually, but they also allow the mean responses to be compared across the four drugs. Thus, it is useful for interpreting and locating any significant interaction that may exist between the two factors.

5.4 A Two-Way Factorial in a Completely Randomized Design

```
data mice;
input  poison 1. @;
   do drug=1 to 4;
      input time 3.2 @;
      output;
   end;
datalines;
1 31 82 43 45
1 45110 45 71
1 46 88 63 66
1 43 72 76 62
2 36 92 44 56
2 29 61 35102
2 40 49 31 71
2 23124 40 38
3 22 30 23 30
3 21 37 25 36
3 18 38 24 31
3 23 29 22 33
;
run;

proc print ;
   title 'Analysis of Survival Times of Mice: Original Data';
run;

proc glm data=mice;
   class poison drug;
   model time = poison drug poison*drug;
   means poison drug/lsd;
run;
```

Fig. 5.36. SAS Example E7: Program

Since there are equal sample sizes for each treatment combination (number of observations in each cell), the Type I and III sums of squares are the same as expected. However, it is recommended that Type III sums of squares be always used in situations where the model does not contain any terms other than those representing fixed classificatory factors and their interactions. From the output from SAS Example E7 shown in Fig. 5.37, the following analysis of variance table is constructed.

SV	df	SS	MS	F	p-value
Treatment	11	2.2043			
Poison	2	1.0330	0.51651	23.22	< .0001
Drug	3	0.9212	0.30707	13.81	< .0001
Poison × Drug	6	0.2501	0.04169	1.87	.1123
Error	36	0.8007	0.02224		
Total	47	3.0051			

Since the interaction between poison and drug is not significant at 5%, one may conclude that these two factors act additively. Thus, it may be reasonable to examine the main effects and test the hypotheses $H_{01}: \bar{\mu}_{1.} = \bar{\mu}_{2.} = \bar{\mu}_{3.}$ and $H_{02}: \bar{\mu}_{.1} = \bar{\mu}_{.2} = \bar{\mu}_{.3} = \bar{\mu}_{.4}$ independently, in order to determine the effects of poisons and drugs, respectively. As seen from the extremely small p-values, both `poison` and `drug` effects are highly significant. The LSD procedure output produced from the `means` statement, shown in Fig. 5.38 for poison and drug means, finds that at $\alpha = .05$, the mean survival times for

(i) Poison 3 is significantly lower than those of Poisons 1 and 2,
(ii) Poison 2 and 3 are not significantly different,

```
              Analysis of Survival Times of Mice: Original Data            3

                              The GLM Procedure

                          Class Level Information

                     Class         Levels    Values

                     poison            3     1 2 3

                     drug              4     1 2 3 4


              Number of Observations Read              48
              Number of Observations Used              48

              Analysis of Survival Times of Mice: Original Data            4

                              The GLM Procedure

Dependent Variable: time

                                  Sum of
Source                    DF     Squares    Mean Square   F Value   Pr > F

Model                     11    2.20435625    0.20039602     9.01   <.0001

Error                     36    0.80072500    0.02224236

Corrected Total           47    3.00508125


             R-Square     Coeff Var      Root MSE     time Mean

             0.733543     31.11108       0.149139     0.479375


Source                    DF   Type I SS    Mean Square   F Value   Pr > F

poison                     2   1.03301250    0.51650625    23.22    <.0001
drug                       3   0.92120625    0.30706875    13.81    <.0001
poison*drug                6   0.25013750    0.04168958     1.87    0.1123


Source                    DF   Type III SS  Mean Square   F Value   Pr > F

poison                     2   1.03301250    0.51650625    23.22    <.0001
drug                       3   0.92120625    0.30706875    13.81    <.0001
poison*drug                6   0.25013750    0.04168958     1.87    0.1123
```

Fig. 5.37. SAS Example E7: Output (pages 3 and 4)

5.4 A Two-Way Factorial in a Completely Randomized Design

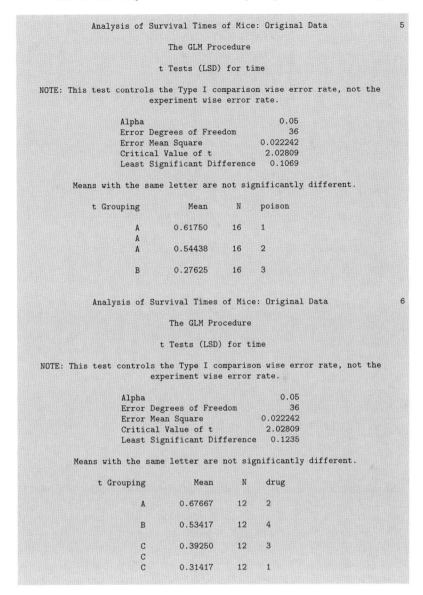

Fig. 5.38. SAS Example E7: Output (pages 5 and 6)

(iii) Drugs 1 and 3 are not significantly different but significantly lower than those of Drugs 2 and 4, respectively, and
(iv) Drug 4 is significantly lower than that of Drug 2.

The analysis would have greatly improved if preplanned comparisons considered important were suggested by the experimenter. Due to the lack of such comparisons, the interpretation of the results relied on multiple comparison procedures. Adjustments for making multiple comparisons can be made by using methods such as those based on Bonferroni or Tukey procedures. These methods will be used in other examples to follow.

5.4.2 Residual analysis and transformations

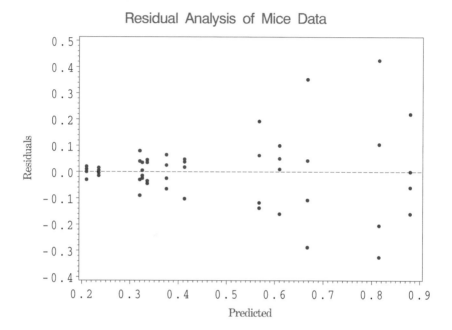

Fig. 5.39. SAS Example E7: Plot of residuals versus predicted values

A residual analysis showed that the variance of the data increases with the expected mean of the observed data. This was clearly evident in the plot of residuals against the predicted values shown in Fig. 5.39. This plot is obtained by modifying the `proc glm` step in the SAS Example E7 program by including the following statement:

```
output out=new r=res p=yhat;
```

This results in the residuals and the predicted values from the fitted model being added to the original data in the SAS data set named `mice` and saved in a new data set named `new`. The new proc steps are shown in Fig. 5.40

This suggested that a *variance stabilizing transformation* may be attempted to increase the sensitivity of the experiment as well as for easier interpretation of the results. If the error variance σ is proportional to a power

5.4 A Two-Way Factorial in a Completely Randomized Design

α of the mean μ, a power transformation of the response of the form y^λ may be attempted to stabilize the variance. The required power transformation can be shown to be given by $\lambda = 1 - \alpha$.

```
proc glm data=mice;
  class poison drug;
  model time = poison drug poison*drug;
  output out=new r=res p=yhat;
run;

symbol1 v=dot c=red i=none;
axis1 c=vipr label=(c=blueviolet h=1.5 a=90 f=centx 'Residuals');
axis2 c=vipr label=(c=blueviolet h=1.5 f=centx 'Predicted');

proc gplot data=new;
plot res*yhat/ vaxis=axis1 haxis=axis2 vref=0 cvref=darkgreen lvref=3;
title c=steelblue h=2 'Residual Analysis of Mice Data';
run;
```

Fig. 5.40. SAS Example E7: Modified version to create residual plot

An empirical method for obtaining an estimate of α for replicated data is to plot the logarithm of the sample standard deviation for each treatment combination against the logarithm of the sample mean. By fitting a straight line, an interval estimate of the *slope* of the line α can be obtained. Estimates of α thus obtained can be used to determine an appropriate transformation λ. Most common such transformations are *reciprocal, inverse square root, logarithmic,* and *square root* respectively for estimates of λ close to $-1, -\frac{1}{2}, 0,$ and $\frac{1}{2}$. The plot of the log s_{ij} versus log $\bar{y}_{ij}$ values for the 12 cells of the mice data is shown in Fig. 5.41.

The estimated slope of the regression is 1.977 and a 95% interval is (1.39, 2.56), suggesting that a reciprocal transformation of survival times (also called an inverse transformation) may be appropriate. Generally, for data measured in time units, an inverse transformation is often found to be appropriate for stabilizing the variance; the data values are transformed into *survival rates*, a natural unit of measure for this study.

SV	df	SS	MS	F	p-value
Treatment	11	56.8622			
Poison	2	34.8771	17.4386	72.63	<.0001
Drug	3	20.4143	6.8048	28.34	<.0001
Poison × Drug	6	1.5708	0.2618	1.09	.3867
Error	36	8.6431	0.2401		
Total	47	65.5053			

The SAS Example E7 program is modified to include the statement `time = 1/time;` in the data step to effect this transformation of the response variable. The analysis of variance obtained from this analysis is shown above.

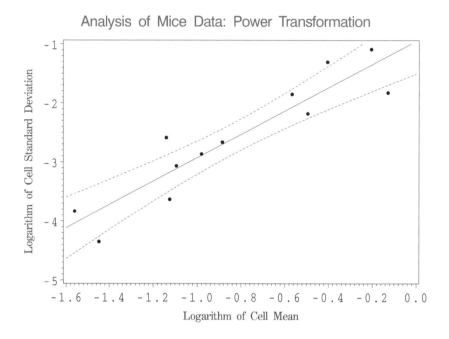

Fig. 5.41. SAS Example E7: Empirical Estimation of Variance Stabilizing Transformation

It is observed that the `poison` × `drug` interaction has become even less significant, thus allowing the experimenter to be more confident of the suitability of an additive model. Further, mean squares for both `poison` and `drug` effects are now much larger relative to the error mean square, implying increased sensitivity compared to the previous analysis. Note that for presentation of the results of the analysis, statistics calculated using the transformed data such as confidence intervals are preferably transformed back to the original units for easier interpretation by the experimenter.

5.5 Two-Way Factorial: Analysis of Interaction

In Section 5.4 it was shown how the treatment sum of squares (SS) with $(ab-1)$ degrees of freedom (df) was subdivided into sums of squares (SS) corresponding to main effects A and B and their interaction effect with $(a-1)$, $(b-1)$,

5.5 Two-Way Factorial: Analysis of Interaction

and $(a-1)(b-1)$ df, respectively. This subdivision was suggested by the effects model and enabled the testing of hypotheses appropriate for determining the presence or absence of interaction and main effects. In Section 5.2 it was shown how the treatment SS in a single-factor experiment may be partitioned into one df SS that are appropriate for making inferences about pre-planned or a priori comparisons among the factor means.

In general, for a factor with a levels, the associated df of $(a-1)$ implies that the SS may be partitioned into a set of $(a-1)$ orthogonal comparisons, and thus into $(a-1)$ SS each with a single degree of freedom. In the case of a two-factor experiment, each of the two main effects and interaction SS may be partitioned into several one df SS corresponding to comparisons of interest. In particular, the $(a-1)(b-1)$ df for interaction may also be partitioned into $(a-1)(b-1)$ one d.f. SS. The partitioning of the interaction SS may be derived on the basis of the main effect comparisons. For example, if a comparison of interest among Factor A means was "Control vs. Others," it might be of interest to the experimenter to examine whether this comparison differs among the levels of Factor B. The resulting comparisons constitute a subset of A × B interaction comparisons.

Generally, one df interaction comparisons that make sense may be formulated by considering one df main effect comparisons of the two factors. For example, in SAS Example E7 (see Section 5.4), the contrast coefficients corresponding to a possible comparison of interest $3\mu_{.1} - \mu_{.2} - \mu_{.3} - \mu_{.4}$ among the Poison means is (3 –1 –1 –1). Possible interaction comparisons may be those obtained by making the same comparison of the cell means at each level of the drug and then comparing them among the levels of the drug.

For example, to test that the above comparison is the same between levels 2 and 3 of the drug factor, the two comparisons of the cell means $3\mu_{21} - \mu_{22} - \mu_{23} - \mu_{24}$ and $3\mu_{31} - \mu_{32} - \mu_{33} - \mu_{34}$ must be compared. This comparison can be written using the contrast coefficients (0 0 0 0 +3 –1 –1 –1 –3 1 1 1), giving an interaction contrast of possible interest. Note that the coefficients may be obtained via the "elementwise product" of the main effect contrasts (0 1 –1) × (3 –1 –1 –1).

SAS Example E8

An example taken from Snedecor and Cochran (1989) illustrates the use of preplanned comparisons in two-way factorial experiments for analyzing interactions. The data shown below are gains in weight of male rats under six feeding treatments in a completely randomized design. The two factors were

A (2 levels) : Level of protein (high, low)

B (3 levels) : Source of protein (beef, cereal, pork)

The data are shown in Table 5.5.

	High Protein			Low Protein		
	Beef	Cereal	Pork	Beef	Cereal	Pork
	73	98	94	90	107	49
	102	74	79	76	95	82
	118	56	96	90	97	73
	104	111	98	64	80	86
	81	95	102	86	98	81
	107	88	102	51	74	97
	100	82	108	72	74	106
	87	77	91	90	67	70
	117	86	120	95	89	61
	111	92	105	78	58	82

Table 5.5. Weight gain in rats under six diets

A standard analysis for a two-way factorial in a completely randomized design would start with computing an analysis variance table as described in Section 5.4. Using the output from a `proc glm` step in SAS, the following analysis of variance table was constructed:

SV	df	SS	MS	F	p-value
Treatments	5	4,612.93	922.59	4.30	.0023
Level	1	3,168.27	3,168.27	14.77	.0003
Source	2	266.53	133.27	0.62	.5411
Level × Source	2	1,178.13	589.07	2.75	.0732
Error	54	11,586.00	214.56		
Total	59	16,198.93			

Since this analysis of variance table shows that the p-value value for the Level×Source interaction is between 10% and 5%, it would not be possible for the experimenter to be entirely comfortable in assuming an additive model, although the null hypothesis of no interaction will be rejected at $\alpha = .1$. This demonstrates the dilemma an experimenter might encounter if one attempts to interpret results from a factorial experiment using the usual partitioning of treatment sum of squares found in an analysis of variance table for two-way classifications. Proceed with an analysis of main effects assuming an additive model or attempt to make sense of the interaction present?

On the other hand, in many experiments of this nature, the structure of the treatment factors may suggest a priori comparisons among levels of factors, which might be more helpful for making useful interpretations. These may lead to a more natural explanation of any interaction that may be present among these factor levels. In the above experiment, comparisons of interest among the sources of protein means may be

- Average of beef and pork with cereal, and
- Beef versus pork.

5.5 Two-Way Factorial: Analysis of Interaction

These comparisons are suggested since beef and pork are animal sources of protein, whereas cereal is a vegetable source. When the main effects sums of squares are subdivided into single-degree of freedom sums of squares, a comparable subdivision of the interaction sum of squares can also be made. To illustrate the procedure, as usual assume the model for observations be

$$y_{ijk} = \mu_{ij} + \epsilon_{ijk}, \quad i = 1, 2; \quad j = 1, 2, 3; \quad k = 1, ..., 10$$

The means in the above model are the population means of the observations obtained from each combination of the two factors Level of Protein and Source of Protein as illustrated in the following table:

		Source			
		Beef	Cereal	Pork	
Level	High	μ_{11}	μ_{12}	μ_{13}	$\bar{\mu}_{1.}$
	Low	μ_{21}	μ_{22}	μ_{23}	$\bar{\mu}_{2.}$
		$\bar{\mu}_{.1}$	$\bar{\mu}_{.2}$	$\bar{\mu}_{.3}$	

The sample cell means and marginal means that are the best estimates of the above parameters are found in the following table:

		Source			
		Beef	Cereal	Pork	
Level	High	100.0	85.9	99.5	95.13
	Low	79.2	83.9	78.7	80.6
		89.6	84.9	89.1	

The usual main effect hypotheses based on the marginal means are $H_{01}: \bar{\mu}_{1.} = \bar{\mu}_{2.}$ and $H_{02}: \bar{\mu}_{.1} = \bar{\mu}_{.2} = \bar{\mu}_{.3}$. However, instead of testing the main effect hypotheses, the corresponding df may be partitioned using orthogonal one df comparisons among the main effect means (marginal means). For example, the 2 df of the Protein sum of squares may be split into two one df sums of squares that represent two orthogonal comparisons. For example, the two comparisons of the protein means stated above may be formulated as the contrast of the marginal means: "Average of beef and pork with cereal" comparison is expressed in the form $(\bar{\mu}_{.1} + \bar{\mu}_{.3})/2 - \bar{\mu}_{.2}$ and "Beef with pork" comparison as $\bar{\mu}_{.1} - \bar{\mu}_{.2}$. Notice that these two comparisons are orthogonal to each other. They are summarized in the following table of contrast coefficients:

Comparison	Coefficients		
Animal vs. Vegetable	+1	-2	+1
Beef vs. Pork	+1	0	-1

The above comparisons are called "main effect" comparisons, as the contrasts involve only the marginal means. The corresponding subdivision of the interaction sum of squares are obtained by the following comparisons of the cell means. These correspond to the comparisons of the cell means $\mu_{11} + \mu_{13} - 2\mu_{12} = \mu_{21} + \mu_{23} - 2\mu_{22}$ and $\mu_{11} - \mu_{13} = \mu_{21} - \mu_{23}$. The comparisons such as $\mu_{11} - \mu_{13}$ and $\mu_{11} + \mu_{13} - 2\mu_{12}$ are called "simple effects." Thus interaction comparisons are comparisons of simple effects.

Comparison	Coefficients
Animal vs. Vegetable × Level of Protein	+1 −2 +1 −1 +2 −1
Beef vs. Pork × Level of Protein	+1 0 −1 −1 0 +1

The first interaction comparison tests if the "Animal vs. Vegetable" effect, if any, is the same at both levels of protein, whereas the second comparison tests if the "Beef vs. Pork" effect, if any, is the same at both levels of protein. The coefficients for the contrast labeled `Animal vs. Vegetable x Level of Protein` are obtained by the elementwise product $(1 \ -1) \times (+1 \ -2 \ 1)$ and those of `Beef vs. Pork x Level of Protein` by the product $(1 \ -1) \times (1 \ 0 \ -1)$.

```
data rats;
input level source @;
    do i=1 to 10;
        input wt @;
        output;
    end;
cards;
1 1  73 102 118 104  81 107 100  87 117 111
1 2  98  74  56 111  95  88  82  77  86  92
1 3  94  79  96  98 102 102 108  91 120 105
2 1  90  76  90  64  86  51  72  90  95  78
2 2 107  95  97  80  98  74  74  67  89  58
2 3  49  82  73  86  81  97 106  70  61  82
;
run;

proc glm ;
    class level source;
    model wt = level source level*source;
    contrast 'animal vs. vegetable'
             source 1 -2 1;
    contrast 'beef vs. pork'
             source 1 0 -1;
    contrast 'an. vs. veg. x level'
             level*source 1 -2 1 -1 2 -1;
    contrast 'bf. vs. pk. x level'
             level*source 1 0 -1 -1 0 1;
run;
```

Fig. 5.42. SAS Example E8: Program

Although the "means model" was used in the discussion making statistical inferences from the two-way classification, the use of the "effects model" is

5.5 Two-Way Factorial: Analysis of Interaction

required for specification of the model as well as contrast coefficients in SAS programs. The "effects model" for the two-way classification is the partitioning of the mean μ_{ij} into main effects and interaction parameters $\mu + \alpha_i + \beta_j + \gamma_{ij}$, as introduced in Section 5.4.

In SAS Example E8 (see program in Fig. 5.42), proc glm is used for the purpose of subdivision of the Source of Protein and the Level x Source sums of squares into single degree of freedom sums of square corresponding to the preplanned comparisons of the protein means discussed earlier. These single degree of freedom sums of squares are obtained using the contrast statement in the proc glm step.

Note that data arranged in this format may be handled conveniently by the use of input, do, and output statements as illustrated in the program. In fact, this structure of the input data is preferable to the more traditional method where a line of input data corresponds to each cell value.

The right-hand side of the model statement

```
model wt = level source level*source;
```

codes the effects model formulation for μ_{ij} given earlier with the terms level and source representing the two classificatory factors "Level of Protein" and "Source of Protein", respectively, and the term level*source representing the interaction effect.

It is important to note that because of the use of the "effects model" parameters in the specification of the single-degree of freedom hypotheses in the above program, it is necessary to ensure that the same ordering of the subscripts for the levels of the factors present in the data be maintained in the program, so that the contrast coefficients (1, –2, 1, etc.) used in defining the comparisons of interest may be used to specify those hypotheses.

This is accomplished by continuing to use the "Level of Protein" as the first factor and the "Source of Protein" as the second factor within the program. This is specified by the order of appearance of these effects in the class statement. In the SAS program (see Fig. 5.42), level appears before source in the class statement. With this class statement in effect, the level*source interaction contrast coefficients are ordered so that the second subscript corresponds to the levels of source and changes faster than the first subscript, which corresponds to the levels of the factor level. Otherwise, the coefficients as given in the previous section may define different comparisons among the means. Using the "effects model" it is seen that

$$\bar{\mu}_{.1} - 2\bar{\mu}_{.2} + \bar{\mu}_{.3}$$
$$= \frac{1}{2}\{\sum_i (\mu + \alpha_i + \beta_1 + \gamma_{i1})\} - \{\sum_i (\mu + \alpha_i + \beta_2 + \gamma_{i2})\}$$
$$+ \frac{1}{2}\{\sum_i (\mu + \alpha_i + \beta_3 + \gamma_{i3})\}$$
$$= \beta_1 - 2\beta_2 + \beta_3 + \bar{\gamma}_{.1} - 2\bar{\gamma}_{.2} + \bar{\gamma}_{.3}$$

Thus, it is clear that a comparison that can simply be specified as $\bar{\mu}_{.1} - 2\bar{\mu}_{.2} + \bar{\mu}_{.3}$ using the cell means involves both the main effect and interaction parameters, in terms of the effects parameters. Fortunately, for making a main effects comparisons using `proc glm`, when the sample sizes are equal, as in this example, one needs to specify only the main effect portion of the contrast. This is because `proc glm` completes the rest of the specification as a convenience for the user. In practice, most users are not aware of this occurrence. Because of this behavior, the statements

```
contrast 'animal vs. vegetable' source 1 -2 1
                        level*source .5 -1 .5 .5 -1 .5 ;
contrast 'animal vs. vegetable' source 1 -2 1 ;
```

will produce identical results. However, when an comparison involves the cell means (as opposed to marginal means), such as the comparison of Beef vs. Pork x Level of Protein, the `level*source` portion of the coefficients is important. Using the "effects model" it may be verified that

$$\mu_{11} - 2\mu_{12} + \mu_{13} - \mu_{21} + 2\mu_{22} - \mu_{23}$$
$$= (\mu + \alpha_1 + \beta_1 + \gamma_{11}) - 2(\mu + \alpha_1 + \beta_2 + \gamma_{12}) + (\mu + \alpha_1 + \beta_3 + \gamma_{13})$$
$$- (\mu + \alpha_2 + \beta_1 + \gamma_{21}) - 2(\mu + \alpha_2 + \beta_2 + \gamma_{22}) + (\mu + \alpha_2 + \beta_3 + \gamma_{23})$$
$$= \gamma_{11} - 2\gamma_{12} + \gamma_{13} + \gamma_{21} - 2\gamma_{22} + \gamma_{23}$$

is a contrast among the interaction parameters and does not involve any main effect parameters.

The following analysis of variance table is constructed from the SAS output shown in Fig. 5.43. A recommended format for the anova table is

SV	DF	SS	MS	F	p-value
Treatments	5	4,612.93	922.59	4.30	.0023
Level	1	3,168.27	3,168.27	14.77	.0003
Animal vs. Vegetable	1	264.03	264.03	1.23	.2722
Beef vs. Pork	1	2.50	2.50	0.01	.9144
(Animal vs. Vegetable)× Level	1	1,178.13	1,178.13	5.49	.0228
(Beef vs. Pork) × Level	1	0.00	0.00	0.00	1.0000
Error	54	11,586.00			
Total	59	16,198.93			

Of the five comparisons tested, two are significant at $\alpha = .05$. There appears to be no difference in the mean gains between Beef and Pork at either level of Protein as well as no difference between the animal and vegetable sources on the average. However, the interaction of animal versus vegetable comparison with level of protein is significant. This indicates that the animal versus vegetable simple effect is different at one level of protein compared to the other. The addition of the estimate statements

5.5 Two-Way Factorial: Analysis of Interaction

```
                         The SAS System                              2

                        The GLM Procedure

Dependent Variable: wt

                                     Sum of
Source                      DF      Squares    Mean Square   F Value   Pr > F

Model                        5    4612.93333     922.58667      4.30   0.0023

Error                       54   11586.00000     214.55556

Corrected Total             59   16198.93333


             R-Square     Coeff Var      Root MSE       wt Mean

             0.284768      16.67039      14.64772      87.86667


Source                      DF     Type I SS    Mean Square   F Value   Pr > F

level                        1    3168.266667    3168.266667    14.77   0.0003
source                       2     266.533333     133.266667     0.62   0.5411
level*source                 2    1178.133333     589.066667     2.75   0.0732


Source                      DF    Type III SS   Mean Square   F Value   Pr > F

level                        1    3168.266667    3168.266667    14.77   0.0003
source                       2     266.533333     133.266667     0.62   0.5411
level*source                 2    1178.133333     589.066667     2.75   0.0732


Contrast                    DF    Contrast SS   Mean Square   F Value   Pr > F

animal vs. vegetable         1     264.033333     264.033333     1.23   0.2722
beef vs. pork                1       2.500000       2.500000     0.01   0.9144
an. vs. veg. x level         1    1178.133333    1178.133333     5.49   0.0228
bf. vs. pk.  x level         1       0.000000       0.000000     0.00   1.0000
```

Fig. 5.43. SAS Example E8: Output

```
estimate 'an. vs. veg. at low'
                source 1 -2 1 level*source 1 -2 1 0 0 0;
estimate 'an. vs. veg. at high'
                source 1 -2 1 level*source 0 0 0 1 -2 1;
```

results in the t-tests shown in Fig. 5.44 for testing the animal versus vegetable simple effect at each level of protein: This clearly shows that the animal versus vegetable simple effect is only significant at the high level of protein.

If preplanned comparisons were not available and interaction turns out to be significant in a two-way classification, one option available is to compare means at each level of one factor. For example, the three protein source means

```
                                                 Standard
  Parameter                      Estimate          Error        t Value      Pr > |t|

  an. vs. veg. at high          27.7000000      11.3460713        2.44        0.0179
  an. vs. veg. at low           -9.9000000      11.3460713       -0.87        0.3868
```

Fig. 5.44. SAS Example E8: Animal versus vegetable at levels of protein

may be compared at each level of protein. The `slice=` option available with the `lsmeans` statement produces F-tests for these hypotheses. In the above example, F-tests for testing the hypotheses $H_0 : \mu_{11} = \mu_{12} = \mu_{13}$ and $H_0 : \mu_{21} = \mu_{22} = \mu_{23}$ are produced by the statement

```
lsmeans level*source/slice=level;
```

and are shown in Fig. 5.45.

One difference in the `lsmeans` statement from the `means` statement is that interaction effects can also be specified as arguments. For example, the statement

```
lsmeans level*source/cl pdiff adjust=tukey;
```

produces tests and confidence intervals for all pairwise differences of the form $\mu_{ij} - \mu_{i'j'}$. The output from this statement is not reproduced here. One caution is that output from such statements usually is quite extensive because of the large number of pairs of differences among the means, even between factors with smaller number of levels such as in the above example.

```
                           The GLM Procedure
                          Least Squares Means

               level*source Effect Sliced by level for wt

                               Sum of
          level       DF       Squares      Mean Square    F Value    Pr > F

            1          2     1280.066667     640.033333      2.98     0.0590
            2          2      164.600000      82.300000      0.38     0.6833
```

Fig. 5.45. SAS Example E8: `slice=` option in the `lsmeans` statement

5.6 Two-Way Factorial: Unequal Sample Sizes

In this section a detailed analysis of a two-way classification with unequal sample sizes is presented. The intent is to show the complexities that arise in the analysis because of the unbalanced data structure compared to the analysis of the complete data set used in the previous analysis of the two-way classification data as discussed in Sections 5.4 and 5.5. The model used for the purpose of the analysis in those sections is

$$y_{ijk} = \mu_{ij} + \epsilon_{ijk}$$

called the "means model." This was extended to an "effects model"

$$\mu_{ij} = \mu + \alpha_i + \beta_j + \gamma_{ij}$$

by partitioning each of the cell means μ_{ij}, as discussed in those sections. In this section, **proc glm** is used for the analysis of the same model and includes several SAS procedure information statements available with **proc glm** for illustrating their use. Some of these statements were used in Section 5.5 for the analysis of interaction comparisons. Although it is instructive to learn the syntax of these statements, an attempt is also made to illustrate typical examples in which these statements may provide useful information to the experimenter.

SAS Example E9

The data set used in SAS Example E8 is modified by deleting some data values to introduce unequal sample sizes ensuring that there are no missing cells. The data used are as shown in Table 5.6.

Poison	Drug			
	A	B	C	D
I	0.31	0.82	0.43	0.71
	0.45	1.10	0.45	0.66
	0.46		0.63	0.62
	0.43		0.76	
II	0.36	0.61	0.44	0.56
	0.40	0.49	0.35	0.71
			0.31	0.38
			0.40	
III	0.22	0.30	0.23	0.30
	0.21	0.37	0.25	0.36
	0.18	0.38	0.22	
	0.23			

Table 5.6. Survival times data: Unequal sample sizes

The following "effects model" is used to specify the model for analysis by **proc glm**,

$$y_{ijk} = \mu + \alpha_i + \beta_j + \gamma_{ij} + \epsilon_{ijk}$$

with $i = 1, 2, 3$ (poison levels), $j = 1, 2, 3, 4$ (drug levels), $k = 1,\ldots,n_{ij}$ (cell sample sizes), and $N = \sum_i \sum_j n_{ij}$ is the total number of observations. Thus, 20 ($= 1+3+4+12$) parameters are utilized to model the mean μ_{ij} as a linear function of parameters representing main effects and interaction. However, only 12 ($= 1+2+3+6$) degrees of freedom are available to be partitioned from the treatment sum of squares for this purpose. Hence, the model is said to be overspecified (or overparameterized); that is, since more parameters are being used than the available degrees of freedom, 8 out of the 20 parameters cannot be estimated. Thus, the normal equations to be solved to obtain the least squares estimates of the 20 parameters will not have a unique solution. To obtain a solution to the normal equations, restrictions (or constraints) on the parameters must be imposed. These may take the form of setting some of the parameters equal to functions of others or simply equating some of them to a baseline value such as zero. In this example, `proc glm` uses the following eight restrictions:

$$\alpha_3 = \beta_4 = \gamma_{14} = \gamma_{24} = \gamma_{31} = \gamma_{32} = \gamma_{33} = \gamma_{34} = 0$$

This implies that the normal equations may be solved to obtain "estimates" of the rest of the parameters and "estimates" of the above parameters will be set to the baseline value of zero. Obviously, this is not a unique solution to the normal equations, as other solutions may be obtained using different sets of restrictions.

In the SAS Example E9 (program shown in Fig. 5.46), the `model` statement in `proc glm` includes options `ss1`, `ss2`, `ss3`, and `ss4`, requesting that all four types of sums of squares computed by `proc glm` be output. In two-way classification models, the Types II, III, and IV sums of squares are all exactly the same in magnitude when the sample sizes are equal, but Type I sums of squares differ from those. When the sample sizes are unequal (with no completely empty cells), Type II sums of squares differ from Types III and IV sums of squares (which are still of the same magnitude). This is observed from page 2 of the output from this program displayed in Fig. 5.47.

Type I sums of squares that correspond to each effect in the model are computed by fitting each term in the order listed in the `model` statement sequentially and measuring the increase in the treatment sum of squares. Thus, Type I sums of squares are appropriate for testing the significance of adding an effect sequentially into the current model. Type III sum of squares, on the other hand, measure the increase in the treatment sum of squares when the effect is added to a model with all other effects already in the model. Thus, Type I sums of squares are appropriate for testing the significance of an effect in the full model. It is recommended that Type III sums of squares be used to construct an analysis of variance table because these correspond to those calculated in Yates' weighted squares of means analysis. That method is recommended for use for unbalanced data when main effects need to be tested in the presence of interaction. In addition, the `solution` option requests that a solution to the normal equations be produced. As expected, parameter estimates corresponding to those eight parameters discussed earlier are equal to zero as printed in the output resulting from this option shown in Fig. 5.48.

The estimates of other parameters shown in this output are not unique (these are flagged with the letter "B" to indicate this) **1**; this means that there is no direct interpretation of the magnitudes of these estimates. However, these values may be used to construct estimates of specific *estimable functions of the parameters* that are useful for interpreting the results of the experiment. Some of these estimates could be obtained directly by using

```
options ls=80 ps=50 nodate pageno=1;
data mice2;
input  poison 1. @;
   do drug=1 to 4;
      input time 3.2 @;
      output;
   end;
datalines;
1 31 82 43  .
1 45110 45 71
1 46   . 63 66
1 43   . 76 62
2 36   . 44 56
2  . 61 35  .
2 40 49 31 71
2  .   . 40 38
3 22 30 23 30
3 21 37 25 36
3 18 38  .  .
3 23   . 22  .
;
run;
proc glm data=mice2;
  class poison drug;
  model time = poison drug poison*drug/ss1 ss2 ss3 ss4 solution;

  means poison drug;
  lsmeans poison/stderr cl pdiff tdiff adjust=tukey;
  lsmeans drug /stderr cl pdiff tdiff adjust=tukey;

  contrast 'poison 1 vs 2' poison -1 1;
  contrast 'poison 1 vs 2 *' poison -1 1
               poison*drug  -.25 -.25 -.25 -.25 .25 .25 .25 .25;
  contrast 'drug A&B vs C&D' drug 1 1 -1 -1;

  estimate 'poison 1 vs 2' poison  -1 1;
  estimate 'poison 1 mean' intercept 1
                           poison 1 0 0
                           drug .25 .25 .25 .25
                           poison*drug .25 .25 .25 .25;
  estimate 'drug A @ poison 1-2' poison -1 1 poison*drug -1 0 0 0 1;

  title 'Analysis of  Two-Way Data : Unequal Sample Sizes';
run;
```

Fig. 5.46. SAS Example E9: Program

`lsmeans`, `contrast`, and `estimate` statements. Results obtained from some of the these statements included in the SAS program are discussed next.

The `means` statement provides the "unadjusted" Poison and Drug means, whereas the `lsmeans` statement produces the "least squares means" or the adjusted means together with their standard errors. The unadjusted means are shown in Fig 5.49. These quantities estimate functions of μ_{ij}. For example, consider the estimates printed for Poison 1. The unadjusted mean 0.6023 with sample size 13 is an estimate of the *weighted* marginal mean $(4\mu_{11} + 2\mu_{12} + 4\mu_{13} + 3\mu_{14})/(4+2+4+3)$, a "weighted average of cell means." As can be observed, this definition makes the marginal mean depend on the cell sample sizes. These are computed by the `means` statement and given on page 4 of the SAS output (see Fig. 5.49). The adjusted mean 0.6508 (see Fig. 5.50), however, is an estimate of $\bar{\mu}_{1\cdot}(= (\mu_{11} + \mu_{12} + \mu_{13} + \mu_{14})/4)$, an "unweighted average of cell means," defined assuming cell sample sizes are all equal as specified in the model definition. These are called *least squares means* and are estimates of cell means and marginal means defined using the population means irrespective of the sample sizes. The results from the `lsmeans` statement are reproduced in Fig. 5.50.

Perhaps estimates provided by the `lsmeans` statement are more appropriate if the interest is in the differences in Poison means of the treatment populations as defined by the model. This is justified from the point of view that for the purpose of comparing factor means, all cell means (treatment means) should be regarded as equally important and therefore, equally weighted, regardless of the sample sizes.

The best estimates of the cell means μ_{ij} and the marginal means $\bar{\mu}_{i\cdot}$ and $\bar{\mu}_{\cdot j}$, respectively, are given by

$$\hat{\mu}_{ij} = \bar{y}_{ij\cdot} = (\sum_k y_{ijk})/n_{ij}$$

$$\hat{\bar{\mu}}_{i\cdot} = \sum_j (\sum_k y_{ijk}/n_{ij})/b = (\sum_j \bar{y}_{ij\cdot})/b$$

$$\hat{\bar{\mu}}_{\cdot j} = \sum_i (\sum_k y_{ijk}/n_{ij})/a = (\sum_i \bar{y}_{ij\cdot})/a$$

where the sample sizes n_{ij} are given in the following table:

		Poison			
		1	2	3	4
	1	4	2	4	3
Drug	2	2	2	4	3
	3	4	3	3	2

The standard error of the difference in the estimates of Factor A means at levels i and i' is

```
                Analysis of Two-Way Data : Unequal Sample Sizes                    1

                                 The GLM Procedure

                              Class Level Information

                       Class         Levels     Values

                       poison             3     1 2 3

                       drug               4     1 2 3 4

                    Number of Observations Read           48
                    Number of Observations Used           36

                Analysis of Two-Way Data : Unequal Sample Sizes                    2

                                 The GLM Procedure

Dependent Variable: time

                                    Sum of
Source                    DF       Squares     Mean Square    F Value    Pr > F

Model                     11    1.22958056      0.11178005      12.70    <.0001

Error                     24    0.21118333      0.00879931

Corrected Total           35    1.44076389

            R-Square     Coeff Var      Root MSE     time Mean

            0.853423      20.98798      0.093805      0.446944


Source                    DF      Type I SS     Mean Square    F Value    Pr > F

poison                     2     0.68676873      0.34338436      39.02    <.0001
drug                       3     0.39430953      0.13143651      14.94    <.0001
poison*drug                6     0.14850230      0.02475038       2.81    0.0323


Source                    DF     Type II SS     Mean Square    F Value    Pr > F

poison                     2     0.72205463      0.36102731      41.03    <.0001
drug                       3     0.39430953      0.13143651      14.94    <.0001
poison*drug                6     0.14850230      0.02475038       2.81    0.0323


Source                    DF    Type III SS     Mean Square    F Value    Pr > F

poison                     2     0.79692581      0.39846290      45.28    <.0001
drug                       3     0.38425215      0.12808405      14.56    <.0001
poison*drug                6     0.14850230      0.02475038       2.81    0.0323


Source                    DF     Type IV SS     Mean Square    F Value    Pr > F

poison                     2     0.79692581      0.39846290      45.28    <.0001
drug                       3     0.38425215      0.12808405      14.56    <.0001
poison*drug                6     0.14850230      0.02475038       2.81    0.0323
```

Fig. 5.47. SAS Example E9: Types I, II, III, and IV sums of squares

```
                    Analysis of Two-Way Data : Unequal Sample Sizes                3

                                  The GLM Procedure

Dependent Variable: time

                                                 Standard
       Parameter               Estimate            Error      t Value    Pr > |t|

       Intercept             0.3300000000 B     0.06632988     4.98       <.0001
       poison      1         0.3333333333 B     0.08563150     3.89       0.0007
       poison      2         0.2200000000 B     0.08563150     2.57       0.0168
       poison      3         0.0000000000 B         .            .          .
       drug        1        -.1200000000 B      0.08123718    -1.48       0.1526
       drug        2         0.0200000000 B     0.08563150     0.23       0.8173
       drug        3        -.0966666667 B      0.08563150    -1.13       0.2701
       drug        4         0.0000000000 B         .            .          .
       poison*drug 1 1      -.1308333333 B      0.10831624    -1.21       0.2389
       poison*drug 1 2       0.2766666667 B     0.12110124     2.28       0.0315
       poison*drug 1 3       0.0008333333 B     0.11164982     0.01       0.9941
       poison*drug 1 4       0.0000000000 B         .            .          .
       poison*drug 2 1      -.0500000000 B      0.11803488    -0.42       0.6756
       poison*drug 2 2      -.0200000000 B      0.12110124    -0.17       0.8702
       poison*drug 2 3      -.0783333333 B      0.11164982    -0.70       0.4897
       poison*drug 2 4       0.0000000000 B         .            .          .
       poison*drug 3 1       0.0000000000 B         .            .          .
       poison*drug 3 2       0.0000000000 B         .            .          .
       poison*drug 3 3       0.0000000000 B         .            .          .
       poison*drug 3 4       0.0000000000 B         .            .          .

NOTE: The X'X matrix has been found to be singular, and a generalized inverse
      was used to solve the normal equations.  Terms whose estimates are
      followed by the letter 'B' are not uniquely estimable.
```

Fig. 5.48. SAS Example E9: Parameter estimates

$$\text{s.e.}(\hat{\hat{\mu}}_{i\cdot} - \hat{\hat{\mu}}_{i'\cdot}) = \frac{s}{b}\sqrt{\sum_j \left(\frac{1}{n_{ij}} + \frac{1}{n_{i'j}}\right)}$$

and the standard error of the difference in the estimates of pair of Factor B means at levels j and j' is

$$\text{s.e.}(\hat{\hat{\mu}}_{\cdot j} - \hat{\hat{\mu}}_{\cdot j'}) = \frac{s}{a}\sqrt{\sum_i \left(\frac{1}{n_{ij}} + \frac{1}{n_{ij'}}\right)}$$

where an estimate of the error variance $\hat{\sigma}^2 = s^2$, where s^2 is the MSE obtained from the anova table, as earlier. Thus, a $(1-\alpha)100\%$ confidence interval for the differences in a pair estimates Factor A and B means are respectively

$$(\hat{\hat{\mu}}_{i\cdot} - \hat{\hat{\mu}}_{i'\cdot}) \pm t_{\alpha/2,\nu} \cdot \frac{s}{b}\sqrt{\sum_j \left(\frac{1}{n_{ij}} + \frac{1}{n_{i'j}}\right)}$$

and

5.6 Two-Way Factorial: Unequal Sample Sizes

$$(\hat{\bar{\mu}}_{.j} - \hat{\bar{\mu}}_{.j'}) \pm t_{\alpha/2,\nu} \cdot \frac{s}{a}\sqrt{\sum_i \left(\frac{1}{n_{ij}} + \frac{1}{n_{ij'}}\right)}$$

where $t_{\alpha/2,\nu}$ is the upper $\alpha/2$ percentage point of the t-distribution with ν degrees of freedom, where $\nu = (N - ab)$ is the degrees of freedom for MSE.

Poison	Drug			
	A	B	C	D
I	0.4125	0.9600	0.5675	0.6633
II	0.3800	0.5500	0.3750	0.5500
III	0.2100	0.3500	0.2333	0.3300

Table 5.7. Survival times cell means

From Table 5.7 the least squares means for Poisons 1 and 2 are respectively $\hat{\bar{\mu}}_{1.} = (0.4125 + 0.9600 + 0.5675 + 0.6633)/4 = 0.6508$ and $\hat{\bar{\mu}}_{2.} = (0.3800 + 0.5500 + 0.3750 + 0.5500)/4 = 0.4638$. The standard error of the difference $\hat{\bar{\mu}}_{2.} - \hat{\bar{\mu}}_{1.}$ is

$$\frac{s}{b}\sqrt{\sum_j \left(\frac{1}{n_{1j}} + \frac{1}{n_{2j}}\right)} = \sqrt{.0088 \times \left(\frac{1}{4} + \frac{1}{2} + \frac{1}{2} + \frac{1}{2} + \frac{1}{4} + \frac{1}{4} + \frac{1}{3} + \frac{1}{3}\right)/4}$$

$$= 0.04$$

Pages 5 and 6 of the SAS output (see Figs. 5.50 and 5.51) resulting from the lsmeans statements contain estimates, tests, and confidence intervals for individual least squares means of poison and drug levels, respectively. These pages also contain estimates, tests and confidence intervals for all pairwise differences of least squares means for the two factors. Moreover, those comparisons incorporate the Tukey adjustment 2 of confidence levels for making multiple comparisons, as requested by using the lsmeans statement option adjust=tukey. Results of the calculations associated with the difference $\hat{\bar{\mu}}_{2.} - \hat{\bar{\mu}}_{1.}$ illustrated earlier can be checked from this output.

The contrast statements are similar to those one might use in the balanced case. For example, the first contrast statement computes an F-statistic to test the hypothesis

$$H_0: \bar{\mu}_{2.} - \bar{\mu}_{1.} = 0$$

and the second to test

$$H_0: (\bar{\mu}_{.1} + \bar{\mu}_{.2})/2 = (\bar{\mu}_{.3} + \bar{\mu}_{.4})/2$$

These functions are *estimable* because they are linear functions of the means μ_{ij}; however, they must first be expressed in terms of effects model parameters

α_i's (poison effects), β_j's (drug effects), and γ_{ij}'s (interaction effects), so that they could be specified in proc glm statements. As discussed in Section 5.5, both main effect and interaction parameters are needed to express the contrast for the means comparison $\bar{\mu}_{2.} - \bar{\mu}_{1.}$. The required expression is obtained by substituting

$$\mu_{ij} = \mu + \alpha_i + \beta_j + \gamma_{ij}$$

in $\bar{\mu}_{2.} - \bar{\mu}_{1.}$ as follows:

$(\mu_{21} + \mu_{22} + \mu_{23} + \mu_{24})/4 - (\mu_{11} + \mu_{12} + \mu_{13} + \mu_{14})/4$
$= \{(\mu + \alpha_2 + \beta_1 + \gamma_{21}) + (\mu + \alpha_2 + \beta_2 + \gamma_{22}) + (\mu + \alpha_2 + \beta_3 + \gamma_{23})$
$\quad + (\mu + \alpha_2 + \beta_4 + \gamma_{24})\}/4 - \{(\mu + \alpha_1 + \beta_1 + \gamma_{11}) + (\mu + \alpha_1 + \beta_2 + \gamma_{12})$
$\quad + (\mu + \alpha_1 + \beta_3 + \gamma_{13}) + (\mu + \alpha_1 + \beta_4 + \gamma_{14})\}/4$
$= (\alpha_2 - \alpha_1) + (\gamma_{21} + \gamma_{22} + \gamma_{23} + \gamma_{24})/4 - (\gamma_{11} + \gamma_{12} + \gamma_{13} + \gamma_{14})/4$

Thus, the contrast statement needed is

```
contrast 'poison 1 vs 2 *' poison -1 1
              poison*drug  -.25 -.25 -.25 -.25 .25 .25 .25 .25;
```

However, SAS allows the user to specify the comparison using only the main effect portion (and so completes the interaction portion of the coefficients needed). Thus, the statement

```
contrast 'poison 1 vs 2' poison -1 1;
```

produces identical results. The contrast statement for the second comparison, $(\mu_{.1} + \mu_{.2})/2 - (\mu_{.3} + \mu_{.4})/2$, that is equivalent to the comparison $\mu_{.1} + \mu_{.2} -$

```
             Analysis of Two-Way Data : Unequal Sample Sizes                 4
                             The GLM Procedure

        Level of                    -------------time------------
        poison         N                Mean              Std Dev

           1          13             0.60230769         0.21358659
           2          11             0.45545455         0.12420657
           3          12             0.27083333         0.06894772


        Level of                    -------------time------------
        drug           N                Mean              Std Dev

           1          10             0.32500000         0.10865337
           2           7             0.58142857         0.28852663
           3          11             0.40636364         0.16776607
           4           8             0.53750000         0.16671190
```

Fig. 5.49. SAS Example E9: Unadjusted means

```
              Analysis of Two-Way Data : Unequal Sample Sizes                 5

                              The GLM Procedure
                            Least Squares Means
                 Adjustment for Multiple Comparisons: Tukey-Kramer

                                Standard                       LSMEAN
       poison     time LSMEAN      Error       Pr > |t|        Number

         1        0.65083333     0.02707906     <.0001           1
         2        0.46375000     0.02950872     <.0001           2
         3        0.28083333     0.02791246     <.0001           3


                       Least Squares Means for Effect poison
                   t for H0: LSMean(i)=LSMean(j) / Pr > |t|

                            Dependent Variable: time

             i/j             1               2               3

              1                            4.67119         9.514176
                                           0.0003          <.0001

              2           -4.67119                         4.503275
                           0.0003                          0.0004

              3           -9.51418        -4.50327
                           <.0001          0.0004


       poison     time LSMEAN       95% Confidence Limits

         1         0.650833         0.594945        0.706722
         2         0.463750         0.402847        0.524653
         3         0.280833         0.223225        0.338442


                         Difference       Simultaneous 95%
                          Between       Confidence Limits for
        i    j             Means          LSMean(i)-LSMean(j)

        1    2            0.187083       0.087066       0.287101
        1    3            0.370000       0.272882       0.467118
        2    3            0.182917       0.081480       0.284353
```

Fig. 5.50. SAS Example E9: Poison comparisons

$\mu_{.3} - \mu_{.4})$ is

contrast 'drug A&B vs C&D' drug 1 1 -1 -1;

The two estimate statements request estimates of the functions of the cell means μ_{ij} : $(\mu_{11} + \mu_{12} + \mu_{13} + \mu_{14})/4$ and $\mu_{21} - \mu_{11}$. The first estimate statement is used to directly compute the Poison 1 least squares mean, in order to illustrate the differences between those resulting from the means and lsmeans statements. The second is simply the difference in cell means between Poison levels II and I at Drug level A. Such differences may be of interest if the interaction is found to be significant.

```
                Analysis of  Two-Way Data : Unequal Sample Sizes                 6
                              The GLM Procedure
                              Least Squares Means
                  Adjustment for Multiple Comparisons: Tukey-Kramer  2

                                        Standard                         LSMEAN
            drug        time LSMEAN      Error         Pr > |t|          Number

             1          0.33416667      0.03126820     <.0001              1
             2          0.62000000      0.03610541     <.0001              2
             3          0.39194444      0.02854383     <.0001              3
             4          0.51444444      0.03377352     <.0001              4

                      Least Squares Means for Effect drug
                      t for H0: LSMean(i)=LSMean(j) / Pr > |t|

                            Dependent Variable: time

           i/j              1              2              3              4

            1                           -5.98441       -1.3647        -3.91691
                                         <.0001         0.5326         0.0034

            2            5.984413                       4.954977       2.13505
                          <.0001                        0.0003         0.1708

            3            1.3647         -4.95498                      -2.77024
                          0.5326         0.0003                        0.0488

            4            3.916906       -2.13505        2.770245
                          0.0034         0.1708         0.0488

                  drug      time LSMEAN      95% Confidence Limits

                   1         0.334167        0.269632       0.398701
                   2         0.620000        0.545482       0.694518
                   3         0.391944        0.333033       0.450856
                   4         0.514444        0.444739       0.584150

                             Difference       Simultaneous 95%
                              Between       Confidence Limits for
                 i    j        Means          LSMean(i)-LSMean(j)

                 1    2       -0.285833      -0.417593      -0.154074
                 1    3       -0.057778      -0.174570       0.059014
                 1    4       -0.180278      -0.307244      -0.053311
                 2    3        0.228056       0.101089       0.355022
                 2    4        0.105556      -0.030828       0.241940
                 3    4       -0.122500      -0.244486      -0.000514
```

Fig. 5.51. SAS Example E9: Drug comparisons

As before, these are expressed as functions of effects model parameters:
(i) First,

$$(\mu_{11} + \mu_{12} + \mu_{13} + \mu_{14})/4$$
$$= \mu + \alpha_1 + (\beta_1 + \beta_2 + \beta_3 + \beta_4)/4 + (\gamma_{11} + \gamma_{12} + \gamma_{13} + \gamma_{14})/4$$
$$= \mu + \alpha_1 + \bar{\beta}_. + \bar{\gamma}_1.$$

The term μ in the above expression requires "intercept 1" to be included in the estimate statement and the term α_1 is coded as "poison 1 0 0" for

5.6 Two-Way Factorial: Unequal Sample Sizes

```
                 Analysis of  Two-Way Data : Unequal Sample Sizes              9

                                The GLM Procedure

Dependent Variable: time

Contrast                       DF    Contrast SS    Mean Square    F Value    Pr > F

poison 1 vs 2                   1     0.19200095     0.19200095      21.82    <.0001
poison 1 vs 2 *                 1     0.19200095     0.19200095      21.82    <.0001
drug A&B vs C&D                 1     0.00474103     0.00474103       0.54    0.4700

                                                 Standard
Parameter                         Estimate        Error      t Value    Pr > |t|

poison 1 vs 2                   -0.18708333     0.04005047    -4.67     <.0001
poison 1 mean                    0.65083333     0.02707906    24.03     <.0001
drug A @ poison 1-2             -0.03250000     0.08123718    -0.40     0.6926
```

Fig. 5.52. SAS Example E9: Output from Contrast and Estimate Statements

representing level 1 of the poison effect. The terms $(\beta_1 + \beta_2 + \beta_3 + \beta_4)/4$ requires the term "drug .25 .25 .25 .25" and $(\gamma_{11} + \gamma_{12} + \gamma_{13} + \gamma_{14})/4$ requires the term "poison*drug .25 .25 .25 .25 0 0 0 0 0 0 0 0." Thus, the complete statement needed to perform this computation is

```
estimate 'poison 1 mean'
            intercept 1
            poison 1 0 0
            drug .25 .25 .25 .25
            poison*drug .25 .25 .25 .25 0 0 0 0 0 0 0 0;
```

The zeros at the end of the estimate statement may be omitted. Of course, the above statement may also be entered in the form

```
estimate 'poison 1 mean' intercept 4
                         poison 4 0 0
                         drug 1 1 1 1
                         poison*drug 1 1 1 1/divisor=4;
```

using the `divisor=` option.

(ii) Similarly,
$$\mu_{21} - \mu_{11} = -\alpha_1 + \alpha_2 + \gamma_{21} - \gamma_{11}$$

The terms $-\alpha_1 + \alpha_2$ requires that "poison –1 1 0" be included in the estimate statement and $\gamma_{21} - \gamma_{11}$ requires that "poison*drug –1 0 0 0 1 0 0 0 0 0 0 0" be included. Thus, the complete statement is

```
'Drug A @ Poison 2 -1' poison -1 1 poison*drug -1 0 0 0 1;
```

again omitting the trailing zeros.

The results of the `contrast` and `estimate` statements used in SAS Example E9 are found in Fig. 5.52.

5.7 Two-Way Classification: Randomized Complete Block Design

When experimental units tend not to be homogeneous (as usually the case in many studies), the experimenter can often employ a different design than a completely randomized design (CRD) that may be more efficient by helping to control the error variance. One such design is called the *randomized complete block design* (RCBD). Recall that the analysis of data from an experiment carried out in a completely randomized design involves the estimation of the error variance by combining (or pooling) the sample variances calculated from responses from experimental units assigned the same treatment. This is called the *within sample variance* or *within treatment variance*.

How different experimental units respond to the application of the same treatment may depend on the nature of the units. For example, the yields of a variety of cereal crop from plots of land may vary widely if the plots are very different from each other in their soil constitution, moisture content, degree of drainage, exposure to sunlight or shade, or in some other way. The yields obtained from such plots planted with the same variety may lead to an inflated estimate of error variance than if the plots were more homogeneous.

By grouping experimental units that are considered similar, the contribution to experimental error due to this variation can be reduced or eliminated. Groups of units that are similar in this way comprise what are called *blocks*. In an agricultural experiment, these might be plots that are contiguously located in the field. Once blocks are formed, a *complete* set of treatments are applied to the experimental units in a block. Thus, the *block size* (the number of experimental units in a block) must be exactly the same as the number of treatments (i.e., the number of levels of the factor under study). This process is repeated for every block available. The treatments are assigned randomly to the experimental units within each block. The number of blocks is determined by the number of experimental units available. For example, if four varieties of corn were under study, the availability of 20 plots would ensure that 5 complete blocks could be formed.

The "blocks' are also called "reps" (for replications) because every treatment is repeated once in each block. A variation is that the complete set of treatments is applied more than once in a block. This would require at least twice the number of experimental units than would be used in a regular RCBD. Another variation is that less than the full set of treatments is used in each block. Such designs are called incomplete block designs and are not discussed in this book. Data from a one-factor experiment performed using a RCBD also form a two-way classification. In the following discussion, the more common practice of each treatment (or combination of treatments, if a factorial arrangement of treatments is used) appearing once in each block is considered

5.7 Two-Way Classification: Randomized Complete Block Design

Model

The model for observations from an RCBD is

$$y_{ij} = \underbrace{\mu + \tau_i}_{\mu_i} + \beta_j + \epsilon_{ij}, \quad i=1,\ldots,t; \quad j=1,\ldots,r$$

It is assumed that the random error ϵ_{ij} is distributed as iid $N(0, \sigma^2)$. In the "means model" μ_i is the mean of ith treatment and τ_i is effect of ith treatment. In this presentation as well as in many textbooks, for the purpose of analyzing results from an RCBD, the effect of the jth block, β_j, is considered a fixed effect. This may not be reasonable, as the blocks are selected randomly and thus are better represented in the model by random effects. In Chapter 6, an analysis of an RCBD as a *mixed model* will be presented. Although blocks are considered fixed effects an additive model is used to represent the observations; that is, an interaction term is not included in the model. Thus, the assumption that treatment differences remain the same from block to block is built into the model. This allows the error variance to be estimated even though treatments are not replicated within each block.

Estimation

The best estimates of μ_i and σ^2 are respectively

$$\hat{\mu}_i = \bar{y}_{i.} = \left(\sum_j y_{ij}\right)/r$$

$$\hat{\sigma}^2 = s^2$$

where s^2 is the error mean square from the analysis of variance table presented below. Since the difference in treatment means of two treatments labeled p and q is $\mu_p - \mu_q = \mu + \tau_p - (\mu - \tau_q) = \tau_p - \tau_q$, it is the same as the difference in the corresponding treatment effects. Similarly, a comparison in treatment means is identical to the same comparison in treatment effects.

The best estimate of the difference between the effects of two treatments labeled p and q is

$$\widehat{\mu_p - \mu_q} = \widehat{\tau_p - \tau_q} = \bar{y}_{p.} - \bar{y}_{q.}$$

with standard error given by

$$s_d = \text{s.e.}(\bar{y}_{p.} - \bar{y}_{q.}) = s\sqrt{\frac{2}{r}}$$

A $(1-\alpha)100\%$ confidence interval for $\mu_p - \mu_q$ (or, equivalently, $\tau_p - \tau_q$) is

$$(\bar{y}_{p.} - \bar{y}_{q.}) \pm t_{\alpha/2,\nu} \cdot s\sqrt{\frac{2}{r}}$$

where $t_{\alpha/2,\nu}$ is the upper $\alpha/2$ percentile point of a t-distribution with $\nu = (t-1)(r-1)$ degrees of freedom.

Testing Hypotheses

An analysis of variance (anova) table corresponding to the above model is

SV	df	SS	MS	F	p-value
Blocks	$r-1$	SS_{Blk}	MS_{Blk}		
Trts	$t-1$	SS_{Trt}	MS_{Trt}	MS_{Trt}/MSE	
Error	$(r-1)(t-1)$	SSE	$MSE(=s^2)$		
Total	$rt-1$	SS_{Tot}			

The above F-statistic tests the hypothesis of equality of treatment means:

$$H_0: \mu_1 = \mu_2 = \cdots = \mu_t \text{ versus } H_a: \text{ at least one inequality}$$

or, equivalently, the hypothesis of equality of treatment effects

$$H_0: \tau_1 = \tau_2 = \cdots = \tau_t \text{ versus } H_a: \text{ at least one inequality.}$$

H_0 is rejected if the observed F-value exceeds the α upper percentile of an F-distribution with $df_1 = t-1$ and $df_2 = (b-1)(t-1)$ or, equivalently, if the computed p-value is less than α, where level α controls the Type I error of the test and is selected by the experimenter prior to the experiment. For a treatment mean $\bar{y}_{i.}$, the model can be used to show that

$$\bar{y}_{i.} = \sum_{j=1}^{b}(\mu + \tau_i + \beta_j + \epsilon_{ij})/b$$
$$= \mu + \tau_i + \bar{\beta} + \bar{\epsilon}_{i.}$$

Thus, the difference $\bar{y}_{p.} - \bar{y}_{q.}$ has the form

$$\bar{y}_{p.} - \bar{y}_{q.} = (\tau_p - \tau_q) + (\bar{\epsilon}_{p.} - \bar{\epsilon}_{q.})$$

Thus, the expected value of $\bar{y}_{p.} - \bar{y}_{q.}$ is

$$E(\bar{y}_{p.} - \bar{y}_{q.}) = \tau_p - \tau_q.$$

It is important to note that the mean difference estimates the difference in treatment effects only since the effect of blocks cancels out. These differences are called 'within block' comparisons because they are averages of treatment differences *within each block* (i.e., $\bar{y}_{p.} - \bar{y}_{q.} = \sum_j (y_{pj} - y_{qj})/r$). Thus, tests and confidence intervals for differences in treatment means $\mu_p - \mu_q$ may be constructed using these sample mean differences. For example, the null hypothesis H_0 is rejected if $t_c > t_{\alpha/2,\nu}$, where

5.7 Two-Way Classification: Randomized Complete Block Design

$$t_c = \frac{|\bar{y}_{p\cdot} - \bar{y}_{q\cdot}|}{s_d}$$

$t_{\alpha/2,\nu}$ is the upper $\alpha/2$ percentile of the t-distribution with $\nu = N-t$ degrees of freedom. Equivalently, a difference τ_p and τ_q is declared significantly different at the α level if

$$|\bar{y}_{p\cdot} - \bar{y}_{q\cdot}| > \text{LSD}_\alpha$$

where $\text{LSD}_\alpha = t_{\alpha/2,\nu} \cdot s_d$. This is used to perform tests of all pairwise differences of treatment means (or treatment effects). Hypotheses of the type $H_0 : \tau_1 - (\tau_2 + \tau_3 + \tau_4 + \tau_5)/4 = 0$ or, equivalently $H_0 : 4\tau_1 - \tau_2 - \tau_3 - \tau_4 - \tau_5 = 0$ may be tested using contrasts of the τ's. These are single df preplanned comparisons discussed in detail in Section 5.2.

5.7.1 Using PROC GLM to analyze a RCBD

SAS Example E10

It is standard agronomic practice to treat seeds chemically to increase germination rate prior to planting. In SAS Example E10, an experiment described in Snedecor and Cochran (1989) in which four seed treatments are compared with a control of no treatment (labeled as "Check" in the data table below) on soybeans is considered. The responses are number of plants that failed to emerge out of 100 seeds planted in each plot. The set of five treatments is replicated five times, each replication representing a block. The data are shown in Table 5.8. The model for the response from the ith treatment in the jth block, y_{ij}, is

$$y_{ij} = \mu + \tau_i + \beta_j + \epsilon_{ij}, \quad i = 1,\ldots,5; \quad j = 1,\ldots,5$$

where τ_i is the effect of ith treatment, β_j is the effect of jth block, and the random error ϵ_{ij} is distributed as iid $N(0, \sigma^2)$.

Treatment	Replication				
	1	2	3	4	5
Check	8	10	12	13	11
Arasan	2	6	7	11	5
Spergon	4	10	9	8	10
Samesan	3	5	9	10	6
Fermate	9	7	5	5	3

Table 5.8. Effect of seed treatments on germination of soybeans (Snedecor and Cochran, 1989)

In the SAS program displayed in Fig. 5.53, an `input` statement with a trailing @ combined with a `iterative do` loop is used to input the data.

```
options ls=80 ps=50 pageno=1 nodate;
data soybean;
input trt \$ @;
do rep=1 to 5;
    input yield @;
    output;
end;
datalines;
check 8 10 12 13 11
arasan 2 6 7 11 5
spergon 4 10 9 8 10
samesan 3 5 9 10 6
fermate 9 7 5 5 3
;
run;

proc glm order=data;
    classes trt rep;
    model yield = trt rep;
    means trt/lsd cldiff alpha=.05;
    contrast 'Control vs Others' trt 4 -1 -1 -1 -1;
run;
```

Fig. 5.53. SAS Example E10: Analysis of Seed Treatments

The trailing @ "holds" the data line after accessing a character value for the variable trt. SAS then reads the next numeric value for the variable yield from the data line. Each time through the loop, the output statement causes an observation to be written containing the current values for trt, rep, and yield to the data set soybean and then returns to read another value for yield. This process is repeated five times for each data line. The first five observation in the SAS data set are thus

Obs	trt	rep	yield
1	check	1	8
2	check	2	10
3	check	3	12
4	check	4	13
5	check	5	11

The class statement specifies the classification variables that will appear in the model statement. Ordinarily, when proc anova or proc glm processes data sets with classification variables, levels of variables listed in a class statement are lexically ordered prior to the processing of the data by the procedure. For levels that are numeric, the values are usually in the increasing order and thus will not affect the analysis. However, when the values of levels are character type, such as the names of chemicals or "check" as values of the variable trt in this example, the ordering might be affected. To determine the ordering of levels used in the procedure, the user must check the output in the SAS Output page titled Class Level Information. In Fig. 5.54 (see page 1), the values shown are check, arasan, spergon, samesan, and fermate. Thus, proc glm retained the same ordering found in the data set for these levels. This is a result of inserting the option order=data in the proc glm statement. If, however, this option is omitted, the proc glm would have reordered the levels to be arasan, check, fermate, samesan, and spergon; as can be observed, the levels are in increasing alphabetical order.

Knowledge of the actual ordering of the levels used in the procedure prior to executing the program is important because the user needs to specify, for example, coefficients of contrasts in the correct order. For example, the coef-

```
                            The SAS System                              1

                           The GLM Procedure

                        Class Level Information

        Class         Levels    Values

        trt              5      check arasan spergon samesan fermate

        rep              5      1 2 3 4 5


                    Number of Observations Read       25
                    Number of Observations Used       25

                            The SAS System                              2

                           The GLM Procedure

Dependent Variable: yield

                                  Sum of
Source                    DF     Squares      Mean Square    F Value    Pr > F

Model                      8   133.6800000    16.7100000       3.09     0.0262

Error                     16    86.5600000     5.4100000

Corrected Total           24   220.2400000


            R-Square     Coeff Var      Root MSE     yield Mean

            0.606974     30.93006       2.325941      7.520000


Source                    DF     Type I SS    Mean Square    F Value    Pr > F

trt                        4   83.84000000   20.96000000       3.87     0.0219
rep                        4   49.84000000   12.46000000       2.30     0.1032


Source                    DF    Type III SS   Mean Square    F Value    Pr > F

trt                        4   83.84000000   20.96000000       3.87     0.0219
rep                        4   49.84000000   12.46000000       2.30     0.1032
```

Fig. 5.54. SAS Example E10: Output (pages 1 and 2)

ficients of the contrast for the comparison of "Check vs Chemicals" specified as 4 –1 –1 –1 –1 with option order=data present (see Fig. 5.53) would have to be changed to –1 4 –1 –1 –1 if this option is omitted.

The analysis of variance table for the seed treatment data constructed using information on page 2 of the SAS output (see Fig. 5.54) and Type III SS is:

SV	df	SS	MS	F	p-value
Treatment	4	83.84	20.96	3.87	.0219
Rep	4	49.84	12.46	2.30	.1032
Error	16	86.56	5.41		
Total	24	220.24			

The hypothesis of no difference among the five seed treatments (the check and four chemicals) $H_0 : \tau_1 = \tau_2 = \cdots = \tau_5$ is rejected in favor of H_a : at least one inequality at $\alpha = .05$ since the p-value is smaller. Since the blocks are considered fixed effects, there is a comparable test for block effects. Rather than performing a standard test for differences in block effects (which is not a meaningful hypothesis considering that the labeling of the blocks is done randomly), some practitioners use the F-statistic as a nominal measure of whether the mean square for blocks is inflated compared to the error mean square. This may indicate whether using a blocked experiment was more efficient compared to a completely randomized design. There are other efficiency measures that may be calculated based on the mean squares from the anova table. It now remains to determine the effects of the seed treatments that are actually different.

The lsd option (or, equivalently, the t option) with the means statement is used to perform all comparisons of pairwise means using the percentage points of the t-distribution. The concurrent use of the cldiff option specifies that these comparisons be given in the form of 95% confidence intervals on pairwise differences of the means. The output is in Fig. 5.55 and shows confidence intervals for 20 pair differences The set of 95% confidence intervals on the relevant pairwise differences of means extracted from the SAS output **1** is

$$\begin{aligned}
\text{check} - \text{spergon} &: (-0.518, 5.718) \\
\text{check} - \text{samesan} &: (1.082, 7.318) \\
\text{check} - \text{arasan} &: (1.482, 7.718) \\
\text{check} - \text{fermate} &: (1.882, 8.118) \\
\text{spergon} - \text{samesan} &: (-1.518, 4.718) \\
\text{spergon} - \text{arasan} &: (-1.118, 5.118) \\
\text{spergon} - \text{fermate} &: (-0.718, 5.518) \\
\text{samesan} - \text{arasan} &: (-2.718, 3.518) \\
\text{samesan} - \text{fermate} &: (-2.318, 3.918) \\
\text{arasan} - \text{fermate} &: (-2.718, 3.518)
\end{aligned}$$

5.7 Two-Way Classification: Randomized Complete Block Design

```
                    The SAS System                        3

                    The GLM Procedure

                   t Tests (LSD) for yield

NOTE: This test controls the Type I comparison wise error rate, not the
      experiment wise error rate.

              Alpha                            0.05
              Error Degrees of Freedom          16
              Error Mean Square               5.41
              Critical Value of t          2.11991
              Least Significant Difference  3.1185

Comparisons significant at the 0.05 level are indicated by ***.

                        Difference
         trt            Between      95% Confidence
       Comparison        Means          Limits

      check   - spergon    2.600    -0.518    5.718
      check   - samesan    4.200     1.082    7.318  ***
      check   - arasan     4.600     1.482    7.718  ***
      check   - fermate    5.000     1.882    8.118  ***
      spergon - check     -2.600    -5.718    0.518
      spergon - samesan    1.600    -1.518    4.718
      spergon - arasan     2.000    -1.118    5.118
      spergon - fermate    2.400    -0.718    5.518
      samesan - check     -4.200    -7.318   -1.082  ***
      samesan - spergon   -1.600    -4.718    1.518
      samesan - arasan     0.400    -2.718    3.518
      samesan - fermate    0.800    -2.318    3.918
      arasan  - check     -4.600    -7.718   -1.482  ***
      arasan  - spergon   -2.000    -5.118    1.118
      arasan  - samesan   -0.400    -3.518    2.718
      arasan  - fermate    0.400    -2.718    3.518
      fermate - check     -5.000    -8.118   -1.882  ***
      fermate - spergon   -2.400    -5.518    0.718
      fermate - samesan   -0.800    -3.918    2.318
      fermate - arasan    -0.400    -3.518    2.718
```

Fig. 5.55. SAS Example E10: Pairwise comparisons and preplanned comparison

$$\text{fermate - arasan} : (-3.518, 2.718)$$

Since zero is included in every interval except (check–samesan), (check–arasan), and (check–fermate), the conclusion that might be drawn is that whereas spergon is not found to be different from the control of no seed treatment on germination rate, samesan, arasan, and fermate are found to produce significantly higher germination rates than the control. Also, there appears to be no significant differences among the chemicals on their effect on germination. It must be noted that that these are individual comparisons and no adjustment is made for making multiple comparisons.

The result of the contrast statement also confirmed that the hypothesis $H_0 : 4\tau_1 - \tau_2 - \tau_3 - \tau_4 - \tau_5 = 0$ is rejected with a p-value of .0028 **2** (see Fig. 5.56), and thus the average effect of the chemicals on emergence is differ-

```
                              The SAS System                            4

                            The GLM Procedure

Dependent Variable: yield

Contrast                   DF    Contrast SS    Mean Square   F Value   Pr > F

Control vs Others           1    67.24000000    67.24000000    12.43    0.0028  2
```

Fig. 5.56. SAS Example E10: F-test for preplanned comparison

ent from the effect of the control of no chemical treatment. Also, note that if the hypothesis tested is one-tailed (e.g., $H_0 : \tau_1 < (\tau_2 - \tau_3 - \tau_4 - \tau_5)/4$ versus $H_a : \tau_1 \geq (\tau_2 - \tau_3 - \tau_4 - \tau_5)/4$), the same set of contrast coefficients may be used. In this instance, the output will be the same but the p-value for the one-tailed test is one-half of the p-value printed, which is .0014. The remainder of the treatment sum of squares after subtracting the sum of squares due to this contrast will be the total sum of squares from any set of three orthogonal comparisons one can make among the four chemicals and, thus, will have three degrees of freedom. The compound hypothesis tested by the corresponding F-statistic is $H_0 : \tau_2 = \tau_3 = \tau_4 = \tau_5$ versus H_a : at least one inequality (i.e., whether there are any differences among the effects of the chemicals). This sum of squares can be obtained by subtraction or directly by including the following contrast statement in the above SAS program (i.e., the SAS program shown in Fig. 5.53):

```
contrast 'Among Chemicals' trt 0 3 -1 -1 -1,
                          trt 0 0  2 -1 -1,
                          trt 0 0  0  1 -1;
```

The results of the partitioning of the treatment sum of squares resulting from the preplanned comparison may be usefully included in the anova table as follows:

SV	df	SS	MS	F	p-value
Treatment	4	83.84	20.96	3.87	.0219
Check vs. Chemical	1	67.24	67.24	12.43	.0028
Among Chemicals	3	16.60	5.53	1.02	.4087
Rep	4	49.84	12.46	2.30	.1032
Error	16	86.56	5.41		
Total	24	220.24			

It is perhaps instructive to note that the sum of squares for the "Among Chemicals" hypothesis may also be obtained by specifying contrasts that are nonorthogonal but distinct as follows:

5.7 Two-Way Classification: Randomized Complete Block Design

```
contrast 'Among Chemicals2' trt 0 1 -1  0  0,
                            trt 0 0  1 -1  0,
                            trt 0 0  0  1 -1;
```

The earlier conclusion of no significant differences among the chemicals on their effect on germination is confirmed.

5.7.2 Using PROC GLM to test for nonadditivity

In Section 5.7.1, an additive model is used to analyze data generated from experiments using RCBDs. As discussed there, this is based on the assumption that differences in responses to the treatments are unaffected by the block effects. In many experiments, this assumption may not be plausible because an interaction may occur between the treatment factor and the blocking factor. This is possible in cases in which the responses are not necessarily obtained from experimental units that are grouped together prior to the experiment to form blocks. For example, a block may consist of responses from a complete replicate of experimental trials performed by a person or (a group of persons such as a team), a time unit (such as a day, a week, or a month), a space unit (such as a lab, a location, a growth chamber, a bench, or an oven) and so forth.

In some instances, it may be that the second factor may not be considered a blocking factor (such as when experimental trials are run within an enclosure to control an environmental factor such as temperature or humidity). If independent replications of the levels of the first factor are not obtained, this arrangement would result in a two-way factorial experiment with only a single response observed per cell. In such cases, a nonadditive model may not be appropriate.

Tukey (1949) proposed a test for nonadditivity in nonreplicated two-way classifications. It was shown later that this Tukey's F-statistic is a test of $H_0: \lambda = 0$ in the model

$$y_{ij} = \mu + \tau_i + \beta_j + \lambda \tau_i \beta_j + \epsilon_{ij}, \quad i = 1, \ldots, t; \quad j = 1, \ldots, r$$

where the interaction is modeled as $\gamma_{ij} = \lambda \tau_i \beta_j$, which is a scalar multiple of the product of the main effects τ_i and β_j. This test is called the *single degree of freedom test of nonadditivity* because the sum of squares due to the null hypothesis has one degree of freedom. This test is popular among practitioners for the important reason that if the data display this type of nonadditivity, it is possible to find a power transformation of the response variable that may restore additivity so that one can proceed with further analysis of the treatment effects. Since this test and associated computations are detailed in many textbooks, they will not be repeated here. However, a method for obtaining Tukey's statistic for testing nonadditivity and an associated p-value using SAS is presented.

SAS Example E11

In the SAS program displayed in Fig. 5.57, the first `proc glm` step is a modified version of the program used in Section 5.7.1. Here, the predicted values from fitting the additive model to the soybean data are saved as a variable name yhat using the `predicted=` (or, equivalently, `p=` as used here) option in a `output` statement. This results in the creation of new SAS data set (named new here), which is the same as the original data set soybeans, augmented by the addition of the new variable named yhat. This data set is used as the input data set to a second `proc glm` step where the model statement

```
insert data step to create the SAS dataset 'soybean' here

proc glm order=data;
    class trt rep;
    model yield = trt rep;
    output out=new p=yhat;
run;

proc glm order=data;
    class trt rep;
    model yield = trt rep yhat*yhat/ss1;
    title "Tukey's Test for Non-additivity";
run;
```

Fig. 5.57. SAS Example E11: Program for Tukey's test of nonadditivity

```
              Tukey's Test for Non-additivity                    4

                      The GLM Procedure

Dependent Variable: yield

                               Sum of
Source                DF      Squares     Mean Square   F Value   Pr > F

Model                  9    135.1756858    15.0195206     2.65    0.0461

Error                 15     85.0643142     5.6709543

Corrected Total       24    220.2400000


          R-Square    Coeff Var     Root MSE    yield Mean

          0.613765    31.66724      2.381377     7.520000


Source            DF    Type I SS    Mean Square   F Value   Pr > F

trt                4   83.84000000   20.96000000    3.70    0.0275
rep                4   49.84000000   12.46000000    2.20    0.1187
yhat*yhat          1    1.49568580    1.49568580    0.26    0.6150
```

Fig. 5.58. SAS Example E11: Output from PROC GLM(page 4)

model yield = trt rep yhat*yhat/ss1; is used to fit a new model. In this model, the term yhat*yhat is equivalent to a regression variable (i.e., a covariate).

The ss1 option requests that only Type I SS be computed and output. From the output in Fig. 5.58, the Tukey's single degree of freedom sum of squares, the F-statistic for performing Tukey's test of nonadditivity, and the corresponding p-value are the Type I SS values pertaining to the regression term yhat*yhat, and are observed to be 1.4956, 0.26, and 0.6150, respectively. Thus, the hypothesis of no interaction (i.e., $H_0 : \lambda = 0$) is not rejected based on this p-value; hence, an additive model is appropriate for this data.

5.8 Exercises

5.1 An experiment was carried out in a *completely randomized design* to compare the differences in the levels of physiologically active polyunsaturated fatty acids (PAPUFA, in percentages) of six different brands of diet margarine, resulting in the following data (Devore, 1982):

Imperial:	14.1	13.6	14.4	14.3	
Parkay:	12.8	12.5	13.4	13.0	12.3
Blue Bonnet:	13.5	13.4	14.1	14.3	
Chiffon:	13.2	12.7	12.6	13.9	
Mazola:	16.8	17.2	16.4	17.3	18.0
Fleischmann's:	18.1	17.2	18.7	18.4	

Prepare and run a SAS program to obtain the output necessary to provide all of the following information. *You must extract numbers from the output and write answers on a separate sheet of paper.*

a. Assuming the *fixed effects one-way classification model* for these data give estimates of true mean PAPUFA percentages $\mu_1, \mu_2, \mu_3, \mu_4, \mu_5$, and μ_6, and the error variance σ^2. Write down the corresponding analysis of variance table including the p-value. State the hypothesis tested by the F-statistic and your decision based on the p-value.

b. Mazola and Fleischmann's are corn based and the others are soybean based. Use an appropriate contrast statement to compute an F-statistic for testing the hypothesis that the average of true mean PAPUFA percentages for corn-based brands is the same as the average for soybean-based brands. Include the corresponding sum of squares and the F-statistic in the above anova table. Based on the p-value, state your conclusions from the experiment in words.

c. Compute Tukey 95% confidence intervals for all pairwise differences in true mean PAPUFA percentages i.e., $(\mu_r - \mu_s)$'s. Explain why these intervals are wider than the t-distribution based 95% confidence intervals (other than the fact that the Tukey percentiles are larger than the corresponding t values).

5.2 Six samples of each of four types of cereal grain grown in a certain region were randomly selected and analyzed to determine the thiamin content (mcg/gm) in an experiment reported in Devore (1982). The data were input to the following SAS program:

```
data ex2;
input cereal $ @;
do i=1 to 6;
    input thiamin @;
    output;
end;
datalines;
Wheat 5.2 4.5 6.0 6.1 6.7 5.8
Barley 6.5 7.0 6.1 7.5 5.9 5.7
Maize 5.8 4.7 6.4 4.9 6.0 5.2
Oats 8.3 6.7 7.8 7.0 5.9 7.2
;
run ;
proc print;
    title ' Thiamin Content in Cereal Grains';
run;
proc glm;
    class cereal ;
    model thiamin = cereal ;
run;
```

Assuming the one-way fixed effects model

$$y_{ij} = \mu + \tau_i + \epsilon_{ij}, \qquad i = 1,\ldots,4; \quad j = 1,\ldots,6$$

where $\mu_i = \mu + \tau_i$ are the population mean thiamin content for each cereal and ϵ_{ij} are iid $N(0, \sigma^2)$ errors, complete the SAS program as needed to answer the following:

a. It is thought that oats are higher in thiamin content than other cereals. Add a contrast statement for testing the appropriate comparison to test this hypothesis.
b. Add an option to the proc statement for the levels of the cereal variable to retain the ordering present in the data.
c. Add an option to the model statement to obtain the estimates of the parameters μ, τ_1, τ_2, τ_3 and τ_4.
d. Compute 95% confidence intervals on all pairwise differences (i.e., $(\mu_p - \mu_q)$'s) adjusted for multiple testing using the Bonferroni method.

5.3 The following are clotting times of plasma (in minutes) for four different levels of a drug compared in a *completely randomized design* (modified from a different experiment reported in Armitage and Berry, 1994). Blood samples were taken from each of eight subjects randomly assigned to each of the four levels of the drug.

	Drug		
0.1%	0.2%	0.3%	0.4%
8.4	9.4	9.8	12.2
12.8	15.2	12.9	14.4
9.6	9.1	11.2	9.8
9.8	8.8	9.9	12.0
8.4	8.2	8.5	8.5
8.6	9.9	9.8	10.9
8.9	9.0	9.2	10.4
7.9	8.1	8.2	10.0

a. Write a SAS data step to read the data (that you enter instream) and create a SAS data set in the form necessary to be used by proc glm.
b. Write a complete SAS proc step to obtain the anova table necessary and include a statement to compute confidence intervals for differences in all pairwise drug means. Report the anova table and the confidence intervals and use them to summarize the results of the experiment.
c. Add a statement to obtain the necessary F-statistic to test whether the average clotting times have an increasing linear trend with the level of the drug. What is your conclusion?

5.4 A marketing consultant conducted an experiment to compare four different package designs for a new breakfast cereal (this problem is described in Kutner et al. (2005); however, the data given are artificial). Twenty-four stores with approximately similar sales volumes were selected and each store was required to carry only one of the package designs. Thus, each package design was randomly assigned to six stores. Other relevant conditions such as price, amount and location of shelf space, and advertising were kept roughly similar for all stores participating in the experiment. Sales, in number of cases, were observed for the study period. The data are

Package Design			
1	2	3	4
12	14	19	24
18	12	17	30
14	13	21	27
15	10	23	28
17	15	16	32
15	12	20	30

Write and execute a SAS program to obtain the output necessary to provide answers to all of the following questions:
a. Construct an analysis of variance using numbers from the SAS output. Test $H_0: \mu_1 = \mu_2 = \mu_3 = \mu_4$ versus H_a: at least one μ_i is different

from the others, using $\alpha = .05$. Use the p-value from the SAS output to make a decision.

b. Use the lsd procedure to test all possible differences $H_0 : \mu_i - \mu_j = 0$ versus $H_a : \mu_i - \mu_j \neq 0$ using $\alpha = .05$. Report the results using the underscoring display. Summarize your conclusions from this procedure in a sentence or two.

c. Use the Tukey procedure to test all possible differences $H_0 : \mu_i - \mu_j = 0$ versus $H_a : \mu_i - \mu_j \neq 0$ using $\alpha = .05$. Report the results using the underscoring display. Point out any conclusions that are different from those made using the lsd procedure. Explain why they are different.

d. It would be more useful to test preplanned comparisons than testing pairwise differences among the four packaging designs given the following additional information on the package designs used.

Package Design	Design Style
1	3 colors, with cartoons
2	3 colors, without cartoons
3	5 colors, with cartoons
4	5 colors, without cartoons

Thus, the experimenter could have planned to

i. compare the the average effect of the three-color designs with the average effect of the five-color designs,

ii. compare the the average effect of the designs with cartoons with the average effect of the designs without cartoons,

iii. compare the the effect of the three-color design with cartoons with the effect of the three-color design without cartoons, and

iv. compare the the effect of the five-color design with cartoons with the effect of the five-color design without cartoons.

Compute appropriate t-statistics to test the comparisons given above controlling the error rate for each comparison at $\alpha = .05$. State your conclusion in each case.

5.5 In an experiment conducted to compare the effects of sleep deprivation on reaction time to onset of light (Kirk, 1982), 32 subjects were randomly divided into 4 groups of 8 subjects each. Four levels of sleep deprivation (12, 24, 36, and 48 hours) were randomly assigned to the four groups. The reaction times of the 32 subjects (in hundredths of a second) are tabulated below. Prepare and run a SAS program to obtain the output necessary to provide all of the following information. You must extract numbers from the output and write answers on a separate sheet of paper. Model this program after SAS Example E3.

	Duration of Sleep Deprivation			
	12hrs	24hrs	36hrs	48hrs
	20	21	25	26
	20	20	23	27
	17	21	22	24
	19	22	23	27
	20	20	21	25
	19	20	22	28
	21	23	22	26
	19	19	23	27

a. Obtain a scatter plot of the data with the levels of duration on the horizontal axis. Overlay line segments connecting the averages. What kind of trend do the averages indicate?

b. Construct an anova table and use the F-statistics to test the hypothesis that the population means of reaction times at each sleep deprivation are all equal against the alternative that there is at least one difference, using $\alpha = .05$

c. Test the hypothesis that there is no linear trend in the population means using $\alpha = .05$

d. Test the hypothesis that the nonlinear components of the trend (i.e., the deviation from linear components) equal zero using $\alpha = .05$. Note that this is the same as a test of lack of fit of a linear trend.

5.6 Researchers conducted an experiment to compare the effectiveness of four new weight-reducing agents to that of an existing agent (Ott and Longnecker, 2001). The researchers randomly divided a random sample of 50 males into 5 equal groups, with preparation A assigned to the first group, B to the second group, and so on. They then gave a prestudy physical to each person in the experiment and told him how many pounds overweight he was. A comparison of the mean number of pounds overweight for the groups showed no significant differences. The researchers then began the study program, and each group took the prescribed preparation for a fixed period of time. The weight losses recorded at the end of the study period are as follows:

Agent	Weight Loss									
A1	12.4	10.7	11.9	11.0	12.4	12.3	13.0	12.5	11.2	13.1
A2	9.1	11.5	11.3	9.7	13.2	10.7	10.6	11.3	11.1	11.7
A3	8.5	11.6	10.2	10.9	9.0	9.6	9.9	11.3	10.5	11.2
A4	12.7	13.2	11.8	11.9	12.2	11.2	13.7	11.8	11.5	11.7
S	8.7	9.3	8.2	8.3	9.0	9.4	9.2	12.2	8.5	9.9

The standard agent is labeled agent S and the four new agents are labeled A1, A2, A3, and A4:

A1: Drug therapy with exercise and counseling
A2: Drug therapy with exercise but no counseling
A3: Drug therapy with counseling but no exercise
A4: Drug therapy with no exercise and no counseling

Denoting the means for the five treated populations by μ_1, μ_2, μ_3, μ_4, and μ_5, respectively, linear combinations for making comparisons among the agent means that will address the following questions can be constructed:

a. Compare the mean for the standard to the average of the four agent means:
$$(\mu_1 + \mu_2 + \mu_3 + \mu_4)/4 - \mu_5$$

b. Compare the average mean for the agents with counseling to average mean for those without counseling (ignoring the standard):
$$(\mu_1 + \mu_3)/2 - (\mu_2 + \mu_4)/2$$

c. Compare the average mean for the agents with exercise to average mean for those without exercise (ignoring the standard):
$$(\mu_1 + \mu_2)/2 - (\mu_3 + \mu_4)/2$$

d. Compare the mean for the agents with counseling to the standard:
$$(\mu_1 + \mu_3)/2 - \mu_5$$

5.7 An experiment, carried out in a *completely randomized design* (Devore, 1982), to compare the effects of five different plate lengths of 4, 6, 8, 10, and 12 inches on axial stiffness of metal plate-connected trusses used for roof support, yielded the following observations on axial stiffness index (kips/in.).

		Plate Length		
4	6	8	10	12
309.2	402.1	392.4	346.7	407.4
409.5	347.2	366.2	452.9	441.8
311.0	361.0	351.0	461.4	419.9
326.5	404.5	357.1	433.1	410.7
349.8	331.0	409.9	410.6	473.4
309.7	348.9	367.3	384.2	441.2
316.8	381.7	382.0	362.6	465.8

Apart from the overall differences of the effects of the five plate lengths on axial stiffness, the experimenter was interested in determining whether the mean stiffness index depends `linearly` on the actual plate length. This may be done using an appropriate orthogonal polynomial of the means. Prepare and run a SAS program to obtain the output necessary to provide all of the following information. You must extract numbers from the output and write answers on a separate sheet of paper.

a. Assuming the fixed effects one-way classification model for the data, give estimates of μ_1, μ_2, μ_3, μ_4, μ_5, and σ^2.

b. Write down the corresponding analysis of variance table including the p-value. State the hypothesis tested by the F-statistic and your decision based on the p-value.

c. Use a `contrast` statement to compute an F-statistic for testing whether the mean stiffness index is linearly related to the plate length. Include the corresponding sum of squares and the F-statistic in an expanded anova table. Based on the p-values, summarize your conclusions.

d. Include an `estimate` statement for the same comparison as in part (c). If the mean stiffness index is shown to have a straight-line relationship with the plate length, then estimate the slope of this line by the estimated value of the comparison divided by the sum of squares of its coefficients. Interpret this slope as an increase or decrease of axial index per inch of plate length.

5.8 A researcher wants to evaluate the difference in mean film thickness of a coating placed on silicon wafers using three different processes (Ott and Longnecker, 2001). Six wafers are randomly assigned to each of the processes. The film thickness and the temperature in the lab during the coating process are recorded for each wafer. The researcher is concerned that fluctuations in temperature may affect the film thickness. The results were analyzed using a one-way covariance model $y_{ij} = \mu_i + \beta(x_{ij} - \bar{x}) + \epsilon_{ij}$ where temperature was used as the covariate. Complete and execute the following SAS program and use the output to answer the questions that follow.

```
data ex8;
input process @;
do i=1 to 6;
  input temperature thickness @@;
  output;
end;
datalines;
1 26 100 35 150 28 106 31  95 29 113 34 144
2 24 118 28 134 29 138 32 147 36 165 35 159
3 37 124 31  95 34 120 27  86 28  98 25  81
;
run;

proc glm data=ex7;
 class ;
 model thickness = ;
 title 'Film thickness adjusted by Temperature';
run;
```

a. Construct an adjusted anova table.

b. Using the above anova table, test the hypothesis of $H_0 : \mu_1 = \mu_2 = \mu_3$ (use the p-value and state decision).
c. Compute 95% confidence intervals for all differences in pairs of means (e.g., $\mu_1 - \mu_2$) adjusted for multiple testing using the Bonferroni method.
d. What does the test of $H_0 : \beta = 0$ tell you? Test this hypothesis using the above adjusted anova table and state your conclusion.
e. Give the sum of squares for the variety effect that is not adjusted for the moisture effect.

5.9 A process engineer is interested in determining if there is a difference in the breaking strength of a mono-filament fiber produced using three different machines for a textile company (Montgomery, 1991). However, the strength of the fiber is also related to its diameter, with thicker fibers being stronger than thinner ones. A random samples of five fiber specimens each were selected from each machine and the breaking strength y (in pounds) and the diameter x (in 10^{-3} inches) (to be used as a covariate) are measured.

Machine 1		Machine 2		Machine 3	
y	x	y	x	y	x
36	20	45	22	35	21
41	25	52	28	37	23
39	24	44	22	42	26
42	25	49	30	34	21
48	32	51	28	32	15

Use proc glm in SAS and *the one-way covariance model* to analyze this data. *Extract numbers from the SAS output and write your own answers to the following questions.*
a. Write an appropriate model for analyzing these data so that it is possible to perform a test of the equality of the slopes from your analysis. Construct an "adjusted" analysis of variance table to test the hypothesis of equal means (or effects) corresponding to the machines. State your conclusion based on the p-value.
b. Provide estimates of the regression coefficients of the straight lines if it is determined that their slopes are different.
c. Calculate 95% confidence intervals on all differences in pair of means (e.g., $\mu_i - \mu_j$) adjusted at the mean value of x.
d. Include a proc step in your SAS program to obtain a scatter plot of y versus x, superimposed by the fitted regression lines.

5.10 The data displayed below are results from an experiment described in Snedecor and Cochran (1989) on the use of drugs in the treatment of leprosy. The drugs were A and D, which were antibiotics and F, an inert drug used as a control. The dependent variable Y was a score of leprosy bacilli measured on each patient after several months of treatment. The covariate X was a pretreatment score of leprosy bacilli.

	Drugs				
A		D		F	
X	Y	X	Y	X	Y
11	6	6	0	16	13
8	0	6	2	13	10
5	2	7	3	11	18
14	8	8	1	9	5
19	11	18	18	21	23
6	4	8	4	16	12
10	13	19	14	12	5
6	1	8	9	12	16
11	8	5	1	7	1
3	0	15	9	12	20

a. Use proc glm and the one-way covariance (equal slopes) model to analyze this data. Construct an adjusted anova table.
b. Using the above anova table, test the hypothesis $H_0 : \mu_1 = \mu_2 = \mu_3$ (use the p-value and state decision).
c. Construct 95% confidence intervals for all differences in pairs of means (e.g., $\mu_1 - \mu_2$) adjusted for multiple testing using the Bonferroni method.
d. What does the test of $H_0 : \beta = 0$ tell you? Test this hypothesis using the above adjusted anova table and state your conclusion.
e. Construct an analysis of variance that is not adjusted for the pre-score. What conclusion can you draw from this Anova table.

5.11 An experiment conducted to study the friction properties of lubricants is described in Mason et al. (1989). A key constituent of lubricants that is of interest to the researchers is the additive that is mixed with the base lubricant. In order to ascertain whether two competing additives produce a different effect on the friction properties of lubricants, 10 mixtures of a base lubricant and each of the additives were made. The mixtures of base lubricant cannot be made sufficiently uniform to ensure that all batches have identical physical properties. Consequently, the plastic viscosity, an important characteristic of the base lubricant that is related to its friction-reducing capability, was measured for each mixture prior to the addition of the additives. This measures the variation among batches due to the base lubricant alone. Analyze the following data to determine whether the additives differ in the mean friction measurements associated with each.

	Additive A		Additive B	
	Plastic Viscosity(x)	Friction Measurement(y)	Plastic Viscosity(x)	Friction Measurement(y)
	12	27.1	15	28.6
	13	26.6	13	37.1
	15	28.9	14	37.9
	14	27.1	14	30.6
	10	23.6	13	33.6
	10	26.4	13	34.9
	13	28.1	13	33.1
	14	26.1	14	34.4
	12	24.4	14	32.6
	14	29.1	12	35.6

Use proc glm and the one-way covariance model to analyze this data. A plot of the data suggests that the slopes of the regression lines for the two additives may be different. The data were analyzed using a one-way covariance model

$$y_{ij} = \alpha_i + \beta_i x_{ij} + \epsilon_{ij}$$

Thus, it is possible to perform a test of the equality of the slopes as a part of your analysis. Extract numbers from the SAS output to write your own answers to the following questions.

a. Construct an "adjusted" analysis of variance table to test the hypothesis $H_0 : \beta_1 = \beta_2$ (or, equivalently, $H_0 : \tau_1 = \tau_2$). Use the p-value to draw a conclusion.
b. If the slopes are found to be unequal, obtain estimates of parameters for the two regressions.
c. Use the lsmeans statement in proc glm to obtain a confidence interval on the difference in the slopes $\alpha_1 - \alpha_2$ adjusted at the mean plastic viscosity.
d. Use the lsmeans statements in proc glm to obtain a confidence interval on the difference in the slopes $\alpha_1 - \alpha_2$ adjusted at plastic viscosity values of 12 and 14.

5.12 In an experiment described in Kirk (1982), four methods for teaching arithmetic were being evaluated. Thirty-two students were randomly assigned to four classrooms, each with eight students. An intelligence test was administered to each student at the beginning of the experiment. The resulting scores (x) are used to adjust the arithmetic achievement scores (y) obtained at the conclusion of the experiment for differences in intelligence among the students. The results are recorded as follows:

Method 1		Method 2		Method 3		Method 4	
y	x	y	x	y	x	y	x
3	42	4	47	7	61	7	65
6	57	5	49	8	65	8	74
3	33	4	42	7	64	9	80
3	47	3	41	6	56	8	73
1	32	2	38	5	52	10	85
2	35	3	43	6	58	10	82
2	33	4	48	5	53	9	78
2	39	3	45	6	54	11	89

a. Use `proc glm` and the one-way covariance (equal slopes) model to analyze this data. Construct an adjusted Anova table.
b. Using the above Anova table, test the hypothesis of $H_0 : \mu_1 = \mu_2 = \mu_3 = \mu_4$ (use the p-value and state decision).
c. Construct 95% confidence intervals for all differences in pairs of means (e.g., $\mu_a - \mu_b$) adjusted for multiple testing using the Tukey method.
d. What does the test of $H_0 : \beta = 0$ tell you? Test this hypothesis using the above adjusted anova table and state your conclusion.
e. Construct an analysis of variance that is not adjusted for the intelligence score. What conclusion can you draw from this Anova table.

5.13 A medical experiment is run to determine the side effects on children when they take various dosages of a drug administered by different methods. A two-way factorial with four dosages (0.5, 1.0, 1.5, and 2.0 milligrams) and three methods of administering (oral, extended release, intravenous) is used in a completely randomized design, with each treatment combination replicated twice. The response variable is the amount of a certain chemical present in the lever after 24 hours. The data are as follows:

Method	Dosage			
	0.5	1.0	1.5	2.0
1	0.414	0.541	0.592	0.672
	0.312	0.423	0.575	0.610
2	0.537	0.513	0.595	0.709
	0.451	0.580	0.573	0.623
3	0.572	0.622	0.613	0.695
	0.554	0.597	0.650	0.751

Use appropriate SAS procedures to analyze these data. Use the output from program to answer all of the following on a separate sheet.
a. Estimate the cell means μ_{ij} and report these in a two-way table. Obtain a graph using `proc gplot` of the cell means, with dosage on the x-axis and using different symbols for each method. Join the points for each method by line segments of different colors and line types.

b. Construct the anova table to test the hypotheses of no interaction and main effects. What are the conclusions from each test? Does the graph in part (a) support your conclusion from the test for interaction. Discuss.

c. Obtain 95% confidence intervals for the pairwise differences in methods means and make an overall conclusion.

d. The hypothesis that the average effects of dosage is linearly related to the level of dosage can be examined by constructing a contrast of the dosage means with appropriate coefficients. Include a `contrast` statement in your program. Partition the dosage sum of squares from the results and determine if there is evidence for such a linear trend.

5.14 A mechanical engineer is comparing the thrust force produced by a drill press w various combinations of drill speed and feed rate. A two-way factorial with four feed rates (0.015, 0.030, 0.045, and 0.060 in./min) and two drill speeds (125 and 200 rpm) is used in a completely randomized design with each treatment combination replicated twice (Montgomery, 1991). The results are as follows:

Drill Speed	Feed Rate			
	0.015	0.030	0.045	0.060
125	2.70	2.45	2.60	2.75
	2.78	2.49	2.72	2.86
200	2.83	2.85	2.86	2.94
	2.86	2.80	2.87	2.88

Use appropriate SAS procedures to analyze these data. Use the output from program to answer all of the following:

a. Use `proc means` to obtain estimates of the cell means μ_{ij}. Construct a two-way table of means showing the Feed Rate and Drill Speed cell means.

b. Obtain a graph using `proc gplot` of the cell means with Feed Rate on the x-axis. Join the points for each Drill Speed by line segments. (This is the interaction plot of means discussed in the text.)

c. Construct the anova table to test the hypotheses of no interaction and zero main effects. What are the conclusions from each test using $\alpha = .05$?

d. Use the graph in part (b) to explain your conclusion from the test for interaction. Comment on the variation in mean response across the levels of Feed Rate at each level of Drill Speed and use it to explain any significant interaction.

e. Compute a 95% confidence interval for the difference in Drill Speed means. Use this interval to say if these means are significantly different.

f. The hypothesis that the average effects of Feed Rate is linearly related to the levels of Feed Rate can be tested by constructing a contrast of the Feed Rate means with coefficients to be obtained from the table of orthogonal polynomial coefficients. Include a contrast statement in your program to do this computation. Insert a line into the anova table to report the results. What do you conclude from this test?

5.15 Ostle (1963) presented an example of an agronomic experiment to assess the effects of date of planting and type of fertilizer on the yield of soybeans. Thirty-two experimental plots were randomly assigned to the treatment combinations so that each combination was replicated four times. The yield data are reported in the following table. Write a SAS program with appropriate procedure steps to obtain the output necessary to answer all of the following questions. Extract or write answers on separate pages and attach any graphs requested.

Date of Planting	Fertilizer Type	Yields from Plots (bushels/acre)			
Early	Check	28.6	36.8	32.7	32.6
	Aero	29.1	29.2	30.6	29.1
	Na	28.4	27.4	26.0	29.3
	K	29.2	28.2	27.7	32.0
Late	Check	30.3	32.3	31.6	30.9
	Aero	32.7	30.8	31.0	33.8
	Na	30.3	32.7	33.0	33.9
	K	32.7	31.7	31.8	29.4

a. Use proc means to obtain estimates of the cell means μ_{ij}. Construct a two-way table showing the date of planting by fertilizer type mean yields.
b. Obtain a scatter plot of the cell means using proc gplot with type of fertilizer on the horizontal axis. Join the points for each date of planting by line segments. (This will produce an interaction plot of treatment means discussed in the text.)
c. Construct the anova table to test the hypotheses of no interaction and zero main effects. What are the conclusions from each test using $\alpha = .05$?
d. Use the graph in part (b) to explain your conclusion from the test for interaction. Comment on the trends in mean response across the levels of type of fertilizer for each date of planting and use it to explain significant interaction, if any.
e. The hypothesis comparing the check with the average of the types of fertilizers (a main effect comparison) can be tested by constructing a contrast of the fertilizer means. Use a contrast statement in your program to do this computation. Add a line to the anova table to report the results. What do you conclude from this test?
f. Use a contrast statement to test the interaction comparison that the comparison in part (e) is the same at both dates of planting. What is

you conclusion and does it support your conclusion for parts (c) and (d)? Add a line to the anova table to report the results.

5.16 The yield (grams per plant) of beet roots grown in pots in response to two crossed factors, wood chips from three different sources and nitrogen at three levels, in a completely randomized design with three replications, are given below (Bliss, 1970). The rate of application of the wood chips was 10 tons/acre and nitrogen levels used were 0, $\frac{1}{2}$, and 1 grams/100 grams of organic matter added as chips. There were nine replications of pots with a control treatment of no chips and, therefore, no additional nitrogen. Use appropriate SAS procedures to analyze the data. Provide answers to the following questions on separate sheets, extracting material from the SAS output as needed.

No Chips			Pine Chips			Oak-Hickory			Aspen-Birch		
0	0	0	0	$\frac{1}{2}$	1	0	$\frac{1}{2}$	1	0	$\frac{1}{2}$	1
16.9	15.5	16.8	13.3	20.8	21.2	3.6	10.0	18.3	1.2	6.0	7.8
17.5	20.0	19.7	16.1	16.5	18.4	3.8	10.5	18.2	1.2	3.3	9.7
20.0	19.0	20.5	14.3	14.0	16.3	1.8	9.9	14.2	4.1	5.1	10.5

a. Estimate the cell means μ_{ij} and report these in a two-way table. Obtain a graph using proc gplot of the cell means with nitrogen levels on the x-axis and using different symbols for each type of wood chip. Join the points for each chip type by line segments of different colors and line types.
b. Construct the anova table to test the hypotheses of no interaction and main effects. What are the conclusions from each test? Does the graph in part (a) support your conclusion from the test for interaction. Discuss.
c. The hypothesis that the average beet yields is linearly related to the levels of nitrogen can be examined by constructing a contrast of the dosage means with appropriate coefficients. Include a contrast statement in your program. Partition the nitrogen main effect sum of squares in to linear and lack-of-fit components and determine if there is evidence for such a linear trend.
d. Partition the wood chips main effect sum of squares into sums of squares corresponding to the following three orthogonal comparisons: chips versus control, hard versus soft woods, and between hard woods. (Note: Pine is a soft wood, whereas the others are all hard woods.) Include contrast statements in your program to obtain F-statistics to test the corresponding hypotheses. Report the results of these tests and use these to make conclusions about the wood chip main effects.
e. Use a contrast statement to extract the interaction sum of squares that correspond to the interaction comparison of linear × between hard woods. What hypothesis does this comparison test? What is your conclusion?

5.17 Use the following data set, which is similar to that used in SAS Example E9 (Fig. 5.46) except that different data values are missing.

Poison	Drug			
	A	B	C	D
I	0.31	0.82	.	0.45
	0.45	.	0.45	0.71
	.	0.88	0.63	0.66
	0.43	0.72	.	0.62
II	0.36	0.92	0.44	0.56
	.	.	0.35	.
	0.40	0.49	0.31	0.71
	.	.	0.40	0.38
III	0.22	0.30	0.23	0.30
	.	0.37	0.25	.
	0.18	0.38	0.24	.
	0.23	0.29	0.22	0.33

You are required to add a proc glm step to this program to do the computations described below. Use the usual two-way classification model for this analysis.

a. Obtain the least squares estimates of μ, α_i, β_j, and γ_{ij} for $i = 1, 2, 3$ and $j = 1, 2, 3, 4$ computed by proc glm.
b. Use lsmeans statements to obtain 95% confidence intervals on pairwise differences in poison and drug means, adjusted for multiple testing.
c. It can be shown that $\bar{\mu}_{.1} - \bar{\mu}_{.2} = \beta_1 - \beta_2 + \bar{\gamma}_{.1} - \bar{\gamma}_{.2}$. Estimate $\bar{\mu}_{.1} - \bar{\mu}_{.2}$ by substituting estimates of the parameters from part (a) in this expression, using hand computation. Include an estimate statement in the proc glm step to verify this estimate and to compute its standard error.
d. Use the results of the estimate statement to compute a 95% confidence interval for $\bar{\mu}_{.1} - \bar{\mu}_{.2}$.
e. Test $H_0 : \bar{\mu}_{.1} = \bar{\mu}_{.2}$ using an appropriate contrast statement.
f. Include a statement to obtain estimates of the cell means μ_{ij} and their standard errors for all i and j.
g. Construct an analysis of variance table to test main effects and interaction for these data and use the p-values to state conclusions. Use the values output from proc glm and write your own complete anova table.

5.18 Kutner et al. (2005) discussed an example in which human growth hormone was administered at a clinical research center to short prepubescent children. The investigator was studying the effects of gender and bone development levels on the rate of growth induced by hormone administration. Three children were randomly selected from each gender-bone development group, but 4 of the 18 children dropped out of the year-long study. The data are as follows:

Gender	Bone Development		
	Severely Depressed	Moderately Depressed	Mildly Depressed
Male	1.4	2.1	0.7
	2.4	1.7	1.1
	2.2		
Female	2.4	2.5	0.5
		1.8	0.9
		2.0	1.3

Use a SAS program with a proc glm step to this program to perform the computations described below. Use the usual two-way classification model for this analysis.

a.) Use proc means to obtain estimates of the cell means μ_{ij}. Construct a two-way table of gender by bone development levels showing the growth rate cell means.

b.) Obtain an interaction plot using proc gplot of the growth rate cell means with bone development levels on the x-axis. Join the points for each gender by line segments. Use colors and line types to identify gender.

c. Obtain the least squares estimates of μ, α_i, β_j, and γ_{ij} for $i = 1, 2$ and $j = 1, 2, 3$ computed by proc glm.

d. Construct an analysis of variance table to test main effects and interaction for these data. Use the values output from proc glm and write your own complete anova table. Using the p-values with $\alpha = .05$, make conclusions from each test.

e. Use lsmeans statements to obtain 90% confidence intervals on pairwise differences in bone development means adjusted for multiple testing using the Tukey method. What do you conclude from these intervals?

f. Kutner et al. (2005) tested the hypothesis whether the average growth rate of children with only mildly depressed bone development is significantly larger than zero. Add an estimate statement to obtain a t-statistic to test this one-sided hypothesis.

g. Test the hypothesis that the growth rates of children with severely depressed bone development is different in males and females using an appropriate contrast statement.

5.19 Four brands of airplane tires are compared to assess the differences in the rate of tread wear. The data were collected on eight planes, with two tires used under each wing. The researcher uses each airplane as a block, mounting four test tires, one of each brand, in random order on each airplane. Thus, a randomized complete block design with "airplane" as a blocking factor is the design used. The amount of tread is measured initially, and after 6 months, the following wear rates obtained. A larger value indicates greater wear.

Airplane	Brand			
	A	B	C	D
1	4.02	2.46	2.06	3.49
2	4.50	3.39	2.91	3.18
3	2.73	1.69	2.37	1.48
4	3.74	1.95	3.39	3.09
5	3.21	1.20	1.72	2.65
6	2.53	1.04	2.52	1.23
7	3.07	2.55	2.42	2.07
8	3.10	1.09	2.22	2.57

Brand A is currently used by the airline and Brands B, C, and D from three different competitors are being evaluated to replace A. Thus, the management is interested in the following:
1. Comparing Brand A with the average wear of Brands B, C, and D,
2. Comparing Brands B, C, and D

Prepare and run a SAS/GLM program necessary and provide answers to the following questions (on a separate sheet) assuming the model SAS Example E10 (see Fig. 5.53).

a. Construct an analysis of variance table and test the hypothesis $H_0: \tau_A = \tau_B = \tau_C = \tau_D$. State your conclusion based on the p-value.
b. Use a contrast statement for making comparison 1 by testing $H_0: \tau_A = (\tau_B + \tau_C + \tau_D)/3$.
c. Use a contrast statement for making the comparison 2 by testing $H_0: \tau_B = \tau_C = \tau_D$. One way to test this hypothesis is to make the comparisons $\tau_B - \tau_C$ and $\tau_B - \tau_D$ simultaneously in a single contrast statement. This results in the computation of a SS with 2 df and therefore, an F-test with 2 df for the numerator. Add the results of (b) and (c) to the anova table as additional lines and summarize conclusions from this analysis.
d. Construct 95% confidence intervals for $\mu_B - \mu_C$ and $\mu_B - \mu_D$, using the output from appropriate means statements used with proc glm.
e. Include the statement output out=new p=fitted r=residual; in the proc glm step. Then use proc gplot and the SAS data set new to obtain scatter plots of Residuals versus Machine and Residuals versus Fitted. Add a reference line at zero value on the residuals axis to each of these plots. Use these plots to comment briefly on whether your model assumptions were reasonable.
f. Perform Tukey's test of nonadditivity using a proc glm step and the SAS data set new created in part (e). What is your conclusion?

5.20 Four machines are compared to assess the differences in the rate of production of a certain part (Part No. Z-15) (Ostle, 1963). The data were collected over five days. All four machines were run each day (in random order), thus using a *randomized complete block design* with "Day" as a blocking factor. The following data are the number of units produced per day.

Day	Machine			
	A	B	C	D
1	293	308	323	333
2	298	353	343	363
3	280	323	350	368
4	288	358	365	345
5	260	343	340	330

Machine A is currently in use in a factory and Machines B, C, and D from three different competitors are being evaluated to replace A. Thus, the management is interested in the following:
1. comparing Machine A with the average production of Machines B, C, and D,
2. comparing B, C, and D

Prepare and run a `proc glm` step necessary and provide complete answers, including hypotheses tested and statistics used, to the following questions (on a separate sheet as before). Use the model shown for the RCBD for analyzing these data.

a. Construct an analysis of variance table and test the hypothesis H_0 : $\tau_A = \tau_B = \tau_C = \tau_D$. State your conclusion based on the p-value.
b. Use a `contrast` statement for making comparison 1 by testing H_0 : $\tau_A = (\tau_B + \tau_C + \tau_D)/3$. What is your conclusion?
c. Use a `contrast` statement for making comparison 2 by testing H_0 : $\tau_B = \tau_C = \tau_D$. One way to test this hypothesis is to make the comparisons $\tau_B - \tau_C$ and $\tau_B + \tau_C - 2\tau_D$ simultaneously, in a single `contrast` statement. This results in the computation of a SS with 2 df and an F-test with 2 df for the numerator. Add these tests as lines in an expanded anova table and summarize the results from your analysis.
d. Construct 95% confidence intervals for $\mu_B - \mu_C$, $\mu_B - \mu_D$, and $\mu_C - \mu_D$, using an appropriate statement in the `proc glm` step. Use the results of parts (c) and (d) to make a statement about the new machines being tried out assuming higher production rate is of interest.
e. Include the statement `output out=stats p=fitted r=residual;` in `proc glm` step. Then use `proc gplot` and the SAS data set `stats` to obtain scatter plots of Residuals versus Machine and Residuals versus Fitted. State the purpose for which these plots may be used. Do these plots identify any problems with your model assumptions?
f.) Perform Tukey's test of nonadditivity using a `proc glm` step and the SAS data set `stats` created in part (e). What is your conclusion?

5.21 A consumer product-testing organization wished to compare the annual power consumption of five different brands of dehumidifier (Devore, 1982). Because power consumption depends on the prevailing humidity level, it was decided to monitor each brand at four different areas of humidity, ranging from moderate to heavy. Within each area, the five brands were randomly assigned to five different locations for testing, resulting in a randomized complete block experiment with the areas as blocks. The resulting power consumption (annual kwh) values are as follows:

Brand	Blocks			
	1	2	3	4
1	685	792	838	875
2	722	806	893	953
3	733	802	880	941
4	811	888	952	1005
5	828	920	978	1023

Use a SAS procedure and the model given for the RCBD for analyzing these data.

a. Construct an analysis of variance table and test the hypothesis $H_0: \tau_1 = \tau_2 = \tau_3 = \tau_4 = \tau_5$. State your conclusion based on the p-value.
b. Use Tukey's underscoring procedure to compare all pairwise treatment effects. Make a concluding statement about the annual power consumption of the five different brands of dehumidifiers.
c. Although comparing the block effects is not of interest use the F-test to comment about the variability among the blocks in this experiment.
d. Add a proc step to your SAS program to perform Tukey's test for nonadditivity.

6

Analysis of Variance: Random and Mixed Effects Models

6.1 Introduction

In Chapter 5, the data sets considered were produced from experiments involving treatment factors that were regarded as *fixed*. The levels of factors studied in such experiments were those that the experimenter was interested in comparing and were not a random sample from a population of all possible levels. As discussed in Section 5.1, *random* factors were defined as those for which the levels of factors in the experiment consisted of a random sample from a population of levels. When random factors are present, the interest of the experimenter is to study the variance of the hypothetical population of factors rather than differences among the effects of different factor levels. Thus, the two types of factors are different not only in the way the treatment levels are selected but also in the way they affect the objectives of the study and, therefore, in the type of inferences made.

Whereas *random models* involve only random effects, *mixed models* are models that incorporate both fixed and random effects. Different variations of these are useful for modeling data generated from many experimental and observational studies. In this chapter, several applications of these models will be discussed and analyzed using SAS software. In the first few sections one-way and two-way random models are considered, followed by several sections presenting different applications of the mixed model. In the latter sections data from an RCBD are reanalyzed and a split-plot experiment presented regarding the blocks as a random factor, instead of a fixed factor. Several SAS procedures will be used in the analyses, primarily `proc glm` and `proc mixed`, and the differences identified and compared.

It is necessary to note some of the differences in the analyses presented here of models that include random effects from those that involve only fixed effects. A primary difference will be the inclusion of a column for *expected mean squares* in the analysis of variance table. An expected mean square is the linear function of the parameters that the particular mean square is expected to estimate unbiasedly and is usually derived mathematically using the computational formula for the mean square (called a *quadratic form*) and the model used for the observations. They are typically used for construction of F-ratios that are used to test whether a particular function of the parameters of interest is zero or not. For example, in the one-way classification model with a fixed effect used in Section 5.2, the expected mean square for the source of variation (SV), labeled Trt, is determined to be $E(\text{MS}_{\text{Trt}}) = \sigma^2 + \sum_i (\alpha_i - \bar{\alpha}_.)^2/(t-1)$ and the expected mean squares for Error is $E(\text{MSE}) = \sigma^2$. Now the fact that $\sum_i (\alpha_i - \bar{\alpha}_.)^2/(t-1)$ is zero if the null hypothesis of $H_0 : \alpha_1 = \alpha_2 = \cdots = \alpha_t$ is true and larger than zero if it is false implies that the ratio of mean squares $\text{MS}_{\text{Trt}}/\text{MSE}$ is an appropriate measure for constructing a statistical test of whether H_0 is true or not.

It was considered unnecessary to include this column as part of the analyses in Chapter 5, partly because the hypothesis being tested using the F-statistic for a particular effect was unambiguous. This was so because the models in that chapter did not contain random effects or nested effects. In experiments using complex designs (e.g., split-plots) that involve only fixed treatment effects, it may not be obvious how F-ratios that test particular hypotheses may be constructed. It may be helpful for the analysis of such experiments to include the expected mean squares column in the anova table.

In experiments that involve random factors, the variance of the response is usually partitioned into parts called *variance components*, explained as variation due to each of the random effects appearing in the model. Apart from determining appropriate test statistics, expected mean squares are also used to estimate variance components using the *method of moments*. This involves equating the expected mean squares of each source of variation in the anova table, to the respective observed mean square and solving the resulting set of linear equations for the variance components. The resulting estimates are called method of moments estimates or anova estimates. These estimates have the useful properties of being unbiased and having minimum variance.

Another difference in the analyses of models that include random effects is that in many situations, closed-form solutions to the normal equations for obtaining maximum likelihood estimates do not exist and, thus, iterative optimization techniques need to be employed to obtain the estimates. Section 6.5 contains an introduction to the mixed model that includes a brief discussion of estimation of fixed effects and variance components that is illustrated with a simplified example. Here, a brief introduction to the iterative methods available is given. The SAS procedure recommended for analyzing mixed models is proc mixed. It incorporates two popular likelihood-based methods: maximum likelihood (ML) and restricted maximum likelihood (REML). A detailed theoretical discussion of these methods is beyond the scope of this book. However, at least a brief explanation will help users of SAS programs like proc mixed understand the basic principles involved. The presentation below (supplemented as needed later) is intended to provide users with a minimal explanation necessary to understand some of the options available in the usage of these procedures as well as help make choices among the possible values that may be specified for them. Readers are urged to obtain additional information from more advanced textbooks on the topic as well as from the detailed descriptions provided in the manuals.

The end result of the models that will be described in this chapter is the specification of the joint distribution of the observations, say y_{ijk}. Generally, it is easier to describe this as the multivariate distribution of an n-dimensional data vector $\mathbf{y}$. In the notation used in Chapters 4 and 5, a regression or a fixed effects anova model was represented by $\mathbf{y} = X\boldsymbol{\beta} + \boldsymbol{\epsilon}$ where the errors (ϵ_i's) were assumed to be iid $N(0, \sigma^2)$ random variables. Another way to express this model is to say that $\mathbf{y}$ is distributed as an n-dimensional multivariate normal with mean vector $\boldsymbol{\mu} = X\boldsymbol{\beta}$ and variance-covariance matrix $\sigma^2 I$ where I is an $n \times n$ identity matrix. In order to allow other variance-covariance structures, this matrix may be represented by the symbol Σ (an element of this matrix is represented by σ_{ij}). For the two-way model used in Section 5.1 in Chapter 5, the vector $\boldsymbol{\beta} = (\mu, \alpha_1, \alpha_2, \alpha_3, \tau_1, \tau_2, \tau_3, \tau_4)'$ and $\Sigma = \sigma^2 I$. Thus, the parameters of that model are $\mu, \alpha_1, \alpha_2, \alpha_3, \tau_1, \tau_2, \tau_3, \tau_4$, and σ^2. The joint density function of the elements of $\mathbf{y}$, using matrix notation is represented by

$$f(\mathbf{y}) = \frac{1}{\sqrt{(2\pi)^n |\Sigma|}} \exp-\frac{1}{2}\left[(\mathbf{y} - X\boldsymbol{\beta})' \Sigma^{-1} (\mathbf{y} - X\boldsymbol{\beta})\right]$$

The likelihood function of the parameters of this model is the same function given on the right-hand side of the above equation but is considered as a function of the parameters (as opposed to a function of the elements in $\mathbf{y}$). It is denoted by $L(\boldsymbol{\theta})$, where $\boldsymbol{\theta}$ is a vector containing all unknown parameters in the density function. Since it is easier to manipulate mathematically, the logarithm of L, called the *log-likelihood* and denoted by $\ell(\boldsymbol{\theta})$, is often used. The log-likelihood for the above model is

$$\ell(\boldsymbol{\theta}) = -\frac{n}{2} \log(2\pi) - \frac{1}{2} \log|\Sigma| - \frac{1}{2}\left[(\mathbf{y} - X\boldsymbol{\beta})' \Sigma^{-1} (\mathbf{y} - X\boldsymbol{\beta})\right]$$

where $\boldsymbol{\theta} = (\boldsymbol{\beta}, \sigma_{11}, \ldots, \sigma_{nn})'$. The *maximum likelihood estimates* (MLEs) are those values of the parameters that maximizes the log-likelihood function $\ell(\boldsymbol{\theta})$ over the *parameter space*. For unbalanced data sets, calculating the MLEs usually require numerical optimization methods that involve iterative procedures. Inference procedures for the parameters based on MLEs usually involve large sample properties of these estimates. Usually, for construction of test statistics and interval estimates, an approximate estimate of the variance-covariance matrix of the estimated parameter vector that results from the optimization procedure is used.

So far this description has only included models that involved regression-type or anova fixed-effects-type parameters. The general model for random and mixed models will be described in other sections of this chapter. In the most general form, the mixed model is given by

$$\mathbf{y} = X\boldsymbol{\beta} + Z\mathbf{u} + \boldsymbol{\epsilon}$$

where the random vectors $\mathbf{u}$ and $\boldsymbol{\epsilon}$ have independent multivariate normal distributions $N(\mathbf{0}, G)$ and $N(\mathbf{0}, R)$, respectively, and the variance-covariance matrices G and R are fixed unknown constants. Using this specification, the variance-covariance matrix of $\mathbf{y}$ is of the form $V = ZGZ' + R$ and the marginal distribution of $\mathbf{y}$ is multivariate normal with mean vector $\boldsymbol{\mu} = X\boldsymbol{\beta}$ and variance-covariance matrix V (i.e., $N(X\boldsymbol{\beta}, V)$).

In the classical variance components model, the random subvectors $\mathbf{u}_1, \mathbf{u}_2, \ldots, \mathbf{u}_k$ of $\mathbf{u}$ (say, corresponding to k random effects) and $\boldsymbol{\epsilon}$ have the multivariate normal distributions $N(\mathbf{0}, \sigma_1^2 I)$, $N(\mathbf{0}, \sigma_2^2 I), \ldots, N(\mathbf{0}, \sigma_k^2 I)$, and $N(\mathbf{0}, \sigma^2 I)$, respectively, where the matrices $\sigma_1^2 I$ and so forth, are diagonal matrices with the diagonal elements all equal to the respective variance components. The variance-covariance matrix of $\mathbf{y}$ is thus given by the matrix V:

$$V = \sum_i Z_i Z_i' \sigma_i^2 + \sigma^2 I.$$

where Z_i is the design matrix for the ith random effect; that is, in this case, both G and R turn out to be diagonal matrices, whose diagonal elements are the variance components $(\sigma_1^2, \sigma_2^2, \ldots, \sigma_k^2, \sigma^2)$.

The log-likelihood function of the parameters in the mixed model is obtained using the marginal distribution of $\mathbf{y}$ given above and is

$$\ell(\boldsymbol{\beta}, V) = -\frac{n}{2}\log(2\pi) - \frac{1}{2}\log|V| - \frac{1}{2}\left[(\mathbf{y} - X\boldsymbol{\beta})'V^{-1}(\mathbf{y} - X\boldsymbol{\beta})\right]$$

In the classical variance components model described above, V is of the form given above and, thus, is a function of the variance components $\boldsymbol{\sigma}^2 = (\sigma_1^2, \sigma_2^2, \ldots, \sigma_k^2, \sigma^2)'$. The MLEs of $\boldsymbol{\beta}$ and the variance components are obtained by equating the first derivatives of $\ell(\boldsymbol{\beta}, V)$ with respect to $\boldsymbol{\beta}$ and V to zero and solving the resulting equations for $\boldsymbol{\beta}$ and $\boldsymbol{\sigma}^2$. The usual strategy is to first solve the set of equations

$$X'V^{-1}X\tilde{\boldsymbol{\beta}} = X'V^{-1}\mathbf{y}$$

for $\tilde{\boldsymbol{\beta}}$, assuming that the variance components are known. The solution can be obtained in closed form as

$$\tilde{\boldsymbol{\beta}} = (X'V^{-1}X)^{-1}X'V^{-1}\mathbf{y}$$

where $\tilde{\boldsymbol{\beta}}$ is a function of the unknown variance components. Substituting $\tilde{\boldsymbol{\beta}}$ in $\ell(\boldsymbol{\beta}, V)$ and using an iterative procedure to maximize the resulting *profile log-likelihood*, $\hat{\boldsymbol{\sigma}}^2$, the maximum likelihood estimate of $\boldsymbol{\sigma}^2$ is obtained. A procedure to ensure that the variance components are in the parameter space (i.e., they are non-negative) is incorporated. The MLE of the variance components $\hat{\boldsymbol{\sigma}}^2$ is then used to compute the MLE of $\boldsymbol{\beta}$ using

$$\hat{\boldsymbol{\beta}} = (X'V(\hat{\boldsymbol{\sigma}}^2)^{-1}X)^{-1}X'V(\hat{\boldsymbol{\sigma}}^2)^{-1}\mathbf{y}$$

where $V(\hat{\boldsymbol{\sigma}}^2)$ is the estimated variance-covariance matrix of $\mathbf{y}$ using the MLE $\hat{\boldsymbol{\sigma}}^2$ of $\boldsymbol{\sigma}^2$.

Even balanced data ML estimates are not identical to the method of moment estimates. This is mainly because the ML estimation method does not "adjust for" using the estimate of the fixed part of the model to estimate the variance components. The so-called restricted maximum likelihood estimation method (REML) overcomes this problem. It is easiest to understand the REML estimation as based on maximizing the log-likelihood of the transformed data vector $K\mathbf{y}$ instead of the log-likelihood of $\mathbf{y}$. The rows of the matrix K, $\mathbf{k}'$, are such that $E(\mathbf{k}'\mathbf{y}) = \mathbf{k}'X\boldsymbol{\beta} = 0$. These linear combinations of the observations are called *error contrasts* in the literature and can be obtained by selecting $n - r$ linearly independent rows of the matrix $I - X(X'X)^{-1}X'$, where r is the rank of X. Once the matrix K is constructed, it can be employed to transform the setup used for the previous maximization problem by transforming $\mathbf{y}$ to $K\mathbf{y}$, $X\boldsymbol{\beta}$ to zero, Z's to KZ, and V to KVK'. Note that the new objective function is the log-likelihood function only of the variance components, but the new observed values are the transformed data values in $K\mathbf{y}$. The results of maximizing the transformed likelihood gives the REML estimates of the variance components, $\hat{\boldsymbol{\sigma}}_R^2$. The estimates of the fixed effect parameters are then obtained from

$$\hat{\boldsymbol{\beta}}_R = (X'V(\hat{\boldsymbol{\sigma}}_R^2)^{-1}X)^{-1}X'V(\hat{\boldsymbol{\sigma}}_R^2)^{-1}\mathbf{y}$$

where $V(\hat{\boldsymbol{\sigma}}_R^2)$ is the estimated variance-covariance matrix using the REML estimate of $\boldsymbol{\sigma}^2$.

6.2 One-Way Random Effects Model

Experiments involving one random factor are considered in this section. This type of experiment is similar to the "Traffic Tickets" example discussed in Chapter 5. The random factor of interest was "Precinct"; that is, precincts in a large city were selected randomly for comparing the number of tickets issued for traffic-related violations. The main interest in this experiment is concerned with making inferences (i.e., estimation and hypothesis tests), about the variance among the precincts in the number of tickets issued. Suppose that a precincts are under study and n officers are randomly sampled in each precinct. Let y_{ij} be the number of tickets issued by the jth officer in the ith precinct.

Model

The one-way random effects model is given by

$$y_{ij} = \mu + A_i + \epsilon_{ij}, \quad i = 1,\ldots,a; \quad j = 1,\ldots,n$$

where the random effects A_i, $i = 1, \ldots, a$, are distributed as iid $N(0, \sigma_A^2)$ random variables independent of the random errors ϵ_{ij}. As usual, the random errors ϵ_{ij}, $i = 1, \ldots, a$; $j = 1, \ldots, n$, are distributed independently as $N(0, \sigma^2)$ random variables. If this is formulated as a "means model" where $\mu_i = \mu + A_i$, then μ_i, $i = 1, \ldots, a$, are assumed to be distributed as iid $N(\mu, \sigma_A^2)$ random variables, where μ represents the mean of the population of the factor levels. In the traffic tickets example, this would correspond to the mean number of tickets issued by all police officers regardless of the precinct. The random effect A_i models the random increment the ith precinct would add to (or subtract from) μ to give the mean number of tickets μ_i issued in that precinct.

It is important to note that the mean of A_i is zero; that is, on average, the incremental mean number of ticket issued in different precincts cancel out. However, σ_A^2, the variance of A_i, and hence of μ_i, measures the variance among the mean numbers of tickets issued in different precincts. If precinct means are all the same, then it will be zero; otherwise, it will be positive and will be larger the more variable the mean numbers of tickets issued among different precincts. On the other hand, the variance among the numbers of tickets issued by officers within each precinct is assumed to be the same for all precincts. This measures the "error" variance σ^2 among the experimental units (police officers) in this study.

Estimation and Hypothesis Testing

An analysis of variance that corresponds to the above model is constructed using the same computational formulas used for the computation of the anova for the one-way fixed effects model. However, as discussed in Section 6.1, an additional column displaying the expected mean squares is included in the anova table for the one-way random effects model:

SV	df	SS	MS	F	E(MS)
A	$a - 1$	SS_A	MS_A	MS_A/MSE	$\sigma^2 + n\sigma_A^2$
Error	$a(n-1)$	SSE	$MSE(= s^2)$		σ^2
Total	$an - 1$				

The computation of the expected mean squares does not require the distributional assumption of normality of the random effects. However, it is required for performing hypothesis tests and constructing confidence intervals using the F-, t-, and the chi-square distributions.

The hypothesis of main interest in the above model is :

$$H_0 : \sigma_A^2 = 0 \quad \text{versus} \quad H_a : \sigma_A^2 > 0$$

The two mean squares needed to construct an appropriate F-ratio are determined so that both the numerator and the denominator mean squares will have the same expectation if the null hypothesis of $\sigma_A^2 = 0$ holds and the numerator will have a larger expectation if $\sigma_A^2 > 0$. By examining the $E(\mathrm{MS})$ column, it can be observed that the F-statistic shown satisfies this requirement. The null hypothesis is rejected if the computed F-statistic exceeds the upper $(1-\alpha)$ percentile of the F-distribution with $a-1$ and $a(n-1)$ degrees of freedom or, equivalently, if the p-value is less than α for an α level selected by the experimenter to control the Type I error rate.

As usual, the estimate of the error variance σ^2 is the MSE from the anova table $\hat{\sigma}^2 = s^2$. If the hypothesis $H_0 : \sigma_A^2 = 0$ is rejected in favor of $H_a : \sigma_A^2 > 0$, the mean squares may also be used to estimate σ_A^2. To do this, the expected mean square (which is a linear combination of the variance components) is set equal to its observed value MS_A from the anova table; that is, set

$$\sigma^2 + n\sigma_A^2 = \mathrm{MS}_A$$

and the resulting equation is solved for σ_A^2 after substituting the estimate s^2 for σ^2. This gives the result

$$\hat{\sigma}_A^2 = \frac{\mathrm{MS}_A - s^2}{n}$$

where the right-hand side consists only of quantities computed from the data and are obtained from the anova table. This method of estimation is called the *method of moments*. These estimates are identical to maximum likelihood estimates when the sample sizes are the same, as is the case in this example.

One approach for obtaining approximate confidence intervals requires the normality assumptions stated in the model definition. A $(1-\alpha)100\%$ confidence interval for σ_A^2 is provided by

$$\frac{\nu\hat{\sigma}_A^2}{\chi^2_{1-\alpha/2,\nu}} < \sigma_A^2 < \frac{\nu\hat{\sigma}_A^2}{\chi^2_{\alpha/2,\nu}} \qquad (6.1)$$

where $\chi^2_{1-\alpha/2,\nu}$ and $\chi^2_{\alpha/2,\nu}$ are the $1-\alpha/2$ and $\alpha/2$ percentile points of the chi-squared distribution with ν degrees of freedom, respectively, and the value of ν is obtained using the Satterthwaite approximation. This approximation is required because, as seen earlier, $\hat{\sigma}_A^2 = \frac{1}{n}\mathrm{MS}_A - \frac{1}{n}\mathrm{MSE}$, a linear combination of two mean squares, and thus it does not have an exact chi-square distribution. The approximation defines ν as

$$\nu = \frac{(n\hat{\sigma}_A^2)^2}{(\mathrm{MS}_A)^2/(a-1) + (s^2)^2/a(n-1)} \qquad (6.2)$$

Note that no approximation is required to obtain a $(1-\alpha)100\%$ confidence interval for σ^2 since the interval is based on a single mean square (i.e. MSE). It is given by:

$$\frac{\nu s^2}{\chi^2_{1-\alpha/2,\nu}} < \sigma^2 < \frac{\nu s^2}{\chi^2_{\alpha/2,\nu}} \qquad (6.3)$$

where $\nu = a(n-1)$, the degrees of freedom for error. Also, note that the above confidence intervals are not symmetrical around the estimate of the corresponding variance component. The confidence intervals for the standard deviations σ_A or σ are found by taking square roots of both end points of each interval.

Although factor levels are independent, the observations from the the same factor are correlated. This correlation is another important quantity that may be estimated from this type of an experiment and is called the *intraclass correlation*:

$$\text{Corr}(y_{ij}, y_{ij'}) = \frac{\sigma_A^2}{\sigma_A^2 + \sigma^2} \quad \text{for } j \neq j'$$

This ratio also measures the proportion of the total variation in y_{ij} only due to the random effect, since obviously, $\text{Var}(y_{ij}) = \sigma_A^2 + \sigma^2$. In plant breeding experiments, for example, investigators might be interested in selecting inbred lines that have large intraclass correlations, as that would indicate variation due to genetic influences of those breeds have a larger effect than, say, environmental effects on the trait being measured.

6.2.1 Using PROC GLM to analyze one-way random effects models

The following example is taken from Snedecor and Cochran (1989). An experiment was conducted at the Iowa Agricultural Experiment Station to determine if there is significant variation of average daily gain of pigs from litter to litter. Average daily gain in weight is an indicator of growth rate in animals. For the study, four litters were chosen at random from a single inbred line of swine. The average daily gains of two animals selected at random from each litter were measured. The data are shown in Table 6.1.

Litter	1	2	3	4
Gain	1.18	1.36	1.37	1.07
	1.11	1.65	1.40	0.90

Table 6.1. Average daily gain of swine

The model for average daily gain is:

$$y_{ij} = \mu + A_i + \epsilon_{ij}, \quad i = 1, \ldots, 4; \quad j = 1, 2$$

where A_i is the effect of the ith litter and is assumed to be an iid $N(0, \sigma_A^2)$ random variable and ϵ_{ij}, the sampling error associated with pigs within each litter, an iid $N(0, \sigma^2)$ random variable.

6.2 One-Way Random Effects Model

```
data hogs;
input litter gain;
datalines;
1   1.18
1   1.11
2   1.36
2   1.65
3   1.37
3   1.40
4   1.07
4   0.90
;
run;
proc glm data=hogs;
  class litter ;
  model gain = litter ;
  random litter/test;
  title 'Average Daily Gain in Swine';
run;
```

Fig. 6.1. SAS Example F1: Program

SAS Example F1

The SAS Example F1 program (see Fig. 6.1) illustrates how `proc glm` can be used to perform the necessary computations. The data may be input using methods used for one-way fixed effects experiments (see, e.g., Fig. 5.1 in Chapter 5). However, in this example, since the sample sizes are small and equal, a straightforward approach can be used. The data are entered exactly in the same format as required by `proc glm`. That is, values for a classification variable `litter` and the response variable `gain` are entered in the lines of data separated by blanks so that they are accessed easily using the *list input* method. The `class` and the `model` statements are exactly as for a fixed effects model; however, an additional statement `random litter/test` is included here. This statement indicates that the effect `litter` in the model is a random effect and also requests that a test be performed to test the hypothesis that the corresponding variance component is significantly different from zero.

From the `proc glm` output reproduced in Fig. 6.2, the construction of the following anova table is straightforward:

SV	df	SS	MS	F	p-value	E(MS)
Litter	3	0.3288	0.1096	7.38	.0416	$\sigma^2 + 2\sigma_A^2$
Boar (Litter)	4	0.0594	0.01485			σ^2
Total	7	0.3882				

Note carefully that information necessary for completing the additional column titled E(MS) containing the expected mean squares is available from page 3 of this output under `Type III Expected Mean Square` **1**. This is a

```
                    Average Daily Gain in Swine                          1

                         The GLM Procedure

                      Class Level Information

                   Class        Levels    Values

                   litter          4      1 2 3 4

               Number of Observations Read        8
               Number of Observations Used        8

                    Average Daily Gain in Swine                          2

                         The GLM Procedure

Dependent Variable: gain

                                  Sum of
Source                DF         Squares     Mean Square    F Value    Pr > F

Model                  3      0.32880000      0.10960000       7.38    0.0416

Error                  4      0.05940000      0.01485000

Corrected Total        7      0.38820000


             R-Square     Coeff Var      Root MSE     gain Mean

             0.846986      9.710006      0.121861      1.255000


Source                DF       Type I SS     Mean Square    F Value    Pr > F

litter                 3      0.32880000      0.10960000       7.38    0.0416


Source                DF     Type III SS     Mean Square    F Value    Pr > F

litter                 3      0.32880000      0.10960000       7.38    0.0416

                    Average Daily Gain in Swine                          3

                         The GLM Procedure

   Source              Type III Expected Mean Square  1

   litter              Var(Error) + 2 Var(litter)

                    Average Daily Gain in Swine                          4

                         The GLM Procedure
             Tests of Hypotheses for Random Model Analysis of Variance  2

Dependent Variable: gain

Source                DF     Type III SS     Mean Square    F Value    Pr > F

litter                 3        0.328800        0.109600       7.38    0.0416

Error: MS(Error)       4        0.059400        0.014850
```

Fig. 6.2. SAS Example F1: Output (pages 1-4)

6.2 One-Way Random Effects Model

simple linear combinations of the variance components σ_A^2 and σ^2. Also, recall that the expected value of mean squares for error, MSE, is always σ^2 under the model assumptions. These two expectations form the equations that are to be used to obtain the method of moments estimates of the variance components.

The results of the F-test of $H_0 : \sigma_A^2 = 0$ is displayed on page 4 titled Tests of Hypotheses for Random Model Analysis of Variance **2** of the output. Note that the F-statistic is the ratio $\mathrm{MS_{Litter}}/\mathrm{MSE}$. The denominator of the F-ratio is the MSE (identified as MS(Error) in the output). Also note that, in the anova table above, the corresponding source of variation is labeled Boar(Litter).

If $\sigma_A^2 = 0$, the expectations of both the numerator and the denominator of the ratio would have the same value of σ^2. If the F-statistic is found to be significantly large, it should lead to the conclusion that σ_A^2 is greater than zero. Thus, in more complex experiments, information from the $E(MS)$ column could be used to identify ratios of mean squares needed to test hypotheses about different variance components.

The p-value from the anova table is less than .05 and, hence, the null hypothesis $H_0 : \sigma_A^2 = 0$ is rejected at $\alpha = .05$; that is, evidence exists in the data from this experiment that there is a significant variation of average daily gain among the litters. Since the litters were a random sample, this result would apply to all litters from the inbred line of swine from which these litters were sampled.

Since the hypothesis $H_0 : \sigma_A^2 = 0$ is rejected, it is useful to quantify this variation by estimating the variance component σ_A^2. The method of moments requires setting the computed values of the mean squares equal to the corresponding expressions found in the $E(MS)$ column and solving the resulting linear equations for the variance components. In this case, the equations are

$$\sigma^2 + 2\sigma_A^2 = 0.1096$$

$$\sigma^2 = 0.01485$$

and solving these equations gives the required estimates

$$\hat{\sigma}^2 = 0.01485$$

$$\hat{\sigma}_A^2 = \frac{0.1096 - 0.01485}{2} = 0.0474$$

Finally, to compute a $(1-\alpha)100\%$ confidence interval for σ_A^2, instead of using Formula (6.1) for hand computation, it may be coded in SAS as shown in the following simple data step:

```
data cint;
   alpha=.05; n=2; a=4;
   msa= 0.1096; s2=0.01485; sa2=.0474;
   nu=(n*sa2)**2/(msa**2/(a-1)+ s2**2/a*(n-1));
```

```
              L= (nu*sa2)/cinv(1-alpha/2,nu);
              U= (nu*sa2)/cinv(alpha/2,nu);
              put nu=    L=     U=;
        run;
```

Note that the first two lines have been completed using information obtained from Fig. 6.2 and that the last three lines use these values for the computation of the confidence interval with the required confidence coefficient. The results are printed on the log page/window instead of the output page/window. The 95% confidence interval for σ_A^2 is (0.01342, 1.3809).

6.2.2 Using PROC MIXED to analyze one-way random effects models

Although the use of the random statement in proc glm gives the user the capability to compute expected mean squares and perform F-tests about variance components, other statements in proc glm do not make use of this information. For example, lsmeans and estimate statements assume that all effects are fixed irrespective of whether the random statement is present or not. Thus, it is recommended that one use proc mixed to analyze both mixed effects models and random effects models. Among other advantages, proc mixed gives the user the option of choosing among several estimation methods in addition to the method of moments as well as the ability to use estimate statements to estimate *best linear unbiased predictors* (BLUPs) (i.e., predictable linear combinations of fixed and random effects).

```
insert data step to create the SAS dataset 'hogs' here

proc mixed data=hogs cl;
   class litter;
   model gain = ;
   random litter/solution;
   title 'Average Daily Gain in Swine';
run;
```

Fig. 6.3. SAS Example F2: Program using PROC MIXED

SAS Example F2

The SAS Example F2 program (see Fig. 6.3) illustrates how proc mixed may be used to perform an analysis of a one-way random model. The essential difference from a proc glm step is that in the model statement in proc mixed, only the fixed part of the model needs to be specified; thus, model gain = ; implies that only an overall mean μ is present in the model in addition to any random effects. The random statement specifies the random effect terms:

6.2 One-Way Random Effects Model

in this case, the term `litter` and a random error term. The variances of the random effects constitute the *variance components* including the `error` variance component that is assumed by default.

The default method of estimation is *restricted maximum likelihood* (commonly known as REML), which assumes that random effects are Normally distributed. Other estimation methods available include *maximum likelihood* and the method known as MIVQUE(0). These can be requested using the proc statement options `method=ml` or `method=mivque0`, respectively. Since the REML estimation is popular among practitioners, that method is set as the default. Finally, the `cl` option specified on the `proc mixed` statement requests the calculation of confidence limits for the variance components.

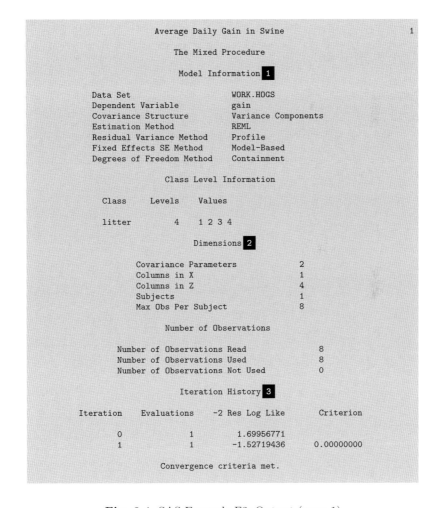

Fig. 6.4. SAS Example F2: Output (page 1)

What follows is a brief explanation of the contents of output page 1 (see Fig. 6.4) from `proc mixed`. In the `Model Information` **1** portion of page 1, the phrase *Variance Components* describing the covariance structure implies that the model specified by the user has been identified as a traditional mixed model in which variances of the random effects parameters (or the variance components) are the only covariance parameters present. The `Dimensions` **2** section gives the sizes of the design matrices (described in Section 6.1 where mixed model theory is introduced). The `noinfo` option on the proc statement may be used to suppress the above two tables. `Iteration History` **3** provides details of the convergence of the iterative procedure used to optimize the objective function used in the case of REML or maximum likelihood methods, respectively. The column labeled "−2 Log Like" for maximum likelihood or "−2 Res Log Like" for REML lists the value of −2 times the log-likelihood or −2 times the log residual likelihood function. This statistic, called the deviance, is used for testing hypotheses about parameters by model comparison. The column labeled "Evaluations" lists the number of times the objective function was evaluated during each iteration. Using the `noitprint` option in the proc statement suppresses the "Iteration History" table.

```
                        Average Daily Gain in Swine                              2

                             The Mixed Procedure

                      Covariance Parameter Estimates 4

             Cov Parm      Estimate     Alpha     Lower      Upper

             litter        0.04738      0.05      0.01341    1.3843
             Residual      0.01485      0.05      0.005331   0.1226

                                Fit Statistics

                  -2 Res Log Likelihood              -1.5
                  AIC (smaller is better)             2.5
                  AICC (smaller is better)            5.5
                  BIC (smaller is better)             1.2

                      Solution for Random Effects 5

                                          Std Err
     Effect    litter    Estimate         Pred       DF    t Value    Pr > |t|

     litter    1         -0.09510         0.1291     4     -0.74      0.5021
     litter    2          0.2161          0.1291     4      1.67      0.1693
     litter    3          0.1124          0.1291     4      0.87      0.4330
     litter    4         -0.2334          0.1291     4     -1.81      0.1448
```

Fig. 6.5. SAS Example F2: Output (page 2)

Output page 2 (see Fig. 6.5) contains the estimates of the variance components given as $\hat{\sigma}_A^2 = 0.04738$ and $\hat{\sigma}^2 = 0.01485$ as `Covariance Parameter`

Estimates **4**. The Satterthwaite approximation introduced previously is used to construct confidence intervals for the variance components appearing here. This is true for all iterative methods, as variance components are constrained to be nonnegative. As will be seen in SAS Example F3, when the method of moments estimation method is used, this constrained is not used so large sample methods are used to calculate these confidence limits (except for the error variance). Solution for Random Effects **5** are predicted values of the `litter` random effects, produced by the `solution` option in the `random` statement. They are estimates of the BLUPs of the random effect for each litter. These predictions may also be obtained using the `estimate` statements:

```
estimate 'Litter 1 Effect' | litter 1 0 0 0;
estimate 'Litter 2 Effect' | litter 0 1 0 0;
estimate 'Litter 3 Effect' | litter 0 0 1 0;
estimate 'Litter 4 Effect' | litter 0 0 0 1;
```

Note that the syntax of the estimate statement is similar to that used in the analysis of fixed effects models with `proc glm`, except that the specification of the random effects must appear after the vertical bar ("|"). For mixed models, both fixed effects and random effects may appear in the same estimate statement: fixed effects before | and random effects after. The output from the above set of estimate statements (not shown) is identical to that of the output from the `solution` option seen in Fig. 6.5.

SAS Example F3

```
insert data step to create the SAS dataset 'hogs' here

proc mixed data=hogs noclprint noinfo method=type3 asycov cl;
  class litter;
  model gain = ;
  random litter/solution;
  title 'Average Daily Gain in Swine';
run;
```

Fig. 6.6. SAS Example F3: Method of moments estimates using PROC MIXED

It is important to note that estimates produced by maximum likelihood estimation methods will be different from the method of moments estimates calculated using `proc glm` *if the sample sizes are not equal*. However, `proc mixed` may be used to also obtain the method of moments estimates as illustrated in SAS Example F3 (see Fig. 6.6). The proc statement options `noclprint` suppresses the class-level information table, and `noinfo` suppresses several other tables that are not relevant here. The option `method=type3` specifies the type of the mean squares (and the corresponding expected values)

```
                          Average Daily Gain in Swine                          1

                               The Mixed Procedure

                           Type 3 Analysis of Variance ■1

                      Sum of
Source          DF    Squares     Mean Square    Expected Mean Square

litter          3     0.328800    0.109600       Var(Residual) + 2 Var(litter)
Residual        4     0.059400    0.014850       Var(Residual)

                           Type 3 Analysis of Variance ■1

                                                 Error
Source     Error Term                            DF      F Value    Pr > F

litter     MS(Residual)                          4       7.38       0.0416
Residual   .                                     .       .          .

                        Covariance Parameter Estimates

           Cov Parm     Estimate    Alpha    Lower      Upper

           litter       0.04738     0.05     -0.04092   0.1357
           Residual     0.01485     0.05     0.005331   0.1226

                  Asymptotic Covariance Matrix of Estimates ■2

                   Row    Cov Parm      CovP1        CovP2

                   1      litter        0.002030     -0.00006
                   2      Residual      -0.00006     0.000110

                          Average Daily Gain in Swine                          2

                               The Mixed Procedure

                            Solution for Random Effects
                                         Std Err
        Effect    litter    Estimate     Pred      DF     t Value    Pr > |t|

        litter    1         -0.09510     0.1291    4      -0.74      0.5021
        litter    2         0.2161       0.1291    4      1.67       0.1693
        litter    3         0.1124       0.1291    4      0.87       0.4330
        litter    4         -0.2334      0.1291    4      -1.81      0.1448
```

Fig. 6.7. SAS Example F3: Output (pages 1 and 2)

that are to be used to estimate the variance components. Usually, Types 1 and 3 are used; however, they will be identical in equal sample sizes case. The option `asycov` requests the asymptotic covariance matrix of the estimated variance components and the option `cl` requests the confidence intervals on the variance components (based on asymptotic standard errors of the estimates of the variance components).

The output is shown in Fig. 6.7. The anova table showing the expected mean squares is given in the table titled `Type 3 Analysis of Variance` ■1.

It is exactly the same as that produced by `proc glm` where the litter effect is tested using the MS(Residual) as the divisor. The confidence intervals for the variance components are calculated using the estimated variance given in the table titled `Asymptotic Covariance Matrix of Estimates` **2**. For example, an asymptotic 95% confidence interval for σ_A^2 is calculated as

$$\hat{\sigma}_A^2 \pm z_{0.025} \times \text{s.e.}(\hat{\sigma}_A^2)$$
$$0.04738 \pm 1.96 \times \sqrt{.00203}$$

giving $(-0.04093, 0.1357)$. Since the number of litters is small, approximating the sampling distribution of $\hat{\sigma}_A^2$ by the normal distribution is questionable. So an interval based on the Satterthewaite approximation may be more appropriate. However, the interval for σ^2 given here is not based on the asymptotic standard error; it is calculated using the formula (6.3). The estimated BLUPs of the litter random effects are the same as those obtained previously.

The model used in the beginning of this section assumed equal sample sizes, for each level of the random factor. The expressions given in the anova table for the expected value of mean square (i.e., $E(\text{MS})$) for effect A was based on this assumption. In the case of unequal sample sizes this expression would be different. However, the investigator need not know this formula, since, as observed previously, when the `random` statement is present `proc glm` provides Type III expected means squares as part of the output, and when the `method=type3` option is present, `proc mixed` computes and outputs this expectation as a part of the Type 3 analysis of variance. Thus, the user may proceed with an analysis based on the method of moments as usual.

SAS Example F4

In SAS Example F4, the program shown in Fig. 6.8 is used to illustrate the analysis of a data set with unequal sample sizes. An experiment on artificial insemination in which semen samples from six different bulls was used to inseminate different numbers of cows is described in Snedecor and Cochran (1989). The data are percentages of conceptions and are recorded in Table 6.2.

The one-way random effects model for the `percent` variable is

$$y_{ij} = \mu + A_i + \epsilon_{ij}, \quad i = 1, \ldots, 6; \quad j = 1, \ldots, n_i$$

where the random effects A_i, $i = 1, \ldots, 6$ are distributed as iid $N(0, \sigma_A^2)$ random variables independent of the random errors ϵ_{ij}, which are distributed independently as $N(0, \sigma^2)$. The n_i's represent the different sample sizes used in the experiment.

In the SAS program, the data are input using `trailing @@` to input pairs of data values for the variables `bull` and `percent`. A `proc mixed` step with a `method=type3` option used in this SAS program is similar to the one used for the analysis of the previous example. The SAS output pages 2 and 3 are reproduced in Fig. 6.9. From page 2, $E(\text{MS})$ for `bull` **1** is seen to be

Table 6.2. Artificial insemination of cows (Snedecor and Cochran, 1989)

Bull 1	Bull 2	Bull 3	Bull 4	Bull 5	Bull 6
46	70	52	47	42	35
31	59	44	21	64	68
37		57	70	50	59
62		40	46	69	38
30		67	14	77	57
		64		81	76
		70		87	57
					29
					60

$\sigma^2 + 5.6686\sigma_A^2$. Thus, the equations to be solved for obtaining method of moments are

$$\sigma^2 + 5.6686\sigma_A^2 = 664.411746$$

$$\sigma^2 = 248.287630$$

and solving these equations give, the estimates $\hat{\sigma}^2 = 248.287630$ and $\hat{\sigma}_A^2 = (664.411746 - 248.287630)/5.6686 = 73.40862$. These values are confirmed from the output that also include asymptotic standard errors and 95% confidence intervals. The interval for σ_A^2 is based on the normal distribution and the asymptotic standard error of its estimate and the interval for σ^2 is calculated using the formula (6.3). See comment made earlier (see SAS Example F3) concerning intervals based on the large sample approximation. More im-

```
data cows;
input bull percent @@;
datalines;
1 46 1 31 1 37  1 62 1 30
2 70 2 59
3 52 3 44 3 57 3 40 3 67 3 64 3 70
4 47 4 21 4 70 4 46 4 14
5 42 5 64 5 50 5 69 5 77 5 81 5 87
6 35 6 68 6 59 6 38 6 57 6 76 6 57 6 29 6 60
;
run;

proc mixed data=cows noclprint noinfo method=type3 asycov cl;
   class bull ;
   model percent = ;
   random bull/solution;
   title 'Artificial Insemination of Cows';
run;
```

Fig. 6.8. SAS Example F4: Program using PROC MIXED

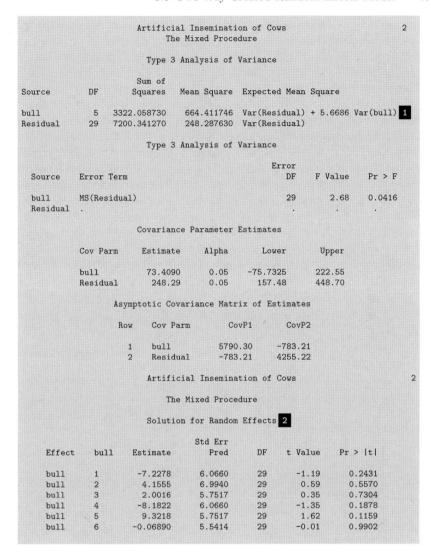

Fig. 6.9. SAS Example F4: Output (pages 1 and 2)

portantly, page 2 displays the estimated BLUPs ❷ of the bull random effects for which the standard errors are now different for each estimate.

6.3 Two-Way Crossed Random Effects Model

An experiment with two random factors that are crossed is considered in this section. This situation is similar to a two-way factorial experiment in a completely randomized design discussed in Section 5.4, except that the levels of the two factors are selected randomly from populations of all possible levels.

Consider an experiment in which two factors that influence the breaking strength of plastic sheeting is under study. Four production machines (Factor A) and five operators (Factor B) are selected for the study. These factor levels are to be considered as random samples from populations of machines and operators used in a typical factory that produces plastic sheeting and every machine will be used by all operators. A machine-operator combination performs a production run that will produce a measurement of breaking strength. As in the fixed effects case, a number of runs are performed by each machine-operator combination in random order, so that replications are available for estimating the random error variance. To have equal sample sizes, the number of replications carried out per each factor combination is kept the same. Assuming that the number of replications is 2, the 40 experimental runs required are performed in a completely randomized design. Again, the main interest in this experiment will be estimation and hypothesis tests about the variance components (i.e., the variances of the random effects).

Model

The two-way crossed random effects model is given by

$$y_{ijk} = \mu + A_i + B_j + AB_{ij} + \epsilon_{ijk}, \quad i = 1, \ldots, a; \quad j = 1, \ldots, b; k = 1, \ldots, n$$

where the effects of Factor A, A_i, are assumed to be iid $N(0, \sigma_A^2)$ random variables, effects of Factor B, B_j, are assumed to be iid $N(0, \sigma_B^2)$ random variables, the effects of the interaction between the two factors, denoted by AB_{ij}, are assumed to be iid $N(0, \sigma_{AB}^2)$, and the random errors ϵ_{ijk} are assumed to be iid $N(0, \sigma^2)$ random variables. In addition, A_i, B_j, AB_{ij}, and ϵ_{ijk} are pairwise independent. If this is formulated as a "means model," where $\mu_{ij} = \mu + A_i + B_j + AB_{ij}$, then μ_{ij}, $i = 1, \ldots, a; j = 1, \ldots, b$. are iid $N(\mu, \sigma_A^2 + \sigma_B^2 + \sigma_{AB}^2 + \sigma^2)$ random variables, where μ represents the mean of the population of the observations (i.e., $E(y_{ijk}) = \mu$).

Thus, the objective of the experiment is to identify the components of variance that contribute significantly to the total variance of the observations. A consequence of the above model is that observations realized from the same level of Factor A (or Factor B or both) are correlated. For example, it can be shown that the covariance between y_{111} and y_{122} is σ_A^2 and that between y_{111} and y_{112} is $\sigma_A^2 + \sigma_B^2 + \sigma_{AB}^2$. Thus, the model defines the "covariance structure" of the observed data vector.

Estimation and Hypothesis Testing

An analysis of variance that correspond to the above model is constructed using the same computational formulas used for the computation of the anova that correspond to the two-way classification model with fixed effects discussed

6.3 Two-Way Crossed Random Effects Model

in Section 5.4. As discussed in Section 6.1, an additional column displaying the expected mean squares is included in the anova table for the two-way random effects model:

SV	df	SS	MS	F	E(MS)
A	$a-1$	SS_A	MS_A	MS_A/MS_{AB}	$\sigma^2 + n\sigma_{AB}^2 + bn\sigma_A^2$
B	$b-1$	SS_B	MS_B	MS_B/MS_{AB}	$\sigma^2 + n\sigma_{AB}^2 + an\sigma_B^2$
AB	$(a-1)(b-1)$	SS_{AB}	MS_{AB}	MS_{AB}/MSE	$\sigma^2 + n\sigma_{AB}^2$
Error	$ab(n-1)$	SSE	MSE($=s^2$)		σ^2
Total	$abn-1$				

Again, the computation of the expected mean squares does not require the distributional assumption of normality of the random effects, but normality is required for performing hypothesis tests and constructing confidence intervals. F-Statistics are constructed for sources of variation A, B, and AB as shown in the analysis of variance table and are used to test the following hypotheses:

(i) $H_0 : \sigma_A^2 = 0$ versus $H_a : \sigma_A^2 > 0$

(ii) $H_0 : \sigma_B^2 = 0$ versus $H_a : \sigma_B^2 > 0$

(iii) $H_0 : \sigma_{AB}^2 = 0$ versus $H_a : \sigma_{AB}^2 > 0$

respectively. Note, carefully, that these ratios are not the same as the F-ratios shown in the two-way fixed effects anova table (see Section 5.4). As in Section 6.2, suitable F-ratios are determined so that both the numerator and the denominator mean squares will have the same expectations if the null hypothesis holds, but the numerator will have a larger expectation under the alternative. For example, by examining the $E(\text{MS})$ column it can be observed that both MS_A and MS_{AB} will have expectation equal to $\sigma^2 + n\sigma_{AB}^2$ if $\sigma_A^2 = 0$; however, the numerator will have expectation equal to $\sigma^2 + n\sigma_{AB}^2 + bn\sigma_A^2$ if $\sigma_A^2 > 0$. Thus, the F-statistic for effect A satisfies this requirement for testing the hypotheses stated in (i). The F-statistics for testing hypotheses (ii) and (iii) are also constructed in a similar fashion.

The estimate of the error variance σ^2 is the MSE from the anova table (i.e., $\hat{\sigma}^2 = s^2$). To estimate the other variance components, the expected mean squares are set equal to the corresponding observed values, and the resulting set of equations is solved. However, if any of the above null hypotheses fail to be rejected, these parameters may be set equal to zero in the above expressions for $E(\text{MS})$ before they are used to estimate the rest of the variance components.

If the hypothesis $H_0 : \sigma_A^2 = 0$ is rejected in favor of $H_a : \sigma_A^2 > 0$, then σ_A^2 may be estimated by solving

$$\sigma^2 + bn\,\sigma_A^2 = \mathrm{MS}_A.$$

Substituting the estimate s^2 for σ^2 gives the estimate

$$\hat{\sigma}_A^2 = \frac{\mathrm{MS}_A - s^2}{bn}$$

where the right-hand side is computed using values obtained from the anova table. As earlier, these are called the *method of moments* estimates.

A $(1-\alpha)100\%$ confidence interval for σ_A^2 is provided by

$$\frac{\nu\hat{\sigma}_A^2}{\chi^2_{1-\alpha/2,\nu}} < \sigma_A^2 < \frac{\nu\hat{\sigma}_A^2}{\chi^2_{\alpha/2,\nu}} \tag{6.4}$$

where $\chi^2_{1-\alpha/2,\nu}$ and $\chi^2_{\alpha/2,\nu}$ are the $1-\alpha/2$ and $\alpha/2$ percentile points of the chi-squared distribution with ν degrees of freedom, respectively. The degrees of freedom for $\hat{\sigma}_A^2 = \frac{1}{bn}\mathrm{MS}_A - \frac{1}{bn}\mathrm{MSE}$ is obtained using the Satterthwaite approximation and is given by

$$\nu = \frac{(bn\hat{\sigma}_A^2)^2}{(\mathrm{MS}_A)^2/(a-1) + (s^2)^2/ab(n-1)}$$

Formulas for constructing confidence intervals for the other variance components can be similarly obtained.

6.3.1 Using PROC GLM and PROC MIXED to analyze two-way crossed random effects models

The data shown in Table 6.3 appear in Kutner et al. (2005). An automobile manufacturer studied the effects of differences between drivers (factor A) and differences between cars (factor B) on gasoline consumption. Four drivers were selected at random and five cars of the same model with manual transmission were also randomly selected from the assembly line. Each driver drove each car twice over a 40-mile test course and the miles per gallon was calculated. The actual trials were run in completely random order.

The interest here is in explaining the variation in gasoline consumption in terms of the variance components and determining whether their contributions to the total variation in the response are significant. The model is

$$y_{ijk} = \mu + A_i + B_j + AB_{ij} + \epsilon_{ijk}, \quad i=1,\ldots,4; \quad j=1,\ldots,5; \quad k=1,2$$

where A_i, B_j, and AB_{ij} are random effects of driver, car, and their interaction, distributed as independent normal random variables with mean zero and variances σ_A^2, σ_B^2, and σ_{AB}^2, respectively, and the random errors ϵ_{ijk} are distributed independently as $N(0,\sigma^2)$ random variables.

6.3 Two-Way Crossed Random Effects Model

Factor A	Factor B (car)				
(driver)	1	2	3	4	5
1	25.3	28.9	24.8	28.4	27.1
	25.2	30.0	25.1	27.9	26.6
2	33.6	36.7	31.7	35.6	33.7
	32.9	36.5	31.9	35.0	33.9
3	27.7	30.7	26.9	29.7	29.2
	28.5	30.4	26.3	30.2	28.9
4	29.2	32.4	27.7	31.8	30.3
	29.3	32.4	28.9	30.7	29.9

Table 6.3. Automobile mileage data

SAS Example F5

The SAS Example F5 program (see Fig. 6.10) illustrates how `proc glm` is used to fit the above model to the gasoline mileage data. The `proc glm` step carries out a standard analysis based on the method of moments for estimation of variance components and F-tests that are valid under the condition that the random effects have independent normal distributions as described earlier. Page 2 of the SAS output (see Fig. 6.11) displays the Types I and III analysis of variance table (which should be identical for balanced data) and are a part

```
data auto;
input  driver 1. @;
    do car=1 to 5;
        input mpg @;
        output;
    end;
datalines;
1  25.3  28.9  24.8  28.4  27.1
1  25.2  30.0  25.1  27.9  26.6
2  33.6  36.7  31.7  35.6  33.7
2  32.9  36.5  31.9  35.0  33.9
3  27.7  30.7  26.9  29.7  29.2
3  28.5  30.4  26.3  30.2  28.9
4  29.2  32.4  27.7  31.8  30.3
4  29.3  32.4  28.9  30.7  29.9
;
run;

proc glm data=auto;
class driver car;
model mpg =driver car driver*car;
random driver car driver*car/test;
title 'Study of Variation in Gasoline Consumption';
run;
```

Fig. 6.10. SAS Example F5: Program

of the standard `proc glm` output, independent of whether the factors were fixed or random.

```
                     Study of Variation in Gasoline Consumption                1

                              The GLM Procedure

                            Class Level Information

                       Class          Levels    Values

                       driver           4       1 2 3 4

                       car              5       1 2 3 4 5

                   Number of Observations Read          40
                   Number of Observations Used          40

                     Study of Variation in Gasoline Consumption                2

                              The GLM Procedure

Dependent Variable: mpg

                                    Sum of
Source                   DF        Squares      Mean Square    F Value    Pr > F

Model                    19      377.4447500    19.8655132     113.03     <.0001

Error                    20        3.5150000     0.1757500

Corrected Total          39      380.9597500


              R-Square     Coeff Var      Root MSE       mpg Mean

              0.990773     1.395209       0.419225       30.04750


Source                   DF       Type I SS     Mean Square    F Value    Pr > F

driver                    3      280.2847500    93.4282500     531.60     <.0001
car                       4       94.7135000    23.6783750     134.73     <.0001
driver*car               12        2.4465000     0.2038750       1.16     0.3715


Source                   DF      Type III SS    Mean Square    F Value    Pr > F

driver                    3      280.2847500    93.4282500     531.60     <.0001
car                       4       94.7135000    23.6783750     134.73     <.0001
driver*car               12        2.4465000     0.2038750       1.16     0.3715
```

Fig. 6.11. SAS Example F5: Output from PROC GLM (pages 1 and 2)

The following anova table is constructed using the information in pages 1-4 of the SAS output. The total and error sums of squares and their respective degrees of freedom are available in Page 2 and Pages 3 and 4 (see Fig. 6.12) show the table of Type III Expected Mean Squares **1** and F-tests

6.3 Two-Way Crossed Random Effects Model

for the variance components driver (σ_A^2), car (σ_B^2), **2** and the driver × car interaction (σ_{AB}^2), **3** respectively.

SV	df	SS	MS	F	p-value	E(MS)
Driver	3	280.285	93.428	458.26	<.0001	$\sigma^2 + 2\sigma_{AB}^2 + 10\sigma_A^2$
Car	4	94.714	23.678	116.14	<.0001	$\sigma^2 + 2\sigma_{AB}^2 + 8\sigma_B^2$
Driver× Car	12	2.446	0.204	1.16	.3715	$\sigma^2 + 2\sigma_{AB}^2$
Error	20	3.515	3.7917			σ^2
Total	39	380.960				

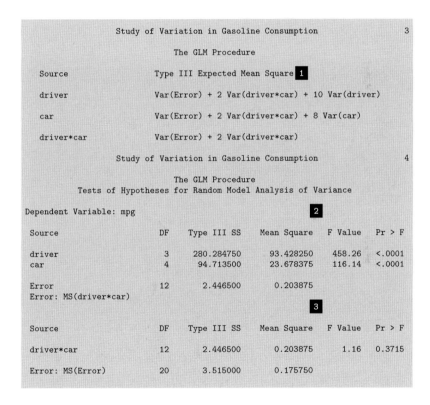

Fig. 6.12. SAS Example F5: Output from PROC GLM (pages 3 and 4)

A p-value of .3715 for the interaction effects leads to the conclusion that σ_{AB}^2 is not different from zero, whereas the p-values of less than .0001 for the driver and car effects respectively show that the corresponding two variance components are significantly larger than zero. Their estimates can be obtained by the method of moments by setting the expressions for expected mean squares found on page 3 (see Fig. 6.12) for each effect equal to its computed mean square (Type I or Type III) from page 2. This gives the set of equations

$$\sigma^2 + 10\sigma_A^2 + 2\sigma_{AB}^2 = 93.428250$$

$$\sigma^2 + 8\sigma_B^2 + 2\sigma_{AB}^2 = 23.678375$$

$$\sigma^2 + 2\sigma_{AB}^2 = 0.203875$$

$$\sigma^2 = 0.175750$$

Solving these gives the estimates

$$\hat{\sigma}^2 = 0.175750$$

$$\hat{\sigma}_{AB}^2 = (0.203875 - 0.175750)/2 = 0.0140625$$

$$\hat{\sigma}_B^2 = (23.678375 - 2 \times 0.0140625 - 0.175750)/8 = 2.9343125$$

$$\hat{\sigma}_A^2 = (93.428250 - 2 \times 0.0140625 - 0.175750)/10 = 9.3224375$$

Confidence intervals for each nonzero variance component based on the Satterthwaite approximation can be computed by hand or by modifying the SAS code given in Section 6.2. For example, a 95% confidence interval for σ_A^2 is given by executing the SAS code

```
data cint;
    alpha=.05; n=2; a=4; b=5;
    msa= 93.428250; msab=0.203875; sa2=9.3224375;
    nu=(n*b*sa2)**2/(msa**2/(a-1)+ msab**2/(a-1)*(b-1));
    L= (nu*sa2)/cinv(1-alpha/2,nu);
    U= (nu*sa2)/cinv(alpha/2,nu);
    put nu=    L=    U=;
run;
```

which results in the interval (2.9864, 130.7911). A confidence interval for σ_B^2 can be similarly calculated. **Proc mixed** produces these intervals by default when an iterative method such as **ml** or **reml** is used along with the proc statement option **cl**.

SAS Example F6

In the SAS Example F6 program (see Fig. 6.13), **proc mixed** is used to fit the above model to the gasoline mileage data. In the proc statement, **method=type3** is specified, so instead of using an iterative algorithm for calculating the likelihood estimates, the method of moments estimators using the Type III expected mean squares are computed.

6.3 Two-Way Crossed Random Effects Model

```
insert data step to create the SAS dataset 'auto' here

proc mixed  data=auto noclprint noinfo method=type3 cl;
class driver car;
model mpg = /;
random driver car driver*car;
run;
```

Fig. 6.13. SAS Example F6: Method of moments estimation using PROC MIXED

The results, both estimates, F-statistics, and their p-values shown in Fig. 6.14 are identical to those obtained using `proc glm`. The variance com-

```
                            The SAS System                              1

                           The Mixed Procedure

                         Type 3 Analysis of Variance

                                      Sum of
                   Source        DF   Squares       Mean Square

                   driver         3   280.284750    93.428250
                   car            4    94.713500    23.678375
                   driver*car    12     2.446500     0.203875
                   Residual      20     3.515000     0.175750

                         Type 3 Analysis of Variance
                                                                  Error
Source      Expected Mean Square                    Error Term    DF  F Value

driver      Var(Residual) + 2 Var(driver*car) + 10 Var(driver)  MS(driver*car) 12  458.26
car         Var(Residual) + 2 Var(driver*car) + 8 Var(car)      MS(driver*car) 12  116.14
driver*car  Var(Residual) + 2 Var(driver*car)                   MS(Residual)   20    1.16
Residual    Var(Residual)                                          .            .       .

                              Type 3 Analysis
                                of Variance

                              Source     Pr > F

                              driver     <.0001
                              car        <.0001
                              driver*car  0.3715
                              Residual      .

                        Covariance Parameter Estimates

               Cov Parm      Estimate    Alpha    Lower      Upper

               driver         9.3224      0.05    -5.6289    24.2738
               car            2.9343      0.05    -1.1677     7.0364
               driver*car     0.01406     0.05    -0.08402    0.1121
               Residual       0.1757      0.05     0.1029     0.3665
```

Fig. 6.14. SAS Example F6: Output

ponent estimates are also the same as those calculated by hand using results from `proc glm`. However, the confidence intervals calculated by `proc mixed` are those based on large sample standard errors and the standard normal percentiles except those for σ^2, which are based on chi-square percentiles. Confidence intervals on the other variance components based on Satterthwaite approximation can be calculated using formulas similar to (6.1) in Section 6.2, as illustrated for SAS Example F5.

SAS Example F7

Finally, the same data are reanalyzed in the SAS Example F7 program (see Fig. 6.15) using REML, the default method in `proc mixed`. In the proc statement, `method` is unspecified, so an iterative algorithm is used to compute the restricted maximum likelihood estimates because the default method is REML.

```
insert data step to create the SAS dataset 'auto' here

proc mixed  data=auto noinfo noitprint covtest cl ;
class driver car;
model mpg = /solution;
random driver car driver*car;
run;
```

Fig. 6.15. SAS Example F7: REML estimation using PROC MIXED

The resulting SAS output (see Fig. 6.16) contains the estimates of the variance components, estimates of their asymptotic standard errors, z-tests, and the associated p-values. These are output as a result of the `covtest` option. Confidence intervals computed using the Satterthwaite approximation are produced as a result of the `cl` option, because in this case the variance components are constrained to be nonnegative. The `solution` option provides the estimates of the fixed effects (see **Solution for Fixed Effects** **1** in Fig. 6.16). Recall that the only fixed effect in the model is $E(y_{ijk}) = \mu$, the population mean of the observations. This is identified as the `Intercept` effect in the output.

The estimates of the variance components are identical to the method of moments estimates, as expected for balanced data. However, the standard errors have been calculated using large-sample results for maximum likelihood estimators, which assume that the numbers of levels for the two factors (i.e., sample sizes) are infinitely large. Thus, the results of the z-tests do not coincide with those of the F-tests based on assuming normal distributions for the random effects.

The confidence interval for σ_A^2, on the other hand, agrees with that computed earlier using the the Satterthwaite approximation. A $(1-\alpha)100\%$ confidence interval for σ_A^2 is given by formula (6.1) in Section 6.2. The computation

6.3 Two-Way Crossed Random Effects Model

```
               Covariance Parameter Estimates

                        Standard     Z
 Cov Parm    Estimate   Error     Value    Pr Z    Alpha    Lower     Upper

 driver      9.3224     7.6284    1.22    0.1108   0.05    2.9864    130.79
 car         2.9343     2.0929    1.40    0.0805   0.05    1.0464     24.9038
 driver*car  0.01406    0.05004   0.28    0.3893   0.05    0.001345   3.592E17
 Residual    0.1757     0.05558   3.16    0.0008   0.05    0.1029     0.3665

                        Fit Statistics

                -2 Res Log Likelihood        86.8
                AIC  (smaller is better)     94.8
                AICC (smaller is better)     96.0
                BIC  (smaller is better)     92.3

                    Solution for Fixed Effects

                          Standard
  Effect        Estimate  Error      DF    t Value    Pr > |t|

  Intercept     30.0475   1.7096      3     17.58      0.0004
```

Fig. 6.16. SAS Example F7: Output

is simplified because degrees of freedom is $\nu = 2z^2$, where z is the Wald statistic, given by $z = \sigma_A^2/\text{s.e.}(\sigma_A^2)$. Substituting the values needed to compute z for each variance component, the confidence intervals given in Fig. 6.16 can be verified.

For example, using the estimate and its standard error (9.3224 and 7.6284, respectively) for the driver variance component in the following SAS data step

```
data;
    alpha=.05; s2= 9.3224; ses2= 7.6284;
    z=s2/ses2;
    nu=2*z**2;
    L= (nu*s2)/cinv(1-alpha/2,nu);
    U= (nu*s2)/cinv(alpha/2,nu);
    put z=  nu=   L=    U=;
run;
```

results in the 95% confidence interval (2.9864, 130.7887).

6.3.2 Randomized complete block design: Blocking when treatment factors are random

In the discussion of the RCBD presented in Section 5.7 of Chapter 5, both the treatment and block effects were considered to be fixed effects. It may be more reasonable to consider the block effects to be random effects. In Section 6.5,

RCBDs with blocks as random effects will be discussed as a special case of the mixed effects model.

In this subsection, a model with both block and treatment effects random is presented. One of the consequences of the way blocks are formed is that, conceptually, it is not feasible for differences in treatment effects to be different from block to block because blocks are formed by grouping experimental units. Therefore, although blocks were considered to be fixed effects, an additive model was used in Section 5.7 to represent the observations; that is, an interaction term was not included in the model for observations from an experiment carried out as an RCBD. The same argument holds for the case when the treatments are random effects thus an interaction term is omitted from the model.

Montgomery (1991) discussed an experiment that uses **subjects** as the *blocking factor* and **analysts** that perform DNA analyses on three samples taken from each subject as *the treatment factor*. As the analysts are a random sample from a population and the samples from each subject are randomly assigned to the analysts, the experimental design is an RCBD and the observed data y_{ij} may be modeled as the additive model

$$y_{ij} = \mu + A_i + B_j + \epsilon_{ij}, \quad i = 1, \ldots, a; \quad j = 1, \ldots, b$$

where A_i, the effects of analysts (Factor A), are iid $N(0, \sigma_A^2)$ random variables, B_j, effects of subjects (Factor B), are iid $N(0, \sigma_B^2)$ random variables, and the random errors ϵ_{ij} are iid $N(0, \sigma^2)$ random variables, and these three set of random variables are pairwise independent. The analysis of the data is similar that of the two-way crossed random effects model except that there is no interaction term in the model. The anova table for this model is

SV	df	SS	MS	F	E(MS)
A	$a-1$	SS_A	MS_A	MS_A/MSE	$\sigma^2 + b\sigma_A^2$
B	$b-1$	SS_B	MS_B	MS_B/MSE	$\sigma^2 + a\sigma_B^2$
Error	$(a-1)(b-1)$	SSE	$MSE(=s^2)$		σ^2
Total	$ab-1$				

A SAS example showing the analysis of data from this model is omitted, as it is straightforward and follows in the same lines as the analysis of data for SAS Example F5, except that determining whether σ_A^2 is nonzero is of interest here.

6.4 Two-Way Nested Random Effects Model

In this section, a two-way random model for responses from an experiment with two random factors when one of the factors is *nested* in the other factor is considered. Consider two factors A and B. Factor B is said to be nested in Factor A if levels of B are different at each level of A. For example, in

6.4 Two-Way Nested Random Effects Model

an extended version of the "Traffic Tickets" example discussed in Chapter 5, suppose that the number of tickets issued by officers in randomly selected precincts in several cities is under study. In this case, suppose also that both cities and precincts are randomly sampled. The factor precinct is nested within the factor city because the levels of precinct are different from city to city. This factor is called the precinct within city, with its levels defined using combinations of the levels of both city and precinct factors. In general, in a two-way nested classification, the sampling of the levels takes place in a *hierarchical* manner: First, the levels of one factor (Factor A) are randomly sampled, and then levels of the *nested* factor (Factor B) are randomly sampled within each level of A. Although in this section both factors are considered random, it is also possible that at least one of them is a fixed factor.

Model

An appropriate model for the situation described above is

$$y_{ijk} = \mu + A_i + B_{ij} + \epsilon_{ijk}, \quad i = 1, \ldots, a;\ j = 1, \ldots, b;\ k = 1, \ldots, n$$

where it is assumed that A_i, the Factor A effect, is iid $N(0, \sigma_A^2)$, B_{ij}, the Factor B within A effect, is iid $N(0, \sigma_B^2)$, and the random error ϵ_{ijk} is iid $N(0, \sigma^2)$. Further, it is assumed that A_i, B_{ij}, and ϵ_{ijk} are pairwise independently distributed. The parameters σ_A^2, σ_B^2, and σ^2 will constitute the "variance components" in this problem. It is important to note that it is assumed that all B_{ij}, irrespective of the level i, have the same variance σ_B^2. For example, in the motivating problem introduced earlier, this is equivalent to assuming that the variances in the mean number of tickets issued among the precincts is the same for all cities. It is possible to examine whether this is a plausible assumption using the observed data.

Estimation and Hypothesis Testing

An analysis of variance that corresponds to the model is constructed using the same computational formulas used for the computation of sums of squares of an anova table if the factors A and B within A are considered to be fixed. These sums of squares would be the same for effect A as in the anova table for the two-way crossed random effects model (given in Section 6.3). For the effect B within A (denoted in the following anova table as B(A)), the sum of squares is obtained by combining (or pooling) the sum of squares for effects B and AB from the anova table for the two-way crossed random effects model. That is $SS_{B(A)} = SS_B + SS_{AB}$ with $df(SS_{B(A)}) = df(SS_B) + df(SS_{AB}) = (b-1) + (a-1)(b-1) = a(b-1)$. As with the random models considered so far, an additional column displaying the expected mean squares is included in the anova table:

SV	df	SS	MS	F	E(MS)
A	$a-1$	SS_A	MS_A	$MS_A/MS_{B(A)}$	$\sigma^2 + n\sigma_B^2 + bn\sigma_A^2$
B(A)	$a(b-1)$	$SS_{B(A)}$	$MS_{B(A)}$	$MS_{B(A)}/MSE$	$\sigma^2 + n\sigma_B^2$
Error	$ab(n-1)$	SSE	$MSE(=s^2)$		σ^2
Total	$abn-1$				

F-statistics are constructed for sources of variation A and B(A) as shown in the analysis of variance table and are used to test the hypotheses:

(i) $\quad H_0: \sigma_A^2 = 0 \quad \text{versus} \quad H_a: \sigma_A^2 > 0$

(ii) $\quad H_0: \sigma_B^2 = 0 \quad \text{versus} \quad H_a: \sigma_B^2 > 0$

respectively. Again, note, carefully, that these ratios are not the same as the F-ratios shown in the two-way crossed random effects anova table (see Section 6.3), although the appropriate F-ratios are determined in the same principle described there. For example, by examining the E(MS) column it can be observed that both MS_A and $MS_{B(A)}$ will have expectation equal to $\sigma^2 + n\sigma_B^2$ if $\sigma_A^2 = 0$, irrespective of the value of σ_B^2. However, the numerator MS_A will have expectation equal to $\sigma^2 + n\sigma_B^2 + bn\sigma_A^2$ if $\sigma_A^2 > 0$. Thus, the F-statistic for effect A meets the requirement for testing the hypotheses stated in (i. The F-statistics for testing hypotheses (ii) is also constructed in a similar manner.

The *method of moments* estimates of variance components are obtained by setting the computed mean squares equal to their corresponding expected values and solving the resulting equations, as usual:

$$MS_A = \sigma^2 + n\sigma_B^2 + bn\sigma_A^2$$
$$MS_{B(A)} = \sigma^2 + n\sigma_B^2$$
$$MSE = \sigma^2$$

The method of moments estimates of σ_A^2, σ_B^2, and σ^2 are thus given by

$$\hat{\sigma}_A^2 = (MS_A - MS_{B(A)})/bn$$
$$\hat{\sigma}_B^2 = (MS_{B(A)} - MSE)/n$$

and $\hat{\sigma}^2 = MSE$, respectively. When the sample sizes are equal (i.e., for balanced data), these estimators are unbiased and have minimum variance. As earlier, a $(1-\alpha)100\%$ confidence interval for σ_A^2 is provided by

$$\frac{\nu\hat{\sigma}_A^2}{\chi^2_{1-\alpha/2,\nu}} < \sigma_A^2 < \frac{\nu\hat{\sigma}_A^2}{\chi^2_{\alpha/2,\nu}}$$

where $\chi^2_{1-\alpha/2,\nu}$ and $\chi^2_{\alpha/2,\nu}$ are the $1-\alpha/2$ and $\alpha/2$ percentile points of the chi-square distribution with ν degrees of freedom, respectively. The degrees of freedom for $\hat{\sigma}_A^2 = \frac{1}{bn}\text{MS}_A - \frac{1}{bn}\text{MS}_{B(A)}$ is obtained using the Satterthwaite approximation and is given by

$$\nu = \frac{(bn\hat{\sigma}_A^2)^2}{(\text{MS}_A)^2/(a-1) + (\text{MS}_{B(A)})^2/a(b-1)}$$

Formulas for constructing confidence intervals for the other variance components can be similarly obtained.

6.4.1 Using PROC GLM to analyze two-way nested random effects models

In order to study the variation of the calcium content in turnip greens, four plants were selected at random. From each plant, three leaves were randomly selected, and from each leaf, two samples of 100 mg each were taken and the calcium content determined. This experiment is described in Snedecor and Cochran (1989). The data appear in Table 6.4.

Plant	Leaf	Determinations of Ca	
1	1	3.28	3.09
	2	3.52	3.48
	3	2.88	2.80
2	1	3.46	2.44
	2	1.87	1.92
	3	2.19	2.19
3	1	2.77	2.66
	2	3.74	3.44
	3	2.55	2.55
4	1	3.78	3.87
	2	4.07	4.12
	3	3.31	3.31

Table 6.4. Calcium content in turnip greens (Snedecor and Cochran, 1989)

The experimenter is interested is verifying whether there is a significant variation in calcium content from plant to plant compared to the variation within a plant. If so, it is also of interest to obtain an estimate of this variation. The model is

$$y_{ijk} = \mu + A_i + B_{ij} + \epsilon_{ijk} \quad i = 1, 4; \; j = 1, 3; \; k = 1, 2$$

where A_i, the effect of Plant i, is assumed to be iid $N(0, \sigma_A^2)$, B_{ij}, the Leaf i within Plant j effect, is assumed to be iid $N(0, \sigma_B^2)$, and ϵ_{ijk}, the samples within Leaf within Plant effect, is iid $N(0, \sigma^2)$. It is also assumed that A_i, B_{ij}, and ϵ_{ijk} are pairwise independent.

SAS Example F8

```
data turnip;
input plant leaf x1-x2;
drop x1-x2;
calcium=x1; output;
calcium=x2; output;
cards;
1 1 3.28 3.09
1 2 3.52 3.48
1 3 2.88 2.80
2 1 2.46 2.44
2 2 1.87 1.92
2 3 2.19 2.19
3 1 2.77 2.66
3 2 3.74 3.44
3 3 2.55 2.55
4 1 3.78 3.87
4 2 4.07 4.12
4 3 3.31 3.31
;
run;

proc glm data=turnip;
   class plant leaf;
   model calcium= plant leaf(plant);
   random plant leaf(plant)/test;
   title 'Analysis of a Two-Way Nested Random Model: Turnip Data';
run;
```

Fig. 6.17. SAS Example F8: Program

The SAS Example F8 program (see Fig. 6.17) illustrates how `proc glm` is used to fit the above model to the calcium in turnip data. The data are entered and input to SAS in a straightforward way. However, note how two separate observations are created in the SAS data set from the two sample values from a leaf entered in the same line of data, using the `output` statements. It is important to recognize that the levels of leaf are specified in the data as if the two factors were crossed; that is, the levels of leaf are labeled 1,2, and 3 for every level of plant.

Both `plant` and `leaf` are declared as classification variables in the `class` statement, but in the `model` statement, a `leaf(plant)` term is used to define the Leaf within Plant effect; that is, the `leaf(plant)` notation represents the "B_{ij}" term in the model. When this model specification is used to code the necessary design matrices that correspond to the random effects, the levels of the factor Leaf within Plant are identified as the levels of leaves within each plant (i.e., 11, 12, 13, 21, 22, ..., etc).

The `random` statement declares that `plant` and `leaf(plant)` are random effects whereas the `test` option requests `proc glm` to construct suitable F-statistics for testing hypotheses about the variance components specified in the `random` statement. The expected values of the mean squares in the anova table determine the ratios of sums of squares to be used to test the two hypotheses of interest: $H_0 : \sigma_A^2 = 0$ versus $H_a : \sigma_A^2 > 0$ and $H_0 : \sigma_B^2 = 0$ versus $H_a : \sigma_B^2 > 0$, as discussed previously.

The Types I and III sums of squares are used to compute the standard output from `proc glm` (see page 2 of Fig. 6.18) and, by default, the error mean square is used as the denominator of the F-ratios constructed to test the above hypotheses **1**. Although this will produce the correct F-statistic to test the `leaf(plant)` effect, the F-statistic calculated for testing the `plant` effect is incorrect. The inclusion of the `test` option in the `random` statement will result in the use of Leaf within Plant mean square as the denominator **2** to test

```
             Analysis of a Two-Way Nested Random Model: Turnip Data              1

                                  The GLM Procedure

                             Class Level Information

                         Class         Levels    Values

                         plant              4    1 2 3 4

                         leaf               3    1 2 3


                    Number of Observations Read          24
                    Number of Observations Used          24

             Analysis of a Two-Way Nested Random Model: Turnip Data              2

                                  The GLM Procedure

Dependent Variable: calcium
                                      Sum of
Source                    DF         Squares     Mean Square    F Value    Pr > F

Model                     11     10.19054583      0.92641326     139.22    <.0001

Error                     12      0.07985000      0.00665417

Corrected Total           23     10.27039583

               R-Square     Coeff Var      Root MSE    calcium Mean

               0.992225      2.708195      0.081573        3.012083

Source                    DF       Type I SS     Mean Square    F Value    Pr > F

plant                      3      7.56034583      2.52011528     378.73    <.0001
leaf(plant)                8      2.63020000      0.32877500      49.41    <.0001

Source                    DF    Type III SS  1   Mean Square    F Value    Pr > F

plant                      3      7.56034583      2.52011528     378.73    <.0001
leaf(plant)                8      2.63020000      0.32877500      49.41    <.0001
```

Fig. 6.18. SAS Example F8: Output from PROC GLM (pages 1 and 2)

```
          Analysis of a Two-Way Nested Random Model: Turnip Data         3
                          The GLM Procedure

Source                    Type III Expected Mean Square

plant                     Var(Error) + 2 Var(leaf(plant)) + 6 Var(plant)

leaf(plant)               Var(Error) + 2 Var(leaf(plant))

          Analysis of a Two-Way Nested Random Model: Turnip Data         4
                          The GLM Procedure
                Tests of Hypotheses for Random Model Analysis of Variance

Dependent Variable: calcium

Source                     DF     Type III SS    Mean Square    F Value    Pr > F

plant                       3       7.560346       2.520115       7.67     0.0097

Error                       8       2.630200       0.328775
Error: MS(leaf(plant))  2


Source                     DF     Type III SS    Mean Square    F Value    Pr > F

leaf(plant)                 8       2.630200       0.328775      49.41     <.0001

Error: MS(Error)           12       0.079850       0.006654
```

Fig. 6.19. SAS Example F8: Output from PROC GLM (pages 3 and 4)

the `plant` effect and this produces the correct F-statistic, as observed from the output shown in Fig. 6.19. The F-statistic for testing the `leaf(plant)` effect is identical to the one on page 2.

The following analysis of variance for the turnip green data is constructed from the Type III sums of squares output from `proc glm`.

SV	df	SS	MS	F	p-value	E(MS)
Plant	3	7.56035	2.52012	7.67	.0097	$\sigma^2 + 2\sigma_B^2 + 6\sigma_A^2$
Leaf(Plant)	8	2.63020	0.32878	49.41	< .0001	$\sigma^2 + 2\sigma_B^2$
Error	12	0.07985	0.00665			
Total	23	10.27040				

The expected mean squares are those derived by `proc glm` and displayed on page 3 in Fig. 6.19 in the table titled "Type III Expected Mean Square." To test $H_0 : \sigma_A^2 = 0$ versus $H_a : \sigma_A^2 > 0$ the statistic $F_1 = 2.52012/0.32878 = 7.67$ is used. Since the p-value is .0097, the null hypothesis is rejected at $\alpha = .05$. The F-statistic $F_2 = 0.32878/0.00665 = 49.41$ is associated with a p-value of $< .0001$. Thus, the null hypothesis of $H_0 : \sigma_B^2 = 0$ versus $H_a : \sigma_B^2 > 0$ is also rejected at $\alpha = .05$. The variance components are estimated by setting the computed mean squares equal to their expected values and solving the resulting set of equations:

$$\sigma^2 + 2\,\sigma_B^2 + 6\,\sigma_A^2 = 2.52012$$
$$\sigma^2 + 2\,\sigma_B^2 = 0.32878$$
$$\sigma^2 = 0.00665$$

The solutions are $\hat{\sigma}^2 = 0.00665$, $\hat{\sigma}_B^2 = (0.32875 - 0.00665)/2 = 0.16105$, and $\hat{\sigma}_A^2 = (2.52012 - 0.32878)/6 = 0.36522$. The conclusion is that the variation in calcium content among the leaves within a plant is about 24 times as large, and among the plants, it is about 55 times as large, as the variation among samples within leaves.

6.4.2 Using PROC MIXED to analyze two-way nested random effects models

In this subsection, `proc mixed` is used to fit the model discussed in Section 6.4.1 to the turnip greens data. The method of moments is used, mainly so that the results can be compared with the analysis obtained previously using `proc glm`. The differences between the `proc mixed` analysis obtained by this method and those obtained using MLE and REML methods are indicated at the end of this subsection.

SAS Example F9

In the SAS Example F9 program (see Fig. 6.20) the `method=type3` used in the proc statement requests that the method of moments estimators using the Type III expected mean squares are to be calculated. The resulting variance component estimates, F-statistics, and their p-values are shown in Fig. 6.21, and they are identical to those from `proc glm`. The variance component estimates are the same as those calculated by hand using results from `proc glm`. The confidence intervals calculated by `proc mixed` are again those based on large sample standard errors and the normal percentiles except those for σ^2, which are based on chi-square percentiles.

```
insert data step to create the SAS dataset 'turnip' here

proc mixed data=turnip noclprint noinfo method=type3 cl;
  class plant leaf;
  model calcium= /solution;
  random plant leaf(plant);
  title 'Analysis of a Two-Way Nested Random Model: Turnip Data';
run;
```

Fig. 6.20. SAS Example F9: Method of moments estimation using PROC MIXED

If the option `method=reml` (the default value) or `method=ml` is specified as the method of estimation, the same values as those obtained from the method

```
              Analysis of a Two way Nested Random Model: Turnip Data                    1

                                  The Mixed Procedure

                                Type 3 Analysis of Variance

                          Sum of
Source              DF   Squares    Mean Square   Expected Mean Square

plant                3   7.560346    2.520115     Var(Residual) + 2 Var(leaf(plant))
                                                  + 6 Var(plant)
leaf(plant)          8   2.630200    0.328775     Var(Residual) + 2 Var(leaf(plant))
Residual            12   0.079850    0.006654     Var(Residual)

                                Type 3 Analysis of Variance

                                                   Error
    Source       Error Term                         DF     F Value    Pr > F

    plant        MS(leaf(plant))                     8        7.67    0.0097
    leaf(plant)  MS(Residual)                       12       49.41    <.0001
    Residual       .                                  .          .         .

                           Covariance Parameter Estimates

         Cov Parm       Estimate     Alpha      Lower       Upper

         plant           0.3652      0.05      -0.3091     1.0395
         leaf(plant)     0.1611      0.05      -0.00006    0.3222
         Residual        0.006654    0.05       0.003422   0.01813

                                  Fit Statistics

                    -2 Res Log Likelihood          2.2
                    AIC (smaller is better)        8.2
                    AICC (smaller is better)       9.4
                    BIC (smaller is better)        6.3

                              Solution for Fixed Effects

                                  Standard
          Effect        Estimate    Error      DF     t Value    Pr > |t|

          Intercept      3.0121     0.3240      3       9.30      0.0026
```

Fig. 6.21. SAS Example F9: Output

of moments will be obtained for balanced data. However, the standard errors will be calculated using large-sample results for maximum likelihood estimators that assume the numbers of levels for the two factors (sample sizes) are large. The inferences made from the resulting z-tests will not be the same as those made from the F-tests based on assuming normal distributions for the random effects. The confidence intervals for the variance components will be those based on the chi-square distribution and the Satterthwaite approximation.

6.5 Two-Way Mixed Effects Model

The mixed model is a linear model that involves both fixed and random effects. In the following subsections, several applications of this model will be discussed. In Chapters 4 and 5, the least squares method was used to obtain the estimates of the parameters of the regression and anova models, respectively, using the matrix form of the respective models:

$$\mathbf{y} = X\boldsymbol{\beta} + \boldsymbol{\epsilon}.$$

In the case of full rank regression models, the solution to the normal equations

$$X'X\boldsymbol{\beta} = X'\mathbf{y},$$

gave the the least squares estimate $\hat{\boldsymbol{\beta}}$ of $\boldsymbol{\beta}$ as

$$\hat{\boldsymbol{\beta}} = (X'X)^{-1}X'\mathbf{y}$$

In Chapter 5, analysis of variance models were also represented in the same matrix model set-up where the X matrix, now called the *design matrix*, was constructed from the linear model describing the responses observed from an experiment and the parameter vector consisted of the model effects. A specific example in Section 5.1 illustrated how the X matrix is constructed for a typical experimental situation.

The matrix representation of a mixed model will be described in this section and the methods of estimation briefly summarized. As an example of a two-factor mixed model with interaction, consider the machine-operator example discussed in Section 6.3. To keep the dimensions of the matrices involved in the example within manageable limits, instead of four production machines (Factor A), consider that the breaking strengths of plastic sheeting from two specific brands of machines were of interest, that three operators were randomly selected, and that two trials were performed by each machine-operator combination in a completely randomized design. Thus, the applicable model may be expressed as the two-way crossed mixed effects model given by

$$y_{ijk} = \mu + \alpha_i + B_j + \alpha B_{ij} + \epsilon_{ijk}, \quad i = 1, \ldots, 2; \quad j = 1, \ldots, 3; \quad k = 1, \ldots, 2$$

where α_i are fixed effects due to the two levels of Factor A (machines), B_j, the random effects of Factor B (operators), are iid $N(0, \sigma_B^2)$ random variables, the interaction effects between the two factors denoted by αB_{ij} are iid $N(0, \sigma_{\alpha B}^2)$, and the random errors ϵ_{ijk} are iid $N(0, \sigma^2)$ random variables. The interaction effects are random because the levels depend on the levels of Factor B, which are randomly sampled. The matrix form of the model is

$$\mathbf{y} = X\boldsymbol{\beta} + Z_1 \mathbf{u}_1 + Z_2 \mathbf{u}_2 + \boldsymbol{\epsilon}$$

where

$$y = \begin{bmatrix} y_{111} \\ y_{112} \\ y_{121} \\ y_{122} \\ y_{131} \\ y_{132} \\ y_{211} \\ y_{212} \\ y_{221} \\ y_{222} \\ y_{231} \\ y_{232} \end{bmatrix}, \quad X = \begin{bmatrix} 1 & 1 & 0 \\ 1 & 1 & 0 \\ 1 & 1 & 0 \\ 1 & 1 & 0 \\ 1 & 1 & 0 \\ 1 & 1 & 0 \\ 1 & 0 & 1 \\ 1 & 0 & 1 \\ 1 & 0 & 1 \\ 1 & 0 & 1 \\ 1 & 0 & 1 \\ 1 & 0 & 1 \end{bmatrix}, \quad \beta = \begin{bmatrix} \mu \\ \alpha_1 \\ \alpha_2 \\ \alpha_3 \end{bmatrix}, \quad Z_1 = \begin{bmatrix} 1 & 0 & 0 \\ 1 & 0 & 0 \\ 0 & 1 & 0 \\ 0 & 1 & 0 \\ 0 & 0 & 1 \\ 0 & 0 & 1 \\ 1 & 0 & 0 \\ 1 & 0 & 0 \\ 0 & 1 & 0 \\ 0 & 1 & 0 \\ 0 & 0 & 1 \\ 0 & 0 & 1 \end{bmatrix}, \quad \mathbf{u}_1 = \begin{bmatrix} B_1 \\ B_2 \\ B_3 \end{bmatrix},$$

$$Z_2 = \begin{bmatrix} 1 & 0 & 0 & 0 & 0 & 0 \\ 1 & 0 & 0 & 0 & 0 & 0 \\ 0 & 1 & 0 & 0 & 0 & 0 \\ 0 & 1 & 0 & 0 & 0 & 0 \\ 0 & 0 & 1 & 0 & 0 & 0 \\ 0 & 0 & 1 & 0 & 0 & 0 \\ 0 & 0 & 0 & 1 & 0 & 0 \\ 0 & 0 & 0 & 1 & 0 & 0 \\ 0 & 0 & 0 & 0 & 1 & 0 \\ 0 & 0 & 0 & 0 & 1 & 0 \\ 0 & 0 & 0 & 0 & 0 & 1 \\ 0 & 0 & 0 & 0 & 0 & 1 \end{bmatrix}, \quad \mathbf{u}_2 = \begin{bmatrix} \alpha B_{11} \\ \alpha B_{12} \\ \alpha B_{13} \\ \alpha B_{21} \\ \alpha B_{22} \\ \alpha B_{23} \end{bmatrix}, \quad \epsilon = \begin{bmatrix} \epsilon_{111} \\ \epsilon_{112} \\ \epsilon_{121} \\ \epsilon_{122} \\ \epsilon_{131} \\ \epsilon_{132} \\ \epsilon_{211} \\ \epsilon_{212} \\ \epsilon_{221} \\ \epsilon_{222} \\ \epsilon_{231} \\ \epsilon_{232} \end{bmatrix}.$$

The observed data vector is arranged so that, as in Section 5.1, the subscripts of the observations are lexically ordered; that is, digits to the right change faster than the ones on the left. For the observation y_{ijk}, the model terms are obtained by scalar products of the appropriate vectors in the corresponding rows of the matrices X, Z_1, and Z_2 and the parameter vectors β, $\mathbf{u}_1$, and $\mathbf{u}_1$ respectively. By locating the 1's in the appropriate rows of X, Z_1, and Z_2 and selecting the terms in the same positions in the parameter vectors β, $\mathbf{u}_1$, and $\mathbf{u}_1$, the relevant model terms can be easily extracted. For example, for observation y_{132}, the relevant elements are found in row 6 of these matrices (these are highlighted), giving the model terms to be μ, α_1, B_3, αB_{13}, and ϵ_{132}. (For compactness, the above model can be reduced to the form

$$y = X\beta + Z\mathbf{u} + \epsilon$$

where Z_1, and Z_2 are combined to form Z and $\mathbf{u}_1$ and $\mathbf{u}_2$ are stacked together to form $\mathbf{u}$, but the expanded form is helpful in expressing the relevant variance-covariance matrices in easily expressible forms.)

In the classical variance component model, the random vectors $\mathbf{u}_1$, $\mathbf{u}_2$, and ϵ have the multivariate normal distributions $N(\mathbf{0}, \sigma_B^2 I)$, $N(\mathbf{0}, \sigma_{\alpha B}^2 I)$, and $N(\mathbf{0}, \sigma^2 I)$, respectively, where the matrices $\sigma_B^2 I$ and so forth. are diagonal matrices with the diagonal elements all equal to the respective variance components. The variance-covariance matrix of $\mathbf{y}$ is thus given by the 12×12 matrix V:

$$V = Z_1 Z_1' \sigma_B^2 + Z_2 Z_2' \sigma_{\alpha B}^2 + \sigma^2 I$$

By performing the necessary matrix multiplications $Z_1 Z_1'$ and $Z_2 Z_2'$, the form of V can be determined, from which important features of the covariance structure of the observations can be obtained. Thus, the variance of an observation y_{ijk} is $\sigma^2 + \sigma_B^2 + \sigma_{\alpha B}^2$ and, covariance between pairs of observations resulting from a replication using

- the same machine but different operators is $\text{Cov}(y_{111}, y_{121}) = 0$
- different machines but the same operator is $\text{Cov}(y_{111}, y_{211}) = \sigma_1^2$
- the same machine and the same operator is $\text{Cov}(y_{111}, y_{112}) = \sigma_B^2 + \sigma_{\alpha B}^2$
- different machines and different operators is $\text{Cov}(y_{111}, y_{221}) = 0$

A clear distinction exists between estimating a fixed parametric function, such as $\mu + \alpha_i$, and *predicting* a random variable such as B_j. To gain a little insight into the theoretical implications, it is necessary to have a minimal understanding of how statistical inferences are made from a mixed model. In general, the mixed model is expressed in the form

$$\mathbf{y} = X\boldsymbol{\beta} + Z\mathbf{u} + \boldsymbol{\epsilon}$$

where the random vectors $\mathbf{u}$ and $\boldsymbol{\epsilon}$ have the multivariate normal distributions $N(\mathbf{0},\ G)$ and $N(\mathbf{0},\ R)$, respectively, where the variance-covariance matrices G and R are fixed unknown constants. Using this form, the variance-covariance matrix of $\mathbf{y}$ is given by $V = ZGZ' + R$. Thus, the covariance structure of $\mathbf{y}$ is determined by the random effects design matrix Z and the covariance structures is defined by the matrices G and R. For the model discussed above, these matrices take simple forms: R is a 12×12 identity matrix and G is a 9×9 diagonal matrix with the diagonal elements $\sigma_B^2, \sigma_B^2, \sigma_B^2, \sigma_{\alpha B}^2, \ldots, \sigma_{\alpha B}^2$. The matrices X and Z consist of constants (usually 0's and 1's) because they are the usual design matrices for the two types of effects. By writing the likelihood function for the parameters, $\boldsymbol{\beta}$, G, and R (i.e., the joint density function of $\mathbf{y}$ and $\mathbf{u}$) and taking derivatives with respect to $\boldsymbol{\beta}$ and $\mathbf{u}$, the following set of equations known as the *mixed model equations* are obtained:

$$\begin{bmatrix} X'R^{-1}X & X'R^{-1}Z \\ Z'R^{-1}X & Z'R^{-1}Z + G^{-1} \end{bmatrix} \begin{bmatrix} \tilde{\boldsymbol{\beta}} \\ \tilde{\mathbf{u}} \end{bmatrix} = \begin{bmatrix} X'R^{-1}\mathbf{y} \\ Z'R^{-1}\mathbf{y} \end{bmatrix}$$

The solutions to mixed model equations are given by $\tilde{\boldsymbol{\beta}} = (X'V^{-1}X)^{-1}X'V^{-1}\mathbf{y}$ and $\tilde{\mathbf{u}} = GZ'V^{-1}(\mathbf{y} - X\tilde{\boldsymbol{\beta}})$. Using these, best linear unbiased estimates of estimable linear functions of $\boldsymbol{\beta}$ as well as best linear unbiased predictors (BLUPs) of *predictable functions* of $\boldsymbol{\beta}$ and $\mathbf{u}$ may be constructed.

A predictable function of $\boldsymbol{\beta}$ and $\mathbf{u}$ is a linear combination of the form $\ell'\boldsymbol{\beta} + \mathbf{m}'\mathbf{u}$, where $\ell'\boldsymbol{\beta}$ is an estimable function of $\boldsymbol{\beta}$. Recall that estimable functions were defined in Section 5.1 of Chapter 5. It is clear that if fixed parameters are not involved, it is possible to predict virtually any linear function of $\mathbf{u}$; however, in practice, predictable functions considered are only those that are interpretable as part of the inference made from a particular experiment. For example, the mixed model equations are due to Henderson (in Henderson et al. (1959); also see Searle et al. (1992)), who developed a procedure for predicting breeding values (defined as a predictable function, say $\mu + B_i$), for randomly selected sires (i.e., sire is the random Factor B) in an animal genetics experiment. See Searle et al. (1992) for a detailed presentation on BLUPs.

In the above machine-operator experiment, $E(y_{ijk}) = \mu + \alpha_i$ is an estimable function of the fixed parameters and estimates the mean strength of plastic sheeting from machine i, averaged over all operators in the population of operators. On the other hand, the function $\mu + \alpha_i + B_{\cdot} + \alpha B_{i\cdot}$ (where $B_{\cdot} = \frac{1}{3}\sum_j B_j$ and $\alpha B_{i\cdot} = \frac{1}{3}\sum_j \alpha B_{ij}$) is the expectation of y_{ijk} averaged over the three operators in the experiment. This is a predictable function different from the estimable function above and estimates the mean strength for Machine i given that the effects of the operators are fixed. Another example of a predictable function is $\alpha_i - \alpha_j + (\alpha B_{i\cdot} - \alpha B_{j\cdot})$, which measures the difference between the two machines i and j.

Although, BLUPs were not discussed for random models considered in Sections 6.2, 6.3, and 6.4, they can also be defined for those models by treating them as mixed models by taking μ as the only fixed effect in each of the models. Thus, for example, in the one-way random model $y_{ij} = \mu + A_i + \epsilon_{ij}$ considered in Section 6.2, the BLUP of $\mu + A_i$ is of the form $\delta \bar{y}_{\cdot\cdot} + (1-\delta)\bar{y}_{i\cdot}$, where $\delta = \sigma^2/(\sigma^2 + n\sigma_A^2)$. From this example, it is clear that the BLUP is a function of the unknown variance components. Thus, for BLUPs to be practically useful, they need to be estimated because the values of the variance components involved in the expressions for the predictors are unknown. In practice, variance components are first estimated and then plugged into the BLUPs to obtain *estimated* BLUPs (eBLUPs).

6.5.1 Two-way mixed effects model: Randomized complete blocks design

In Section 5.7, the analysis of a randomized blocks design with block effects considered as fixed effects was discussed. However, in practice, block effects need to be considered as random effects because statistical inferences that will be made about the differences in treatment effects from such a design must be valid regardless of the choice of blocks. By considering the blocks used in the experiment as a random sample from a hypothetical population of blocks, the effects of the blocks can be specified in the model as random effects. From such a model, inferences regarding differences in the treatment effects can be made using the (unconditional) means of the observations, with the variance of the block effects then accounting for the variability among blocks present.

With random block effects, using the same arguments presented when blocks were considered fixed, the model may still be specified as an additive model since the block effects do not interact with the treatment effects; thus, no interaction term is necessary in this case, as well.

Model

An appropriate model for a RCBD with random block effects is

$$y_{ij} = \mu + \tau_i + B_j + \epsilon_{ij}, \quad i=1,\ldots,t;\ j=1,\ldots,r$$

where τ_i is the effect of the ith treatment, the effect of the jth block, B_j, is assumed to be iid $N(0, \sigma_B^2)$, and the random error ϵ_{ij} iid $N(0, \sigma^2)$ distributed independently of B_j. As a consequence of this model, the mean of a response to treatment i is $E(y_{ij}) = \mu + \tau_i$ and the variance is $\text{Var}(y_{ij}) = \sigma^2 + \sigma_B^2$. Further, the covariance between two observations in the same block is $\text{Cov}(y_{ij}, y_{i'j}) = \sigma_B^2$, but observations from different blocks are uncorrelated.

Estimation and Hypothesis Testing

An analysis of variance that corresponds to the model is constructed using the same computational formulas as when blocks were considered fixed but the expected mean squares must be calculated using the assumptions described above. As with the random models, an additional column displaying the expected mean squares is included in the anova table for a mixed model:

SV	df	SS	MS	F	E(MS)
Blocks	$r-1$	SS_A	MS_A	MS_A/MSE	$\sigma^2 + r\sigma_B^2$
Trts	$t-1$	SS_{Trt}	MS_{Trt}	MS_{Trt}/MSE	$\sigma^2 + r\dfrac{\sum_i(\tau_i - \bar{\tau})^2}{(t-1)}$
Error	$(r-1)(t-1)$	SSE	$MSE (= s^2)$		σ^2
Total	$rt-1$				

The F-statistic for Trts tests the hypothesis of equality of treatment effects:

$$H_0: \tau_1 = \tau_2 = \cdots = \tau_t \text{ versus } H_a: \text{ at least one inequality}$$

or, equivalently,

$$H_0: \mu_1 = \mu_2 = \cdots = \mu_t \text{ versus } H_a: \text{ at least one inequality}$$

where $\mu_i = \mu + \tau_i$, the ith treatment mean. H_0 is rejected if the observed F-value exceeds the α upper percentile of an F-distribution with $df_1 = t-1$ and $df_2 = (r-1)(t-1)$. The best estimate of the difference between the effects of two treatments labeled p and q is

$$\widehat{\mu_p - \mu_q} = \widehat{\tau_p - \tau_q} = \bar{y}_{p.} - \bar{y}_{q.}$$

with standard error given by

$$s_d = \text{s.e.}(\bar{y}_{p.} - \bar{y}_{q.}) = s\sqrt{\dfrac{2}{r}}$$

where $s = \sqrt{MSE}$. A $(1-\alpha)100\%$ confidence interval for $\mu_p - \mu_q$ (or, equivalently, $\tau_p - \tau_q$) is

$$(\bar{y}_{p\cdot} - \bar{y}_{q\cdot}) \pm t_{\alpha/2,\nu} \cdot s\sqrt{\frac{2}{r}}$$

where $t_{\alpha/2,\nu}$ is the upper $\alpha/2$ percentile point of a t-distribution with $\nu = (r-1)(t-1)$ degrees of freedom. Thus, none of these results regarding the treatment effects is different from the fixed block effects case. Similarly, standard errors for linear comparisons of treatment means and the corresponding t-tests may be calculated.

SAS Example F10

The data from the experiment comparing five seed treatments in five replications described in Snedecor and Cochran (1989) and used in SAS Example E11 (see Section 5.7) is re-analyzed in this example, but considering the blocks as random effects. The data were shown in Table 5.8 in Section 5.7. The model is thus

$$y_{ij} = \mu + \tau_i + B_j + \epsilon_{ij}, \quad i = 1, \ldots, 5; \; j = 1, \ldots, 5$$

where τ_i is the effect of the ith seed treatment, the effect of the jth block, B_j, is assumed to be iid $N(0, \sigma_B^2)$, and the random error ϵ_{ij} iid $N(0, \sigma^2)$ distributed independently of B_j. The proc glm step (see Fig. 6.22) is similar to that used in SAS Example E11 except for the inclusion of the random statement and use of the lsmeans statement instead of the means statement. This random statement leads to the computation of the expected mean squares for terms in the model statement. The Q option causes the matrix for the quadratic forms (described below) that appear in the expected mean squares for fixed effects to be explicitly displayed. In this example, there is a single such quadratic form for the treatment effect.

The lsmeans statement requests 95% confidence intervals for pairwise differences in seed treatment means (effects) that are adjusted for simultaneous inference using the Tukey method. The results would be exactly the same as those resulting from the use of the means trt/tukey cldiff; statement. Note also that the contrast statement is placed ahead of the random statement so that the expected value of the mean square for testing the contrast hypothesis is also calculated.

Edited forms of the output from the SAS Example F10 program appears in Figs. 6.23, 6.24, and 6.25. Page 2 (see Fig. 6.23) contains the default Type III F-statistics for the trt effects and the results of the test of the contrast hypothesis **1**. For the RCBD (in the mixed model case), these F-statistics are the correct statistics for testing for fixed effects. The contrast hypothesis is rejected (p-value$= .0028$); thus, the average effect of the seed treatments on germination is found to be different from the effect of the control of no treatment.

On page 4 (see Fig. 6.24) of the output, the Type III expected mean squares for the model effects and the contrast are displayed. Both the trt and the contrast expected mean square expressions **2** contain a term with

```
data soybean;
input trt $ @;
do rep=1 to 5;
    input yield @;
    output;
end;
datalines;
check   8 10 12 13 11
arasan  2  6  7 11  5
spergon 4 10  9  8 10
samesan 3  5  9 10  6
fermate 9  7  5  5  3
;
run;

proc glm order=data;
    class trt rep;
    model yield = trt rep;
    contrast 'Check vs Chemicals' trt 4 -1 -1 -1 -1;
    random rep/Q;
    lsmeans trt/pdiff cl adjust=tukey;
run;
```

Fig. 6.22. SAS Example F10: Analysis of seed treatments

a quadratic form (labeled Q(trt)). For the RCBD, Q(trt) is of the form $r\frac{\sum_i(\tau_i - \bar{\tau})^2}{(t-1)}$ (as given in the anova table given earlier in this subsection).

This can be verified using the matrix of the quadratic form output on page 3 **3**. To obtain the form of Q for an effect, one needs to calculate the quadratic form $\tau' A \tau$ and divide by the degrees of freedom for the effect. Here, A is the matrix the columns of which are printed and τ is the vector of fixed effects parameters $\tau = (\tau_1, \tau_2, \tau_3, \tau_4, \tau_5)'$. Thus, the computation requires the matrix multiplication

$$\begin{bmatrix} \tau_1 & \tau_2 & \tau_3 & \tau_4 & \tau_5 \end{bmatrix} \begin{bmatrix} 4 & -1 & -1 & -1 & -1 \\ -1 & 4 & -1 & -1 & -1 \\ -1 & -1 & 4 & -1 & -1 \\ -1 & -1 & -1 & 4 & -1 \\ -1 & -1 & -1 & -1 & 4 \end{bmatrix} \begin{bmatrix} \tau_1 \\ \tau_2 \\ \tau_3 \\ \tau_4 \\ \tau_5 \end{bmatrix} = 5(\sum_i (\tau_i - \bar{\tau})^2)$$

giving Q(trt)$=5(\sum_i(\tau_i-\bar{\tau})^2)/4$. This expected mean square is not different from the case where blocks were considered fixed effects.

In a similar fashion, the Q for the contrast expected mean square may be calculated using the corresponding matrix given on page 3 **4**. In balanced data situations for common experimental designs, such as the RCBD, it is not necessary to use the Q option since the form of the expected mean squares for the fixed effects is available from many textbooks. It was used in this example for illustrating how the output matrix from the Q option is used to construct the quadratic form. This option is mainly useful for determining the expected mean squares in complex situations.

Pages 5 (and 6, not shown) contain the results of the lsmeans statement (see Fig. 6.25). The pdiff option produced the second portion of this page of the SAS output, which gives p-values **5** associated with testing pairwise differences in means (i.e., hypotheses of the form $H_0 : \mu_i = \mu_j$ versus $H : \mu_i \neq \mu_j$

```
                         The SAS System                              2

                        The GLM Procedure

Dependent Variable: yield

                                    Sum of
Source                    DF       Squares     Mean Square    F Value    Pr > F

Model                      8    133.6800000     16.7100000       3.09    0.0262

Error                     16     86.5600000      5.4100000

Corrected Total           24    220.2400000


           R-Square     Coeff Var       Root MSE     yield Mean

           0.606974      30.93006        2.325941       7.520000


Source                    DF      Type I SS     Mean Square    F Value    Pr > F

trt                        4    83.84000000     20.96000000       3.87    0.0219
rep                        4    49.84000000     12.46000000       2.30    0.1032


Source                    DF    Type III SS     Mean Square    F Value    Pr > F

trt                        4    83.84000000     20.96000000       3.87    0.0219
rep                        4    49.84000000     12.46000000       2.30    0.1032


Contrast 1                DF    Contrast SS    Mean Square    F Value    Pr > F

Check vs Chemicals         1    67.24000000     67.24000000      12.43    0.0028
```

Fig. 6.23. SAS Example F10: Output (page 2)

for all pairs (i,j)). The 95% confidence intervals for pairwise differences 6 were produced as a result of the `cl` option. The confidence intervals are adjusted for multiple comparisons using Tukey's procedure.

Zero is included in every interval except 1–2 (i.e., check–arasan), and 1–5 (i.e., check–fermate). This represents one less pair of means not found to be different than when t-based confidence intervals (i.e., those unadjusted for multiple comparisons) were used in Section 5.7. Thus, this procedure is slightly more conservative than using t-based intervals. The conclusions are similar to those drawn in Section 5.7 except that both *spergon* and *samesan* are found to be different from the control.

SAS Example F11

In this program (displayed in Fig. 6.26) `proc mixed` is used to analyze the data from the experiment comparing five seed treatments using Type 3 sums of squares and the method of moments. As discussed in Section 6.2, there are

```
                          The SAS System                                3

                         The GLM Procedure
          Quadratic Forms of Fixed Effects in the Expected Mean Squares

                Source: Type III Mean Square for trt  3

                      trt check         trt arasan        trt spergon

   trt check         4.00000000         -1.00000000       -1.00000000
   trt arasan       -1.00000000          4.00000000       -1.00000000
   trt spergon      -1.00000000         -1.00000000        4.00000000
   trt samesan      -1.00000000         -1.00000000       -1.00000000
   trt fermate      -1.00000000         -1.00000000       -1.00000000

                Source: Type III Mean Square for trt

                           trt samesan       trt fermate

          trt check        -1.00000000       -1.00000000
          trt arasan       -1.00000000       -1.00000000
          trt spergon      -1.00000000       -1.00000000
          trt samesan       4.00000000       -1.00000000
          trt fermate      -1.00000000        4.00000000


           Source: Contrast Mean Square for Check vs Chemicals  4

                      trt check         trt arasan        trt spergon

   trt check         4.00000000         -1.00000000       -1.00000000
   trt arasan       -1.00000000          0.25000000        0.25000000
   trt spergon      -1.00000000          0.25000000        0.25000000
   trt samesan      -1.00000000          0.25000000        0.25000000
   trt fermate      -1.00000000          0.25000000        0.25000000

             Source: Contrast Mean Square for Check vs Chemicals

                           trt samesan       trt fermate

          trt check        -1.00000000       -1.00000000
          trt arasan        0.25000000        0.25000000
          trt spergon       0.25000000        0.25000000
          trt samesan       0.25000000        0.25000000
          trt fermate       0.25000000        0.25000000


                          The SAS System                                4

                         The GLM Procedure

   Source                    Type III Expected Mean Square

   trt                       Var(Error) + Q(trt)  2

   rep                       Var(Error) + 5 Var(rep)


   Contrast                  Contrast Expected Mean Square

   Check vs Chemicals        Var(Error) + Q(trt)  2
```

Fig. 6.24. SAS Example F10: Quadratic form for the expected treatment mean square)

```
                        The SAS System                          5

                        The GLM Procedure
                      Least Squares Means
             Adjustment for Multiple Comparisons: Tukey

                                              LSMEAN
              trt         yield LSMEAN        Number

              check        10.8000000            1
              arasan        6.2000000            2
              spergon       8.2000000            3
              samesan       6.6000000            4
              fermate       5.8000000            5

             Least Squares Means for effect trt
             Pr > |t| for H0: LSMean(i)=LSMean(j)  5

                      Dependent Variable: yield

i/j            1            2            3            4            5

 1                       0.0443       0.4242       0.0740       0.0261
 2         0.0443                     0.6602       0.9987       0.9987
 3         0.4242       0.6602                     0.8102       0.4999
 4         0.0740       0.9987       0.8102                     0.9812
 5         0.0261       0.9987       0.4999       0.9812

              trt       yield LSMEAN     95% Confidence Limits

              check      10.800000       8.594891      13.005109
              arasan      6.200000       3.994891       8.405109
              spergon     8.200000       5.994891      10.405109
              samesan     6.600000       4.394891       8.805109
              fermate     5.800000       3.594891       8.005109

               Least Squares Means for Effect trt

                          Difference        Simultaneous 95%
                           Between         Confidence Limits for
               i   j        Means           LSMean(i)-LSMean(j)  6

               1   2       4.600000        0.093171       9.106829
               1   3       2.600000       -1.906829       7.106829
               1   4       4.200000       -0.306829       8.706829
               1   5       5.000000        0.493171       9.506829
               2   3      -2.000000       -6.506829       2.506829
               2   4      -0.400000       -4.906829       4.106829
               2   5       0.400000       -4.106829       4.906829
               3   4       1.600000       -2.906829       6.106829
               3   5       2.400000       -2.106829       6.906829
               4   5       0.800000       -3.706829       5.306829
```

Fig. 6.25. SAS Example F10: Output (edited pages 5 and 6)

several advantages to using `proc mixed` instead of `proc glm` even when iterative estimation methods are not used. Although using the `random` statement in `proc glm` expected mean squares and F-tests about variance components can be computed, other statements in `proc glm` do not make use of this information. For example, the standard anova tables `proc glm` produces still regard all effects as fixed effects, whereas in `proc mixed`, separate tests are performed for the variance components. Moreover, `proc mixed` allows the user to choose among several estimation methods as well as the ability to use estimate statements to estimate BLUPs involving fixed and random effects.

```
insert data step to create the SAS data set 'soybean' here

proc mixed order=data method=type3 covtest cl;
   class trt rep;
   model yield = trt;
   random rep;
   lsmeans trt/diff cl adjust=tukey;
   contrast 'Check vs Chemicals' trt 4 -1 -1 -1 -1;
run;
```

Fig. 6.26. SAS Example F11: Analysis of seed treatments using PROC MIXED

The `method=type3` option in the `proc` statement specifies that the variance components are to be estimated by the method of moments using the Type 3 sums of squares. The standard errors of the estimated variance components and associated confidence intervals are computed as a result of the `covtest` and the `cl` options. As discussed in Section 6.2, the model statement in `proc mixed` requires only the fixed part of the model to be specified; thus, `model yield = trt;` implies an overall mean μ and the fixed effect `trt` are in the model. The `random` statement specifies the random portion of the model: here, the random effect `rep` and a random error term. The variance of this effect and the error variance constitute the two `variance components` specified by this model. Thus, the `model` and `random` statements (in addition to the `class` statement) are needed to specify a mixed model in `proc mixed`. The `lsmeans` statement requests 95% confidence intervals for pairwise differences in seed treatment effects adjusted for simultaneous inference using the Tukey method. The `diff` option is redundant here as `adjust=` option implies the `diff` option.

The information on page 1 (see Fig. 6.27) is the same as that described earlier for SAS Examples F2 and F3 in Section 6.2. The `Type 3 Analysis of Variance` table **1** provides the expected means squares for all effects and is exactly the same as those produced by `proc glm`. The actual F-tests for both fixed and random effects are shown on page 2 **2** (see Fig. 6.28). These are the same as those in the standard `Type III SS` table from `proc glm`. However, a separate table for the F-tests of the fixed effects is also provided lower in page 2 **3**. The table titled `Covariance Parameter Estimates` **4** gives estimates of variance components obtained via the method of moments. As in Section 6.2, these are obtained by solving

$$\sigma^2 + 5\sigma_B^2 = 12.46$$

```
                         The SAS System                                1

                       The Mixed Procedure

                       Model Information

      Data Set                    WORK.SOYBEAN
      Dependent Variable          yield
      Covariance Structure        Variance Components
      Estimation Method           Type 3
      Residual Variance Method    Factor
      Fixed Effects SE Method     Model-Based
      Degrees of Freedom Method   Containment


                     Class Level Information

         Class     Levels    Values

         trt          5      check arasan spergon samesan
                             fermate
         rep          5      1 2 3 4 5


                          Dimensions

              Covariance Parameters      2
              Columns in X               6
              Columns in Z               5
              Subjects                   1
              Max Obs Per Subject       25


                    Number of Observations

              Number of Observations Read       25
              Number of Observations Used       25
              Number of Observations Not Used    0


                  Type 3 Analysis of Variance

                  Sum of
Source      DF   Squares    Mean Square  Expected Mean Square

trt          4   83.840000   20.960000   Var(Residual) + Q(trt)
rep          4   49.840000   12.460000   Var(Residual) + 5 Var(rep)
Residual    16   86.560000    5.410000   Var(Residual)
```

Fig. 6.27. SAS Example F11: Output (page 1)

$$\sigma^2 = 5.41$$

which give $\hat{\sigma}^2 = 5.41$ and $\hat{\sigma}_B^2 = (12.46 - 5.41)/5 = 1.41$. The confidence intervals calculated by `proc mixed` for the variance components are those based on large-sample standard errors and the normal percentiles except those for σ^2. For example, the Wald statistic z is given by $z = \hat{\sigma}_B^2/\text{s.e.}(\hat{\sigma}_B^2)$ and a 95% confidence interval for σ_B^2 is thus $11.4478 \pm (1.96)(8.7204) = (-5.6442, 28.5398)$.

6.5 Two-Way Mixed Effects Model

```
                        The SAS System                                    2

                       The Mixed Procedure

                  Type 3 Analysis of Variance 2

                                         Error
Source       Error Term                   DF        F Value    Pr > F

trt          MS(Residual)                 16          3.87     0.0219
rep          MS(Residual)                 16          2.30     0.1032
Residual     .                             .            .        .

                   Covariance Parameter Estimates 4

                      Standard      Z
Cov Parm   Estimate    Error      Value    Pr Z    Alpha    Lower    Upper

rep         1.4100    1.8032      0.78    0.4342   0.05   -2.1241   4.9441
Residual    5.4100    1.9127      2.83    0.0023   0.05    3.0008  12.5310

                          Fit Statistics

                 -2 Res Log Likelihood        101.9
                 AIC (smaller is better)      105.9
                 AICC (smaller is better)     106.6
                 BIC (smaller is better)      105.1

                   Type 3 Tests of Fixed Effects 3

                         Num     Den
               Effect    DF      DF     F Value    Pr > F

                trt       4      16       3.87     0.0219

                            Contrasts

                         Num     Den
          Label          DF      DF     F Value    Pr > F

          Check vs Chemicals  1   16     12.43     0.0028
```

Fig. 6.28. SAS Example F11: Output (page 2)

If a $(1 - \alpha)100\%$ confidence interval based on the Satterthwaite approximation is desired, a formula similar to (6.1) in Section 6.2 could be used. The relevant SAS data step is

```
data cint;
    alpha=.05; a=5; r=5;
    msb= 12.46; s2=5.41; sb2=1.41;
    nu=(a*sb2)**2/(msb**2/(r-1)+ s2**2/((a-1)*(r-1)));
    L= (nu*sb2)/cinv(1-alpha/2,nu);
    U= (nu*sb2)/cinv(alpha/2,nu);
    put nu=    L=    U=;
run;
```

```
                    The SAS System                                3

                   The Mixed Procedure

                   Least Squares Means

                      Standard
Effect  trt      Estimate    Error   DF  t Value  Pr > |t|  Alpha    Lower    Upper

trt     check    10.8000    1.1679   16    9.25   <.0001    0.05    8.3242  13.2758
trt     arasan    6.2000    1.1679   16    5.31   <.0001    0.05    3.7242   8.6758
trt     spergon   8.2000    1.1679   16    7.02   <.0001    0.05    5.7242  10.6758
trt     samesan   6.6000    1.1679   16    5.65   <.0001    0.05    4.1242   9.0758
trt     fermate   5.8000    1.1679   16    4.97    0.0001   0.05    3.3242   8.2758
```

Fig. 6.29. SAS Example F11: Estimates and standard errors of means

Executing the above gives the interval (0.3075, 430.516). However, since the Wald statistic z and the degrees of freedom $\nu = 2z^2$ are already available, the code given at the end of SAS Example F7 is simpler to use.

Note, however, that the variance component σ_B^2 is not of major interest in this experiment, but an estimate and a valid hypothesis test of the variation among the blocks are available to the experimenter. The output from `lsmeans` is of main interest and appears in pages 3 and 4 (see Figs. 6.29, 6.30, and 6.31 for extracted parts from these pages). In Fig. 6.29, estimates, standard errors, and confidence intervals for the treatment means $\mu_i = \mu + \tau_i$ are given. It is important to note that the standard errors are computed (correctly) using the formula $\sqrt{(\hat{\sigma}^2 + \hat{\sigma}_B^2)/5} = \sqrt{(5.41 + 1.41)/5} = 1.1679$. Note that in SAS Example F10, `proc glm` would have used the formula $\sqrt{(\hat{\sigma}^2/5}$ (if the `stderr` option was specified requesting it) because rep is considered fixed in `proc glm` for the purpose of this computation.

```
               Differences of Least Squares Means

                                   Standard
Effect  trt       _trt      Estimate  Error   DF  t Value  Pr > |t|  Adj P

trt     check    arasan      4.6000  1.4711   16    3.13    0.0065   0.0443
trt     check    spergon     2.6000  1.4711   16    1.77    0.0962   0.4242
trt     check    samesan     4.2000  1.4711   16    2.86    0.0115   0.0740
trt     check    fermate     5.0000  1.4711   16    3.40    0.0037   0.0261
trt     arasan   spergon    -2.0000  1.4711   16   -1.36    0.1928   0.6602
trt     arasan   samesan    -0.4000  1.4711   16   -0.27    0.7892   0.9987
trt     arasan   fermate     0.4000  1.4711   16    0.27    0.7892   0.9987
trt     spergon  samesan     1.6000  1.4711   16    1.09    0.2929   0.8102
trt     spergon  fermate     2.4000  1.4711   16    1.63    0.1223   0.4999
trt     samesan  fermate     0.8000  1.4711   16    0.54    0.5941   0.9812
```

Fig. 6.30. SAS Example F11: t-Tests for pairwise differences of means

6.5 Two-Way Mixed Effects Model

The results of the t-tests are shown in Fig. 6.30. Whereas the p-values computed for the standard t-tests are given, an additional column (titled Adj P 5) provides the p-values adjusted for multiple testing. Here, the adjustment is based on Tukey's studentized range distribution, the p-values being calculated are probabilities that studentized range random variable $q(t, \nu)$ exceeds $|\bar{y}_i - \bar{y}_j|/\sqrt{s^2/r}$, where $\nu = (a-1)(r-1)$ and $\bar{y}_i$ and $\bar{y}_j$ are a pair of trt means. The SAS function probmc can be used to verify this by executing a data step such as

```
data pval;
   a=5; r=5; diff=4.6; nu=(a-1)*(r-1); s2=5.41;
   q=diff/sqrt(s2/r);
   p=1-probmc("Range", q, ., nu, 5);
   put df= q= p= ;
run;
```

which gives $q = 4.42226$, $p = 0.04429$. The adjusted p-values do not change the results obtained previously for these comparisons.

```
                          The SAS System                              4

                        The Mixed Procedure

                   Differences of Least Squares Means

Effect trt    _trt     Adjustment      Alpha   Lower    Upper    Adj      Adj
                                                                Lower    Upper
trt    check  arasan   Tukey-Kramer    0.05   1.4815   7.7185   0.09317  9.1068
trt    check  spergon  Tukey-Kramer    0.05  -0.5185   5.7185  -1.9068   7.1068
trt    check  samesan  Tukey-Kramer    0.05   1.0815   7.3185  -0.3068   8.7068
trt    check  fermate  Tukey-Kramer    0.05   1.8815   8.1185   0.4932   9.5068
trt    arasan spergon  Tukey-Kramer    0.05  -5.1185   1.1185  -6.5068   2.5068
trt    arasan samesan  Tukey-Kramer    0.05  -3.5185   2.7185  -4.9068   4.1068
trt    arasan fermate  Tukey-Kramer    0.05  -2.7185   3.5185  -4.1068   4.9068
trt    spergon samesan Tukey-Kramer    0.05  -1.5185   4.7185  -2.9068   6.1068
trt    spergon fermate Tukey-Kramer    0.05  -0.7185   5.5185  -2.1068   6.9068
trt    samesan fermate Tukey-Kramer    0.05  -2.3185   3.9185  -3.7068   5.3068
```

Fig. 6.31. SAS Example F11: Confidence intervals for pairwise differences of means

Two sets of confidence intervals are shown in Fig. 6.31, one set based on the t-distribution and the second set adjusted for multiple comparisons using Tukey's studentized range statistic. The results are exactly the same as those obtained from the previous analysis using proc glm.

6.5.2 Two-way mixed effects model: Crossed classification

The machine-operator example at the beginning of Section 6.5, introducing two-factor mixed models, discussed an experiment in which a fixed factor (two brands of machines) and a random factor (three randomly selected operators) were used. The general set up of such experiments will have "a" levels of a fixed factor A and "b" levels of a random factor B, where each factorial combination of levels of A and B is replicated n times in a completely randomized design. The interaction effect between A and B is random because the levels of the interaction effect involve the levels of Factor B which are randomly selected.

Model

The two-way crossed mixed effects model is given by

$$y_{ijk} = \mu + \alpha_i + B_j + \alpha B_{ij} + \epsilon_{ijk}, \quad i = 1, \ldots, a; \quad j = 1, \ldots, b; \quad k = 1, \ldots, n$$

where α_i are fixed effects (levels of Factor A), B_j are random effects (levels of Factor B) that are iid $N(0, \sigma_B^2)$ random variables, the interaction effects between the two factors denoted by αB_{ij} are iid $N(0, \sigma_{\alpha B}^2)$, and the random errors ϵ_{ijk} are iid $N(0, \sigma^2)$ random variables. The random variables B_j, αB_{ij}, and ϵ_{ijk} are pairwise independent. The mean of the responses is $E(y_{ijk}) = \mu + \alpha_i$ and the variance $\text{Var}(y_{ijk}) = \sigma^2 + \sigma_B^2 + \sigma_{\alpha B}^2$. The responses from the same level of random factor B have covariance σ_B^2 if they are from different levels of factor A and $\sigma_B^2 + \sigma_{\alpha B}^2$ if they are from the same level of Factor A. If they are from different levels of Factor B, they are uncorrelated.

A Special Comment The above model, sometimes called the *unconstrained parameters* (UP) model in the literature, is adopted by several authors as well as SAS software for the analysis of data from two-way crossed mixed effects experiments. However, other authors favor an alternative form of the model called the *constrained parameters* (CP) model, where the so-called summation restrictions are imposed on the fixed effects as well as the fixed-by-random interaction parameters. Using these constraints, in effect, imposes a covariance structure among the observations that is different from the one prescribed by the UP model. Although a relationship exists between the two sets of variance components, the meaning assigned to the parameters by experimenters, hence the interpretation of statistical inferences made, may be different. The two major differences between the two models are (i) the CP model allows for negative covariance among observations from different levels of Factor A whereas the UP model does not and (ii) expected mean squares for effect B are different for the two models. The second fact results in two different denominators for the F-statistic for testing the variance of effect B. The UP model is adopted for the rest of the discussion in this book.

Estimation and Hypothesis Testing

The usual format of the anova table for computing the required F-statistics for testing hypotheses of interest in a two-way classification is:

SV	df	SS	MS	F	E(MS)
A	$a-1$	SS_A	MS_A	MS_A/MS_{AB}	$\sigma^2 + n\sigma^2_{\alpha B} + bn\frac{\sum_i(\alpha_i-\bar{\alpha})^2}{(a-1)}$
B	$b-1$	SS_B	MS_B	MS_B/MS_{AB}	$\sigma^2 + n\sigma^2_{\alpha B} + an\sigma^2_B$
AB	$(a-1)(b-1)$	SS_{AB}	MS_{AB}	MS_{AB}/MSE	$\sigma^2 + n\sigma^2_{\alpha B}$
Error	$ab(n-1)$	SSE	MSE		σ^2
Total	$abn-1$	SS_{Tot}			

The F-statistic for A tests the hypothesis of equality of treatment effects:

$$H_0: \alpha_1 = \alpha_2 = \cdots = \alpha_t \text{ versus } H_a: \text{ at least one inequality}$$

or, equivalently,

$$H_0: \mu_1 = \mu_2 = \cdots = \mu_t \text{ versus } H_a: \text{ at least one inequality}$$

where $\mu_i = \mu + \alpha_i$, the ith treatment mean. H_0 is rejected if the observed F-value exceeds the α upper percentile of an F-distribution with $df_1 = a-1$ and $df_2 = (a-1)(b-1)$. It is informative to compare the expected mean squares of the denominator and numerator of this F-ratio and note that they will have the same expectation if the null hypothesis above holds and the numerator will have a larger expectation if the null hypothesis is not true. It can be verified that $bn\frac{\sum_i(\alpha_i-\bar{\alpha})^2}{(t-1)}$ is zero if $\alpha_1 = \alpha_2 = \cdots = \alpha_t$ and is positive otherwise.

Inferences about estimable functions of the fixed effects can be made using the appropriate mean squares to calculate their standard errors. The best estimate of μ_i, the ith treatment mean, and its standard error are

$$\hat{\mu}_i = \bar{y}_{i..} = (\sum_j \sum_k y_{ijk})/bn,$$
$$\text{s.e.}(\bar{y}_{i..}) = s_{AB}/\sqrt{bn}$$

where $s^2_{AB} = MS_{AB}$ and $i = 1, \ldots, a$. This result follows since it can be shown that the variance of $\bar{y}_{i..}$ is $(\sigma^2 + n\sigma^2_{\alpha B})/bn$ and $E(MS_{AB}) = \sigma^2 + n\sigma^2_{\alpha B}$; that is, MS_{AB} estimates $\sigma^2 + n\sigma^2_{\alpha B}$. Note that the above means would be identical to the "Factor A means" calculated in the two-way classification with fixed effects case discussed in Section 5.4, but the standard errors are obviously not the same.

The best estimate of the difference between the effects of two treatments labeled p and q is

$$\widehat{\mu_p - \mu_q} = \widehat{\alpha_p - \alpha_q} = \bar{y}_{p..} - \bar{y}_{q..}$$

with standard error given by

$$s_d = \text{s.e.}(\bar{y}_{p..} - \bar{y}_{q..}) = s_{AB}\sqrt{\frac{2}{bn}}$$

where $s_{AB}^2 = \text{MS}_{AB}$. The standard error of linear contrasts of the effects (or the means) is similarly obtained by replacing s^2 with s_{AB}^2 in the fixed effects model formulas.

A $(1-\alpha)100\%$ confidence interval for $\mu_p - \mu_q$ (or equivalently, $\alpha_p - \alpha_q$) is

$$(\bar{y}_{p..} - \bar{y}_{q..}) \pm t_{\alpha/2,\nu} \cdot s_{AB}\sqrt{\frac{2}{bn}}$$

where $t_{\alpha/2,\nu}$ is the upper $\alpha/2$ percentile point of a t-distribution with $\nu = (a-1)(b-1)$ degrees of freedom. Thus, these results regarding the treatment effects are similar to those for the fixed effects case except that here s is replaced with s_{AB}. Similarly, standard errors for other linear comparisons of treatment means and the corresponding t-tests may be calculated.

F-Statistics shown in the anova table for sources of variation B and AB are used to test the hypotheses

(i) $H_0: \sigma_B^2 = 0$ versus $H_a: \sigma_B^2 > 0$

(ii) $H_0: \sigma_{\alpha B}^2 = 0$ versus $H_a: \sigma_{\alpha B}^2 > 0$

respectively. The null hypotheses in (i) or (ii) are rejected if the corresponding F-values exceed the α upper percentiles of F-distributions with $df_1 = b - 1$ and $df_2 = (a-1)(b-1)$ or $df_1 = (a-1)(b-1)$ and $df_2 = ab(n-1)$, respectively. Note that these F-statistics have different denominators: For the test of (i) the denominator is MS_{AB}, whereas for (ii) it is MSE. Comparing the expected mean squares of the denominator and numerator of these F-ratios allows one to verify whether these are the appropriate F-ratios. (With respect to the above special comment regarding the alternative model, note that the expected mean square for effect B under the CP model is $\sigma^2 + an\sigma_B^2$, suggesting that the appropriate F-statistic for testing hypothesis (i) is MS_B/MSE under that model. Thus, the denominator for testing (i) under the CP model will be different.)

Method of moments estimators are obtained as usual by setting the expected mean squares to their respective computed values. Thus, estimates of σ_B^2, $\sigma_{\alpha B}^2$, and σ^2 are given by

$$\hat{\sigma}_B^2 = (\text{MS}_B - \text{MS}_{AB})/an$$
$$\hat{\sigma}_{\alpha B}^2 = (\text{MS}_{AB} - \text{MSE})/n$$

and $\hat{\sigma}^2 = \text{MSE}$, respectively. When the data are balanced, these estimators are unbiased and have minimum variance. Confidence intervals for the variance components are obtained as described in Section 6.4. For example, a $(1-\alpha)100\%$ confidence interval for σ_B^2 is provided by

6.5 Two-Way Mixed Effects Model

$$\frac{\nu\hat{\sigma}_B^2}{\chi_{1-\alpha/2,\nu}^2} < \sigma_B^2 < \frac{\nu\hat{\sigma}_B^2}{\chi_{\alpha/2,\nu}^2}$$

where $\chi_{1-\alpha/2,\nu}^2$ and $\chi_{\alpha/2,\nu}^2$ are the $1-\alpha/2$ and $\alpha/2$ percentile points of the chi-square distribution with ν degrees of freedom, respectively. The degrees of freedom for $\hat{\sigma}_B^2 = \frac{1}{an}\text{MS}_B - \frac{1}{an}\text{MS}_{AB}$ is obtained using the Satterthwaite approximation and is given by

$$\nu = \frac{(an\hat{\sigma}_B^2)^2}{(\text{MS}_B)^2/(b-1) + (\text{MS}_{AB})^2/(a-1)(b-1)}$$

SAS Example F12

The two most crucial factors that influence the strength of solders used in cementing computer chips into the mother board of guidance systems of airplanes are identified to be the machine used to insert the solder and the operator of the machine. Four types of solder machine used in the plant were selected for a study planned to examine this dependence. Each of three operators selected at random from the operators available at the company's plants made two solders on each of the four machines in random order. The data, taken from Ott and Longnecker (2001), appear in Table 6.5.

Operator	Machine			
	1	2	3	4
1	204	205	203	205
	205	210	204	203
2	205	205	206	209
	207	206	204	207
3	211	207	209	215
	209	210	214	212

Table 6.5. Strength of solder in computer chips

From the description, it is clear that `machine` is a fixed factor with four levels, that the factor `operator` is random with three levels, and that the two factors are crossed. The two-way crossed mixed effects model used for the analysis of these data is given by

$$y_{ijk} = \mu + \alpha_i + B_j + \alpha B_{ij} + \epsilon_{ijk}, \quad i=1,\ldots,4;\ j=1,\ldots,3;\ k=1,\ldots,2$$

where α_i is the effect of machine i, B_j is the effect of operator j distributed as iid $N(0,\sigma_B^2)$ random variables, and the interaction effects between the two factors are denoted by αB_{ij} distributed as iid $N(0, \sigma_{\alpha B}^2)$. The random errors ϵ_{ijk} are distributed as iid $N(0, \sigma^2)$ random variables and the random variables B_j, αB_{ij}, and ϵ_{ijk} are pairwise independent.

```
data solder;
   infile "C:\Documents and Settings\...\ex17-22.txt";
   input operator machine strength;
run;

proc glm data=solder;
   class machine operator;
   model strength = machine operator machine*operator;
   random operator machine*operator/test;
   lsmeans machine/stderr;
run;
```

Fig. 6.32. SAS Example F12: Analysis of solder strength using PROC GLM

In the SAS Example F12 program, displayed in Fig. 6.32, `proc glm` is used to perform a conventional analysis of a mixed model. The data are read from a text file using the `list` input style. The `model` statement contains both the fixed and random effects as usual. The `random` statement identifies the `operator` and the `machine*operator` interaction as random effects. The `test` option requests `proc glm` to construct appropriate F-statistics for testing hypotheses about the variance components corresponding to these effects. The expected values of the mean squares in the anova table (see below) determines the ratios of sums of squares that need to be used for testing the two hypotheses of interest regarding the variances of `operator` and `machine*operator` effects: $H_0: \sigma_B^2 = 0$ vs. $H_a: \sigma_B^2 > 0$ and $H_0: \sigma_{\alpha B}^2 = 0$ vs. $H_a: \sigma_{\alpha B}^2 > 0$ as discussed previously. The anova table is:

SV	df	SS	MS	F	p-value	$E(\text{MS})$
Machine	3	12.458	4.1528	0.56	.6619	$\sigma^2 + 2\sigma_{\alpha B}^2 + 2\sum_i (\alpha_i - \bar{\alpha})^2$
Operator	2	160.333	80.1667	10.77	.0103	$\sigma^2 + 2\sigma_{\alpha B}^2 + 8\sigma_B^2$
Machine× Operator	6	44.667	7.4444	1.96	.1507	$\sigma^2 + 2\sigma_{\alpha B}^2$
Error	12	45.500	3.7917			σ^2
Total	23	262.958				

The `lsmeans` statement illustrates the estimation of an interesting BLUP. BLUPs for a mixed model were discussed in the introduction to this section. A BLUP consists of the sum of an estimable linear function for the fixed parameters (e.g., here $\mu + \alpha_i$) and a different function of the random effect parameters. The `lsmeans` statement in `proc glm` estimates the expectation of y_{ijk} averaged over the levels of the random factor (operator): $\mu + \alpha_i + \bar{B}_. +$

```
              Analysis of Strength of Solder in Computer Chips using PROC GLM        3
                                 The GLM Procedure
  Source                    Type III Expected Mean Square

  machine                   Var(Error) + 2 Var(machine*operator) + Q(machine)

  operator                  Var(Error) + 2 Var(machine*operator) + 8 Var(operator)

  machine*operator          Var(Error) + 2 Var(machine*operator)


              Analysis of Strength of Solder in Computer Chips using PROC GLM        4
                                 The GLM Procedure
                   Tests of Hypotheses for Mixed Model Analysis of Variance

Dependent Variable: strength

Source                         DF     Type III SS  ❶   Mean Square   F Value   Pr > F

machine                         3      12.458333         4.152778      0.56    0.6619
operator                        2     160.333333        80.166667     10.77    0.0103

Error                           6      44.666667         7.444444
Error: MS(machine*operator)  ❷


Source                         DF     Type III SS       Mean Square   F Value   Pr > F

machine*operator                6      44.666667         7.444444      1.96    0.1507

Error: MS(Error)  ❸            12      45.500000         3.791667


              Analysis of Strength of Solder in Computer Chips using PROC GLM        5

                                 Least Squares Means

                              strength       Standard
                machine        LSMEAN          Error        Pr > |t|

                   1         206.833333      0.794949       <.0001
                   2         207.166667      0.794949       <.0001
                   3         206.666667      0.794949       <.0001
                   4         208.500000      0.794949       <.0001
```

Fig. 6.33. SAS Example F12: Output pages 3, 4, and 5

$\overline{\alpha B}_{i.}$ (i.e., the conditional expectation conditioned on the operator and the interaction effects), where $\bar{B}_{.} = \frac{1}{3}\sum_j B_j$ and $\overline{\alpha B}_{i.} = \frac{1}{3}\sum_j \alpha B_{ij}$. Here, the interest is only in the effects of operators used in the study; conditioning on them is equivalent to considering their effects to be "fixed." (The lsmeans statement options pdiff, cl, and adj= may be used to obtain confidence intervals for pairs of differences of the above BLUPs, adjusted for multiple testing using a method of choice.)

The output pages 1 and 2 contain the standard output from `proc glm` (factor level information and Type I and III SS and associated F-tests considering all factors to be fixed); thus they are omitted. Pages 3 and 4 (see Fig. 6.33) contain the Type III Expected Mean Squares **1** and F-tests for all three effects in the model. These are constructed using the appropriate denominator mean squares. The denominator for the F-statistics for testing both the `machine` effect hypothesis (i.e., $H_0 : \alpha_1 = \alpha_2 = \alpha_3$) and the hypothesis about the variance of the `operator` random effect (i.e., $H_0 : \sigma_B^2 = 0$), respectively, is the mean square for the interaction effect `machine*operator` **2**. The denominator for the F-statistic, for testing the hypothesis about the variance of the `machine*operator` random effect (i.e., $H_0 : \sigma_{\alpha B}^2 = 0$) is the mean square for error **3**. The tests fail to reject either the machine main effect hypothesis or the interaction hypothesis at $\alpha = .05$; however, the operator variance component, σ_B^2, is found to be significantly different from zero. As demonstrated previously, the method of moments estimates can be obtained as usual. They are $\hat{\sigma}^2 = 3.792$, $\hat{\sigma}_B^2 = (7.444 - 3.792)/2 = 1.826$, and $\hat{\sigma}_{\alpha B}^2 = (80.16667 - 7.444)/8 = 9.0903$. The interaction plot of the mean solder strengths, shown in Fig. 6.34, is discussed later.

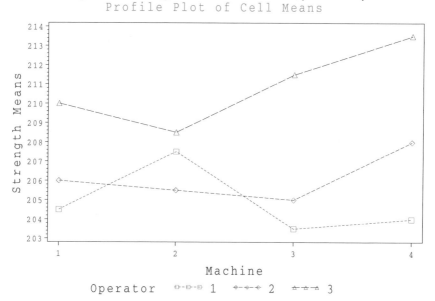

Fig. 6.34. SAS Example F12: Interaction plot of solder strength data

SAS Example F13

In this program (displayed in Fig. 6.35), `proc mixed` is used to analyze the solder strength using Type 3 sums of squares and the method of moments so that the results may be compared with the previous analysis using `proc glm`. In addition to the advantages discussed in Section 6.2, `proc mixed` allows the use of `estimate` statements to obtain estimates of BLUPs with the correct standard errors. The only term required on the right-hand side of the `model` statement is the fixed effect term `machine`. The `random` statement declares the `operator` and `machine*operator` effects to be random effects.

The option `ddfm=` is required to specify the method that must be used by `proc mixed` for the computation of denominator degrees of freedom F-tests, t-tests, confidence intervals, etc. for fixed effects or any function of fixed effects such as contrasts or BLUPs. Here, the value specified is `satterth`. To understand what this means, recall that the Satterthwaite approximation was used in Sections 6.3 and 6.4 for the construction of confidence intervals of variance components in random models. In those sections, this approximation was required when the denominator happened to be a linear combination of mean squares (rather than a single mean square).

```
data solder;
infile "C:\Documents and Settings\...\ex17-22.txt";
input operator machine strength;
run;

proc mixed data=solder noclprint noinfo method=type3 cl;
    class machine operator;
    model strength = machine /ddfm=satterth;
    random  operator machine*operator;
    lsmeans machine;
    estimate 'BLUP_1: Oper 3'
             intercept 4
             machine 1 1 1 1|
             operator 0 0 4
             machine*operator  0 0 1 0 0 1 0 0 1 0 0 1/divisor=4;

    estimate 'BLUP_2: Oper 3'
             intercept 4
             machine 1 1 1 1|
             operator 0 0 4/divisor=4;

    estimate 'LSMEAN for Mach 1'
             intercept 3
             machine 3 0 0 0|
             operator 1 1 1
             machine*operator  1 1 1 0 0 0 0 0 0 0 0 0/divisor=3;
title 'Analysis of Strength of Solder in Computer Chips using PROC MIXED';
run;
```

Fig. 6.35. SAS Example F13: Analysis of solder strength using PROC MIXED

In the case of the two-way mixed model, if the sample sizes are unequal, the denominator of tests associated with fixed effects will be a linear combination of mean squares. Thus, the Satterthwaite approximation is needed for inference associated with the fixed effects. It could be, in the balanced data case, that this approximation may never be needed.

It is recommended that ddfm=kr be used for models with multiple random effects when the sample sizes are unequal. The reason is that the kr option, which stands for Kenward-Roger, employs a method developed by Kenward and Roger (1997), which adjusts the standard errors as well as the degrees of freedom when an approximation is needed. In many balanced models, no standard error adjustment is required. When it is needed, not only the degrees of freedom associated with the respective statistics but also the value of the t-statistics (or F-statistics) themselves are affected.

The profiles of mean solder strengths in Fig. 6.34 indicate several attributes of the random effects in the model. First, the profiles are roughly parallel, indicating that there is no appreciable machine-operator interaction. Second, there is a variation of the profiles from an average, an indication of the variability among operators. The management of the company, always interested in efficiency, might be intrigued by Operator 3, who appears to have a higher performance level than the other operators in the study over all four machines, as observed from this graph. By conditioning on both the operator and machine*operator random effects, a predictable function measuring Operator 3's expected mean strength is obtained as $\mu + \bar{\alpha}_{\cdot} + B_3 + \overline{\alpha B}_{\cdot 3}$, where $\bar{\alpha}_{\cdot} = \frac{1}{4}\sum_i \alpha_i$ and $\overline{\alpha B}_{\cdot 3} = \frac{1}{4}\sum_i \alpha B_{i3}$. The estimate statement labeled "BLUP_1: Oper 3" requests that an estimate and a standard error of this BLUP be computed. Note that the option divisor=4 used with the estimate statement enables the user to specify the linear combination of the parameters needed by entering the coefficients as whole numbers (instead of fractions). The coefficients needed to specify the estimation of $\mu + \bar{\alpha}_{\cdot} + B_3 + \overline{\alpha B}_{\cdot 3}$ are

μ
1

α_1	α_2	α_3	α_4
1/4	1/4	1/4	1/4

B_1	B_2	B_3
0	0	1

αB_{11}	αB_{12}	αB_{13}	αB_{21}	αB_{22}	αB_{23}	αB_{31}	αB_{32}	αB_{33}	αB_{41}	αB_{42}	αB_{43}
0	0	1/4	0	0	1/4	0	0	1/4	0	0	1/4

Since the two random effects are independent, it is possible to condition only on one of the random effects. By conditioning on the operator effect alone (i.e., averaging the interaction over the population of all operators), another measure of mean strength which can be constructed for Operator 3 is the expected mean $\mu + \bar{\alpha}_{\cdot} + B_3$. The estimate statement labeled "BLUP_2:

6.5 Two-Way Mixed Effects Model

```
Analysis of Strength of Solder in Computer Chips using PROC MIXED      1
                           The Mixed Procedure
                         Type 3 Analysis of Variance

                                        Sum of
             Source            DF       Squares         Mean Square
             machine            3      12.458333         4.152778
             operator           2     160.333333        80.166667
             machine*operator   6      44.666667         7.444444
             Residual          12      45.500000         3.791667

                         Type 3 Analysis of Variance

Source            Expected Mean Square                    Error Term

machine           Var(Residual) + 2                       MS(machine*operator)
                  Var(machine*operator) + Q(machine)
operator          Var(Residual) + 2                       MS(machine*operator)
                  Var(machine*operator)
                  + 8 Var(operator)
machine*operator  Var(Residual) + 2                       MS(Residual)
                  Var(machine*operator)
Residual          Var(Residual)                              .

                         Type 3 Analysis of Variance

                                Error
             Source              DF       F Value     Pr > F
             machine              6        0.56       0.6619
             operator             6       10.77       0.0103
             machine*operator    12        1.96       0.1507
             Residual             .         .           .

                      Covariance Parameter Estimates   1

          Cov Parm           Estimate   Alpha    Lower      Upper

          operator            9.0903    0.05   -10.5784    28.7590
          machine*operator    1.8264    0.05    -2.6505     6.3032
          Residual            3.7917    0.05     1.9497    10.3320
```

Fig. 6.36. SAS Example F13: Output page 1 from PROC MIXED

Oper 3" results in the computation of its estimate and standard error. One would expect estimates of these two BLUPs to be similar, given that the interaction is not significant. As noted earlier, these BLUPs cannot be estimated using the `estimate` statement in `proc glm`. Finally, the `estimate` statement labeled "LSMEAN for Mach 1" requests that an estimate and a standard error of the predictable function $\mu + \alpha_1 + \bar{B}_{.} + \overline{\alpha B}_{1.}$ be computed. See the discussion on the output from this example below for an explanation for the inclusion of this statement.

Except for fit statistics and the confidence intervals for the variance components, all results that appear on pages 1 and 2 of the output from SAS

Example F13 (shown in Figs. 6.36 and 6.37) agree with those from proc glm. The estimates of the variance components **1** are also the same as those obtained from proc glm. The confidence intervals, except those for σ^2, are based on large sample results and may not be appropriate for small sample sizes used in this example. Intervals based on Satterthwaite approximation can be constructed as illustrated earlier in this section. SAS code given previously may be used for this purpose; however, it is recommended that the covtest option be added to the proc statement to obtain the required standard errors of the variance component estimates.

The estimates of the two BLUPs for Operator 3 are given on page 3 **2** (shown in Fig. 6.37), and as conjectured, they are similar in value. However, the standard errors differ and, the reason for this can be surmised from observing that their degrees of freedom are fractions. The fractions result from the fact that the standard errors are calculated using the Satterthwaite approximation. The two standard errors are obtained using different linear com-

```
       Analysis of Strength of Solder in Computer Chips using PROC MIXED        2

                                 The Mixed Procedure

                                    Fit Statistics

                        -2 Res Log Likelihood          100.7
                        AIC (smaller is better)        106.7
                        AICC (smaller is better)       108.2
                        BIC (smaller is better)        104.0

                              Type 3 Tests of Fixed Effects

                                 Num      Den
                      Effect      DF       DF     F Value     Pr > F

                      machine      3        6        0.56      0.6619

                                     Estimates

                                       Standard
          Label              Estimate    Error      DF    t Value   Pr > |t|

          BLUP_1: Oper 3      210.71    0.6775     12.7    311.00     <.0001
          BLUP_2: Oper 3      210.54    0.9343      6.78   225.34     <.0001
          LSMEAN for Mach 1   206.83    0.7949     12      260.18     <.0001

                                 Least Squares Means

                                           Standard
          Effect    machine    Estimate      Error     DF   t Value  Pr > |t|

          machine      1         206.83      2.0666   3.19   100.08   <.0001
          machine      2         207.17      2.0666   3.19   100.25   <.0001
          machine      3         206.67      2.0666   3.19   100.00   <.0001
          machine      4         208.50      2.0666   3.19   100.89   <.0001
```

Fig. 6.37. SAS Example F13: Output page 2 from PROC MIXED

binations of mean squares and their degrees of freedom are computed using approximations as illustrated in previous examples.

The "LSMEAN for Mach 1" is the BLUP $\mu + \alpha_1 + \bar{B}_. + \overline{\alpha B}_{1.}$. This is a BLUP for Machine 1 and is the expectation of the observations for Machine 1 conditioned on the observed operators. The `lsmeans machine;` statement, on the other hand, estimates the mean for Machine 1 as $\mu + \alpha_1$, the unconditional expectation of the observations for Machine 1. Thus, both estimated values are the same, 206.83, but the standard errors of the two estimators are different. The `lsmeans` statement gives the standard error as 2.0666 **3** and the `estimate` statement calculates it as 0.7949 **4**. Both are correct since they are estimating standard errors of different BLUPs. In comparison, the `lsmeans` statement in `proc glm` (see Fig. 6.33) computes the standard error as 0.7949 by (incorrectly) considering both `operator` and `machine*operator` as fixed effects.

The above analysis was repeated using the default estimation method of REML as well. The covariance estimates and their standard errors results of the F-tests for the variance components, the t-tests for the estimates of BLUPs, and the lsmeans estimates all remain unchanged, as the data set is balanced. The output includes an F-test for the fixed effect (machine) using Type 3 anova which is also exactly the same as obtained earlier. The only difference is that confidence intervals for the variance components calculated by `proc mixed` with `method=type3` are those based on large-sample standard errors and normal percentiles, whereas with REML, these are based on the chi-square distribution and the Satterthwaite approximation as displayed in Fig. 6.38.

Covariance Parameter Estimates

Cov Parm	Estimate	Alpha	Lower	Upper
operator	9.0903	0.05	2.2609	719.61
machine*operator	1.8264	0.05	0.4064	441.64
Residual	3.7917	0.05	1.9497	10.3320

Fig. 6.38. SAS Example F13: Confidence Intervals for the Variance Components with `method=reml` in PROC MIXED

6.5.3 Two-way mixed effects model: Nested classification

A two-way random model for an experiment with two random factors where one of the factors is *nested* in the other factor was considered in Section 6.4. In this subsection, a mixed model for an experiment with two factors A and B, where Factor B is nested in Factor A is discussed. Recall that the sampling of the levels of a nested factor takes place in a *hierarchical* manner: Thus, levels of Factor B are randomly sampled within each level of A, which are fixed. Suppose that in the machine-operator example, the primary interest is in the performance characteristics of, say, 4 types of machine and that it is possible to randomly sample 12 operators from all available operators. If three operators are assigned randomly to work on each machine and each of them perform the production operation on the assigned machine only, then the operators are nested within each machine.

This type of mixed effects model occurs naturally in animal and plant breeding experiments. As an example, consider the weight gain data from a pig-raising experiment described in Snedecor and Cochran (1989). The breeding values of five sires (bulls) are being evaluated. Each sire is mated to a random group of dams. Each mating produced a litter of pigs whose average daily gain was measured. Thus, the dams are nested within each sire (i.e., levels of "dam" are different for each level of "sire"). Here, the Sire genetic effect is a fixed effect and the additive Dam within Sire genetic effect is considered a random effect. Perhaps the sires are selected from several breeding lines being evaluated. A trait such as weight gain of offspring produced from each mating is usually used to compare genetic merits of breeding lines. The information gained from the variation of the Dam within Sire effect will lead to more precise estimation of the sire effects.

Model

An appropriate model for the situation described above is

$$y_{ijk} = \mu + \alpha_i + B_{ij} + \epsilon_{ijk}, \quad i = 1, \ldots, a;\ j = 1, \ldots, b;\ k = 1, \ldots, n$$

where α_i is the effect of level i Factor A, and it is assumed that the Factor B within A effect, B_{ij}, is an iid $N(0, \sigma_B^2)$ random variable, the random error ϵ_{ijk} is an iid $N(0, \sigma^2)$ random variable, and B_{ij} and ϵ_{ijk} are independently distributed. The parameters σ_B^2, and σ^2 are the "variance components" to be estimated in this model.

Estimation and Hypothesis Testing

An anova table that corresponds to the model is constructed using the same computational formulas used for the computation of the anova if the factors A and B within A were considered fixed. The sums of squares would be the same for effect A as in the anova table for the two-way crossed random effects model (given in Section 6.3). As noted in Section 6.4, the sum of squares for the effect B(A) is obtained by pooling the sum of squares for the effects B and AB and thus has $a(b-1)$ degrees of freedom. As with other models containing random effects considered so far, an additional column displaying the expected mean squares is included in the anova table:

SV	df	SS	MS	F	E(MS)
A	$a-1$	SS_A	MS_A	$MS_A/MS_{B(A)}$	$\sigma^2 + n\sigma_B^2 + bn\frac{\sum_i(\alpha_i-\bar{\alpha})^2}{(a-1)}$
B(A)	$a(b-1)$	$SS_{B(A)}$	$MS_{B(A)}$	$MS_{B(A)}/MSE$	$\sigma^2 + n\sigma_B^2$
Error	$ab(n-1)$	SSE	$MSE(=s^2)$		σ^2
Total	$abn-1$				

The F-statistic for the source A in the anova table tests the hypothesis of equality of factor A effects:

$$H_0: \alpha_1 = \alpha_2 = \cdots = \alpha_a \text{ versus } H_a: \text{ at least one inequality.}$$

or, equivalently,

$$H_0: \mu_1 = \mu_2 = \cdots = \mu_a \text{ versus } H_a: \text{ at least one inequality}$$

where $\mu_i = \mu + \alpha_i$, the mean of an observation at the ith level of A. The F-ratios shown in the analysis of variance table for sources of variation A and B(A) are constructed using the same principle described in Section 6.3 or 6.5.2. The expected mean squares of the denominator and numerator of the F-ratio for A will have the same expectation if the null hypothesis above holds and the numerator will have a larger expectation if it is not true since $\sum_i(\alpha_i - \bar{\alpha})^2$ is zero if $\alpha_1 = \alpha_2 = \cdots = \alpha_a$ and is positive otherwise. H_0 is rejected if the observed F-value exceeds the α upper percentile of an F-distribution with $df_1 = a - 1$ and $df_2 = a(b-1)$. The standard errors for means, pairwise mean comparisons, and linear contrasts of the fixed effects (or the means) are obtained by replacing s^2 with $s^2_{B(A)}$ (and the corresponding degrees of freedom with $a(b-1)$ where needed) in the fixed effects model formulas, where $s^2_{B(A)} = MS_{B(A)}$. For example, a $(1-\alpha)100\%$ confidence interval for $\mu_p - \mu_q$ (or, equivalently, $\alpha_p - \alpha_q$) is

$$(\bar{y}_{p..} - \bar{y}_{q..}) \pm t_{\alpha/2,\nu} \cdot s_{B(A)}\sqrt{\frac{2}{bn}}$$

where $t_{\alpha/2,\nu}$ is the upper $\alpha/2$ percentile point of a t-distribution with $\nu = a(b-1)$.

To test the hypothesis $H_0: \sigma_B^2 = 0$ versus $H_a: \sigma_B^2 > 0$, observe that the mean square for the B(A) effect, $MS_{B(A)}$, has expectation equal to σ^2 if $\sigma_B^2 = 0$ and, thus, MSE is the appropriate denominator for testing the nested effect. The null hypothesis is rejected for F-values that exceed the α upper percentile of an F-distribution with $df_1 = a(b-1)$ and $df_2 = ab(n-1)$. The *method of moments* estimates of variance components are obtained the usual way by setting the computed mean squares equal to their corresponding expected values. Solving the resulting equations

$$\text{MS}_{\text{B(A)}} = \sigma^2 + n\sigma_B^2$$
$$\text{MSE} = \sigma^2$$

give, the estimates $\hat{\sigma}^2 = \text{MSE}$ and $\hat{\sigma}_B^2 = (\text{MS}_{\text{B(A)}} - \text{MSE})/n$.

SAS Example F14

The data for the animal breeding experiment discussed in the introduction are given in Table 6.6. Recall that each of five sires from different breeding lines is mated to two randomly chosen dams, each mating producing a litter of pigs whose average daily gain was measured. Thus, the dams are nested within each sire. The Sire effect is a fixed effect and the Dam within Sire effect is a random effect.

Sire	Dam	Average Daily Gain	
1	1	2.77	2.38
	2	2.58	2.94
2	1	2.28	2.22
	2	3.01	2.61
3	1	2.36	2.71
	2	2.72	2.74
4	1	2.87	2.46
	2	2.31	2.24
5	1	2.74	2.56
	2	2.50	2.48

Table 6.6. Average daily gain of two pigs in each litter (Snedecor and Cochran, 1989)

An appropriate model for analyzing the average daily gain is given by

$$y_{ijk} = \mu + \alpha_i + B_{ij} + \epsilon_{ijk} \quad i = 1, \ldots, 5; \ j = 1, \ldots, 2; \ k = 1, \ldots, 2.$$

where the fixed effect of Sire i is α_i, the effect of Dam j within Sire i, B_{ij}, is assumed to be iid $N(0, \sigma_B^2)$, the random error ϵ_{ijk} is assumed to be iid $N(0, \sigma^2)$, and B_{ij} and ϵ_{ijk} are independent. Proc glm is used in the SAS Example F14 program, displayed in Fig. 6.39, to fit the above model to the average daily gain data, using the method of moments. The data are entered with the values for each sire-dam combination in separate data lines. Thus, each line can be held with a trailing @ modifier for the two replications to be read. They are written, along with the current sire-dam values, as two separate records in the SAS data set using a do-end structure and an output statement. In the proc step, both sire and dam are declared in the class

6.5 Two-Way Mixed Effects Model

```
data pigs;
input sire dam @;
do rep=1 to 2;
  input gain @;
  output;
end;
datalines;
1  1  2.77 2.38
1  2  2.58 2.94
2  1  2.28 2.22
2  2  3.01 2.61
3  1  2.36 2.71
3  2  2.72 2.74
4  1  2.87 2.46
4  2  2.31 2.24
5  1  2.74 2.56
5  2  2.50 2.48
;
run;

proc glm data=pigs;
   class sire dam;
   model gain = sire dam(sire);
   random  dam(sire)/test;
   lsmeans sire/stderr;
   lsmeans sire/stderr e=dam(sire);
   title 'Analysis of Average Daily Gain using PROC GLM';
run;
```

Fig. 6.39. SAS Example F14: Analysis of average daily gain using PROC GLM

statement. The nested effect B_{ij} is represented in the model statement as dam(sire), the notation used in the SAS language to represent a nested effect. It is also identified as a random effect by including it in the random statement as any other random effect (see SAS Example F8 in Section 6.4).

Pages 1 and 2 of the output from the SAS Example F14 program contain factor level information, total SS, and the standard output that include Types I and III SS and corresponding F-statistics calculated by proc glm and are not shown. The Type III sums of squares are used to compute the expected mean squares (see page 3 of Fig. 6.40 **1**) of random effects. The test option in the random statement will result in the use of dam(sire) mean square as the denominator to obtain the appropriate statistic to test the sire effect, as seen on the output page 4 **2**. This will produce the correct F-statistic to test the sire effects hypothesis $H_0 : \alpha_1 = \alpha_2 = \cdots = \alpha_5$ versus H_a : at least one inequality. The anova table is

SV	df	SS	MS	F	p-value	E(MS)
Sire	4	0.09973	0.02493	0.22	0.9155	$\sigma^2 + 2\sigma_B^2 + \frac{10}{3}\sum_i(\alpha_i - \bar{\alpha})^2$
Dam(Sire)	5	0.56355	0.11271	2.91	0.0707	$\sigma^2 + 2\sigma_B^2$
Error	10	0.38700	0.03870			
Total	19	1.05028				

```
              Analysis of Average Daily Gain using PROC GLM                  3

                             The GLM Procedure

     Source                Type III Expected Mean Square  ▮1

     sire                  Var(Error) + 2 Var(dam(sire)) + Q(sire)

     dam(sire)             Var(Error) + 2 Var(dam(sire))

              Analysis of Average Daily Gain using PROC GLM                  4
                             The GLM Procedure
              Tests of Hypotheses for Mixed Model Analysis of Variance

Dependent Variable: gain

Source                       DF    Type III SS    Mean Square    F Value    Pr > F

sire                          4       0.099730       0.024932       0.22    0.9155

Error: MS(dam(sire))          5       0.563550       0.112710 ▮2


Source                       DF    Type III SS    Mean Square    F Value    Pr > F

dam(sire)                     5       0.563550       0.112710       2.91    0.0707

Error: MS(Error)             10       0.387000       0.038700
```

Fig. 6.40. SAS Example F14: Analysis of average daily gain using PROC GLM (output pages 3 and 4)

The p-value of .92 indicates that there are no significant differences among the sire effects. However, the p-value for the test of dam(sire) variance component $H_0 : \sigma_B^2 = 0$ versus $H_a : \sigma_B^2 > 0$ is less than $\alpha = .10$ so it is rejected at 1%. The estimates of the variance components are obtained using the method of moments as usual. These are obtained by solving

$$\sigma^2 + 2\sigma_B^2 = 0.11271$$

$$\sigma^2 = 0.0387$$

which give $\hat{\sigma}^2 = 0.0387$ and $\hat{\sigma}_B^2 = (0.112710 - 0.0387)/2 = 0.037$. Finally, the first lsmeans statement produces estimates of the BLUPs $\mu + \alpha_i + \bar{B}_{i\cdot}$ (same as the *conditional means* of the five sires) for $i = 1, \ldots, 5$ and their standard errors ▮3 . By overriding the default error term using the e=dam(sire) option, the second lsmeans statement produces the estimates of $\mu + \alpha_i$ and the correct standard errors; ▮4 that is, these will be identical to the estimates of unconditional means and their standard errors. Note that for the general case, the unconditional variance of $\bar{y}_{i\cdot\cdot}$ is equal to $\sigma_B^2/b + \sigma^2/bn$. Thus, the standard error of $\bar{y}_{i\cdot\cdot}$ is $\sqrt{0.037/2 + 0.0387/4} = 0.1679$ as given in this output (see Fig. 6.41).

```
                Analysis of Average Daily Gain using PROC GLM                5
                              Least Squares Means  3

                                         Standard
            sire      gain LSMEAN           Error      Pr > |t|

             1          2.66750000       0.09836158     <.0001
             2          2.53000000       0.09836158     <.0001
             3          2.63250000       0.09836158     <.0001
             4          2.47000000       0.09836158     <.0001
             5          2.57000000       0.09836158     <.0001


                Analysis of Average Daily Gain using PROC GLM                6
                              Least Squares Means  4

       Standard Errors and Probabilities Calculated Using the Type III MS for
                            dam(sire) as an Error Term

                                         Standard
            sire      gain LSMEAN           Error      Pr > |t|

             1          2.66750000       0.16786155     <.0001
             2          2.53000000       0.16786155     <.0001
             3          2.63250000       0.16786155     <.0001
             4          2.47000000       0.16786155     <.0001
             5          2.57000000       0.16786155     <.0001
```

Fig. 6.41. SAS Example F14: Analysis of Average Daily Gain using PROC GLM (output pages 5 and 6)

A $(1-\alpha)100\%$ confidence interval based on the Satterthwaite approximation can be constructed for σ_B^2 using a formula similar to (6.1) in Section 6.2. This is implemented in the SAS data step

```
data cint;
   alpha=.05; a=5; b=2; n=2;
   msb=0.112710; s2=0.0387; sb2=0.037;
   nu=(n*sb2)**2/(msb**2/(a*(b-1))+ s2**2/(a*b*(n-1)));
   L= (nu*sb2)/cinv(1-alpha/2,nu);
   U= (nu*sb2)/cinv(alpha/2,nu);
   put nu=    L=     U=;
run;
```

and gives the 95% confidence interval (0.010106, 1.38352) for this example.

A SAS proc mixed step (see Fig. 6.42) is added to the SAS Example F14 program to illustrate the use of the estimate statements in proc mixed to obtain the best estimates of the linear function $\mu + \alpha_i$ and the predictable function $\mu + \alpha_i + \bar{B}_{i\cdot}$. Only those parts of the output that are relevant are reproduced here (see Fig. 6.43). First, note that, by default, proc mixed uses the REML method and calculates the confidence intervals on variance components based in the Satterthwaite approximation 5 . The 95% interval for σ_B^2 is close to the one calculated previously using the results from proc glm. Second, note that lsmeans produces the same estimates as with proc

```
proc mixed data=pigs noclprint noinfo cl;;
   class sire dam;
   model gain = sire/ddfm=satterth;
   random dam(sire);
   lsmeans sire;
   estimate 'Sire 1 BLUE'
            intercept 1 sire 1 0 0 0 0;
   estimate 'Sire 1 BLUP'
            intercept 2
            sire    2 0 0 0 0|
            dam(sire) 1 1 0 0 0 0 0 0 0 0/divisor=2;
   title 'Analysis of Average Daily Gain using PROC MIXED';
run;
```

Fig. 6.42. SAS Example F14: Analysis of average daily gain using PROC MIXED

glm; that is, the standard error is that of the unconditional estimate of $\mu + \alpha_1$, 0.1679 **7**. The two estimate statements further clarify how the two BLUPs differ: The standard error of the BLUP for Sire 1 $\mu + \alpha_1 + \bar{B}_{1\cdot}$ is 0.09836 **6** and is different from that of the above. This estimate gives the performance

```
                    Covariance Parameter Estimates 5

            Cov Parm      Estimate    Alpha     Lower    Upper

            dam(sire)     0.03701     0.05      0.01011  1.3825
            Residual      0.03870     0.05      0.01889  0.1192

                              Estimates 6

                              Standard
    Label         Estimate    Error       DF   t Value   Pr > |t|

    Sire 1 BLUE   2.6675      0.1679       5   15.89     <.0001
    Sire 1 BLUP   2.6675      0.09836     10   27.12     <.0001

                         Least Squares Means 7

                                 Standard
    Effect   sire   Estimate     Error       DF   t Value   Pr > |t|

    sire     1      2.6675       0.1679       5   15.89     <.0001
    sire     2      2.5300       0.1679       5   15.07     <.0001
    sire     3      2.6325       0.1679       5   15.68     <.0001
    sire     4      2.4700       0.1679       5   14.71     <.0001
    sire     5      2.5700       0.1679       5   15.31     <.0001

                              Estimates 8

            Label             Lower     Upper

            Sire 1 BLUE       2.2360    3.0990
            Sire 1 BLUP       2.4483    2.8867
```

Fig. 6.43. SAS Example F14: Analysis of average daily gain data with PROC MIXED

of a particular sire averaged only over the set of dams he was mated with. If the `estimate` statement option `cl` is included as in the following

```
estimate 'Sire 1 BLUP'
         intercept 2
         sire       2 0 0 0 0|
         dam(sire)  1 1 0 0 0 0 0 0 0 0/divisor=2 cl;
```

a t-based confidence interval (95%, by default) is calculated instead of the t-test. Recall that the `divisor=` option allows the user to avoid entering fractional coefficients when formulating estimate statements. In this example, every coefficient specified must be divided by two to obtain the actual linear function of the parameters estimated. This output (obtained in separate run) is reproduced at the bottom **8** of Fig. 6.43. Obviously, the interval for the BLUP is narrower because the estimate is an average only over the three dams nested in sire 1 and not over the entire population of dams.

SAS Example F15

Four manufacturing processes of comparable costs are being tried out for obtaining increased surface quality of a precision machine part. The aim is to obtain the best roughness average values (known as R_a in the industry) at the lowest possible cost. The costs depend on the milling, grinding, and polishing activities involved in each of the four processes. Each of the processes is being run at each of four different plants by three different teams selected at each plant at random from the plant's workforce. Three specimens of the part are produced by each team using a process in random order and the surface finish is measured in microinches. The data are displayed in Table 6.7.

Process	Team		
	1	2	3
1	29	34	16
	12	24	17
	21	29	24
2	42	35	36
	37	39	18
	32	30	23
3	38	23	16
	35	36	27
	28	30	19
4	16	10	12
	13	16	17
	25	20	18

Table 6.7. Surface Finish (in μ-inches)

The model is

$$y_{ijk} = \mu + \alpha_i + B_{ij} + \epsilon_{ijk}, \quad i = 1, \ldots, 4; \ j = 1, \ldots, 3; \ k = 1, \ldots, 3$$

where α_i represents the fixed effect of Process i, B_{ij} is the effect of Team j within each Process i assumed to be iid $N(0, \sigma_B^2)$, and the random error ϵ_{ijk} is assumed to be iid $N(0, \sigma^2)$. Also B_{ij} and ϵ_{ijk} are assumed to be independent. The SAS Example F15 program, displayed in Fig. 6.44, is used to fit the above model to the surface finish data, using the default method REML. Since the data set is balanced, the estimates will be the same as those obtained from the method of moments.

```
data parts;
input process $ @;
do team =1 to 3;
input finish @;
output;
end;
datalines;
1 29 34 16
1 12 24 17
1 21 29 24
2 42 35 36
2 37 39 18
2 32 30 23
3 38 23 16
3 35 36 27
3 28 30 19
4 16 10 12
4 13 16 17
4 25 20 18
;
run;

proc mixed data=parts noclprint noinfo cl covtest;
    class process team;
    model finish = process /ddfm=satterth;
    random team(process);
    lsmeans process/diff cl adj=bon;
    estimate 'Process 4 mean:'
            intercept 3
            process 0 0 0 3|
            team(process) 0 0 0 0 0 0 0 0 1 1 1/divisor=3;
    title 'Analysis of Surface Finish by Process using PROC MIXED';
run;
```

Fig. 6.44. SAS Example F15: Analysis of surface finish data with PROC MIXED

The first part of the output is shown in Fig. 6.45. The estimates of the variance components and confidence intervals based on the Satterthwaite approximation are provided **1**. According to the p-value (.1317), the variance component team(process) is not significantly different from zero. However, the z-tests for variance components based on large-sample results, may not be valid for the numbers of levels of random factors used in this experiment.

6.5 Two-Way Mixed Effects Model

```
                Analysis of Surface Finish by Process using PROC MIXED                    1

                               The Mixed Procedure

                                 Iteration History

         Iteration    Evaluations    -2 Res Log Like         Criterion

             0             1           222.99895630
             1             1           220.51004767          0.00000000

                              Convergence criteria met.

                          Covariance Parameter Estimates  ①

                              Standard        Z
Cov Parm         Estimate      Error        Value      Pr Z    Alpha    Lower     Upper

team(process)    15.9722      14.2804       1.12      0.1317   0.05    4.7596    335.95
Residual         35.3056      10.1918       3.46      0.0003   0.05   21.5255    68.3270

                                   Fit Statistics

                    -2 Res Log Likelihood                220.5
                    AIC (smaller is better)              224.5
                    AICC (smaller is better)             224.9
                    BIC (smaller is better)              225.5

                         Type 3 Tests of Fixed Effects  ②

                               Num      Den
                Effect         DF       DF      F Value     Pr > F

                process         3        8       5.19       0.0279

                                    Estimates  ③

                                Standard
      Label            Estimate   Error      DF     t Value    Pr > |t|

      Process 4 mean:  16.3333   1.9806      24      8.25      <.0001
```

Fig. 6.45. SAS Example F15: Analysis of surface finish data with PROC MIXED (page 1)

From an analysis (not shown) using `proc mixed` using the `method=type3` option, the following anova table was constructed:

SV	df	SS	MS	F	p-value	E(MS)
Process	3	1295.639	431.880	5.19	0.0279	$\sigma^2 + 3\sigma_B^2 + 3\sum_i(\alpha_i - \bar{\alpha})^2$
Team(Process)	8	665.778	83.222	2.36	0.0498	$\sigma^2 + 3\sigma_B^2$
Error	24	847.333	35.306			σ^2
Total	35	2808.750				

The resulting F-test for the variance component `team(process)` rejects $\sigma_B^2 = 0$ at $\alpha = .05$.

464   6 Analysis of Variance: Random and Mixed Effects Models

The four process means are found to be significantly different at $\alpha = .05$ using the F-test based on Type 3 SS ❷ (p-value=0.0279). The coefficients specified in the `estimate` statement correspond to the linear combination of parameters needed to estimate the conditional mean $\mu + \alpha_4 + \bar{B}_{4\cdot}$ for process 4 conditioned on the nested effect ❸. The standard error of this estimate is $\sqrt{35.3056/9} = 1.9806$. This is different from the standard error estimate of the process 4 unconditional mean as illustrated below.

```
             Analysis of Surface Finish by Process using PROC MIXED         2

                              The Mixed Procedure

                             Least Squares Means ❹

                       Standard
Effect  process Estimate  Error  DF t Value Pr > |t| Alpha   Lower    Upper

process   1     22.8889   3.0409  8   7.53  <.0001   0.05  15.8766  29.9012
process   2     32.4444   3.0409  8  10.67  <.0001   0.05  25.4322  39.4567
process   3     28.0000   3.0409  8   9.21  <.0001   0.05  20.9877  35.0123
process   4     16.3333   3.0409  8   5.37  0.0007   0.05   9.3211  23.3456

                    Differences of Least Squares Means ❺

                                   Standard
Effect  process _process Estimate   Error   DF  t Value  Pr > |t|  Adj P

process   1        2     -9.5556   4.3004    8   -2.22   0.0570   0.3420
process   1        3     -5.1111   4.3004    8   -1.19   0.2687   1.0000
process   1        4      6.5556   4.3004    8    1.52   0.1659   0.9955
process   2        3      4.4444   4.3004    8    1.03   0.3316   1.0000
process   2        4     16.1111   4.3004    8    3.75   0.0057   0.0339
process   3        4     11.6667   4.3004    8    2.71   0.0265   0.1592

                    Differences of Least Squares Means ❻
                                                                Adj       Adj
Effect  process _process Adjustment  Alpha   Lower    Upper    Lower     Upper

process   1        2     Bonferroni   0.05 -19.4724   0.3613  -24.5163   5.4052
process   1        3     Bonferroni   0.05 -15.0280   4.8057  -20.0718   9.8496
process   1        4     Bonferroni   0.05  -3.3613  16.4724   -8.4052  21.5163
process   2        3     Bonferroni   0.05  -5.4724  14.3613  -10.5163  19.4052
process   2        4     Bonferroni   0.05   6.1943  26.0280    1.1504  31.0718
process   3        4     Bonferroni   0.05   1.7498  21.5835   -3.2941  26.6274
```

Fig. 6.46. SAS Example F15: Analysis of surface finish data with PROC MIXED (page 2)

The second part of the output, shown in Fig. 6.46, displays the results of the `lsmeans` statement. It shows the standard errors, t-tests, and t-based confidence intervals of the estimates of the unconditional means $\mu + \alpha_i$ ❹. The `diff cl adj=bon` options request t-tests and 95% confidence intervals for pairwise differences in process means (or effects). The p-values reported for pairwise t-tests ⑤ and confidence intervals for the differences ⑥ are both adjusted for simultaneous inference using the Bonferroni method.

Recall that the standard errors are computed using the formula

$$\sqrt{(\hat{\sigma}^2 + 3\hat{\sigma}_B^2)/9} = \sqrt{83.222/9} = 3.0409$$

since it is easily shown that $\text{Var}(\bar{y}_{4..}) = (\sigma^2 + 3\sigma_B^2)/9$ and $\text{MS}_{\text{B(A)}}$ estimates $\sigma^2 + 3\sigma_B^2$. In the part of the output in Fig. 6.46 containing the t-tests, the Adj P column contains the p-values Bonferroni-adjusted for multiple testing. This is done by simply multiplying the standard p-values by the number of comparisons made (i.e., $a(a-1)/2 = 6$, in this example). Similarly, for calculating the Bonferroni adjusted confidence intervals, the upper tail $(1-\alpha/12)100$ percentile of the t-distribution with eight degrees of freedom replaces the $(1-\alpha/2)100$ percentile used for calculating the usual one-at-time t-based intervals. The SAS function call quantile('T',1-alpha,df) can be used to compute these percentiles in a data step.

Using the adjusted p-values, it is found that only processes 2 and 4 means are significantly different. The Bonferroni procedure controls the *maximum experimentwise error rate* (i.e., the probability of making at least one type 1 error at the specified α value). This is somewhat conservative, but the same result is obtained with Tukey's method. If no adjustment is made for multiple testing, processes 3 and 4 are significantly different in addition to processes 2 and 4. The conclusion that may be made from this experiment is that process 4 produced a significantly better surface finish than process 2, which produced the worst result. There is not enough evidence in the data to differentiate the other two process means from those of either processes 2 or 4.

6.6 Models with Random and Nested Effects for More Complex Experiments

Several random and mixed models commonly used for the analysis of experiments were discussed in previous sections. The levels of factors studied in such experiments were combinations of fixed, random, or nested effects. In order to keep the introduction to these models to a reasonable level of complexity but be still informative, the examples of experiments considered were somewhat straightforward. In this section, several experiments that require more complicated models than those discussed so far are considered in order to build on the knowledge acquired from that introduction. In particular, several experiments that involve different combinations of factors or different randomization procedures from those discussed previously are introduced in several subsections. The presentation is slightly different from the previous sections in that instead of introducing the general model and inference procedures for a class of models and then illustrating with an example, the discussion in the following subsections is motivated by a specific example provided to illustrate the class of models.

6.6.1 Models for nested factorials

The so-called nested factorial experiments involve various combinations of crossed and nested factors that are either fixed or random. In the simplest set up, two factors are crossed in a factorial arrangement and a third factor is nested within combinations of those two factors. Instead, the third factor may be nested within only one of the two factors, or the third factor may be nested in a completely different factor. In any case, using the arrangement of the treatment factors and the experimental design structure, one should be able to formulate an appropriate model using the principles illustrated in previous sections of this chapter. Once the appropriate model is determined and the status of each effect in the model, whether fixed or random, is declared, **proc mixed** may be used to perform the computations necessary to analyze the data. The following experiment discussed in Montgomery (1991), provides a typical example of this type of model.

SAS Example F16

A study is designed to find ways to improve the speed of the assembly operation involved in the manual insertion of electronic components on printed circuit boards. An industrial engineer is comparing three assembly fixtures and two workplace layouts. It was decided to select four operators randomly to perform the assembly operation for each fixture-layout combination. Since the workplaces are in different locations within the plant, it is not feasible to use the *same* four operators for each layout. Thus, four different operators are chosen for each of the layouts. Two replications are obtained for each

Fixture	Layout							
	1				2			
	Operator				Operator			
	1	2	3	4	1	2	3	4
1	22	23	28	25	26	27	28	24
	24	24	29	23	28	25	25	23
2	30	29	30	27	29	30	24	28
	27	28	32	25	28	27	23	30
3	25	24	27	26	27	26	24	28
	21	22	25	23	25	24	27	27

Table 6.8. Circuit assembly time data

6.6 Models with Random and Nested Effects for More Complex Experiments

treatment combination. The assembly times are measured in seconds and are shown in Table 6.8.

In this experiment, operators are nested within levels of `layouts` (i.e., different set of operators for each layout). The effect of `operator within layout` is random. The two factors `fixture` and `layout` are crossed because levels of layout occur with each level of fixture. The effects of fixture, layout, and their interaction `fixture`×`layout` are fixed effects. Thus, it is clear that operators are nested in one (layout) of those two crossed factors.

Since operators are nested in layouts, `layout` × `operator(layout)` interaction is not tenable; however, it is possible for `fixture` to interact with `operator(layout)`; hence, to be able to test whether it is significant, a term corresponding to this interaction is included in the model. Thus, the appropriate model is

$$y_{ijk\ell} = \mu + \alpha_i + \beta_j + G_{k(j)} + \gamma_{ij} + \alpha G_{ik(j)} + \epsilon_{ijk\ell}$$

with $i = 1, 2, 3$ (fixtures), $j = 1, 2$ (layouts), $k = 1, 2, 3, 4$ (operators), $\ell = 1, 2$ (reps), where α_i is the effect of the ith operator, β_j is the effect of the jth layout, $G_{k(j)}$ is the effect of the operator k within layout j, assumed to be iid $N(0, \sigma_G^2)$ random variables, γ_{ij} is the interaction effect of the ith level of fixture and the jth level of layout, $\alpha G_{ik(j)}$ is the interaction effect of the ith level of fixture and the operator k within layout j, assumed to be iid $N(0, \sigma_{\alpha G}^2)$ random variables, and $\epsilon_{ijk\ell}$ are the usual random errors assumed to be iid $N(0, \sigma^2)$ random variables. In addition, the three random effects are mutually independent. Thus, the unconditional mean $\mu_{ij} = E(y_{ijk\ell}) = \alpha_i + \beta_j + \gamma_{ij}$ contains the fixed effects only, and the covariance matrix of $\mathbf{y} = (y_{ijk\ell})$ is defined using the variance components.

The SAS Example F16 program, displayed in Fig. 6.47, is used to fit the above model to the circuit assembly data, using the method of moments. The data are input using the method illustrated in several examples previously. The set of responses for the four operators for each fixture-layout combination, are read from a data line using a do-loop and each combination of fixture, layout, operator, and the response values written to the SAS data set using an output statement. The trailing @ symbol is used to hold the line after accessing the fixture and layout values as well as for repeatedly reading the responses for the four operators. The `method=type3` option in the proc statement requests the method of moments to be used. The `cl` option calculates confidence intervals for variance components based on the Satterthwaite approximation. The class statement in `proc mixed` step declares `fixture`, `layout`, and `operator` as classification variables, in that order. The `model` statement includes the fixed effects `fixture`, `layout` and `fixture*layout` and the `random` statement, the random effects `operator(layout)` and `fixture*operator(layout)`. Explanations of the `lsmeans` and `estimate` statements included in the proc step are provided in the discussion of the output produced from this program.

The following anova table is constructed using the output from SAS Example F16 displayed in Fig. 6.48.

```
data circuit;
input fixture layout @;
do operator=1 to 4;
  input time @;
  output;
end;
datalines;
1  1  22  23  28  25
1  2  26  27  28  24
1  1  24  24  29  23
1  2  28  25  25  23
2  1  30  29  30  27
2  2  29  30  24  28
2  1  27  28  32  25
2  2  28  27  23  30
3  1  25  24  27  26
3  2  27  26  24  28
3  1  21  22  25  23
3  2  25  24  27  27
;
run;

proc mixed data=circuit noclprint noinfo method=type3 cl;
    class fixture layout operator;
    model time = fixture layout fixture*layout/ddfm=satterth;
    random operator(layout) fixture*operator(layout);
    lsmeans fixture/pdiff cl adj=tukey;
    estimate 'F1-F2 for Op 4(1)'
            fixture 1 -1 0|
            fixture*operator(layout) 0 0 0 1   0 0 0 0
                                     0 0 0 -1  0 0 0 0
                                     0 0 0 0   0 0 0 0;
    estimate 'F1-F2 for Op 1(2)'
            fixture 1 -1 0|
            fixture*operator(layout) 0 0 0 0   1 0 0 0
                                     0 0 0 0  -1 0 0 0
                                     0 0 0 0   0 0 0 0;
    title 'Analysis of Nested Factorials using PROC MIXED';
run;
```

Fig. 6.47. SAS Example F16: Analysis of circuit assembly time data with PROC MIXED

SV	df	SS	MS	F	p-value	$E(\text{MS})$
F	2	82.7917	41.3958	7.55	.0076	$\sigma^2 + 2\sigma^2_{\alpha G} + Q(\alpha, \gamma)$
L	1	4.0833	4.0833	0.34	.5807	$\sigma^2 + 2\sigma^2_{\alpha G} + 6\sigma^2_G + Q(\beta, \gamma)$
F×L	2	19.0417	9.5208	1.74	.2178	$\sigma^2 + 2\sigma^2_{\alpha G} + Q(\gamma)$
O(L)	6	71.9167	11.9861	2.18	.1174	$\sigma^2 + 2\sigma^2_{\alpha G} + 6\sigma^2_G$
F×O(L)	12	65.8333	5.4861	2.35	.0360	$\sigma^2 + 2\sigma^2_{\alpha G}$
Error	24	56.0000	2.3333			σ^2
Total	47	299.6667				

6.6 Models with Random and Nested Effects for More Complex Experiments

The Type 3 sums of squares **1**, Expected Mean Squares **2**, F-statistics and p-values **3** available in page 1 of the output are used for this purpose. The method of moments estimates for the variance components can be obtained using the equations

$$\sigma^2 + 2\,\sigma^2_{\alpha G} + 6\,\sigma^2_G = 11.9861$$

```
               Analysis of Nested Factorials using PROC MIXED             1
                           The Mixed Procedure
                       Type 3 Analysis of Variance  1

                                        Sum of
             Source              DF    Squares       Mean Square

             fixture              2    82.791667      41.395833
             layout               1     4.083333       4.083333
             fixture*layout       2    19.041667       9.520833
             operator(layout)     6    71.916667      11.986111
             fixtu*operat(layout) 12   65.833333       5.486111
             Residual            24    56.000000       2.333333

                       Type 3 Analysis of Variance  2

Source                Expected Mean Square

fixture               Var(Residual) + 2 Var(fixtu*operat(layout))
                      + Q(fixture,fixture*layout)
layout                Var(Residual) + 2 Var(fixtu*operat(layout)) + 6
                      Var(operator(layout)) + Q(layout,fixture*layout)
fixture*layout        Var(Residual) + 2 Var(fixtu*operat(layout))
                      + Q(fixture*layout)
operator(layout)      Var(Residual) + 2 Var(fixtu*operat(layout))
                      + 6 Var(operator(layout))
fixtu*operat(layout)  Var(Residual) + 2 Var(fixtu*operat(layout))
Residual              Var(Residual)

                       Type 3 Analysis of Variance  3
                                                 Error
Source                Error Term                  DF   F Value  Pr > F

fixture               MS(fixtu*operat(layout))    12    7.55    0.0076
layout                MS(operator(layout))         6    0.34    0.5807
fixture*layout        MS(fixtu*operat(layout))    12    1.74    0.2178
operator(layout)      MS(fixtu*operat(layout))    12    2.18    0.1174
fixtu*operat(layout)  MS(Residual)                24    2.35    0.0360
Residual              .                            .     .        .

                   Covariance Parameter Estimates  4

        Cov Parm             Estimate   Alpha    Lower    Upper

        operator(layout)      1.0833    0.05   -1.2927   3.4593
        fixtu*operat(layout)  1.5764    0.05   -0.7156   3.8684
        Residual              2.3333    0.05    1.4226   4.5157
```

Fig. 6.48. SAS Example F16: Analysis of circuit assembly time data (page 1)

$$\sigma^2 + 2\sigma_{\alpha G}^2 = 5.4861$$

$$\sigma^2 = 2.3333$$

giving the estimates $\hat{\sigma}^2 = 2.3333$, $\hat{\sigma}_{\alpha G} = 1.5764$, and $\hat{\sigma}_G^2 = 1.0833$, agreeing with the values on page 1. Large-sample confidence intervals are available from page 1 **4**; intervals based on Satterthwaite approximations can be constructed as usual.

From the anova table, the effects of assembly fixtures are significantly different at $\alpha = .05$. However, the workplace layouts have no significant effects on mean assembly times and there is no significant interaction between layouts and fixtures. The variance component $\sigma_{\alpha G}^2$ measuring the interaction between fixtures and operators(layout) is found to be significantly different from zero. This indicates that the differences in effects of fixtures varies among operators within layouts although when averaged over the fixtures it is not significantly different from zero.

```
          Analysis of Nested Factorials using PROC MIXED              2

                       The Mixed Procedure

                         Fit Statistics

          -2 Res Log Likelihood            187.3
          AIC (smaller is better)          193.3
          AICC (smaller is better)         194.0
          BIC (smaller is better)          193.6

                 Type 3 Tests of Fixed Effects

                        Num      Den
          Effect         DF       DF    F Value   Pr > F

          fixture         2       12     7.55     0.0076
          layout          1        6     0.34     0.5807
          fixture*layout  2       12     1.74     0.2178

                         Estimates  5
                              Standard
    Label            Estimate    Error      DF   t Value   Pr > |t|

    F1-F2 for Op 4(1)  -3.3340   1.4441    11.7   -2.31    0.0402
    F1-F2 for Op 1(2)  -2.8605   1.8744     6.79  -1.53    0.1721

                      Least Squares Means
                          Standard
    Effect  fixture  Estimate   Error   DF  t Value  Pr > |t|  Alpha  Lower  Upper

    fixture    1     25.2500   0.6916  15.5  36.51   <.0001    0.05  23.78  26.72
    fixture    2     27.9375   0.6916  15.5  40.40   <.0001    0.05  26.47  29.41
    fixture    3     25.0625   0.6916  15.5  36.24   <.0001    0.05  23.59  26.53
```

Fig. 6.49. SAS Example F16: Analysis of circuit assembly time data (page 2)

6.6 Models with Random and Nested Effects for More Complex Experiments

Although it is possible to compare the differences among fixtures unconditionally (i.e., averaging over the population of operators), the significant interaction suggests that some operators may be more effective than others. Thus, to compare mean performance using the assembly fixtures p and q (say) for each operator $k(j)$ in the experiment, predictable functions of the following type are needed:

$$\alpha_p - \alpha_q + \bar{\gamma}_{p.} - \bar{\gamma}_{q.} + \alpha G_{pk(j)} - \alpha G_{qk(j)}$$

Two examples of this type of BLUP are included in the SAS program for illustration. Note that the estimable function of the fixed effects parameters includes the interaction terms $\bar{\gamma}_{p.} - \bar{\gamma}_{q.}$; however, specifying this part can be omitted from the `estimate` statements, as SAS automatically includes the coefficients for the interaction term. Note that the combination $k(j)$ signifies a different operator for all combinations of values of k (levels of operator) and j (levels of layout). Again, the subscripts change in accordance with lexical ordering. Here, note that the effect `layout` occurs before `operator` in the `class` statement; thus, operator subscripts change faster. Thus, the values of $k(j)$ for the random effect parameter $\alpha G_{ik(j)}$ occur in the order 1(1), 2(1), 3(1), 4(1), 1(2), 2(2), 3(2), 4(2) for each $i = 1, 2, 3$; that is, the `layout × operator(layout)` interaction parameter vector will have 24 elements. These 24 positions are coded as either $0, 1$, or -1 when coding the $\alpha G_{14(1)} - \alpha G_{24(1)}$ part in the comparison:

```
estimate 'F1-F2 for Op 4(1)'
         fixture 1 -1 0|
         fixture*operator(layout) 0 0 0 1   0 0 0 0
                                  0 0 0 -1  0 0 0 0
                                  0 0 0 0   0 0 0 0;
```

The t-statistics and p-values **5** from the estimate statements used in SAS Example F16 program are shown in Fig. 6.49. These tests show that there is a significant difference (at $\alpha = .05$) between the effects of fixtures 1 and 2 on the average speed of assembly for Operator 4 in Layout 1 but not for Operator 1 in Layout 2. Other simple effects of this type may be similarly tested. Fig. 6.50 contains the results of the `lsmeans` statement that produce t-tests **6** and confidence intervals **7** for pairwise differences of fixture means (or effects) adjusted for multiple testing using Tukey's method. As usual these are estimates of differences of the unconditional means for pairs of fixtures p and q averaged over the layouts:

$$\alpha_p - \alpha_q + \bar{\gamma}_{p.} - \bar{\gamma}_{q.}.$$

`Proc mixed` calculates the standard errors of these differences correctly. Since the $F \times L$ interaction is not significant, the above results can be usefully interpreted. They show that Fixture 2 mean is significantly larger than the means of both Fixtures 1 and 3 but that those of Fixtures 1 and 3 are similar.

```
              Analysis of Nested Factorials using PROC MIXED               3

                             The Mixed Procedure

                     Differences of Least Squares Means  6

                                      Standard
 Effect    fixture  _fixture  Estimate    Error    DF   t Value   Pr > |t|

 fixture     1         2      -2.6875    0.8281    12    -3.25     0.0070
 fixture     1         3       0.1875    0.8281    12     0.23     0.8247
 fixture     2         3       2.8750    0.8281    12     3.47     0.0046

                     Differences of Least Squares Means  7

 Effect    fixture                                        Adj     Adj
          _fixture  Adjustment   Adj P  Alpha  Lower    Upper   Lower   Upper

 fixture  1    2   Tukey-Kramer  0.0179  0.05  -4.4918  -0.8832  -4.8967  -0.4783
 fixture  1    3   Tukey-Kramer  0.9722  0.05  -1.6168   1.9918  -2.0217   2.3967
 fixture  2    3   Tukey-Kramer  0.0119  0.05   1.0707   4.6793   0.6658   5.0842
```

Fig. 6.50. SAS Example F16: Analysis of circuit assembly time data (page 3)

6.6.2 Models for split-plot experiments

The split-plot design is often used when one factor is more readily applied to large experimental units, called whole-plots (or main plots), and another factor can be applied to smaller experimental units within the whole-plot called subplots. Another frequent use of a split-plot design is when more precision is needed for comparisons among the levels of one factor than for the other factor. To ensure that a factor is more accurately estimated, its levels are applied to the subplots so that it will naturally have more replications. Note that in this introduction, only two-way factorials are considered, though treatments at each level (whole-plots or subplots) may be in any arrangement. For example, the whole-plot treatments (or sub-plot treatments or both) themselves may be factorial combinations of other factors.

A typical example is a field experiment in which irrigation levels are applied to larger plots and a factor like crop varieties or levels of fertilizer are randomized among smaller plots within each larger plot assigned a type of irrigation. The proper analysis of a split-plot design recognizes that treatments applied to whole-plots are subject to a different experimental errors than treatments applied to subplots. This results in the use of different mean squares as denominators for the F-ratios used to test the respective treatment effects.

Although the split-plot design originated in field experiments, it has found useful applications in many other areas. Consider the following experiment described in Montgomery (1991). A paper manufacturer is interested in studying the effects of three different pulp preparation methods and four different cooking temperatures on the tensile strength of paper. If a randomized complete

6.6 Models with Random and Nested Effects for More Complex Experiments

block design (RCBD) is to be used for this experiment, the 12 method by treatment combinations (a 3 × 4 factorial), would need to be applied within each block (or replication). This requires that 12 pulp batches be prepared for each block, using each of the 3 methods of pulp preparation and each batch assigned to one of the 4 cooking temperatures.

However, the experiment was actually conducted in the following manner. Since the pilot plant is only capable of making 12 runs per day, the experimenter ran 1 replicate on each of 3 days and considered the runs performed per day as a block. On each day, he prepared three large batches of pulp using each of the three preparation methods (in random order). Then he divided each batch into four smaller samples and randomly assigned each sample to be cooked using one of the four temperatures. This is repeated for each of the three large batches. This procedure is repeated in each of the 3 days thus producing 36 tensile strength measurements.

The treatment arrangement continues to be a 3 × 4 factorial; however, the design is no longer a randomized block design. Rather, it is a split-plot design because the 12 treatment combinations are not assigned completely at random to 12 pulp batches; the 3 methods are assigned to the 3 large batches of pulp (whole-plots), which are then subdivided into 4 smaller samples (subplots) and assigned the temperatures. Thus, a split-plot design also introduces a different randomization scheme for factorial treatment combinations.

The preparation method is the whole-plot factor, whereas the cooking temperature is the subplot factor. The experimental procedure is simplified because the experimenter makes 3 large batches of pulp using the 3 methods of preparation instead of making 12 small batches. The experimental plan is sketched out in Fig. 6.51, in which the levels of method are denoted by A1, A2, and A3 and the levels of temperature are denoted by B1, B2, B3, and B4.

Rep I			Rep II			Rep III		
A3	A1	A2	A1	A3	A2	A3	A2	A1
B1	B3	B2	B3	B3	B4	B2	B1	B4
B4	B1	B4	B2	B1	B1	B3	B2	B3
B3	B2	B1	B4	B2	B2	B4	B3	B1
B2	B4	B3	B1	B4	B3	B1	B4	B2

Fig. 6.51. SAS Example F17: Plan for the strength of paper experiment

The model for observations from this experiment reflects the fact that there are two types of experimental unit in the experiment and therefore two types of experimental error. The model and the analysis of variance for a general split-plot experiment similar to the above example is described below. It is easier to consider the model as representing two distinct parts of the design.

The whole-plot design is an RCBD because the whole-plot treatment levels (Factor A) are replicated r times; thus, it is easier to visualize the model for the whole-plot part. The subplot part is also analyzed as an RCBD, where the whole-plots serve as blocks for the levels of the subplot treatment (Factor B). Assume that Factor A and Factor B have a and b levels, respectively. Let y_{ijk} represent an observation from the ith replication, the jth level of Factor A, and the kth level of Factor B. Then the model that describes this observation is given by

$$y_{ijk} = \mu + \beta_i + \tau_j + \epsilon_{ij} + \alpha_k + \delta_{jk} + \epsilon^*_{ijk}$$

where μ is the overall mean, β_i is the effect of the ith block (or replication), $i = 1, \ldots, r$, τ_j is the effect of the jth whole-plot factor (i.e., Factor A), $j = 1, \ldots, a$, ϵ_{ij} is the experimental error associated with the ijth whole-plot, α_k is the effect of the kth subplot factor (i.e., Factor B), $k = 1, \ldots, b$, δ_{jk} is the interaction between method j and temperature k, and ϵ^*_{iij} is the experimental error associated with ijkth subplot. As usual the random errors ϵ_{ij} are assumed to be iid $N(0, \sigma^2_W)$ and ϵ^*_{ijk} are assumed to be iid $N(0, \sigma^2_S)$ random variables. The anova table associated with this model is

SV	df	SS	MS	F
Rep	$r-1$	SS_{REP}	MS_{REP}	
A	$a-1$	SS_A	MS_A	MS_A/MSE_W
Error A	$(r-1)(a-1)$	SSE_W	MSE_W	
B	$(b-1)$	SS_B	MS_B	MS_B/MSE_S
AB	$(a-1)(b-1)$	SS_{AB}	MS_{AB}	MS_{AB}/MSE_S
Error B	$a(r-1)(b-1)$	SSE_S	MSE_S	

Note carefully that whereas the subplot part of the experiment is, by necessity, an RCBD, the whole-plot part need not be so. In the above experiment, it was decided to use "days" as a blocking factor and, thus, an RCBD was used for the whole-plot part of the expriment. However, it is common to use a completely randomized design for the whole-plot part of the experiment. For example, if the a levels of the whole-plot factor were applied in a completely random manner to the ar whole-plots, then the whole-plot design would be a CRD. In this case, the source Rep will not appear in the above anova table and the degrees of freedom for Error A will change to $a(r-1)$.

6.6.3 Analysis of split-plot experiments using PROC GLM

SAS Example F17

The observed data from the tensile strength of paper experiment described in Section 6.6.2 are found in Table 6.9. In the SAS Example F17 program, shown in Fig. 6.52 the data are input by the nesting of three do loops to create the

6.6 Models with Random and Nested Effects for More Complex Experiments

values of the variables, `temp`, `rep` and `method`. Then it uses the list input style to read each of nine values for each temperature contained in a line of data. Again a trailing @ symbol is used to read the data values repeatedly from the same data line. The `output` statement writes a record into the SAS data set after reading each data value along with the current values of each of `temp`, `rep`, and `method` variables. Thus only the data values need to be entered in the appropriate sequence in the four lines of data.

Temperature	Replication								
	I			III			III		
	Method			Method			Method		
	1	2	3	1	2	3	1	2	3
200	30	34	29	28	31	31	31	35	32
225	35	41	26	32	36	30	37	40	34
250	37	38	33	40	42	32	41	39	39
275	36	42	36	41	40	40	40	44	45

Table 6.9. Tensile strength of paper data

```
data paper;
do temp=200,225,250,275;
  do rep=1 to 3;
    do method=1 to 3;
      input strength @;
      output;
    end;
  end;
end;
datalines;
30 34 29 28 31 31 31 35 32
35 41 26 32 36 30 37 40 34
37 38 33 40 42 32 41 39 39
36 42 36 41 40 40 40 44 45
;
run;

proc glm data=paper;
  class rep method temp;
  model strength = rep method rep*method temp method*temp;
  test h=method e=rep*method;
  lsmeans method*temp/slice=method;
  lsmeans method/pdiff cl adj=tukey e=rep*method;
  contrast 'Temp:Linear Trend' temp -3 -1 1 3;
  title 'Analysis of a Split-Plot Experiment using PROC GLM';
run;
```

Fig. 6.52. SAS Example F17: Analysis of strength of paper data with PROC GLM

In SAS Example F17, `proc glm` is used to perform a traditional analysis in which the block (or replication) effect is considered a fixed effect in the model. Since the `random` statement cannot be used because there are no random effects in the model, the `test` option used for testing fixed effects is not available The user must provide an appropriate effect (or combination of effects) to be used as the error term (or denominator) of the F-ratio appropriate for testing hypotheses (and for constructing confidence intervals) about the effects of the whole-plot factor. This is the most important difference from an analysis of data from a similar experiment performed as a randomized blocks design from that of a split-plot experiment. Recall that under standard models with only fixed effects, SAS does not require the user to specify a term to represent the random error term in the `model` statement. This implies that the degrees of freedom remaining after sum of squares for the terms specified in the model are taken into account are automatically used to compute the residual or error sum of squares.

However, in the model shown for the split-plot experiment, there are two error terms. Thus, the user is required to determine how the sum of squares for the whole-plot error (labeled Error A) is to be calculated and specify it in a `test` statement available in `proc glm`. Usually the error term must be specified as a function of effects aleady present in the model statement. Recall that in this example, the whole-plot part of the design is an RCBD with replications as blocks and the levels of the whole-plot factor `method` as treatment. Thus, the appropriate error sum of squares for the whole-plot design is equivalent to the sum of squares that correspond to the replication by method interaction. Thus, as illustrated in Fig. 6.52, by including a term that corresponds to the replication by method interaction in the model statement one is able to specify it as the error term to be used for testing the `method` effects.

The data are classified according to `rep`, `method`, and `temp`, so these are included in the `class` statement as usual. The `model` statement is specified as

```
strength = rep method rep*method temp method*temp;
```

The effects in the `model` statement correspond to the terms in the model definition above except for the term that corresponds to Error A. As noted above, the term `rep*method` is included because it is intended that the sum of squares corresponding to this effect will be used as Error A. It is seen that Error B is the error (or residual) sum of squares after the other terms in the model are accounted for. The `test` statement

```
test h=method  e=rep*method;
```

specifies that the F-ratio for testing the `method` effect to be constructed using the mean square corresponding to the term `rep*method` as the denominator (or error term). Recall that, ordinarily, the mean square corresponding to the

6.6 Models with Random and Nested Effects for More Complex Experiments

```
             Analysis of a Split-Plot Experiment using PROC GLM            1
                            The GLM Procedure
                          Class Level Information

                  Class          Levels     Values

                  rep               3       1 2 3

                  method            3       1 2 3

                  temp              4       200 225 250 275


                  Number of Observations Read          36
                  Number of Observations Used          36

             Analysis of a Split-Plot Experiment using PROC GLM            2

                            The GLM Procedure

Dependent Variable: strength

                                   Sum of
Source                  DF         Squares    Mean Square    F Value   Pr > F

Model                   17      751.4722222    44.2042484      11.13   <.0001

Error                   18       71.5000000     3.9722222

Corrected Total         35      822.9722222


              R-Square     Coeff Var     Root MSE    strength Mean

              0.913120      5.531963     1.993043         36.02778


Source                  DF       Type I SS    Mean Square    F Value   Pr > F

rep                      2      77.5555556    38.7777778       9.76    0.0013
method                   2     128.3888889    64.1944444      16.16    <.0001  1
rep*method               4      36.2777778     9.0694444       2.28    0.1003
temp                     3     434.0833333   144.6944444      36.43    <.0001
method*temp              6      75.1666667    12.5277778       3.15    0.0271


Source                  DF     Type III SS    Mean Square    F Value   Pr > F

rep                      2      77.5555556    38.7777778       9.76    0.0013
method                   2     128.3888889    64.1944444      16.16    <.0001  1
rep*method               4      36.2777778     9.0694444       2.28    0.1003
temp                     3     434.0833333   144.6944444      36.43    <.0001
method*temp              6      75.1666667    12.5277778       3.15    0.0271


  Tests of Hypotheses Using the Type III MS for rep*method as an Error Term  2

Source                  DF     Type III SS    Mean Square    F Value   Pr > F

method                   2     128.3888889    64.1944444       7.08    0.0485
```

Fig. 6.53. SAS Example F17: Analysis of strength of paper data (Page 1-2)

residual (i.e., Error B here), would be used to construct this F-ratio in the Type I or III analysis of variance table in `proc glm`.

The user therefore must recognize that there will be two F-ratios for `method` (and associated p-values) that will appear in the output, and that the proper F ratio to be used for `method` is the one that results from the `test` statement, appearing separately **2** in page 2 of the output shown in Fig. 6.53. The F ratio for `method` appearing in the usual Type I or III anova tables **1** will be incorrect for testing the `method` effects in the split-plot experiment. The anova table is thus completed using the values on page 2:

SV	df	SS	MS	F	p-value
Rep	2	77.55	38.78		
Method	2	128.39	64.20	7.08	.0485
Error A	4	36.28	9.07		
Temp	3	434.08	144.69	36.45	.0001
Method×Temp	6	75.17	12.53	3.15	.0271
Error B	18	71.50	3.97		
Total	35	822.97			

From the anova table `method`× `temp` interaction as well as the two main effects `method` and `temp` are significant at $\alpha = .05$. Further analysis of these results depend on the intentions of the experimenter. Making conclusions about

```
insert data step to create the SAS dataset 'paper' here

proc sort data=paper;
  by temp method;
run;

proc means data=paper noprint mean;
  by temp method;
  var strength;
  output out=meandat mean=cellmean;
run;

title1 c=firebrick h=2 'Analysis of Strength of Paper';
title2 c=cornflowerblue h=1.5 f=centx 'Profile Plot of Cell Means';

symbol1 c=crimson i=join v=square h=1.5 l=2;
symbol2 c=darkorange i=join v=diamond h=1.5 l=3;
symbol3 c=cadetblue i=join v=triangle h=1.5 l=4;

axis1 v=(c=stb h=1) label=(c=dodgerblue h=1.5 f=swissu a=90 'Cell Means');
axis2 v=(c=stb h=1) offset=(.2 in) label=(c=dodgerblue h=1.5
                              f=swissu 'Levels of Temperature');

proc gplot data=meandat;
  plot cellmean*temp=method/vaxis=axis1 haxis=axis2 hm=4;
run;
```

Fig. 6.54. SAS Example F17: Program for interaction plot using PROC GPLOT

6.6 Models with Random and Nested Effects for More Complex Experiments

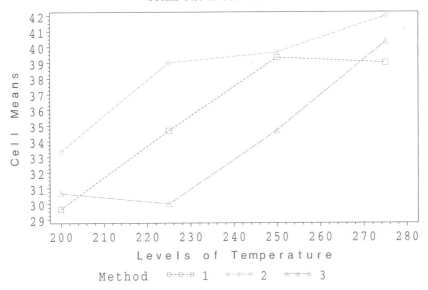

Fig. 6.55. SAS Example F17: Plot of mean strength versus temperature by method

method and temperature effects is complicated by the fact that interaction is significant. An interaction plot was obtained by the execution of the SAS program in Fig. 6.54. The plot, displayed in Fig. 6.55, suggests that, on the average, the mean strength appears to increase with temperature. However, the pattern of increase is different for each method. For example, as the temperature increases from 200 to 225, the mean strength increases for methods 1 and 2, whereas for method 3, the mean strength stays about the same. However, for the temperature range over 225, a similar rate of increase in mean

```
              Analysis of a Split-Plot Experiment using PROC GLM                4

                              The GLM Procedure
                              Least Squares Means

                 method*temp Effect Sliced by method for strength 3

                                Sum of
             method      DF     Squares      Mean Square    F Value    Pr > F

               1          3    184.666667      61.555556     15.50    <.0001
               2          3    121.666667      40.555556     10.21     0.0004
               3          3    202.916667      67.638889     17.03    <.0001
```

Fig. 6.56. SAS Example F17: Analysis of strength of paper data (Page 4)

strength is seen for method 3 also. However, for method 2, mean strength does not show an appreciable change as the temperature increases from 225 to 250, as also observed for method 1 as the temperature increases from 250 to 275. These differences in the effect of temperature for each method manifests itself as a significant interaction effect. One may use the `proc glm` statement

$$\text{lsmeans method*temp/slice=method;}$$

to examine temperature effect at each level of method. The output in Fig. 6.56 confirms the interaction plot: that the mean strengths are significantly different among the temperature levels for each method. Note that each test is associated with three degrees of freedom 3 that corresponds to comparing the four temperatures for each method; that is, each slice tests a hypothesis of the form $H_0 : \mu_{j1} = \mu_{j2} = \mu_{j3} = \mu_{j4}$ versus H_a : at least one inequality, for each j, where $\mu_{jk} = \mu + \tau_j + \alpha_k + \delta_{jk}$.

However, this finding by itself is not sufficient to make a useful conclusion. One may need to use multiple comparisons to determine, say, the best method to use at each temperature. Thus, the type of inferences one may make depend on the purpose of the experiment. If the intention was to find the best method over all temperatures, one could compare the effects of method averaged over temperatures (thus ignoring the interaction).

```
                        strength
        method          LSMEAN          95% Confidence Limits

          1            35.666667        33.252936      38.080397
          2            38.500000        36.086269      40.913731
          3            33.916667        31.502936      36.330397


              Least Squares Means for Effect method

                       Difference           Simultaneous 95%
                        Between            Confidence Limits for
         i    j          Means             LSMean(i)-LSMean(j)

         1    2        -2.833333         -7.215118       1.548451
         1    3         1.750000         -2.631784       6.131784
         2    3         4.583333          0.201549       8.965118
```

Fig. 6.57. SAS Example F17: Analysis of strength of paper data (Page 5)

The statement

$$\text{lsmeans method/pdiff cl adj=tukey e=rep*method;}$$

compares the method marginal means using 95% confidence intervals adjusted using Tukey's method. It is important to note that in `proc glm`, the error term used for calculating the standard errors needs to be specified for the whole-

6.6 Models with Random and Nested Effects for More Complex Experiments

plot effect: in this case, `method`. Otherwise, the error mean square will be used by default. Thus, the option `e=rep*method` is included in the `lsmeans` statement.

The confidence intervals in Fig. 6.57 (displaying part of output page 5) show that only methods 2 and 3 are significantly different and that method 2 produces the strongest paper averaged over the temperature range, a finding that appears to be confirmed by the graph in Fig. 6.55. Finally, the experimenter may have selected equispaced levels for `temperature` so that the trend in the increase or decrease in mean strength could be examined using orthogonal polynomials. Again, the interaction may be ignored and the effects averaged over the three methods for this purpose. (This trend in temperature could also be examined for each method using appropriate contrast statements; however, that is a useful option only if the experimenter is interested in a particular method or if the trend appears to be different for each method.)

```
              Analysis of a Split-Plot Experiment using PROC GLM        6

                             The GLM Procedure

Dependent Variable: strength

Contrast                DF      Contrast SS    Mean Square    F Value    Pr > F

Temp:Linear Trend        1      432.4500000    432.4500000     108.87    <.0001
```

Fig. 6.58. SAS Example F17: Analysis of strength of paper data (Page 6)

The contrast statement in Fig. 6.52, tests the hypothesis of linear trend in the marginal means averaged over the methods. The results of this statement are shown in Fig. 6.58. The appropriate partitioning of the `temp` SS is thus given by

SV	df	SS	MS	F	p-value
Temp	3	434.08	144.69	36.45	.0001
Linear	1	432.45	432.45	108.87	$<.0001$
Lof	2	1.63	0.82	0.21	≈ 1
Error B	18	71.50	3.97		

The results of the F-tests confirms a linear trend in the mean strength (averaged over the methods) as the cooking temperature increases.

6.6.4 Analysis of split-plot experiments using PROC MIXED

SAS Example F18

In previous discussions of the RCBD (in Section 6.5.1 and others), reasons were provided for regarding block effects as random effects in experiments

482   6 Analysis of Variance: Random and Mixed Effects Models

that make use of blocks. In the analysis of the paper data in Section 6.6.3, neither the rep effect nor the rep*method effects were considered random effects. Recall that for computing the standard errors of method mean comparisons in SAS Example F17, the whole-plot error term was required to be explicitly identified in the lsmeans statement using the e=rep*method option. As mentioned previously, the random statement in proc glm does not result in the correct standard errors, as it does not set up the variance structure required for analyzing a general mixed model. Thus, proc mixed is the appropriate SAS procedure available for analyzing data from split-plot experiments. In SAS Example F18, the analysis of the strength of paper data is repeated using proc mixed. A discussion of the changes needed if the whole-plot part of the experiment is conducted as a completely randomized design appears following this example.

```
insert data step to create the SAS dataset 'paper' here

proc mixed data=paper noclprint noinfo method=type3 cl;;
   class rep method temp;
   model strength = method  temp method*temp/ddfm=satterth;
   random rep rep*method;
   lsmeans method*temp/slice=method;
   lsmeans method/pdiff cl adj=tukey;
   contrast 'Temp:Linear Trend' temp -3 -1 1 3;
   title 'Analysis of a Split-Plot Experiment using PROC MIXED';
run;
```

Fig. 6.59. SAS Example F18: Analysis of strength of paper data with PROC MIXED

The specification of the model for the observations from a split-plot experiment with blocks as a random factor is slightly different. It is restated again with only the differences from the previous model highlighted:

$$y_{ijk} = \mu + B_i + \tau_j + \epsilon_{ij} + \alpha_k + \delta_{jk} + \epsilon^*_{ijk}$$

where B_i is the effect of the ith block (or replication) assumed to be iid $N(0, \sigma_B^2)$ ϵ_{ij} is the experimental error associated with the ijth whole-plot assumed to be iid $N(0, \sigma_W^2)$, and ϵ^*_{iij} is the experimental error associated with ijkth subplot assumed to be iid $N(0, \sigma_S^2)$. The SAS Example F18 program is displayed in Fig. 6.59. As earlier, the fixed effects are specified in the model statement and the random effects in the random statement. The user is again required to determine how the whole-plot error is calculated and then declare it as a random effect. Note also that the method of computation of variance components is specified as type3. The specification of the option ddfm=satterth requests that the Satterthwaite approximation be calculated to obtain degrees of freedom for any statistic for which it is required. Com-

6.6 Models with Random and Nested Effects for More Complex Experiments

```
                Analysis of a Split-Plot Experiment using PROC MIXED           1

                                 The Mixed Procedure

                                Type 3 Analysis of Variance

                                         Sum of
                Source           DF      Squares          Mean Square

                method            2      128.388889        64.194444
                temp              3      434.083333       144.694444
                method*temp       6       75.166667        12.527778
                rep               2       77.555556        38.777778
                rep*method        4       36.277778         9.069444
                Residual         18       71.500000         3.972222

                                Type 3 Analysis of Variance
                                                                              Error
Source       Expected Mean Square                         Error Term           DF

method       Var(Residual) + 4 Var(rep*method)            MS(rep*method)        4
             + Q(method,method*temp)
temp         Var(Residual) + Q(temp,method*temp)          MS(Residual)         18
method*temp  Var(Residual) + Q(method*temp)               MS(Residual)         18
rep          Var(Residual) + 4 Var(rep*method)            MS(rep*method)        4
             + 12 Var(rep)
rep*method   Var(Residual) + 4 Var(rep*method)            MS(Residual)         18
Residual     Var(Residual)                                    .                 .

                                Type 3 Analysis of Variance

                        Source        F Value     Pr > F

                        method          7.08      0.0485
                        temp           36.43      <.0001
                        method*temp     3.15      0.0271
                        rep             4.28      0.1016
                        rep*method      2.28      0.1003
                        Residual          .          .

                            Covariance Parameter Estimates

                Cov Parm        Estimate   Alpha     Lower      Upper

                rep              2.4757    0.05     -3.9439    8.8953
                rep*method       1.2743    0.05     -1.9343    4.4829
                Residual         3.9722    0.05      2.2679    8.6869
```

Fig. 6.60. SAS Example F18: Analysis of strength of paper data using PROC MIXED(Page 1)

putation of the denominator sum of squares statistics for certain statistics require this approximation, as will be illustrated later.

As can be observed, every statement used with `proc glm` is available for use with `proc mixed`; however, the user is not required to specify the error terms to be used for tests of fixed effects or for computation of standard errors for comparisons of fixed effects. Page 1 of the output from `proc mixed` (note that the options `noclprint` and `noinfo` are in effect) is shown in Fig. 6.60. It contains the Type 3 sums of squares and F-tests for all effects, expected mean squares, estimates, and asymptotic confidence intervals for variance components. The following anova table that includes an expected mean squares column is assembled using this information:

SV	df	SS	MS	F	p-value	$E(\text{MS})$
Rep	2	77.55	38.78	4.28	.1016	$\sigma_S^2 + 4\sigma_W^2 + 12\sigma_b^2$
Method	2	128.39	64.20	7.08	.0485	$\sigma_S^2 + 4\sigma_W^2 + Q(\beta,\tau)$
Error A	4	36.28	9.07	2.28	.1003	$\sigma_S^2 + 4\sigma_W^2$
Temp	3	434.08	144.69	36.45	.0001	$\sigma_S^2 + Q(\alpha,\delta)$
Method×Temp	6	75.17	12.53	3.15	.0271	$\sigma_S^2 + Q(\delta)$
Error B	18	71.50	3.97			σ_S^2
Total	35	822.97				

As seen here, the F-statistics for the whole-plot effects, subplot effects, and their interactions agree with those produced by `proc glm`. However, for unbalanced data, this will be not the case. The interest here is in comparison of the fixed effects and not the estimation or testing of the variance components. However, the magnitudes of the variance components should give the experimenter an idea about the precision of the experiment. For example, the

```
                    Type 3 Tests of Fixed Effects  1

                      Num      Den
         Effect        DF       DF    F Value   Pr > F

         method        2        4      7.08     0.0485
         temp          3       18     36.43     <.0001
         method*temp   6       18      3.15     0.0271


                            Contrasts  2

                      Num      Den
         Label         DF       DF    F Value   Pr > F

         Temp:Linear Trend   1   18   108.87    <.0001
```

Fig. 6.61. SAS Example F18: Analysis of strength of paper data using PROC MIXED(page 2)

subplot error variance with 18 degrees of freedom is 3.9722, quite large compared to the whole-plot error variance of 1.2743 with four degrees of freedom. Thus, there is substantial variation among the smaller batches that is not accounted for by the different cooking temperatures. Confidence intervals based on the chi-square percentiles can be constructed for these as usual using the Satterthwaite approximation.

6.6 Models with Random and Nested Effects for More Complex Experiments 485

A separate table showing the F-tests for the fixed effects **1** extracted from page 2 of the output is shown in Fig. 6.61. The result of the contrast statement is also extracted from page 2 and shown here **2**. These results are identical to those from SAS Example F17 using `proc glm`. Relevant portions of results from the `lsmeans` statement for making pairwise comparisons of method means, extracted from output pages 3, 4, and 5, are edited and included in Fig. 6.62. The results from the `lsmeans` statement with the `slice=` is also shown here. Again, these results are identical to those obtained from `proc glm`.

```
                    Differences of Least Squares Means
                            Standard
Effect  method  _method  Estimate  Error  DF  t Value  Pr > |t|  Adjustment    Adj P
method    1       2      -2.8333  1.2295   4   -2.30    0.0825   Tukey-Kramer  0.1660
method    1       3       1.7500  1.2295   4    1.42    0.2277   Tukey-Kramer  0.4129
method    2       3       4.5833  1.2295   4    3.73    0.0203   Tukey-Kramer  0.0434

                    Differences of Least Squares Means
                                                               Adj      Adj
Effect        method  _method  Alpha   Lower    Upper   Lower    Upper

method          1        2     0.05   -6.2469  0.5802  -7.2151  1.5485
method          1        3     0.05   -1.6635  5.1635  -2.6318  6.1318
method          2        3     0.05    1.1698  7.9969   0.2015  8.9651

                    Tests of Effect Slices
                                    Num    Den
      Effect          method         DF     DF    F Value   Pr > F

      method*temp       1            3      18     15.50    <.0001
      method*temp       2            3      18     10.21    0.0004
      method*temp       3            3      18     17.03    <.0001
```

Fig. 6.62. SAS Example F18: Analysis of strength of paper data using PROC MIXED(Extracted from pages 3, 4, and 5)

Other contrast statements can be used to analyze the interactions by making comparisons among the interaction means. Use of the `slice=` option produced an F-test to compare the four `temp` means for a given method simultaneously (i.e. to test $H_0 : \mu_{j1} = \mu_{j2} = \mu_{j3} = \mu_{j4}$), as discussed previously. A more detailed analysis involves pairwise comparison of interaction means. For example, the contrast statements

```
contrast 'T1 vs T2 @ M1' temp 1 -1  0  0
                    method*temp  1 -1  0  0  0 0 0 0  0 0 0 0;
contrast 'T1 vs T3 @ M1' temp 1  0 -1  0
                    method*temp  1  0 -1  0  0 0 0 0  0 0 0 0;
contrast 'T1 vs T4 @ M1' temp 1  0  0 -1
                    method*temp  1  0  0 -1  0 0 0 0  0 0 0 0;
```

are each single-degree of freedom comparisons that compare cell means μ_{ij} for pairs of temperature levels at method level 1. Similarly, comparison made earlier of the linear trend of temp means can also be performed for each method using similar contrast statements. The F-statistics produced for these contrasts using proc glm are identical to those produced by proc mixed as the denominator uses the subplot error mean square in both procedures.

However, for comparisons of cell means μ_{ij} for pairs of method levels at fixed temperature levels, the two SAS procedures produce different F-statistics. This is because the F-statistics in these cases are obtained via the Satterthwaite approximation. More precisely, the denominator of the required F-statistic, in general, is an estimate of a linear combination of σ_S^2 and σ_W^2. This is a weighted average of the two mean squares MSE$_W$ and MSE$_W$, the distribution of which is approximated by a chi-square distribution with degrees of freedom approximately obtained using the method of Satterthwaite, as illustrated previously. Thus, proc mixed, along with the ddfm=satterth option in the model statement is required for these F-statistics to be correctly computed. As mentioned elsewhere, it is recommended that the ddfm=kr be used for models with multiple random effects when the sample sizes are unequal. In SAS Example F17 and SAS Example F18 programs, the following contrast statements are included to illustrate this computation:

```
contrast 'M1 vs M2 @ T1' method 1 -1  0
                        method*temp 1 0 0 0  -1 0 0 0   0 0 0 0;
contrast 'M1 vs M3 @ T1' method 1  0 -1
                        method*temp 1 0 0 0   0 0 0 0  -1 0 0 0;
```

These compare the cell means for Method 1 with Method 2 at Temperature 1 and cell means for Method 1 with Method 3 at Temperature 1. The two sets of F-statistics produced by proc glm and proc mixed, respectively, for these contrasts are produced in Figs. 6.63 and 6.64. Note carefully that the degrees of freedom calculated for the denominator by proc mixed are obtained using the Satterthwaite approximation.

Contrast	DF	Contrast SS	Mean Square	F Value	Pr > F
M1 vs M2 @ T1	1	20.16666667	20.16666667	5.08	0.0370
M1 vs M3 @ T1	1	1.50000000	1.50000000	0.38	0.5466

Fig. 6.63. Simple effect contrasts of whole-plot factor with PROC GLM

Using the model for the split-plot experiments introduced in Section 6.6.2, it can be shown, in general, that the variance of the sample cell mean of Method 1 at Temperature 1, $\bar{y}_{.11}$, is $(\sigma_W^2 + \sigma_S^2)/r$. Thus the mean square appropriate for the denominator of F-statistics for making comparisons of this type of means is an estimate of $\sigma_W^2 + \sigma_S^2$. By examining the anova table

6.6 Models with Random and Nested Effects for More Complex Experiments

for this experiment, it is easy to observe that a mean square with an expected value of $\sigma_W^2 + \sigma_S^2$ is not directly available. However, it is possible to *synthesize* a mean square (say, MS*) by using a linear combination of MSE_W and MSE_S. To derive this linear combination by examining the expected values of these mean squares, note that

$$\frac{E(\text{MSE}_W) + 3E(\text{MSE}_S)}{4} = \sigma_W^2 + \sigma_S^2.$$

Thus, the linear combination needed is

$$\text{MS}^* = \frac{1}{4}\text{MSE}_W + \frac{3}{4}\text{MSE}_S$$

because the $E(\text{MS}^*)$ will then be equal to $\sigma_W^2 + \sigma_S^2$. The Satterthwaite approximation gives the degrees of freedom for a synthesized mean square of this type. Using this approximation, the degrees of freedom for $\text{MS}^* = \frac{1}{4}\text{MSE}_W + \frac{3}{4}\text{MSE}_S = .25\text{MSE}_W + .75\text{MSE}_S$ is given by

$$\nu = \frac{(\text{MS}^*)^2}{\frac{(.25\text{MSE}_W)^2}{df_1} + \frac{(.75\text{MSE}_S)^2}{df_2}},$$

where df_1 and df_2 are the degrees of freedom for MSE_W and MSE_S, respectively. The calculation can be done in a SAS data step similar to those used for computations of confidence intervals for components of variance in the previous sections. The execution of the data step results in the values $\text{MS}^* = 5.245$ and $\nu = 15.47$. This value for the denominator degrees of freedom, rounded to 15.5, is identical to the value reported in Fig. 6.64.

```
data df;
    mse1=9.07; mse2=3.97; df1=4; df2=18;  p=.25; q = .75;
    ms_star= p*mse1 +q*mse2;
    nu=ms_star**2/((p*mse1)**2/df1+ (q*mse2)**2/df2);
    put ms_star=  nu=  ;
run;
```

	Contrasts			
Label	Num DF	Den DF	F Value	Pr > F
M1 vs M2 @ T1	1	15.5	3.84	0.0682
M1 vs M3 @ T1	1	15.5	0.29	0.6005

Fig. 6.64. Simple effect contrasts of whole-plot factor with PROC MIXED

Finally, if the whole-plot design is a CRD instead of an RCBD as used in the strength of paper experiment described in SAS Example F17, the model is modified as follows:

$$y_{ijk} = \mu + \tau_j + \epsilon_{ij} + \alpha_k + \delta_{jk} + \epsilon^*_{ijk}.$$

The whole-plot error is now estimated by the *replication within method* mean square usually denoted by the term `rep(method)` in the model statement. The partitioning of the degrees of freedom for the whole-plot analysis is adjusted as follows:

Whole Plot Design: RCBD

SV	df
Rep	$r-1$
A	$a-1$
Error A	$(r-1)(a-1)$

Whole Plot Design: CRD

SV	df
A	$a-1$
Error A	$a(r-1)$

In this case, for use in SAS procedures such as `proc glm` or `proc mixed`, a variable denoting the *replication number* is also input as part of the data. Levels of this variable usually identify experimental units used for the whole-plot treatments. Suppose that the variable name `rep` is used for this purpose in the paper example. Then the terms `rep` and `rep*method` are replaced by the single term `rep(method)` in the model statement and the error term for testing the method main effect using `proc glm` (as in SAS Example F17) becomes `rep(method)`. Thus, the test statement changes to `test h=method e=rep(method);`. The random statement is modified to `random rep(method);` in the `proc mixed` step.

6.7 Exercises

6.1 A textile mill weaves a fabric on a large number of looms (Montgomery, 1991). To investigate whether there is an appreciable variation among the output of cloth per minute by the looms, the process engineer selects five looms at random and measured their output on five randomly chosen days. The following data are obtained:

Loom	Output (lbs/min)				
1	14.0	14.1	14.2	14.0	14.1
2	13.9	13.8	13.9	14.0	14.0
3	14.1	14.2	14.1	14.0	13.9
4	13.6	13.8	14.0	13.9	13.7
5	13.8	13.6	13.9	13.8	14.0

a. Write the *one-way random model* you will use to analyze this data stating assumptions about each parameter in the model and tell what each parameter represents. Construct the corresponding analysis of variance using `proc glm`. Write the anova table including a column for Expected Mean Square ($E(\text{MS})$).

b. Express the hypothesis that there is no variability in output among the looms, in terms of the model parameters. Perform a test of this hypothesis using the analysis in part (a) using $\alpha = .05$.

c. If the hypothesis in part (b) is rejected, estimates of the variance components associated with the model in part (a) may be desired. Use the method of moments to obtain these estimates from the results of parts (a) and (b).

d. Calculate 95% confidence intervals for the variance components that are found to be nonzero.

6.2 A sugar manufacturer wants to determine whether there is significant variability in purity of batches of raw cane among batches obtained from different suppliers as well as among different batches obtained from the same supplier (Montgomery, 1991). Four batches of raw cane are obtained at random from each of three randomly selected suppliers. Three determinations of purity are made using random samples from each batch. The data are given as follows (note that the original data were given in coded form):

Batch	Supplier		
	1	2	3
1	94	94	95
	92	91	97
	93	90	93
2	91	93	91
	90	97	93
	89	95	95
3	91	92	94
	93	93	92
	94	91	95
4	94	93	96
	97	96	95
	93	95	94

a. Use a two-way random model to analyze these data. Write an appropriate model explaining what each term in the model represents and stating any assumptions made. Prepare and run a `proc mixed` program necessary to test hypotheses and estimate variance components using this model.

b. Construct an analysis of variance table on a separate sheet (including expected mean squares) using the output from the program. Test all hypotheses concerning variance components of interest and interpret the results of these tests.

c. Provide estimates of parameters of interest (variance components) depending on the outcome of each of the hypotheses tested in part (b); that is, only nonzero variance components need to be estimated. Show work.

d. Calculate 95% confidence intervals for the variance components that are found to be nonzero.

6.3 A manufacturer of diet foods suspects that the batches of raw material furnished by her supplier differ significantly in sodium content. There is a large number of batches currently in the warehouse and the variability of sodium content among these batches is of interest. Five of these are randomly selected for study. Determinations of sodium in five samples taken from each batch were made and the data obtained are reported in the following table.

Batch 1	Batch 2	Batch 3	Batch 4	Batch 5
23.36	23.59	23.51	23.28	23.29
23.48	23.46	23.64	23.40	23.46
23.56	23.42	23.46	23.37	23.37
23.39	23.49	23.52	23.46	23.32
23.40	23.50		23.39	23.38
	23.48			23.41
				23.35

a. Write the one-way random model you will use to analyze this data stating assumptions made about each parameter in the model and what each parameter represents. Construct the corresponding analysis of variance table using `proc glm`, including an additional column for Expected Mean Square, $E(MS)$. You must extract numbers from the SAS output to write down your own table.

b. Using model parameters, express the hypothesis that there is no variability in sodium content among batches, using model parameters. Perform a test of this hypothesis using your analysis of variance table from part (a). Use the p-value for making the decision.

c. If the hypothesis in part (b) is rejected, estimates of the variance components associated with the model in part (a) can be computed. Obtain these estimates depending on the results of part (b).

d. Calculate 95% confidence intervals for the variance components that are found to be nonzero.

6.4 In an experiment to study variability of a blood pH measurements among animals from different dams as well as from different sires described in Sokal and Rohlf (1995), 10 female mice (dams) were successfully mated over a period of time to 2 males (sires). Different sires were employed for the 10 dams, implying that a total of 20 sires were used in the experiment.

Dam	Sire	Blood pH readings			
1	1	48	48	52	54
	2	48	53	43	39
2	1	45	43	49	40
	2	50	45	43	36
3	1	40	45	42	48
	2	45	33	40	46
4	1	44	51	49	51
	2	49	49	49	50
5	1	54	36	36	40
	2	44	47	48	48
6	1	41	42	36	47
	2	47	36	43	38
7	1	40	34	37	45
	2	42	37	46	40
8	1	39	31	30	41
	2	50	44	40	45
9	1	52	54	52	56
	2	56	39	52	49
10	1	50	45	43	44
	2	52	43	38	33

Four mice were selected at random from each of the resulting litters and the blood pH of each mouse was determined. The following data (which have been coded) were extracted from the original data to produce equal sample sizes.

a. Write the two-way nested effects model for these observations, explaining each term and stating any assumptions made. Prepare and run a `proc mixed` program necessary to analyze the data using this model.
b. Construct an analysis of variance table (including expected mean squares) using the output from the program. Test all hypotheses concerning the variance components of interest and interpret the results of these tests.
c. Provide estimates of the variance components depending on the outcome of each of the hypotheses tested in part (b); that is, you need to estimate only those variance components that are determined to be nonzero as a result of the above tests. State results of your analysis in a summary statement.
d. Calculate 95% confidence intervals for the variance components that are found to be nonzero.

6.5 In an experiment described in Dunn and Clark (1987), four brands of airplane tires are compared to assess the differences in the rate of tread wear. The data were collected on eight planes, with two tires used under each wing. The researcher uses each airplane as a block, mounting four test tires, one of each brand, in random order on each airplane. Thus, a randomized complete block design with "airplane" as a blocking factor is the design used. The amount of tread is measured initially and after 6 months and the following wear rates obtained: Note that a larger value

Airplane	Brand			
	A	B	C	D
1	4.02	2.46	2.06	3.49
2	4.50	3.39	2.91	3.18
3	2.73	1.69	2.37	1.48
4	3.74	1.95	3.39	3.09
5	3.21	1.20	1.72	2.65
6	2.53	1.04	2.52	1.23
7	3.07	2.55	2.42	2.07
8	3.10	1.09	2.22	2.57

indicates greater wear. Brand A is currently used by the airline and Brands B, C, and D from three different competitors are being evaluated to replace A. Thus, the management is interested in
  i. comparing Brand A with the average wear of Brands B C, and D,
  ii. comparing Brands B, C, and D

Prepare and run a proc mixed program necessary and provide answers to the following questions (on a separate sheet) assuming the model for a randomized complete block design.

a. Construct an analysis of variance table and test the hypothesis $H_0 : \tau_A = \tau_B = \tau_C = \tau_D$. State your conclusion based on the p-value.

b. Use a contrast statement for making comparison (i) by testing $H_0 : \tau_A = (\tau_B + \tau_C + \tau_D)/3$.

c. Use a contrast statement for making comparison (ii) by testing $H_0 : \tau_B = \tau_C = \tau_D$. One way to test this hypothesis is to make the comparisons $\tau_B - \tau_C$ and $\tau_B - \tau_D$ simultaneously in a single contrast statement. This results in the computation of a sum of squares with 2 df and an F-test with 2 df for the numerator. Add these results to this anova table as additional lines and summarize conclusions from this analysis.

d. Construct 95% confidence intervals for $\mu_B - \mu_C$ and $\mu_B - \mu_D$, using the output from appropriate estimate statements used with proc mixed.

6.6 A compound is sent to five randomly selected laboratories in the United States for a routine analysis. At each laboratory, four chemists are chosen at random and each chemist makes three chemical determinations on the compound using the same method of chemical analysis. The object is to study the variation of this method from laboratory to laboratory and also among chemists within each laboratory. The data obtained are as follows:

Chemist	Laboratory				
	1	2	3	4	5
1	2.24	2.44	1.97	2.54	2.44
	2.51	2.40	2.05	2.49	2.36
	2.37	2.51	2.13	2.42	2.45
2	2.65	2.26	2.23	2.46	2.67
	2.57	2.37	2.20	2.39	2.61
	2.48	2.41	2.27	2.40	2.59
3	2.41	2.38	2.25	2.71	2.64
	2.37	2.19	2.28	2.70	2.58
	2.40	2.35	2.31	2.78	2.55
4	2.25	2.75	2.37	2.62	2.38
	2.38	2.58	2.30	2.55	2.41
	2.40	2.62	2.44	2.59	2.35

Use a SAS program with **proc mixed** to answer the following questions:
a. Write the appropriate model for the analysis of these data, stating the effect each term used in the model represents and the assumptions made about these effects.
b. Construct an appropriate anova table including the required F-statistics, p-values and expected mean squares.
c. State the two hypotheses of interest for this experiment using the model parameters in part (a). Use the F-statistics and p-values above to make conclusions. Use $\alpha = .05$.
d. Estimate parameters of interest in this experiment. Note that these estimates depend on the outcomes of the hypotheses tested above.

6.7 The objective of a case study discussed in Ott and Longnecker (2001) was to determine whether the pressure drop across the expansion joint in electric turbines was related to gas temperature. Also, the researchers wanted to assess the variation in readings from various types of pressure gauge and whether they were consistent across different gas temperatures. Three levels of gas temperatures that cover the operational range of the turbine were selected 15°C, 25°C, and 35°C. Four types of gauge were randomly chosen for use in the study from the hundreds of different types pressure gauges used to monitor pressure in the lines. Six replications of each of the 12 temperature-gauge factorial combinations were run and the pressure measured.

Use a SAS program with **proc mixed** to answer the following questions:
a. Construct an appropriate anova table including appropriate F-statistics, p-values and expected mean squares.
b. Use the expected mean squares in the anova table to determine which ratios of sums of squares are to be used to test the two hypotheses

Temperature (°C)	Gauge			Pressure			
15	G1	40	40	37	47	42	41
	G2	43	34	38	42	39	35
	G3	42	35	35	41	43	36
	G4	47	47	40	36	41	47
25	G1	57	57	65	67	63	59
	G2	49	43	51	49	45	43
	G3	44	45	49	45	46	43
	G4	36	49	38	45	38	42
35	G1	35	35	35	46	41	42
	G2	41	43	44	36	42	41
	G3	42	41	34	35	39	36
	G4	41	44	35	46	44	46

of interest regarding variance components. Check your answers with those provided by `proc glm`

c. Construct an interaction plot appropriate for studying any significant interaction.

d. If there is significant variation among the gauges, suggest some BLUPs that might be useful for comparing the performance of gauges at each temperature. Use `estimate` statements to make appropriate comparisons.

6.8 A manufacturing company wishes to study the variation in tensile strength of yarns produced on four different looms (Sahai and Ageel, 2000). In an experiment designed for this purpose, 12 machinists were selected and each loom was assigned to 3 different machinists at random. Samples from two different runs by each machinist were obtained and tested. The data in standard units are given as follows:

Loom											
1			2			3			4		
Machinist			Machinist			Machinist			Machinist		
1	2	3	1	2	3	1	2	3	1	2	3
38.2	53.5	15.3	61.3	41.5	35.3	47.1	22.5	14.7	15.5	19.3	21.6
21.6	51.5	26.7	58.3	38.5	27.3	34.3	25.7	26.3	32.3	35.7	26.5

a. Write the appropriate model for the analysis of these data, stating the effect each term used in the model represents and the assumptions made about these effects.

b. Construct an analysis of variance table needed to analyze this data. Include a column of expected mean squares.
c. Use the anova table to test whether the mean tensile strength of yarn produced by the four looms are significantly different using $\alpha = .05$.
d. Test an appropriate hypothesis about the variability of tensile strength of yarn among the machinists within each loom using $\alpha = .05$.
e. Estimate the variance components of relevant effects of this model by constructing 95% confidence intervals for them.
f. Carry out comparisons of pairwise differences among the mean tensile strength of yarn produced by the looms adjusted for multiple comparisons using the Bonferroni adjustment.

6.9 In a study reported in Dunn and Clark (1987), each of three different sprays were applied to four trees selected at random. After 1 week, the concentration of nitrogen was measured on each of six leaves obtained in a random way from each tree. The data are given in the following table.

Spray	Leaf	Tree			
		1	2	3	4
1	1	4.50	5.78	13.32	11.59
	2	7.04	7.69	15.05	8.96
	3	4.98	12.68	12.67	10.95
	4	5.48	5.89	12.42	9.87
	5	6.54	4.07	10.03	10.48
	6	7.20	4.08	13.50	12.79
2	1	15.32	14.53	10.89	15.12
	2	14.97	14.51	10.27	14.79
	3	14.81	12.61	12.21	15.32
	4	14.26	16.13	12.77	11.95
	5	15.88	13.65	10.45	12.56
	6	16.01	14.78	11.44	15.31
3	1	7.18	6.70	5.94	4.08
	2	7.98	8.28	5.78	5.46
	3	5.51	6.99	7.59	5.40
	4	7.48	6.40	7.21	6.85
	5	7.55	4.96	6.12	7.74
	6	5.64	7.03	7.13	6.81

a. Write the appropriate model for the analysis of these data, stating the effect each term used in the model represents and the assumptions made about these effects.
b. Construct an analysis of variance table needed to analyze these data. Include a column of expected mean squares.

c. Use the anova table to test whether the mean nitrogen content resulting from the three sprays are significantly different using $\alpha = .05$.
d. Test an appropriate hypothesis about the variability of nitrogen content among the trees within each spray using $\alpha = .05$.
e. Estimate the variance components of relevant effects of this model by constructing 95% confidence intervals for them.
f. Carry out comparisons of pairwise differences among the mean nitrogen content resulting from the three sprays adjusted for multiple comparisons using the Tukey adjustment.

6.10 In a health awareness study (Kutner et al., 2005), each of three states independently devised a health awareness program. Three cities within each state of similar demographics were selected at random and five households within each city were randomly selected to evaluate the effectiveness of the program. A composite index based on responses of members of the selected households who were interviewed before and after participation in the program was used for measuring the impact of the health awareness program. The data on health awareness are given as follows (the larger the index, the greater the awareness):

	State								
	1			2			3		
	City			City			City		
Household	1	2	3	1	1	3	1	2	3
1	42	26	34	47	56	68	19	18	16
2	56	38	51	58	43	51	36	40	28
3	35	42	60	39	65	49	24	27	45
4	40	35	29	62	70	71	12	31	30
5	28	53	44	65	59	57	33	23	21

a. Write the appropriate model for the analysis of these data, stating the effect each term used in the model represents and the assumptions made about these effects.
b. Construct an analysis of variance table needed to analyze these data. Include a column of expected mean squares.
c. Use the anova table to test whether the mean awareness is significantly different among the three states using $\alpha = .1$.
d. Test an appropriate hypothesis about the variability of awareness among cities within states using $\alpha = .1$.
e. Construct 90% confidence intervals for pairwise comparisons between state means, using the Tukey procedure.
f. Construct 90% confidence interval for the variance component measuring variability of awareness among cities within states.

6.11 Consider an experiment to examine the variation in the effects of different analysts on chemical analyses for the DNA content of plaque (Montgomery, 1991). Three female subjects (ages 18-20 years) were chosen for the study. Each subject was allowed to maintain her usual diet, supplemented with 30 mg (15 tablets) of sucrose per day. No toothbrushing or mouthwashing was allowed during the study. At the end of the week, plaque was scraped from the entire dentition of each subject and divided into three samples. The three samples of plaque from each of the subjects were randomly assigned to three analysts chosen at random. They performed an analysis for the DNA content (in micrograms). The data obtained are as follows:

Analyst	Subject		
	1	2	3
1	13.2	10.6	8.5
2	12.5	9.6	7.9
3	13.0	9.9	8.3

a. Write the appropriate model for the analysis of these data, stating the effect each term used in the model represents and the assumptions made about these effects.
b. Construct an appropriate anova table including appropriate F-statistics, p-values and expected mean squares.
c. State the hypothesis of interest for this experiment using the model parameters in part (a). Use the F-statistic and p-value above to make conclusions. Use $\alpha = .05$.

6.12 An engineer is designing a battery for use in a device that will be subjected to extreme variations in temperature (Montgomery, 1991). He considers a two-way factorial with plate material and temperature as the two factors but has a large number of feasible choices for plate material and temperatures. Suppose that three plate materials and three temperatures were chosen, both at random, for use in the study. Four batteries were tested at each combination of plate material and temperature and the resulting 36 tests are run in a random sequence. The data, observed battery life, are in the following table.

Use a SAS `proc mixed` program to analyze these data. Provide answers to the following questions:

a. Write the appropriate model for the analysis of these data, explaining each term in the model and the assumptions made about these.
b. Construct an analysis of variance table needed to analyze this data. Include a column of expected mean squares.

| Material | Temperature | | | | | |
Type	15°F		70°F		125°F	
1	130	155	34	40	20	70
	74	180	80	75	82	58
2	150	188	136	122	25	70
	159	126	106	115	58	45
3	138	110	174	120	96	104
	168	160	150	139	82	60

c. Use the anova table to test whether temperature, type of material, or their interaction contribute significantly to the variation in battery life using $\alpha = .1$.

d. Estimate significant variance components by providing point estimates and by calculating 95% confidence intervals.

6.13 In a lab experiment carried out in a completely randomized design, a soil scientist studied the growth of barley plants under three different levels of salinity (control, 6 bars, 12 bars) in a controlled growth medium (Kuehl, 2000). Two replications of each treatment were obtained and three plants were measured in each replication. The data on the dry weight of plants in grams are as follows:

Salinity	Container	Weight(g)		
Control	1	11.29	11.08	11.10
	2	7.37	6.55	8.50
6 Bars	1	5.64	5.98	5.69
	2	4.20	3.34	4.21
12 Bars	1	4.83	4.77	5.66
	2	3.28	2.61	2.69

Use a SAS `proc mixed` program to analyze this data and provide answers to the following questions:

a. Write the appropriate model for the analysis of these data, explaining each term in the model and the assumptions made about these.

b. Construct an analysis of variance table needed to analyze these data. Include a column of expected mean squares.

c. Use the anova table to test whether the mean dry weights are significantly different among the three salinity levels using $\alpha = .05$. What is the standard error of a salinity level mean?

d. Test an appropriate hypothesis about the variability of weight among containers within salinity levels using $\alpha = .1$.

e. Partition the sum of squares for salinity effect using two orthogonal polynomials corresponding to linear and quadratic effects, each with one degree of freedom. Interpret the results of the F-tests.

6.14 An experiment, conducted in a split-plot design to determine the effect of three bacterial inoculation treatments applied to two cultivars of grasses on dry weight yields is discussed in Littell et al. (1991). The cultivar is the whole-plot factor and inoculi, the subplot factor. Each of the two cultivars are replicated four times. The data are as follows:

	Replication							
	I		II		III		IV	
Inoculi	Cultivar		Cultivar		Cultivar		Cultivar	
	A	B	A	B	A	B	A	B
Control	27.4	29.4	28.9	28.7	28.6	27.2	26.7	26.8
Live	29.7	32.5	28.7	32.4	29.7	29.1	28.9	28.6
Dead	34.5	34.4	33.4	36.4	32.9	32.6	31.8	30.7

Use a SAS `proc mixed` program to analyze this data and provide answers to the following questions:

a. Write the appropriate model for the analysis of these data, explaining each term in the model and the assumptions made about these.

b. Construct an analysis of variance table needed to analyze these data. Include a column of expected mean squares.

c. Use the anova table to test whether the mean dry weights are significantly different among the three inoculation levels using $\alpha = .05$. What is the standard error of a inoculation level mean?

d. Use the anova table to test whether the mean dry weights are significantly different among the three two cultivars using $\alpha = .05$. What is the standard error of a cultivar mean?

e. Construct confidence intervals for comparing the cultivar means of the two inoculi with the control, adjusted for multiple testing.

6.15 Soy protein isolates (SPI), widely used in the food industry, are usually stored in dry powder form produced via spray-drying or freeze-drying, to enhance shelf life and make them easier to distribute. A study was conducted at Iowa State University (Deak and Johnson, 2007) to determine how various properties of SPI are affected by the method used to dry them and to compare those of dried SPI to fresh (undried) or frozen-thawed SPI. Another factor that may affect the properties of SPI is the temperature used in the extraction process to create SPI. Thus, a two-factor experiment was conducted in which the two factors and their levels are Temperature at levels 25, 40, 60, 80 °C and Method with levels 1 = fresh, 2 = frozen and then thawed, 3 = freeze dried, 4 = spray dried.

Twelve batches of SPI were created so that the four temperature levels were assigned to three batches at random. Each batch was split into four parts and the four methods were assigned to the four parts of each SPI completely at random. Many response variables were measured for each part of each SPI, but data for emulsion capacity (EC, grams of oil emulsified by 1 gram of product) are reported in the following table (data were graciously provided by the first author of the above reference; the EC values were rounded to the nearest whole number).

		Temperature										
		25°C			40°C			60°C			80°C		
	Batch	Method	EC	Batch	Method	EC	Batch	Method	EC	Batch	Method	EC
1	1	1	549	4	1	568	7	1	478	10	1	442
		2	531		2	595		2	503		2	433
		3	573		3	557		3	501		3	473
		4	600		4	591		4	512		4	480
2	2	1	551	5	1	584	8	1	481	11	1	449
		2	640		2	632		2	526		2	496
		3	559		3	608		3	458		3	448
		4	587		4	602		4	485		4	475
3	3	1	538	6	1	582	9	1	485	12	1	473
		2	591		2	606		2	524		2	503
		3	557		3	591		3	469		3	458
		4	584		4	583		4	497		4	471

Use `proc mixed` (and other procedures if needed) in a SAS program to perform the following analyses:

a. Write the appropriate model for the analysis of these data, explaining each term in the model and the assumptions made about these.
b. Construct an analysis of variance table needed to analyze these data. Include a column of expected mean squares.
c. Is there a significant interaction between temperature and method? Conduct an appropriate test to answer this question. Provide a test statistic, its degrees of freedom, a p-value, and a brief conclusion.
d. Construct an interaction plot to study the interaction between temperature and method. Use the levels of temperature on the x-axis drawn to scale. Comment.
e. Use the anova table to test whether the mean EC are significantly different among the four temperature levels using $\alpha = .05$. What is the standard error of a temperature level mean?
f. Use the anova table to test whether the mean EC are significantly different among the four methods using $\alpha = .05$. What is the standard error of a method mean?

g. Construct 95% confidence intervals for pairwise differences among the the four temperature means adjusted for multiple testing using the Tukey adjustment.
h. Calculate a t-statistic for testing whether there is a difference between the effects of spray-dry and freeze-dry methods. Perform the test using $\alpha = .05$.
i. Calculate an F-statistic for testing whether there is a difference between the effects of temperatures $25°C$ and $40°C$ when the freeze-dry method is used. Perform the test using $\alpha = .05$.

A
SAS/GRAPH

A.1 Introduction

Although SAS/GRAPH procedures produce graphical output that are acceptable with minimal specifications, the quality of the graphs may be vastly improved by the use of a few additional options and statements. These include statements that give the user the ability to control the appearance of lines of text, the axes, plot symbols, lines connecting points, fill patterns, etc. For example, by using the options available in the `axis` statement, the scaling of an axis, color, font, and values that appear at each major tick mark on the axis, and the number, length, and placement of major and minor tick marks on the axis can be modified.

Because of the flexibilty and generality of SAS/GRAPH software, it is a burdensome and time-consuming process to master every capability available in the entire SAS/GRAPH system. However, a basic knowledge of a few SAS/GRAPH statements and procedures will enable even a novice to use SAS/GRAPH to construct presentation-quality statistical graphics displays.

Some knowledge of the flow of operations in a SAS/GRAPH program and a basic idea of a few important features available in SAS/GRAPH software and their relationship to various parameter settings is helpful for the beginner. The following titled paragraphs contain a short definition of each feature, followed by a brief explanation of how each affects the appearance and production of a SAS graph.

SAS/GRAPH Device Drivers

A *SAS device driver* consists of an executable program and a *device entry*, which contains settings for a group of *device parameters* (e.g., `hsize=`, etc.) that determine the default appearance of graphics output on a particular device. If a device is not explicitly specified in a SAS/GRAPH program, it will select a device driver appropriate for the display window. If the graphics output is to be reproduced on a particular device, the user may specify the

A SAS/GRAPH

name of the appropriate device driver goptions statement using the device= parameter. SAS/GRAPH software contains device driver definitions for a large number of graphics devices supported by SAS/GRAPH stored in a *device catalog*. The SAS/GRAPH procedure gdevice may be executed to examine the contents of the *device entry* for a selected device.

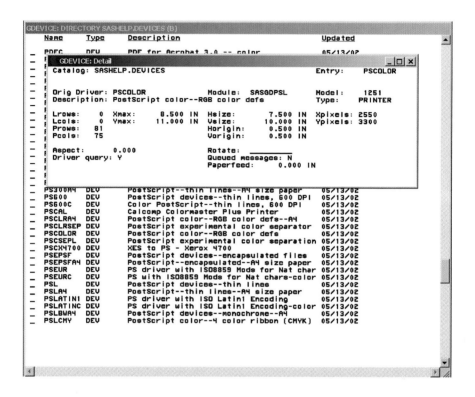

Fig. A.1. View of proc gdevice output showing the Directory window superimposed by the Detail window for the pscolor device driver.

In a windowing environment, these contents are accessible to the user through a series of windows. The main window, the Directory window, is displayed when the user first submits a proc gdevice; step. In the Directory window, the user can scroll through the list of device entries that are stored in the default catalog. When the appropriate device driver is located, the cursor is placed on the first column (i.e., *the selection-field*) and the right mouse button is clicked. By selecting the browse option in the resulting pop-up dialog, a new window (called the Detail window) appears. A screen-shot of the Directory window superimposed by the Detail window for the pscolor device driver is shown in Fig. A.1.

As seen in Fig. A.1, the Detail window displays a set of basic features, such as hsize and vsize, of a device. Additional device parameters are available through several subsidiary windows, which are available to be viewed by clicking on the yellow diamond-shaped Next Screen tool (which is present on the View toolbar when the Detail window is selected, but not shown here). This tool allows the user to navigate among the set of subsidiary windows associated with the device driver currently selected. The next window, the Parameters window, shown in Fig. A.2, includes additional device parameters that control the plotting of the graph such as the default *hardware font* (identified as Chartype) or the number of colors available (Maxcolors). The Gcolors window shows the *colors list* that the device driver uses by default. The Gcolors window for the pscolor device is displayed in Fig. A.3. The Chartype window, shown in Fig. A.4, lists the set of *hardware fonts* available for the current device.

The entire contents of a device entry can be listed in the Output window by submitting the following SAS code, illustrated here for the pscolor device:

```
proc gdevice catalog=sashelp.devices nofs;
    list pscolor;
run;
```

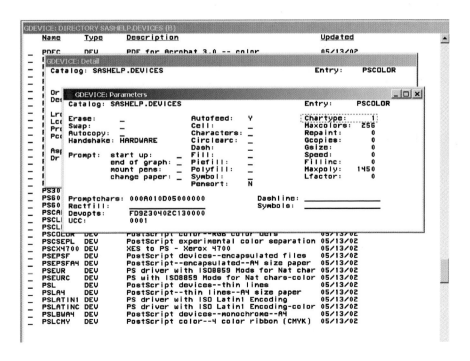

Fig. A.2. View of proc gdevice output showing the Parameters window for the pscolor device

A more compact summary of the settings of the device parameters for a specified device driver can be obtained by executing the `gtestit` procedure. After choosing a device, say, by using a `goptions device=pscolor;` statement, execute `proc gtestit; run;`. In a SAS windowing environment, selecting Solutions → Accessories → Graphics Test Pattern items from the main drop-down menu system will result in the same output.

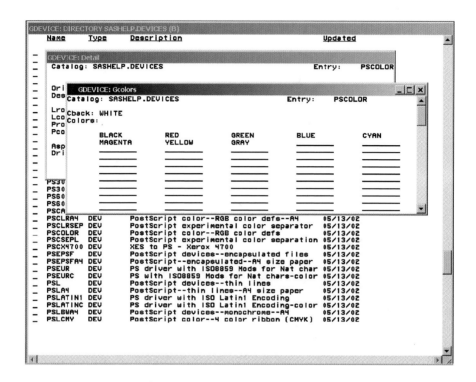

Fig. A.3. View of `proc gdevice` output showing the `Gcolors` window for the `pscolor` device

Some useful examples of SAS/GRAPH device drivers are discussed briefly here. The generic printer drivers `winprtm` (for black-and-white (monochrome) printers), `winprtg` (for gray-scale printers), `winprtc` (for color printers), or `winplot` (for plotters) allow SAS/GRAPH to send generic graphics commands to the Windows printer driver for a specific device. The Windows printer driver (usually supplied with Windows or with the output device) then converts the commands to the printer's own format for producing the output.

When a Windows printer driver is not available for a device or when options such as `hsize=` or `vsize=` are to be used to customize the size specifications of a graph, the use of a SAS/GRAPH native driver, such as `pscolor`, is recommended. Some other examples of native drivers are `hpljs3`, which

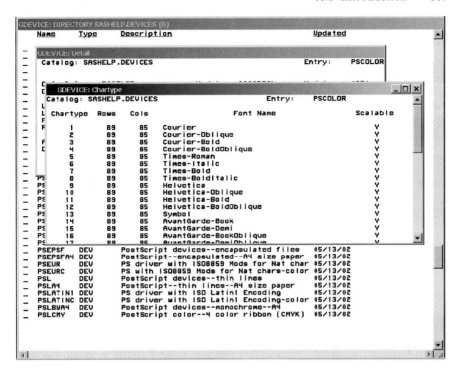

Fig. A.4. View of proc gdevice output showing the Chartype window for the pscolor device

produces output in the PCL5 language that is used by Hewlett Packard LaserJet III printers, and hp7550, which produces HPGL output that is used by Hewlett-Packard 7550 plotters.

SAS/GRAPH Graphics Output

SAS/GRAPH procedures that create graphics output require a device driver to display the output on computer monitors or printers. Many default features of the graphics output are determined by parameter values that are stored in the device entry. SAS/GRAPH obtains these values from the device driver for the default device (or the device driver for the the device specified in a SAS/GRAPH program) for producing graphics output appropriate for displaying on that device. Graphics options available in the goptions statement, SAS/GRAPH *global statements* (e.g., the axis statement), and SAS procedure statements allow the user to override these device parameter settings for the duration of a SAS session. Graphics parameters set thus hold until the user *resets* them or modifies them by executing another goptions statement or a SAS/GRAPH global statement. For an example of a goptions statement, see Fig. 3.2 in Chapter 3. The *executable module* of a device driver converts *device-independent graphics output* produced by a user-written SAS/GRAPH program into *device-dependent graphics output*; that is, commands that are recognized by the particular device that enables it to display the user's graph. See Fig. A.5 for a diagram showing the flow of these activities during the execution of a SAS/GRAPH program.

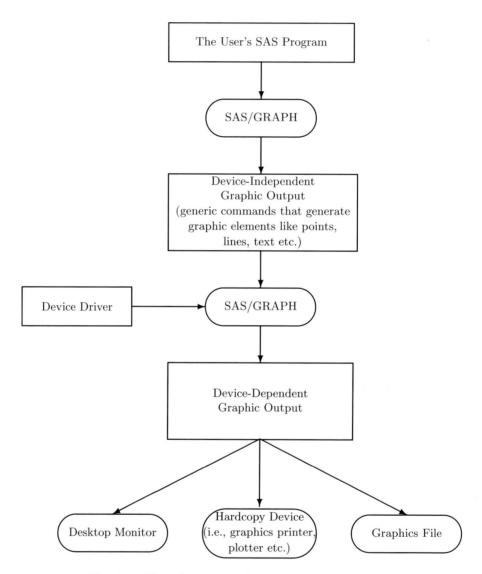

Fig. A.5. Flow of operations for producing SAS graphics

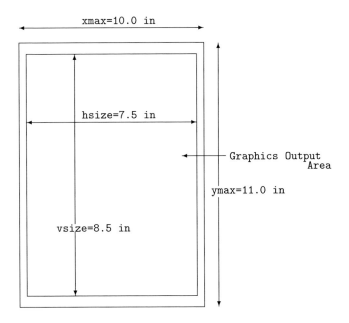

Fig. A.6. Graphics output area of the `pscolor` device

SAS/GRAPH Catalogs and Graphics Stream Files

Graphics output produced by SAS/GRAPH procedures are stored in SAS *catalogs* that can be either temporary or permanent. Temporary catalogs are deleted at the end of the current SAS session. Each graph is stored as a *catalog entry* of the type GSEG. SAS/GRAPH procedure `greplay` enables graphics stored as catalog entries to be redisplayed without using the procedure that produced the graph. This facility is useful for creating multiple plots to be displayed on the same window or printed page. By default, SAS/GRAPH stores graphics produced by procedures in a temporary catalog named `work.gseg`. Catalog entries may be device dependent or device independent. A *graphics stream file* (GSF) is a file that contains device-specific graphics commands generated by the device driver and thus can be directly sent to a device or transported to another location, for example, for generating hardcopy output.

Graphics Output Area

Device parameters `xmax=` and `ymax=` determine the size of *the entire physical display area* for a device. For hardcopy devices, such as high-resolution printers and plotters, the parameters `hsize=` and `vsize=` may be used to define a smaller area as the *graphics output area*, so a margin is possible. For example, see the `Detail` window in Fig. A.1 for settings of these parameters for the `pscolor` device driver. For a PC monitor, the *graphics output area* is identical

510   A SAS/GRAPH

to the entire window. Figure A.6 displays a sketch the graphics output area for a `pscolor` device.

The user can change the size of the *graphics output area* by specifying new values for `hsize=` and `vsize=` in a `goptions` statement. For a monitor display device, `hsize=` and `vsize=` are set to zero in the device entry because the entire area of a graphics window is used to display the graphics output. By default, various graphic elements will be placed in the graphic output area in locations indicated in Fig. A.7.

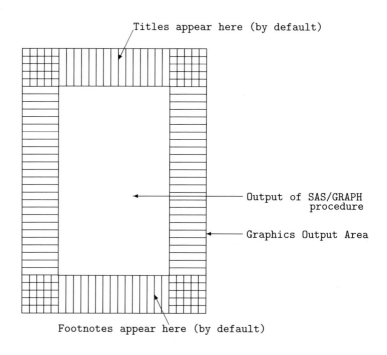

Fig. A.7. Placement of graphics elements in the graphics output area

Device Resolution and Cell Size

Device resolution is the number of pixels/inch defined for a particular device, in both horizontal and vertical directions (i.e., `xpixels/inch` and `ypixels/inch`, respectively). For example, for the above `pscolor` device driver, settings of `ypixels=3300` and `ymax=11.00 in` set in the device entry defines the y-resolution to be 300 pixels/inch (300 `dpi` or 300 dots per inch) (i.e., 3300 pixels/11.00 inches). Changing `hsize=` and/or `vsize=` will not affect the resolution, because the number of `xpixels` and `ypixels` are fixed for a device (i.e., for specific set of values of `xmax` and `ymax`).

Cells control the default sizes of graphic elements (i.e., `titles`, `symbols`, `labels`, etc.). The size of a cell is defined by the settings for device parameters (`lcols`, `lrows`) for landscape mode and (`pcols`, `prows`) for portrait mode. These settings are usually overridden by setting `hpos` (which overrides `lcols`, `pcols`) and `vpos` (which overrides `prows`, `lrows`) by specifying values for these parameters in the `goptions` statement.

Thus, using `hpos=` and `vpos=` to set alternate values for these parameters, the user may change the default cell size. As seen from the device entry for the `pscolor` device driver, device parameters `hsize=7.5 in`, `vsize=10.0 in`, `pcols=75`, and `prows=81` determine the default cell size to be $(7.5/75) \times (10.0/81)$ inches (i.e., 0.1×0.12345 inches). Now, for example, specifying `hpos=100` and `vpos=100` in a `goptions` statement changes the cell size to $(7.5/100) \times (10.0/100)$ inches (i.e., 0.075×0.1 inches).

By default, SAS/GRAPH uses units measured in *cell* size to draw graphic elements. For example, character heights for plotting a title uses two *cell* units as character heights if `h=2` is specified (i.e., if units are omitted) in a `title` statement. Since changing cell size will affect the appearance of graphic elements, absolute units such as `in` or `cm` may be preferred for preserving the appearance of a graph on different devices.

SAS/GRAPH Colors

The user can specify colors to be used for plotting various graphics elements in SAS/GRAPH. Colors can be specified as values for certain options in procedure action statements for plotting graphics elements in all SAS/GRAPH procedures that produce graphics output (e.g., `caxis=` option to specify axis color in a `plot` statement).

Colors specified in procedure action statements override those specified as values of the `color=` option in SAS/GRAPH global statements. The `color=` option in the `axis` statement or the `note` statement are examples. Colors can also be specified in some options available in the `goptions` statement, such as `ctext=` or `csymbol=`, for specifying colors for text or symbols, respectively, but these are overridden by the values given in procedure statements.

If SAS/GRAPH cannot find color specifications in the user's program for plotting the respective graphics elements, default values are taken in sequential fashion from a list specified in the `colors=` option in the `goptions` statement. This list is called the *colors list*. If a color list is not provided in the user's `goptions` statement, SAS/GRAPH uses the device's *default colors list* (see Fig. A.3) for plotting graphics elements.

SAS/GRAPH Color Names

SAS/GRAPH supports several color-naming schemes and allows advanced users to define their own color-naming schemes. However, for the purposes of this text, only those color names such as BLUE, PURPLE, and VIYPK ("vivid yellowish pink") predefined by SAS/GRAPH are recommended to be used in SAS/GRAPH programs. An abbreviated list along with their RGB values (coded in hexadecimal) is provided in Table B.13 in Appendix B. A predefined list of the SAS color names and their accompanying RGB values are contained in the SAS Registry and may be viewed using the Registry Editor. On a windowing environment, selecting `Solutions` → `Accessories` → `Registry Editor` from the to menu bar will result in the appearance of the Registry Editor. Double-clicking on the COLORNAMES\HTML icon enables one to view the contents of this subkey of the SAS Registry. Executing

```
proc registry list startat="colornames\html";run;
```

writes the contents of the registry into the SAS Log. In addition to viewing predefined SAS color names and RGB values, the SAS Registry Editor also can be used to create and define the user's own color names and RGB values.

SAS/GRAPH Fonts

Users control the type style of text and symbols in graphics output by specifying values for the `font=` option available with SAS/GRAPH global statements for defining various graphics elements. The value specified for this option can be the name of an existing SAS/GRAPH *software font*, a user-generated font, or a *hardware font*. For purposes of this text, only software and hardware fonts will be discussed. If `font=none` is specified, the default hardware font for the current device is used. In the `Parameters` window for the `pscolor` device, shown in Fig. A.2, the default *hardware font* is identified by the parameter `Chartype`. The `Chartype` window (see Fig. A.4) lists the *hardware fonts* that are available for the current device. The default hardware font for the `pscolor` device is thus the `Courier` font because the value for Chartype is 1. In some situations, when the device's hardware font cannot be used, a software font named `simulate` font is substituted. This font simulates the selected device's set of characters.

Either a name of a software font or `none` may be specified as the value for a keyword option such as `ftext=` in the `goptions` statement or the `font=` option in SAS/GRAPH global statements (including the `note` statement). The specifications in the `goptions` statement are overridden by a value specified in a `font=` option. If no value is found for the font, SAS/GRAPH will use `font=none` as the default for plotting all text, except for `title1` statements, for which `font=complex` will be used as the default. SAS/GRAPH software fonts are physically located in the SAS/GRAPH font catalog named SASHELP.FONTS. A subset of SAS software fonts for text in Roman alphabet is listed in Table A.1. SAS software fonts are also available for non-Roman text characters (e.g., `arabic` and `greek`). The character set for the `greek` font is displayed in Fig. A.8

Type Style	Font Name	Type Sample
Century Bold	CENTB	A B C a b c 1 2 3
Century Bold Empty	CENTBE	A B C a b c 1 2 3
Century Italic	CENTBI	A B C a b c 1 2 3
Century Italic Empty	CENTBIE	A B C a b c 1 2 3
Century Expanded	CENTX	A B C a b c 1 2 3
Century Expanded Italic	CENTXI	A B C a b c 1 2 3
Hershey Sans Serif	SIMPLEX	A B C a b c 1 2 3
Hershey Sans Serif Bold	DUPLEX	A B C a b c 1 2 3
Hershey Serif	COMPLEX	A B C a b c 1 2 3
Hershey Serif Bold	TRIPLEX	A B C a b c 1 2 3
Hershey Serif Italic	ITALIC	A B C a b c 1 2 3
Swiss	SWISS	A B C a b c 1 2 3
Swiss Empty	SWISSE	A B C a b c 1 2 3
Swiss Bold	SWISSB	A B C a b c 1 2 3
Swiss Bold Empty	SWISSBE	A B C a b c 1 2 3
Swiss Expanded	SWISSX	A B C a b c 1 2 3
Swiss Expanded Empty	SWISSXE	A B C a b c 1 2 3
Swiss Expanded Bold	SWISSXB	A B C a b c 1 2 3
Swiss Italic	SWISSI	A B C a b c 1 2 3
Swiss Italic Empty	SWISSIE	A B C a b c 1 2 3
Zapf	ZAPF	A B C a b c 1 2 3
Zapf Bold	ZAPFB	A B C a b c 1 2 3
Zapf Italic	ZAPFI	A B C a b c 1 2 3
Zapf Bold Italic	ZAPFBI	A B C a b c 1 2 3
Old English	OLDENG	A B C a b c 1 2 3
Script	SCRIPT	A B C a b c 1 2 3
Cscript	CSCRIPT	A B C a b c 1 2 3

Table A.1. Selected SAS/GRAPH fonts for Roman alphabet text

A.2 SAS/GRAPH Statements

SAS/GRAPH statements are of two kinds: statements that appear anywhere in the program and action statements that appear within proc steps. Statements of the first kind, called global statements, define parameters globally; that is, these definitions can be referenced in any proc step action statement and remain in effect until their settings are overridden by other global statements or until the end of the current SAS session.

In global definition statements, the SAS/GRAPH keyword (e.g., `axis`) suffixed by an integer that uniquely identifies the definition is followed by a

	!	"	#	$	ϑ	&	'	(	)	*	+	,	−
	!	"	#	$	%	&	'	(	)	*	+	,	−
.	/	0	1	2	3	4	5	6	7	8	9	:	;
.	/	0	1	2	3	4	5	6	7	8	9	:	;
∅	=	ς	?	@	Α	Β	Ξ	Δ	Ε	Φ	Γ	Η	Ι
<	=	>	?	@	A	B	C	D	E	F	G	H	I
Ε	Κ	Λ	Μ	Ν	Ο	Π	Θ	Ρ	Σ	Τ	Υ	∇	Ω
J	K	L	M	N	O	P	Q	R	S	T	U	V	W
Χ	Ψ	Ζ	_	α	β	ξ	δ	ε	φ	γ	η	ι	ϵ
X	Y	Z	_	a	b	c	d	e	f	g	h	i	j
κ	λ	μ	ν	ο	π	θ	ρ	σ	τ	υ	∂	ω	χ
k	l	m	n	o	p	q	r	s	t	u	v	w	x
ψ	ζ	}	|	{									
y	z	{	|	}									

Fig. A.8. Character set for the `greek` font

set of arguments that consist of one or more options that are separated by at least one blank. Some of these options may be a list of parameters that are separated by at least one blank and enclosed in parentheses. Both options and parameters may consist of keyword literals (e.g., `frame`) or specifications of the form `keyword=value`.

Options provided via global statements take precedence over default settings in a `goptions` statement. If parameter values are not available from either of these statements, then SAS/GRAPH will use the parameter settings from a device entry for the purpose of producing the graph.

Action statements that appear in a proc step (e.g., `plot` statement) provide information specific to the graphs being drawn by the particular procedure. They may designate specific global definitions that are to be used for drawing various elements of the graph. In action statement syntax, a list of options may be specified following the action statement arguments preceded by a slash. These options may be in the form of keyword literals (e.g., overlay) or specifications of the form keyword=value. The action statement plot logpres*bpoint=1 logpres*bpoint=2/vaxis=axis1 haxis=axis2 overlay; used in Fig. 3.2 in Chapter 3 illustrates the use of options. Here, vaxis=axis1 and haxis=axis2 specify that the axis definitions axis1 and axis2 are to be used for drawing the vertical and the horizontal axes, respectively, and the keyword overlay specifies that the two plots requested be drawn on the same graph.

GOPTIONS *options* ;	
cback=	specify background color, for use with terminal/monitor (usually ignored by hardcopy device)
colors=( )	default *colors list*: colors used for various elements of a graph separated by blanks. This list will be used if colors for these elements are not specified explicitly in SAS/GRAPH statements. This list overrides the colors list of the current device driver (see, e.g., Fig. A.3)
ftext=	default font to be used for all text
ctext=	default color to be used for all text in the graph
gunit=	*pct, in, cm or cell*; unit for text height in title, note, footnote, symbol, axis, and legend statements; *cell* is the default unit if not specified here
vpos=	number of cells in vertical axis (10 to 300)
hpos=	number of cells in horizontal axis (10 to 300)
hsize= vsize=	in inches in inches } specify size of display area.
device=	device-driver-name (e.g., ps300, pscolor, hp7480a, hplj4si, x4045p, etc.) (Execute proc gdevice to view the complete list of device drivers supported in SAS/GRAPH.)
targetdevice=	specify intended hardcopy device to display graph on the monitor as it would appear on that device (may not be an exact reproduction, but SAS will duplicate it as close as possible)
gaccess=	specifies format and destination of graphics stream (e.g., gsasfile)
gsfname=	specifies graphics stream file name

Table A.2. Summary of goptions parameters

A.2.1 Goptions statement

The SAS/GRAPH `goptions` statement can be used to specify certain graphics options globally as well as the name of the device driver to be used to produce device-dependent graphics output. Since all available options are too numerous to be described here, a useful subset is summarized in Table A.2. Executing `proc goptions;` will cause a full list of the available options to be displayed. An important graphics option is the `device=` option. The device-specific graphics output will be produced using the device driver specified using these options. Graphics options may also be set interactively during a SAS session using the System Options window but this is omitted from this discussion.

A.2.2 SAS/GRAPH global statements

SAS/GRAPH global statements define different elements of a graph. They can appear anywhere in a SAS program, typically before they are actually used in SAS proc steps. These definitions remain in effect until they are changed or overridden by new definitions and may be used in appropriate procedure action statements in any proc step.

A new definition may be inserted at any point in the program before it is used in an action statement to obtain a desired effect in a graph. Thus, groups of global statements and action statements followed by a `run` statement may appear within the same proc step. The keywords for global definition statements are `title`, `footnote`, `axis`, `legend`, `pattern`, and `symbol`. These keywords are followed by a definition number (e.g., `title1`).

TITLE, FOOTNOTE, and NOTE statements

A useful selection of options that may be specified on the `title`, `footnote`, and `note` statements are summarized in Table A.3. The `note` statement, although included here, is not a global statement; that is, `note` statements must be included in proc steps put together to draw the graphs on which the notes are to be placed.

AXIS statement

Action statements in proc steps assign `axis` definitions as values of keyword options such as `vaxis=` or `haxis=`, as illustrated in the SAS/GRAPH program in Fig. 3.2 in Chapter 3. As in other global definition statements, axis definitions are identified by a definition number appended as a suffix (e.g., `axis1`). Options available with the axis statement control various properties of an axis of a graph, such as the number of major tick marks and the text that is to appear at each of the tick marks. A subset of options available with the axis statement are summarized in Table A.4. These descriptions are supplemented by the following additional notes:

A.2 SAS/GRAPH Statements   517

```
TITLEn    <options> <text-string> <options> <text-string> ...;
FOOTNOTEn <options> <text-string> <options> <text-string> ...;
NOTEn     <options> <text-string> <options> <text-string> ...;
```

color= c=	specify color for the text-string that follows; `red`, `blue`, `green`, `black`, `violet`, etc. or SAS color names like `lightpink`, `steelblue`, or `libr`. *Default*: first color in *colors list* or first default color for device
font= f=	specify font for the text-string that follows; `swiss`, `triplex`, `greek`, `math`, `italic` etc. *Default:* `complex` for `title1` and *none* for all other text. Specifying *none* causes the default hardware font for the output device to be used
height= h=	text height specified as *n in*, *n cm*, *n pct*, *n cells*, or *n* with units unspecified. *Default:* the unit specified in `gunit=` in `goptions` or *cells*
justify j=	alignment of text-string `l`, `r`, or, `c`, for left-, right-, or center-justify *Default:* c
angle= a=	angle in degrees displayed in the range -90 to 90
move= m=	(x,y) coordinates of starting location of text string, specified using units, as in `h=` (e.g., `(1,2) in`, `(2 cm, 3 pct)`); can be specified as relative coordinates, $(\pm x, \pm y)$, starting from the end of last text line drawn in the current statement.
lspace= ls=	line space; space above a title or below footnotes to be left blank; specified using units as in `h=`
underlin= u=	underline text string; 0, 1, 2, or, 3 specifies thickness of line.
box=	draw box around text-string; 1, 2, 3, or, 4 specifies thickness of line.

Table A.3. Summary of `title`, `footnote`, and `note` statement options

order=   The `order=` option is especially useful for specifying a list of numeric values that represent major tick marks on an axis. Some examples of several possible forms are

        `order=(10,20,30,40,50,60)`
        `order=1 to 9`
        `order=10 to 60 by 10`
        `order=60 to 10 by -10`
        `order=15,25,30 to 60 by 10,75`
        `order=('Iowa','Nebraska','Ohio','Wisconsin','Illinois')`

As in SAS Example C1 (see Fig. 3.2 in Chapter 3), use of this option overrides the default values for placement of the major tick marks calculated by SAS/GRAPH for the axis, using its own scaling algorithm. In either case, the *values that are displayed at the tick marks* may be defined by a value= option, discussed below. As in the last order= example, the order= option may also be used for character-valued axis variables, in which case the values for order= is a list of character strings enclosed in quotes.

label= A value for label= option is a list of text description parameters along with a text-string in quotes to be used as the axis label, all enclosed in parentheses. If the text-string is omitted, the description parameters are applied to the default label, which is the name of the variable plotted on the axis or the label assigned to that variable. The parameters c= f= h= a= j= respectively specify the color, font, height,

AXISn *options* ;	
order=	specify list of data values in the order they will appear on the axis. See below for various formats for this list
label=	specify *none* or a label on the axis in the form: (*text-description* 'text') where *text-description* represent text description parameters c= f= h= a= j= . *Default:* the variable name or the variable label
value=	specify *none* or values associated with major tick marks in the form: (*text-description* 'text1' *text-description* 'text2'...) where *text-description* represent text description parameters c= f= h= a= j= . *Defaults:* the values of the variable to appear at major tick marks as calculated by SAS/GRAPH procedure
major=	specify *none* or major tick mark attributes in the form (c= n= h=) for color, number, and height
minor=	(specify *none* or minor tick mark attribute in the form (c= n= h=) for color, number, and height
origin=	specify as (x, y) coordinate with or without units, in which case default unit will be used
length=	n *in,* n *cm,* n *pct,*n *cell,* or n in default units
style=	kind of line drawn for axis, specified using line option number
color=	specifies color of axis line, all tick marks, and all text. Color for axis label can be changed using the label= option and the color for the major tick mark value using the option value=
width=	specifies width of axis line in multiples of 1. Default is 1

Table A.4. Summary of axis statement options

angle, and justification, of the text-string to be plotted. Values for these are to be specified in a similar way to options on other global statements are specified (e.g., see title). If a value for any of these is not specified (by omitting the parameter specification entirely), a default value for the parameter will be assumed as follows:

- for height: 1 in default units, unless a value is specified in the htext= option in a goptions statement. The default unit is *cell* unless a value is specified in the gunit= option in a goptions statement.
- for angle: 0; that is, text is to be plotted horizontally.
- for font: *none*, in which case the default hardware font for the output device will be used.
- for color: the first color in the *colors list*, unless a value is specified in the ctext= option in a goptions statement.
- for justification: *right* (or *r*) for the left vertical axis and *center* (or *c*) for the horizontal axis.

value= The value= option defines the values that *appear* at the major tick marks, whether the actual values that represent the major tick marks are calculated by SAS/GRAPH by default or those specified in an order= option. It is of the form

value= <*text-description*> <*text*> <*text-description*> <*text*> ... ;

where *text-description* represents the group of text description parameters c=   f=   h=   a=   j=   , as discussed earlier for the label= option.

style= This option specifies a line type for the axis line and takes integer values in the range 0 to 46. The default is 1, a solid line. See Fig. A.9 for an abbreviated list of line types.

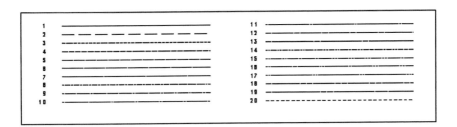

Fig. A.9. Line types for the style= and line= options

SYMBOL statement

Symbol definitions are referenced in SAS/GRAPH procedure action statements using the the keyword symbol suffixed with a sequence number to

SYMBOLn *options* ;		
color= c=	color for symbols, lines, confidence limit (CL) lines, outlines	
ci=	color for lines	
co=	color for CL lines, staffs, bars, and outlines	
cv=	color for symbols	
value= v=	text-string, special symbol, or **none** to specify plot symbol; Setting for **value** depends on the choice of **font**	
font= f=	specifies the font for the plot symbol. See below for details.	
height= h=	height of plot symbol, in units as described for the **title** statement	
width= w=	thickness of interpolated lines in units of 1. Default is 1.	
repeat= r=	number of times symbol statement is to be reused.	
line= l=	line option from line option table (Fig. A.9).	
interpol= i=	type of interpolation of plot points. Various settings for this parameter and their options are summarized in Table A.7	

Table A.5. Summary of **symbol** statement options

indicate that a particular symbol definition be used to represents the points plotted. The **symbol** statements not only define what plot symbols are to be used for the points plotted but also how the points are to be interpolated. Using the **overlay** option in the plot statement in a SAS/GRAPH proc step, several plots representing different views of the same data set may be plotted on the same graph, by making use of a different **symbol** definition in each plot. In the SAS Example C1 program shown in Fig. 3.2 in Chapter 3, this technique is used to overlay a regression line fitted to a set of data on a scatter plot of the data.

SAS/GRAPH plots a symbol using the specifications supplied for the options **value=** and/or **font=** in a **symbol** statement. A subset of options available with the symbol statement are summarized in Table A.5. These descriptions are supplemented by the following additional notes:

A.2 SAS/GRAPH Statements

value= To plot symbols like those shown in Table A.6, the value= option is set to the value corresponding to the selected symbol exactly as shown (e.g., value=plus or value=+ for plotting + or ⊕ as a symbol, respectively) with the font specification omitted. The font= option was introduced previously with respect to its use with other global statements, in particular, relative to text fonts. The software text fonts discussed, such as simplex, swissb and italic, are also available for use as plotting symbols. Any text-string or single character can be plotted as a symbol using any text font available in SAS/GRAPH, by setting the text-string or the single character as a value for the value= option (e.g., value='City' and font=swissb).

font= For plotting nontext symbols, an appropriate font name is selected (e.g., math, music or special) and value= option is set to appropriate character code obtained from the corresponding font map. The font map for the math font, for example, is displayed in Fig. A.10. For Roman text, a font name is selected from Table. A.1. For plotting non-Roman text characters, the appropriate font name is specified (e.g., greek) and the value= is set to an appropriate Roman text-string or character, using *font maps* such as the map shown in Fig. A.8. For example, to plot the symbol Ω, use font=greek and value=C. These font maps can be obtained by executing proc gfont (see the description of the gfont procedure in SAS/GRAPH documentation for an example). Symbols appearing in Fig. A.11 can be produced by setting the value= option to the character code that maps to the symbol and setting font= option to special.

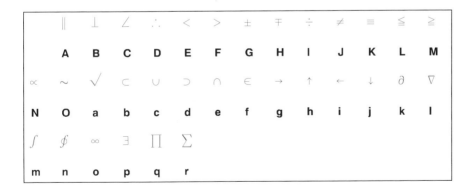

Fig. A.10. Character set for the math font

repeat= This option allows the user to specify the number of times a symbol definition is reused before the next symbol definition is applied. If the repeat= option is omitted, each symbol definition is applied only once. A slight complication to this behavior arises, however, if no symbol color op-

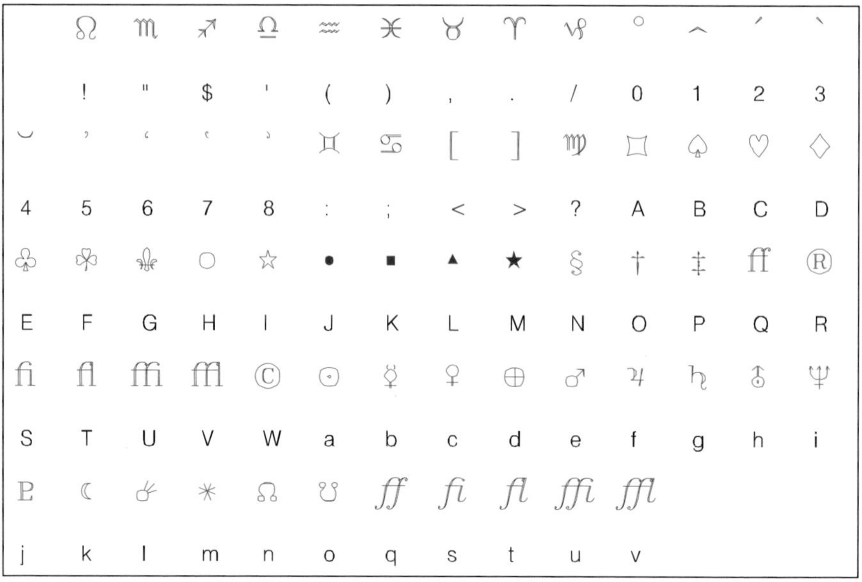

Fig. A.11. Character set for the special font

value=	Plot Symbol	value=	Plot Symbol
plus	+	hash	#
x	×	point	.
star	*	dot	●
square	□	circle	○
diamond	◇	- (i.e.,hyphen)	⊙
triangle	△	+ (i.e.,plus)	⊕

Table A.6. A subset of special symbols for plotting points: set `value=` option exactly as shown and omit the font specification

tion (such as `ci=`, `cv=`, `co=`, `color=` or the graphics option `csymbol=`) are used. In this case, the symbol definition is first cycled through each color in the colors list each time the definition is used. Then the resulting set of definitions are reused the number of times specified in the `repeat=` option.

`interpol=` This option specifies the type of interpolation among the plot points. A selection of interpolations available is summarized in Table. A.7. The simplest example occurs when the keyword *join* is specified as the value (i.e., `i=join`): The points are connected by line segments in the order the pairs of values each plotted point represents occur in the data set. Thus, a meaningful interpretation is to be made from the plot, the data values must first be arranged in increasing order of magnitude of the values for the variable plotted on the horizontal axis. A more complex interpolation request is when a value `rl`, `rq`, or `rc` is specified, in which case, a regression line representing a linear, quadratic, or cubic polynomial, respectively, fitted to the data is plotted as the interpolated line. Other options can be combined with these letters to request confidence bands corresponding to a selected confidence coefficient to be plotted around the fitted line.

A `hilo`, `std`, and `box` (suffixed with associated suboptions some of which are summarized in Table. A.7, if desired) allow a graphical summary of the distribution of the responses (y-values) at each of the x-values to be displayed. Obviously, such summaries are useful for comparing sample distributions across different values of the x variable if replications are available at each x-value. Thus these interpolations make sense only when the x-variable is a category variable. These may include nominal variables such as "gender," "state," "region," etc., constructed ordinal variables such as "income group" or "age group," or classificatory variables such as "treatments" or "blocks."

PATTERN statement

The `pattern` global statement defines the *fill pattern* and the color of areas in a graph, such as bars and blocks, underneath and between plotted lines, slices of pie charts, and areas in maps. The fill patterns can be empty, solid, or composed of parallel lines or cross-hatched lines, where the density and angle of the lines can be controlled.

Table A.8 summarizes some options available for pattern definitions and some of these are described in more detail:

> value= Options that may be specified for fill patterns in bars or blocks are `empty` or `e`, `solid` or `s`, or a pattern specified in the form
>
> *style<density>*
>
> where *style* and *density* are one-character codes obtained from the pattern guide in Fig. A.12. For example, in a bar or a block, `v=L1 c=red` will produce a red-colored, slanted line pattern similar to the fill pattern shown in the lower left corner of Fig. A.12. If the option is used to specify a fill pattern for a map or areas under curves, the values `mempty` or `me`, `msolid` or `ms`, or a pattern specified in the form
>
> *Mdensity<style<angle>>*
>
> where *style* and *density* are one-character codes obtained from the pattern guide in Fig. A.12, and *angle* is a value in degrees $0, \ldots, 360$ measured anticlockwise from the horizontal, may be used. For example, using `v=M3X45`

```
interpolate=        (type of interpolation of plot points)
i=
    none            unconnected points
    join            connected in order of occurrence
    box             box-plots
    box <options>
        h       fill
        j       join medians
        t       tops and bottoms
    hilo            vertical line connecting min to max
    hilo <options>
        b       bar
        j       join means
        t       tops and bottoms
        bj
        tj
    r <option>  fits a regression line
        l       linear fit
        q       quadratic fit
        c       cubic fit
    r <fit type><0><option< 50|...|99 >>
                    fits a regression line with fit type l, q, or c
            cli with prediction bands for individual prediction
            clm with confidence bands for estimated mean
    spline          fits a smoothed line using spline method (details omitted)
    std < 1|2|3 ><var><option>
                    vertical line connecting mean ± 1, 2, or 3 times $s_y$,
                    where $s_y$=std. dev. of each sample
                        if var=m, use s.e. of mean $s_{\bar{y}}$ instead of $s_y$
                        if var=p, $s_y$=std. dev. using pooled variance
                        if var=mp, $s_{\bar{y}}$ computed with pooled variance
        b       connects with bar instead of vertical line
        j       connects means across bars with a line
        t       adds tops and bottoms to each vertical line
    step            step function (details omitted)
```

Table A.7. Selection of settings available for `interpolate=` in the `symbol` statement

`c=blue` will produce the cross-hatching fill pattern shown in the center of Fig. A.8 in blue in a map or an area specified under a curve.

`repeat=` This option has a similar effect as when used in a `symbol` statement described earlier. It specifies the number of times a pattern definition is applied before the next pattern definition is used. By default, each pattern definition is applied only once. However, if no pattern statement color

PATTERNn *options* ;	
color= c=	color for the bar or subgroup; if color is not specified the `pattern` statement will be repeated for each color in the `colors` list
value= v=	pattern for the bar/block, map/plot or pie/star as described in the text
repeat= r=	number of times same `pattern` statement is to be reused

Table A.8. Summary of pattern statement options

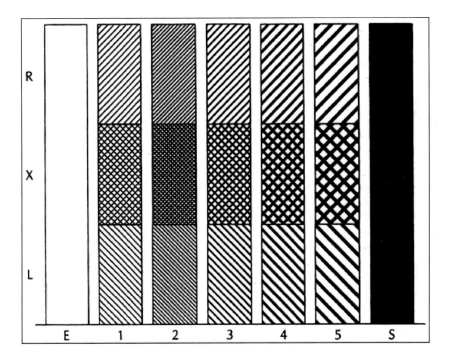

Fig. A.12. Values for style and density parameters in the value= option in the pattern statement

option (color=) or graphics pattern color option (cpattern=) is present, the symbol definition is first cycled through each color in the colors list. Then the resulting set of patterns is repeated a number of times specified in the repeat= option.

A.3 Printing and Exporting Graphics Output

Printing SAS/GRAPH output depends on the operating environment, the output device, and whether the output device is connected to the computer or accessible via a network. In a SAS windowing environment, the default device is usually a SAS graph window and an appropriate device driver (e.g., WIN) is specified as the default. SAS/GRAPH output may be printed directly from the SAS graph window using the `File` drop down menu by selecting the `Print` item while the Graph window is active (i.e., highlighted). At the outset, if it is intended to send the graph directly from the SAS graph window to the output device, `targetdevice=` option must be used, say, in a `goptions` statement in the SAS program, to specify a printer driver. This will ensure that the graph to appear in the output device will be duplicated in the display window. If a target-device has not been specified, the printed output will be automatically produced using an appropriate Windows generic driver (e.g., `winprtc`).

Device-dependent graphics commands produced by a device driver selected in the SAS/GRAPH program can be directly sent to a device for plotting or printing or saved in an external file. For sending graphics output directly to a printer that has been previously configured in the Windows system, SAS/GRAPH provides several generic printer drivers (e.g., `winprtc`; see Section A.1 at the end of the paragraph *SAS/GRAPH Device Drivers*). These drivers send generic graphics commands to the Windows printer driver for the output device. The Windows printer driver (usually supplied with Windows or with the output device) then converts the commands to the printer's own format for producing the output.

When a Windows printer driver is not available for a device or when options such as `hsize=` or `vsize=` are to be used to customize the size specifications of a graph, the use of a SAS/GRAPH native driver, such as `pscolor`, is recommended. As discussed in the paragraph entitled *SAS/GRAPH Device Drivers* in Section A.1, device driver definitions of both of the above types are available in a SAS/GRAPH catalog and are accessible via `proc gdevice`.

Alternatively, the *graphics stream file* created by the a device driver may be exported to an external file. Typically, the device deriver selected for this purpose produces the graphics output in one of several *graphics file formats* (e.g., `bmp`, `ps`, `eps`, `gif`, `tiff`, `jpeg`, `png`, `pdf`, etc.). These files may be printed using host commands, (e.g., `lpr` command in UNIX), viewed with an appropriate software or a browser (e.g., `ghostview` or `MS Photoshop`), or imported into other software (e.g., `latex` or `MS Word`). For example, the output (two graphs) from the program for SAS Example C1 will be saved in a file named `c1.ps` as postscript output if the `goptions` statement in Fig. 3.2 of Chapter 3 is replaced by the two statements in Fig. A.13. The `c1.ps` file produced by executing the modified program can then be used in any of the ways described earlier.

In a windowing environment, SAS/GRAPH output may be exported directly from the Graph window using the `File` drop-down menu by selecting

```
filename gsasfile "C:\Documents and Settings\...\Chapter3\c1.ps";
goptions rotate=landscape gaccess=gsasfile device=pscolor
                        gsfmode=append hsize=8 in vsize=6 in;
```

Fig. A.13. SAS Example C1: Amending the program to save postcript file

the `Export as Image...` item while the Graph window is active (i.e., highlighted). In the resulting `Export` window, a `file type` (e.g., `jpg`) is selected and a *file name* entered for saving the exported graphics file in a chosen location. The file types available are those supported by the operating system. Although this method can be used to create graphics stream files of many format types, a much larger choice of device drivers is available when the `goptions` statement is used to create a GSF.

B

Tables

530   B Tables

Table B.1. Fuel consumption data for the 48 contiguous states (Weisberg, 1985)

(1) STATE: State FIPS code
(2) POP: 1971 Population, in thousands
(3) TAX: 1972 Motor Fuel Tax Rate, in cents per gallon
(4) NUMLIC: 1971 Number of Licensed Drivers, in thousands
(5) INCOME: 1972 Per Capita income, in thousands of dollars
(6) ROADS: 1971 Miles of Federal-Aid Primary Highways, in thousands
(7) FUELC: 1972 Fuel Consumption, in millions of gallons

STATE	POP	TAX	NUMLIC	INCOME	ROADS	FUELC
ME	1,029	9.0	540	3.571	1.976	557
NH	771	9.0	441	4.092	1.250	404
VT	462	9.0	268	3.865	1.586	259
MA	5,787	7.5	3,060	4.870	2.351	2,396
RI	968	8.0	527	4.399	0.431	397
CT	3,082	9.0	1,760	5.342	1.333	1,408
NY	18,366	8.0	8,278	5.319	11.868	6,312
NJ	7,367	8.0	4,074	5.126	2.138	3,439
PA	11,926	8.0	6,312	4.447	8.577	5,528
OH	10,783	7.0	5,948	4.512	8.507	5,375
IN	5,291	8.0	2,804	4.391	5.939	3,068
IL	11,251	7.5	5,903	5.126	14.186	5,301
MI	9,082	7.0	5,213	4.817	6.930	4,768
WI	4,520	7.0	2,465	4.207	6.580	2,294
MN	3,896	7.0	2,368	4.332	8.159	2,204
IA	2,883	7.0	1,689	4.318	10.340	1,830
MO	4,753	7.0	2,719	4.206	8.508	2,865
ND	632	7.0	341	3.718	4.725	451
SD	579	7.0	419	4.716	5.915	501
NE	1,525	8.5	1,033	4.341	6.010	976
KS	2,258	7.0	1,496	4.593	7.834	1,466
DE	565	8.0	340	4.983	0.602	305
MD	4,056	9.0	2,073	4.897	2.449	1,883
VA	4,764	9.0	2,463	4.258	4.686	2,604
WV	1,781	8.5	982	4.574	2.619	819
NC	5,214	9.0	2,835	3.721	4.746	2,953
SC	2,665	8.0	1,460	3.448	5.399	1,537
GA	4,720	7.5	2,731	3.846	9.061	2,979
FL	7,259	8.0	4,084	4.188	5.975	4,169
KY	3,299	9.0	1,626	3.601	4.650	1,761
TN	4,031	7.0	2,088	3.640	6.905	2,301
AL	3,510	7.0	1,801	3.333	6.594	1,946
MS	2,263	8.0	1,309	3.063	6.524	1,306
AR	1,978	7.5	1,081	3.357	4.121	1,242
LA	3,720	8.0	1,813	3.528	3.495	1,812
OK	2,634	7.0	1,657	3.802	7.834	1,695
TX	11,649	5.0	6,595	4.045	17.782	7,451
MT	719	7.0	421	3.897	6.385	506
ID	756	8.5	501	3.635	3.274	490
WY	345	7.0	232	4.345	3.905	334
CO	2,357	7.0	1,475	4.449	4.639	1,384
NM	1,065	7.0	600	3.656	3.985	744
AZ	1,945	7.0	1,173	4.300	3.635	1,230
UT	1,126	7.0	572	3.745	2.611	666
NV	527	6.0	354	5.215	2.302	412
WA	3,443	9.0	1,966	4.476	3.742	1,757
OR	2,182	7.0	1,360	4.296	4.083	1,331
CA	20,468	7.0	12,130	5.002	9.794	10,730

Table B.2. Daily maximum ozone concentrations at Stamford, Connecticut (Stmf) and Yonkers, New York (Ykrs), during the period May 1, 1974 to September 30, 1974, recorded in parts per billion (ppb), reproduced from Chambers et al. (1983)

May		June		July		August		September	
Stmf	Ykrs	Stmf	Ykrs	Stmf	Ykrs	Stmf	Ykrs	Stmf	Ykrs
66	47	61	36	152	76	80	66	113	66
52	37	47	24	201	108	68	82	38	18
–	27	–	52	134	85	24	47	38	25
–	37	196	88	206	96	24	28	28	14
–	38	131	111	92	48	82	44	52	27
–	–	173	117	101	60	100	55	14	9
49	45	37	31	119	54	55	34	38	16
64	52	47	37	124	71	91	60	94	67
68	51	215	93	133	–	87	70	89	74
26	22	230	106	83	50	64	41	99	74
86	27	–	49	–	27	–	67	150	75
52	25	69	64	60	37	–	127	146	74
43	–	98	83	124	47	170	96	113	42
75	55	125	97	142	71	–	56	38	–
87	72	94	79	124	46	86	54	66	38
188	132	72	36	64	41	202	100	38	23
118	–	72	51	76	49	71	44	80	50
103	106	125	75	103	59	85	44	80	34
82	42	143	104	–	53	122	75	99	58
71	45	192	107	46	25	155	86	71	35
103	80	–	56	68	45	80	70	42	24
240	107	122	68	–	78	71	53	52	27
31	21	32	19	87	40	28	36	33	17
40	50	114	67	27	13	212	117	38	21
47	31	32	20	–	25	80	43	24	14
51	37	23	35	73	46	24	27	61	32
31	19	71	30	59	62	80	77	108	51
47	33	38	31	119	80	169	75	38	15
14	22	136	81	64	39	174	87	28	21
–	67	169	119	–	70	141	47	–	18
71	45			111	74	202	114		

Note: Missing observations are shown as "--".
Source: Stamford, Connecticut Department of Environmental Protection.
Yonkers, Boyce Thompson Institute.

Table B.3. Rainfall in acre-feet from 52 clouds, of which 26 were chosen randomly and seeded with silver oxide, reproduced here from Chambers et al. (1983)

Rainfall from Control Clouds	Rainfall from Seeded Clouds
1,202.6	2,745.6
830.1	1,697.8
372.4	1,656.0
345.5	978.0
321.2	703.4
244.3	489.1
163.0	430.0
147.8	334.1
95.0	302.8
87.0	274.7
81.2	274.7
68.5	255.0
47.3	242.5
41.1	200.7
36.6	198.6
29.0	129.6
28.6	119.0
26.3	118.3
26.1	115.3
24.4	92.4
21.7	40.6
17.3	32.7
11.5	31.4
4.9	17.5
4.9	7.7
1.0	4.1

Source: Simpson et al. (1975)

Table B.4. Demographic data for 60 countries reproduced from Ott et al. (1987)

(1) COUNTRY:	country name (20 characters max.)
(2) BIRTHRAT:	crude birth rate
(3) DEATHRAT:	crude death rate
(4) INF_MORT:	infant mortality rate
(5) LIFE_EXP:	life expectancy in years
(6) POPURBAN:	percent population in urban areas
(7) PERC_GNP:	per capita GNP in U.S. dollars
(8) LEV_TECH:	level of technology (100 is maximum)
(9) CIVILLIB:	degree of civil liberties (1 = minimal denial of civil liberties, 7 = maximal denial)

(1)	(2)	(3)	(4)	(5)	(6)	(7)	(8)	(9)
ALGERIA	45	12	109	60	52	2,400	17	6
ARGENTINA	24	8	35	70	83	2,030	23	3
AUSTRALIA	16	7	10	75	86	9,210	71	1
AUSTRIA	12	12	12	73	56	9,210	50	1
BOLIVIA	42	16	124	51	46	510	10	3
BRAZIL	31	8	71	63	68	1,890	15	3
BULGARIA	14	11	17	72	65	3,900	44	7
CANADA	15	7	9	75	76	12,000	75	1
CHILE	24	6	24	70	83	1,870	22	5
COLOMBIA	28	7	53	64	67	1,410	15	3
CZECH.	15	12	16	71	74	5,800	72	6
DENMARK	10	11	8	74	83	11,490	71	1
EGYPT	37	10	80	57	44	700	13	5
FINLAND	14	9	6	74	60	10440	57	2
FRANCE	14	10	9	75	73	11,390	62	2
GHANA	47	15	107	52	40	320	10	5
GREECE	14	9	15	74	70	3,970	23	2
HUNGARY	12	14	19	70	54	2,150	49	5
ITALY	11	10	12	74	72	6,350	41	2
INDIA	34	13	118	53	23	260	6	3
IRAQ	46	13	72	59	68	3,400	12	7
IRELAND	19	9	11	73	56	4,810	48	1
ISRAEL	24	6	14	74	87	5,360	33	2
IVORY COAST	46	18	122	47	42	720	9	5
JAPAN	13	6	6	77	76	10,100	53	1
KENYA	54	13	80	53	16	340	11	5
MADAGSCR	45	17	67	50	22	290	12	6
MALAWI	52	20	165	45	12	210	12	7
MALAYSIA	29	7	29	67	32	1,870	14	4
MOROCCO	41	12	99	58	42	750	12	5

(continued)

Table B.4. continued

(1) COUNTRY:	country name (20 characters max.)
(2) BIRTHRAT:	crude birth rate
(3) DEATHRAT:	crude death rate
(4) INF_MORT:	infant mortality rate
(5) LIFE_EXP:	life expectancy in years
(6) POPURBAN:	percent population in urban areas
(7) PERC_GNP:	per capita GNP in U.S. dollars
(8) LEV_TECH:	level of technology (100 is maximum)
(9) CIVILLIB:	degree of civil liberties (1 = minimal denial of civil liberties, 7 = maximal denial)

(1)	(2)	(3)	(4)	(5)	(6)	(7)	(8)	(9)
NETHERLANDS	12	8	8	76	88	9,910	68	1
NEW ZEALAND	16	8	13	74	83	7,410	66	1
NIGERIA	48	17	105	50	28	760	8	3
NORWAY	12	10	8	76	71	13,820	63	1
PAKISTAN	43	15	120	50	19	390	8	5
PERU	35	10	99	59	65	1,040	12	3
PHILIPPINES	32	7	50	64	37	760	15	5
POLAND	20	10	19	71	59	4,200	53	5
PORTUGAL	14	9	20	71	30	2,190	22	2
ROMANIA	15	10	28	71	49	2,200	33	6
SENEGAL	50	19	141	43	42	440	11	4
SOUTH AFRICA	35	14	92	54	56	2,450	33	6
SPAIN	13	7	10	74	91	4,800	28	2
SRI LANKA	27	6	34	68	22	330	9	4
SWEDEN	11	11	7	76	83	12,400	81	1
SWITZRLND	11	9	8	76	58	16,390	57	1
SYRIA	47	7	57	64	47	1,680	16	7
THAILAND	25	6	51	63	17	810	12	4
TOGO	45	17	113	49	20	280	15	6
TUNISIA	33	10	85	61	52	1,290	15	5
TURKEY	35	10	110	63	45	1,230	14	5
U.S.S.R.	20	10	32	69	64	6,350	59	7
UNITED KINGDOM	13	12	10	73	76	9,050	61	1
UNITED STATES	16	9	11	75	74	14,090	100	1
URUGUAY	18	9	32	69	84	2,490	20	4
VENEZUELA	33	6	39	69	76	4,100	25	2
WEST GERMANY	10	11	10	74	94	11,420	66	2
YUGOSLOVIA	17	10	32	70	46	2,570	23	5
ZAIRE	45	16	106	50	34	160	10	7
ZAMBIA	48	15	101	51	43	580	12	6

Table B.5. Hydrocarbon (HC) emissions at idling speed, in parts per million (ppm), for automobiles of various years of manufacture (Koopmans, 1987)

Pre- 1963	1963-1967		1968-1969		1970-1971		1972-1974	
2351	620	900	1088	241	141	190	140	220
1293	940	405	388	2999	359	140	160	400
541	350	780	111	199	247	880	20	217
1058	700		558	188	940	200	20	58
411	1150		294	353	882	223	223	235
570	2000		211	117	494	188	60	1880
800	823		460		306	435	20	200
630	1058		470		200	940	95	175
905	423		353		100	241	360	85
347	270		71		300	223	70	

Note: The data were extracted via random sampling from that of an extensive study of the pollution control existing in automobiles in current service in Albuquerque, New Mexico.

Table B.6. Pollution and demographic data measured on SMSAs in 1960 (McDonald and Schwing, 1973)

(1) PREC:	Average annual precipitation in inches
(2) JANTEMP:	Average January temperature in degrees Fahrenheit
(3) JULTEMP:	Average July temperature in degrees Fahrenheit
(4) OVER65:	% of 1960 SMSA population aged 65 or older
(5) HOUSEHOLD:	Average household size
(6) EDUC:	Median school years completed by those over 22
(7) HOUSING:	% of housing units which are sound & with all facilities
(8) POPN:	Population per square mile in urbanized areas, 1960
(9) NONWHT:	% nonwhite population in urbanized areas, 1960
(10) WHTCOL. :	% employed in white collar occupations
(11) POOR :	% of families with income $< \$3000$
(12) HC :	Relative hydrocarbon pollution potential
(13) NOX :	Relative nitric oxides pollution potential
(14) SO2 :	Relative sulfur dioxide pollution potential
(15) HUMID :	Annual average % relative humidity at 1 PM
(16) MORT :	Total age-adjusted mortality rate per 100,000

(1)	(2)	(3)	(4)	(5)	(6)	(7)	(8)	(9)	(10)	(11)	(12)	(13)	(14)	(15)	(16)
36	27	71	8.1	3.34	11.4	81.5	3243	8.8	42.6	11.7	21	15	59	59	921.870
35	23	72	11.1	3.14	11.0	78.8	4281	3.5	50.7	14.4	8	10	39	57	997.875
44	29	74	10.4	3.21	9.8	81.6	4260	0.8	39.4	12.4	6	6	33	54	962.354
47	45	79	6.5	3.41	11.1	77.5	3125	27.1	50.2	20.6	18	8	24	56	982.291
43	35	77	7.6	3.44	9.6	84.6	6441	24.4	43.7	14.3	43	38	206	55	1071.289
53	45	80	7.7	3.45	10.2	66.8	3325	38.5	43.1	25.5	30	32	72	54	1030.380
43	30	74	10.9	3.23	12.1	83.9	4679	3.5	49.2	11.3	21	32	62	56	934.700
45	30	73	9.3	3.29	10.6	86.0	2140	5.3	40.4	10.5	6	4	4	56	899.529
36	24	70	9.0	3.31	10.5	83.2	6582	8.1	42.5	12.6	18	12	37	61	1001.902
36	27	72	9.5	3.36	10.7	79.3	4213	6.7	41.0	13.2	12	7	20	59	912.347
52	42	79	7.7	3.39	9.6	69.2	2302	22.2	41.3	24.2	18	8	27	56	1017.613
33	26	76	8.6	3.20	10.9	83.4	6122	16.3	44.9	10.7	88	63	278	58	1024.885
40	34	77	9.2	3.21	10.2	77.0	4101	13.0	45.7	15.1	26	26	146	57	970.467
35	28	71	8.8	3.29	11.1	86.3	3042	14.7	44.6	11.4	31	21	64	60	985.950
37	31	75	8.0	3.26	11.9	78.4	4259	13.1	49.6	13.9	23	9	15	58	958.839
35	46	85	7.1	3.22	11.8	79.9	1441	14.8	51.2	16.1	1	1	1	54	860.101
36	30	75	7.5	3.35	11.4	81.9	4029	12.4	44.0	12.0	6	4	16	58	936.234
15	30	73	8.2	3.15	12.2	84.2	4824	4.7	53.1	12.7	17	8	28	38	871.766
31	27	74	7.2	3.44	10.8	87.0	4834	15.8	43.5	13.6	52	35	124	59	959.221
30	24	72	6.5	3.53	10.8	79.5	3694	13.1	33.8	12.4	11	4	11	61	941.181
31	45	85	7.3	3.22	11.4	80.7	1844	11.5	48.1	18.5	1	1	1	53	891.708
31	24	72	9.0	3.37	10.9	82.8	3226	5.1	45.2	12.3	5	3	10	61	871.338
42	40	77	6.1	3.45	10.4	71.8	2269	22.7	41.4	19.5	8	3	5	53	971.122
43	27	72	9.0	3.25	11.5	87.1	2909	7.2	51.6	9.5	7	3	10	56	887.466
46	55	84	5.6	3.35	11.4	79.7	2647	21.0	46.9	17.9	6	5	1	59	952.529
39	29	75	8.7	3.23	11.4	78.6	4412	15.6	46.6	13.2	13	7	33	60	968.665
35	31	81	9.2	3.10	12.0	78.3	3262	12.6	48.6	13.9	7	4	4	55	919.729
43	32	74	10.1	3.38	9.5	79.2	3214	2.9	43.7	12.0	11	7	32	54	844.053
11	53	68	9.2	2.99	12.1	90.6	4700	7.8	48.9	12.3	648	319	130	47	861.833
30	35	71	8.3	3.37	9.9	77.4	4474	13.1	42.6	17.7	38	37	193	57	989.265
50	42	82	7.3	3.49	10.4	72.5	3497	36.7	43.3	26.4	15	18	34	59	1006.490
60	67	82	10.0	2.98	11.5	88.6	4657	13.5	47.3	22.4	3	1	1	60	861.439
30	20	69	8.8	3.26	11.1	85.4	2934	5.8	44.0	9.4	33	23	125	64	929.150
25	12	73	9.2	3.28	12.1	83.1	2095	2.0	51.9	9.8	20	11	26	58	857.622
45	40	80	8.3	3.32	10.1	70.3	2682	21.0	46.1	24.1	17	14	78	56	961.009
46	30	72	10.2	3.16	11.3	83.2	3327	8.8	45.3	12.2	4	3	8	58	923.234
54	54	81	7.4	3.36	9.7	72.8	3172	31.4	45.5	24.2	20	17	1	62	1113.156
42	33	77	9.7	3.03	10.7	83.5	7462	11.3	48.7	12.4	41	26	108	58	994.648
42	32	76	9.1	3.32	10.5	87.5	6092	17.5	45.3	13.2	29	32	161	54	1015.023
36	29	72	9.5	3.32	10.6	77.6	3437	8.1	45.5	13.8	45	59	263	56	991.290
37	38	67	11.3	2.99	12.0	81.5	3387	3.6	50.3	13.5	56	21	44	73	893.991
42	29	72	10.7	3.19	10.1	79.5	3508	2.2	38.8	15.7	6	4	18	56	938.500
41	33	77	11.2	3.08	9.6	79.9	4843	2.7	38.6	14.1	11	11	89	54	946.185

(continued)

Table B.6. (continued)

(1) PREC:	Average annual precipitation in inches
(2) JANTEMP:	Average January temperature in degrees Fahrenheit
(3) JULTEMP:	Average July temperature in degrees Fahrenheit
(4) OVER65:	% of 1960 SMSA population aged 65 or older
(5) HOUSEHOLD:	Average household size
(6) EDUC:	Median school years completed by those over 22
(7) HOUSING:	% of housing units which are sound & with all facilities
(8) POPN:	Population per square mile in urbanized areas, 1960
(9) NONWHT:	% nonwhite population in urbanized areas, 1960
(10) WHTCOL. :	% employed in white collar occupations
(11) POOR :	% of families with income < $3000
(12) HC :	Relative hydrocarbon pollution potential
(13) NOX :	Relative nitric oxides pollution potential
(14) SO2 :	Relative sulfur dioxide pollution potential
(15) HUMID :	Annual average % relative humidity at 1 PM
(16) MORT :	Total age-adjusted mortality rate per 100,000

(1)	(2)	(3)	(4)	(5)	(6)	(7)	(8)	(9)	(10)	(11)	(12)	(13)	(14)	(15)	(16)
44	39	78	8.2	3.32	11.0	79.9	3768	28.6	49.5	17.5	12	9	48	53	1025.502
32	25	72	10.9	3.21	11.1	82.5	4355	5.0	46.4	10.8	7	4	18	60	874.281
34	32	79	9.3	3.23	9.7	76.8	5160	17.2	45.1	15.3	31	15	68	57	953.560
10	55	70	7.3	3.11	12.1	88.9	3033	5.9	51.0	14.0	144	66	20	61	839.709
18	48	63	9.2	2.92	12.2	87.7	4253	13.7	51.2	12.0	311	171	86	71	911.701
13	49	68	7.0	3.36	12.2	90.7	2702	3.0	51.9	9.7	105	32	3	71	790.733
35	40	64	9.6	3.02	12.2	82.5	3626	5.7	54.3	10.1	20	7	20	72	899.264
45	28	74	10.6	3.21	11.1	82.6	1883	3.4	41.9	12.3	5	4	20	56	904.155
38	24	72	9.8	3.34	11.4	78.0	4923	3.8	50.5	11.1	8	5	25	61	950.672
31	26	73	9.3	3.22	10.7	81.3	3249	9.5	43.9	13.6	11	7	25	59	972.464
40	23	71	11.3	3.28	10.3	73.8	1671	2.5	47.4	13.5	5	2	11	60	912.202
41	37	78	6.2	3.25	12.3	89.5	5308	25.9	59.7	10.3	65	28	102	52	967.803
28	32	81	7.0	3.27	12.1	81.0	3665	7.5	51.6	13.2	4	2	1	54	823.764
45	33	76	7.7	3.39	11.3	82.2	3152	12.1	47.3	10.9	14	11	42	56	1003.502
45	24	70	11.8	3.25	11.1	79.8	3678	1.0	44.8	14.0	7	3	8	56	895.696
42	33	76	9.7	3.22	9.0	76.2	9699	4.8	42.2	14.5	8	8	49	54	911.817
38	28	72	8.9	3.48	10.7	79.8	3451	11.7	37.5	13.0	14	13	39	58	954.442

Table B.7. Heat evolved Y (in cal/g) from cement as a function of percentages in weight of tricalcium aluminate (X_1), tricalcium silicate (X_2), tricalcium aluminoferrite (X_1), and dicalcium silicate (X_1) in the clinkers. Data reproduced from Draper and Smith (1981)

X_1	X_2	X_3	X_4	Y
7.0	26.0	6.0	60.0	78.5
1.0	29.0	15.0	52.0	74.3
11.0	56.0	8.0	20.0	104.3
11.0	31.0	8.0	47.0	87.6
7.0	52.0	6.0	33.0	95.9
11.0	55.0	9.0	22.0	109.2
3.0	71.0	17.0	6.0	102.7
1.0	31.0	22.0	44.0	72.5
2.0	54.0	18.0	22.0	93.1
21.0	47.0	4.0	26.0	115.9
1.0	40.0	23.0	34.0	83.8
11.0	66.0	9.0	12.0	113.3
10.0	68.0	8.0	12.0	109.4

Table B.8. 5% critical values for test of discordancy for a single outlier in a general linear model with normal error structure, using the studentized residual as test statistic (reproduced from Lund (1975))

q \ n	1	2	3	4	5	6	8	10	16	25
5	1.92									
6	2.07	1.93								
7	2.19	2.08	1.94							
8	2.28	2.20	2.10	1.94						
9	2.35	2.29	2.21	2.10	1.95					
10	2.42	2.37	2.31	2.22	2.11	1.95				
12	2.52	2.49	2.45	2.39	2.33	2.24	1.96			
14	2.61	2.58	2.55	2.51	2.47	2.41	2.25	1.96		
16	2.68	2.66	2.63	2.60	2.57	2.53	2.43	2.26		
18	2.73	2.72	2.70	2.68	2.65	2.62	2.55	2.44		
20	2.78	2.77	2.76	2.74	2.72	2.70	2.64	2.57	2.15	
25	2.89	2.88	2.87	2.86	2.84	2.83	2.80	2.76	2.60	
30	2.96	2.96	2.95	2.94	2.93	2.93	2.90	2.88	2.79	2.17
35	3.03	3.02	3.02	3.01	3.00	3.00	2.93	2.97	2.91	2.64
40	3.08	3.08	3.07	3.07	3.06	3.06	3.05	3.03	3.00	2.84
45	3.13	3.12	3.12	3.12	3.11	3.11	3.10	3.09	3.06	2.96
50	3.17	3.16	3.16	3.16	3.15	3.15	3.14	3.14	3.11	3.04
60	3.23	3.23	3.23	3.23	3.22	3.22	3.22	3.21	3.20	3.15
70	3.29	3.29	3.28	3.28	3.28	3.28	3.27	3.27	3.26	3.23
80	3.33	3.33	3.33	3.33	3.33	3.33	3.32	3.32	3.31	3.29
90	3.37	3.37	3.37	3.37	3.37	3.37	3.36	3.36	3.36	3.34
100	3.41	3.41	3.40	3.40	3.40	3.40	3.40	3.40	3.39	3.38

Note: n = number of observations; q = number of independent variables (including count for intercept if fitted)

Table B.9. 1% critical values for test of discordancy for a single outlier in a general linear model with normal error structure, using the studentized residual as test statistic (reproduced from Lund (1975))

q \ n	1	2	3	4	5	6	8	10	16	25
5	1.98									
6	2.17	1.98								
7	2.32	2.17	1.98							
8	2.44	2.32	2.18	1.98						
9	2.54	2.44	2.33	2.18	1.99					
10	2.62	2.55	2.45	2.33	2.18	1.99				
12	2.76	2.70	2.64	2.56	2.46	2.34	1.99			
14	2.86	2.82	2.78	2.72	2.65	2.57	2.35	1.99		
16	2.95	2.92	2.88	2.84	2.79	2.73	2.58	2.35		
18	3.02	3.00	2.97	2.94	2.90	2.85	2.75	2.59		
20	3.08	3.06	3.04	3.01	2.98	2.95	2.87	2.76	2.20	
25	3.21	3.19	3.18	3.16	3.14	3.12	3.07	3.01	2.75	
30	3.30	3.29	3.28	3.26	3.25	3.24	3.21	3.17	3.04	2.21
35	3.37	3.36	3.35	3.34	3.34	3.33	3.30	3.25	3.19	2.81
40	3.43	3.42	3.42	3.41	3.40	3.40	3.38	3.36	3.30	3.05
45	3.48	3.47	3.47	3.46	3.46	3.45	3.44	3.43	3.38	3.23
50	3.52	3.52	3.51	3.51	3.51	3.50	3.49	3.48	3.45	3.34
60	3.60	3.59	3.59	3.59	3.58	3.58	3.57	3.56	3.54	3.48
70	3.65	3.65	3.65	3.65	3.64	3.64	3.64	3.63	3.61	3.57
80	3.70	3.70	3.70	3.70	3.69	3.69	3.69	3.68	3.67	3.64
90	3.74	3.74	3.74	3.74	3.74	3.74	3.73	3.73	3.72	3.70
100	3.78	3.78	3.78	3.77	3.77	3.77	3.77	3.77	3.76	3.74

Note: n = number of observations; q = number of independent variables (including count for intercept if fitted)

Table B.10. 5% critical values based on the Bonferroni bounds for the t-test for a single outlier using externally studentized residual in a linear regression model.

k / n	1	2	3	4	5	6	7	8	9	10	11	12
5	9.92	63.66										
6	6.23	10.89	76.39									
7	5.07	6.58	11.77	89.12								
8	4.53	5.26	6.90	12.59	101.86							
9	4.22	4.66	5.44	7.18	13.36	114.59						
10	4.03	4.32	4.77	5.60	7.45	14.09	127.32					
11	3.90	4.10	4.40	4.88	5.75	7.70	14.78	140.05				
12	3.81	3.96	4.17	4.49	4.98	5.89	7.94	15.44	152.79			
13	3.74	3.86	4.02	4.24	4.56	5.08	6.02	8.16	16.08	165.52		
14	3.69	3.79	3.91	4.07	4.30	4.63	5.16	6.14	8.37	16.69	178.25	
15	3.65	3.73	3.83	3.95	4.12	4.36	4.70	5.25	6.25	8.58	17.28	190.98
16	3.62	3.68	3.77	3.87	4.00	4.17	4.41	4.76	5.33	6.36	8.77	17.85
17	3.59	3.65	3.72	3.80	3.90	4.04	4.21	4.46	4.82	5.40	6.47	8.95
18	3.57	3.62	3.68	3.75	3.83	3.94	4.08	4.26	4.51	4.88	5.47	6.57
19	3.56	3.60	3.65	3.71	3.78	3.86	3.97	4.11	4.30	4.55	4.93	5.54
20	3.54	3.58	3.62	3.67	3.73	3.81	3.89	4.00	4.15	4.33	4.59	4.98
21	3.53	3.57	3.60	3.65	3.70	3.76	3.83	3.92	4.03	4.18	4.37	4.64
22	3.52	3.55	3.59	3.63	3.67	3.72	3.78	3.86	3.95	4.06	4.21	4.40
23	3.52	3.54	3.57	3.61	3.65	3.69	3.75	3.81	3.88	3.98	4.09	4.24
24	3.51	3.53	3.56	3.59	3.63	3.67	3.71	3.77	3.83	3.91	4.00	4.12
25	3.50	3.53	3.55	3.58	3.61	3.65	3.69	3.73	3.79	3.85	3.93	4.02
26	3.50	3.52	3.54	3.57	3.60	3.63	3.66	3.70	3.75	3.81	3.87	3.95
27	3.50	3.52	3.54	3.56	3.58	3.61	3.65	3.68	3.72	3.77	3.83	3.89
28	3.50	3.51	3.53	3.55	3.58	3.60	3.63	3.66	3.70	3.74	3.79	3.84
29	3.49	3.51	3.53	3.55	3.57	3.59	3.62	3.64	3.68	3.71	3.76	3.81
30	3.49	3.51	3.52	3.54	3.56	3.58	3.60	3.63	3.66	3.69	3.73	3.77
31	3.49	3.50	3.52	3.54	3.55	3.57	3.59	3.62	3.64	3.67	3.71	3.74
32	3.49	3.50	3.52	3.53	3.55	3.57	3.59	3.61	3.63	3.66	3.69	3.72
33	3.49	3.50	3.52	3.53	3.54	3.56	3.58	3.60	3.62	3.64	3.67	3.70
34	3.49	3.50	3.51	3.53	3.54	3.56	3.57	3.59	3.61	3.63	3.66	3.68
35	3.49	3.50	3.51	3.52	3.54	3.55	3.57	3.58	3.60	3.62	3.64	3.67
36	3.49	3.50	3.51	3.52	3.54	3.55	3.56	3.58	3.60	3.61	3.63	3.66
37	3.49	3.50	3.51	3.52	3.53	3.55	3.56	3.57	3.59	3.61	3.62	3.65
38	3.49	3.50	3.51	3.52	3.53	3.54	3.56	3.57	3.58	3.60	3.62	3.64
39	3.49	3.50	3.51	3.52	3.53	3.54	3.55	3.57	3.58	3.59	3.61	3.63
40	3.49	3.50	3.51	3.52	3.53	3.54	3.55	3.56	3.58	3.59	3.60	3.62
50	3.51	3.51	3.52	3.53	3.53	3.54	3.54	3.55	3.56	3.57	3.57	3.58
60	3.53	3.53	3.54	3.54	3.54	3.55	3.55	3.56	3.56	3.57	3.57	3.58
70	3.55	3.55	3.55	3.56	3.56	3.56	3.56	3.57	3.57	3.57	3.58	3.58
80	3.57	3.57	3.57	3.57	3.58	3.58	3.58	3.58	3.58	3.59	3.59	3.59
90	3.59	3.59	3.59	3.59	3.59	3.59	3.60	3.60	3.60	3.60	3.60	3.60
100	3.60	3.60	3.60	3.61	3.61	3.61	3.61	3.61	3.61	3.61	3.62	3.62

(continued)

Table B.10. (continued)

k\n	13	14	15	16	17	18	19	20	25	30	35
15											
16	203.72										
17	18.40	216.45									
18	9.13	18.93	229.18								
19	6.67	9.30	19.46	241.91							
20	5.60	6.76	9.46	19.96	254.65						
21	5.03	5.67	6.85	9.62	20.46	267.38					
22	4.68	5.08	5.73	6.93	9.78	20.94	280.11				
23	4.44	4.71	5.12	5.78	7.02	9.93	21.41	292.84			
24	4.27	4.47	4.75	5.17	5.84	7.10	10.07	21.87			
25	4.14	4.30	4.50	4.79	5.21	5.89	7.17	10.21			
26	4.05	4.17	4.32	4.53	4.82	5.25	5.95	7.25			
27	3.97	4.07	4.19	4.35	4.56	4.85	5.29	6.00			
28	3.91	3.99	4.09	4.21	4.37	4.59	4.88	5.33	356.51		
29	3.86	3.93	4.01	4.11	4.24	4.40	4.61	4.91	24.05		
30	3.82	3.88	3.95	4.03	4.13	4.26	4.42	4.64	10.87		
31	3.79	3.84	3.90	3.97	4.05	4.15	4.28	4.44	7.59		
32	3.76	3.80	3.85	3.91	3.98	4.07	4.17	4.30	6.23		
33	3.74	3.77	3.82	3.87	3.93	4.00	4.08	4.19	5.50	420.17	
34	3.71	3.75	3.79	3.83	3.88	3.94	4.01	4.10	5.06	26.05	
35	3.70	3.73	3.76	3.80	3.85	3.90	3.96	4.03	4.76	11.45	
36	3.68	3.71	3.74	3.77	3.81	3.86	3.91	3.97	4.55	7.90	
37	3.67	3.69	3.72	3.75	3.79	3.83	3.87	3.92	4.39	6.43	
38	3.66	3.68	3.70	3.73	3.76	3.80	3.84	3.88	4.27	5.65	483.83
39	3.65	3.67	3.69	3.71	3.74	3.77	3.81	3.85	4.18	5.18	27.90
40	3.64	3.66	3.68	3.70	3.73	3.75	3.79	3.82	4.10	4.86	11.98
50	3.59	3.60	3.61	3.62	3.63	3.65	3.66	3.67	3.77	3.92	4.22
60	3.58	3.59	3.59	3.60	3.61	3.61	3.62	3.63	3.68	3.74	3.84
70	3.59	3.59	3.59	3.60	3.60	3.61	3.61	3.62	3.65	3.68	3.73
80	3.60	3.60	3.60	3.60	3.61	3.61	3.61	3.62	3.64	3.66	3.69
90	3.61	3.61	3.61	3.61	3.62	3.62	3.62	3.62	3.64	3.65	3.67
100	3.62	3.62	3.62	3.63	3.63	3.63	3.63	3.63	3.64	3.66	3.67

Note: n = number of cases; k = number of explanatory variables

Table B.11. 1% critical values based on the Bonferroni bounds for the t-test for a single outlier using externally studentized residual in a linear regression model

k \ n	1	2	3	4	5	6	7	8	9	10	11	12
5	22.33	318.31										
6	10.87	24.46	381.97									
7	7.84	11.45	26.43	445.63								
8	6.54	8.12	11.98	28.26	509.3							
9	5.84	6.71	8.38	12.47	29.97	572.96						
10	5.41	5.96	6.87	8.61	12.92	31.6	636.62					
11	5.12	5.50	6.07	7.01	8.83	13.35	33.14	700.28				
12	4.91	5.19	5.58	6.17	7.15	9.03	13.75	34.62	763.94			
13	4.76	4.97	5.25	5.66	6.26	7.27	9.22	14.12	36.03	827.61		
14	4.64	4.81	5.02	5.32	5.73	6.35	7.39	9.40	14.48	37.40	891.27	
15	4.55	4.68	4.85	5.08	5.37	5.80	6.43	7.50	9.57	14.82	38.71	954.93
16	4.48	4.59	4.72	4.90	5.12	5.43	5.86	6.51	7.60	9.73	15.15	39.98
17	4.41	4.51	4.62	4.76	4.94	5.17	5.48	5.92	6.59	7.70	9.88	15.46
18	4.36	4.44	4.54	4.66	4.80	4.98	5.21	5.53	5.98	6.66	7.80	10.03
19	4.32	4.39	4.47	4.57	4.69	4.83	5.01	5.25	5.57	6.03	6.72	7.89
20	4.29	4.35	4.42	4.50	4.60	4.72	4.86	5.05	5.29	5.62	6.08	6.79
21	4.26	4.31	4.37	4.44	4.52	4.62	4.74	4.89	5.08	5.33	5.66	6.13
22	4.23	4.28	4.33	4.39	4.46	4.55	4.65	4.77	4.92	5.11	5.36	5.70
23	4.21	4.25	4.30	4.35	4.41	4.49	4.57	4.67	4.80	4.95	5.14	5.40
24	4.19	4.22	4.27	4.32	4.37	4.43	4.51	4.59	4.70	4.82	4.98	5.17
25	4.17	4.20	4.24	4.28	4.33	4.39	4.45	4.53	4.62	4.72	4.85	5.00
26	4.15	4.18	4.22	4.26	4.30	4.35	4.41	4.47	4.55	4.64	4.74	4.87
27	4.14	4.17	4.20	4.24	4.27	4.32	4.37	4.43	4.49	4.57	4.66	4.76
28	4.13	4.15	4.18	4.21	4.25	4.29	4.33	4.38	4.44	4.51	4.59	4.68
29	4.12	4.14	4.17	4.20	4.23	4.26	4.30	4.35	4.40	4.46	4.53	4.60
30	4.11	4.13	4.15	4.18	4.21	4.24	4.28	4.32	4.36	4.42	4.47	4.54
31	4.10	4.12	4.14	4.17	4.19	4.22	4.26	4.29	4.33	4.38	4.43	4.49
32	4.09	4.11	4.13	4.15	4.18	4.21	4.24	4.27	4.31	4.35	4.39	4.45
33	4.08	4.10	4.12	4.14	4.17	4.19	4.22	4.25	4.28	4.32	4.36	4.41
34	4.08	4.09	4.11	4.13	4.15	4.18	4.20	4.23	4.26	4.29	4.33	4.37
35	4.07	4.09	4.11	4.12	4.14	4.16	4.19	4.21	4.24	4.27	4.31	4.34
36	4.07	4.08	4.10	4.12	4.13	4.15	4.18	4.20	4.22	4.25	4.28	4.32
37	4.06	4.08	4.09	4.11	4.13	4.14	4.16	4.19	4.21	4.24	4.26	4.29
38	4.06	4.07	4.09	4.10	4.12	4.13	4.15	4.17	4.20	4.22	4.25	4.27
39	4.06	4.07	4.08	4.10	4.11	4.13	4.14	4.16	4.18	4.21	4.23	4.26
40	4.05	4.06	4.08	4.09	4.10	4.12	4.14	4.15	4.17	4.19	4.22	4.24
50	4.03	4.04	4.05	4.06	4.07	4.07	4.08	4.09	4.10	4.12	4.13	4.14
60	4.03	4.04	4.04	4.05	4.05	4.06	4.06	4.07	4.08	4.08	4.09	4.10
70	4.03	4.04	4.04	4.05	4.05	4.05	4.06	4.06	4.07	4.07	4.08	4.08
80	4.04	4.04	4.05	4.05	4.05	4.06	4.06	4.06	4.07	4.07	4.07	4.08
90	4.05	4.05	4.05	4.06	4.06	4.06	4.06	4.07	4.07	4.07	4.07	4.08
100	4.06	4.06	4.06	4.06	4.07	4.07	4.07	4.07	4.07	4.08	4.08	4.08

(continued)

Table B.11. (continued)

k\n	13	14	15	16	17	18	19	20	25	30	35
15											
16	1018.59										
17	41.21	1082.25									
18	15.76	42.41	1145.92								
19	10.17	16.05	43.57	1209.58							
20	7.98	10.31	16.33	44.70	1273.24						
21	6.85	8.06	10.44	16.60	45.81	1336.90					
22	6.18	6.91	8.14	10.56	16.86	46.89	1400.56				
23	5.74	6.22	6.97	8.22	10.68	17.11	47.94	1464.23			
24	5.43	5.78	6.27	7.02	8.29	10.80	17.36	48.97			
25	5.20	5.46	5.81	6.31	7.07	8.36	10.92	17.60			
26	5.03	5.23	5.49	5.85	6.35	7.13	8.43	11.03			
27	4.89	5.05	5.26	5.52	5.88	6.39	7.17	8.50			
28	4.78	4.91	5.08	5.28	5.55	5.91	6.43	7.22	1782.54		
29	4.69	4.80	4.94	5.10	5.31	5.58	5.94	6.47	53.84		
30	4.62	4.71	4.82	4.96	5.12	5.33	5.60	5.97	18.71		
31	4.56	4.64	4.73	4.84	4.97	5.14	5.35	5.63	11.53		
32	4.50	4.57	4.65	4.75	4.86	4.99	5.16	5.37	8.81		
33	4.46	4.52	4.59	4.67	4.76	4.88	5.01	5.18	7.44	2100.85	
34	4.42	4.47	4.53	4.60	4.68	4.78	4.89	5.03	6.64	58.3	
35	4.39	4.43	4.49	4.55	4.62	4.70	4.79	4.91	6.11	19.7	
36	4.36	4.40	4.45	4.50	4.56	4.63	4.71	4.81	5.75	11.99	
37	4.33	4.37	4.41	4.46	4.51	4.57	4.64	4.73	5.48	9.09	
38	4.31	4.34	4.38	4.42	4.47	4.52	4.59	4.66	5.27	7.64	2419.15
39	4.28	4.32	4.35	4.39	4.43	4.48	4.54	4.60	5.11	6.79	62.44
40	4.27	4.29	4.33	4.36	4.40	4.44	4.49	4.55	4.98	6.23	20.60
50	4.15	4.17	4.18	4.20	4.22	4.23	4.25	4.28	4.42	4.65	5.11
60	4.11	4.12	4.12	4.13	4.14	4.15	4.17	4.18	4.25	4.34	4.49
70	4.09	4.09	4.10	4.11	4.11	4.12	4.13	4.13	4.17	4.23	4.30
80	4.08	4.09	4.09	4.09	4.10	4.10	4.11	4.11	4.14	4.17	4.22
90	4.08	4.08	4.09	4.09	4.09	4.10	4.10	4.10	4.12	4.15	4.18
100	4.08	4.09	4.09	4.09	4.09	4.10	4.10	4.10	4.12	4.13	4.15

Note: n = number of cases; k = number of explanatory variables

Table B.12. Table of coefficients for orthogonal polynomials: equally spaced factor levels

Number of Levels	Degree of Polynomial	Factor Level					
		1	2	3	4	5	6
2	1	−1	+1				
3	1	−1	0	+1			
	2	+1	−2	+1			
4	1	−3	−1	+1	+3		
	2	+1	−1	−1	+1		
	3	−1	+3	−3	+1		
5	1	−2	−1	0	+1	+2	
	2	+2	−1	−2	−1	+2	
	3	−1	+2	0	−2	+1	
	4	+1	−4	+6	−4	+1	
6	1	−5	−3	−1	+1	+3	+5
	2	+5	−1	−4	−4	−1	+5
	3	−5	+7	+4	−4	−7	+5
	4	+1	−3	+2	+2	−3	+1
	5	−1	+5	−10	+10	−5	+1

Table B.13. A subset of SAS color names and RGB values

Name	RGB Value	Description
BGR	CX83838C	Bluish gray
BIBG	CX4DBFBC	Brilliant bluish green
BIG	CX4DBF81	Brilliant green
BIGB	CX4D7EBF	Brilliant greenish blue
BIGY	CXAEE554	Brilliant greenish yellow
BIO	CXD9892B	Brilliant orange
BIP	CXA030B2	Brilliant purple
BRGR	CX595753	Brownish gray
BRPK	CXBFB9A6	Brownish pink
BWH	CXDEDDED	Bluish white
DABGR	CX535359	Dark bluish gray
DABR	CX191714	Dark brown
DAG	CX364C40	Dark green
DAGB	CX2A3440	Dark greenish blue
DAOL	CX161911	Dark olive
DAP	CX423045	Dark purple
DAPK	CX995C67	Dark pink
DAR	CX40262B	Dark red
DAY	CX899952	Dark yellow
DAYBR	CX26251F	Dark yellowish brown
DAYPK	CX99615C	Dark yellowish pink
DEBR	CX261C0F	Deep brown
DEOLG	CX15260D	Deep olive green
DEP	CX3E1745	Deep purple
DER	CX4C1923	Deep red
DEY	CX839938	Deep yellow
GRB	CX5C5C73	Grayish blue
GRG	CX63736A	Grayish green
GRGY	CXA9BF86	Grayish greenish yellow
GRR	CX73545A	Grayish red
GRY	CXB5BF93	Grayish yellow
GWH	CXECEDEC	Greenish white
LIB	CX5A58A6	Light blue
LIBG	CX6EA6A4	Light bluish green
LIBR	CX8C7962	Light brown
LIG	CX6EA688	Light green
LIGB	CX6E86A6	Light greenish blue
LIGRR	CX997078	Light grayish red
LIO	CXD9A465	Light orange
LIOL	CX628033	Light olive
LIP	CX9B58A6	Light purple
LIPGR	CXBDB2BF	Light purplish gray
LIPK	CXE599A7	Light pink
LIRBR	CX8C7367	Light reddish brown
LIRP	CX99528E	Light reddish purple
LIY	CXCDE57A	Light yellow
LIYG	CX80BF88	Light yellowish green

(continued)

Table B.13. (continued)

Name	RGB Value	Description
MOB	CX3E3D73	Moderate blue
MOBG	CX4C7372	Moderate bluish green
MOBR	CX594E41	Moderate brown
MOGY	CX9DBF66	Moderate greenish yellow
MOLG	CX769966	Moderate yellow green
MOP	CX6B3D73	Moderate purple
MOPK	CXBA7C87	Moderate pink
MOY	CXABBF66	Moderate yellow
MOYG	CX5D8C64	Moderate yellowish green
OLGR	CX575953	Olive gray
PAB	CX8585A6	Pale blue
PAG	CX90A69A	Pale green
PAPK	CXE5BFC6	Pale pink
PAY	CXD9E5B0	Pale yellow
PWH	CXEBDDED	Purplish white
STB	CX201F73	Strong blue
STBG	CX2E7371	Strong bluish green
STBR	CX593B18	Strong brown
STG	CX2E734E	Strong green
STGY	CX8DBA44	Strong greenish yellow
STP	CX671F73	Strong purple
STPB	CX3F1F73	Strong purplish blue
STPK	CXD9576E	Strong pink
STR	CX731727	Strong red
STRO	CX8C411C	Strong reddish orange
STYG	CX388C43	Strong yellowish green
VIB	CX090766	Vivid blue
VIBG	CX138C89	Vivid bluish green
VIGB	CX13478C	Vivid greenish blue
VIGY	CX80BF1A	Vivid greenish yellow
VILG	CX44A616	Vivid yellow green
VIO	CXB26306	Vivid orange
VIP	CX6F0980	Vivid purple
VIPB	CX2B0766	Vivid purplish blue
VIPR	CX4C052C	Vivid purplish red
VIR	CX33070F	Vivid red
VIRO	CX803009	Vivid reddish orange
VIRP	CX59064C	Vivid reddish purple
VIV	CX53098C	Vivid violet
VIY	CX99BF1A	Vivid yellow
VIYG	CX16A629	Vivid yellowish green
VLIB	CX7674D9	Very light blue
VLIGB	CX90B0D9	Very light greenish blue
VLIP	CXCB74D9	Very light purple
VLIV	CXAC74D9	Very light violet
VPAG	CXBCD9C5	Very pale green
YWH	CXE8EDD5	Yellowish white

References

Armitage, P. and Berry G. (1994) *Statistical Methods in Medical Research*, Third Edition. Blackwell, Malden, MA.

Bliss, C.I. (1970) *Statistics in Biology*, Volume 2. McGraw-Hill, New York.

Bowerman, B.L. and O'Connell, R.T. (2004) *Business Statistics in Practice*, Fourth Edition. McGraw-Hill/Irwin, Chicago.

Box, G.E.P., Hunter, W.G., and Hunter, J.S. (1978) *Statistics for Experimenters*. Wiley, New York.

Chambers, J.M., Cleveland, W.S., Kleiner, B., and Tukey, P.A. (1983) *Graphical Methods in Data Analysis*. Wadsworth, Belmont, CA.

Deak, N.A. and Johnson, L.A. (2007) Effects of extraction temperature and preservation method on functionality of soy protein. *J. Am. Oil Chem. Soc.*, 84, 259-268.

Devore, J.L. (1982) *Probability and Statistics for Engineering and the Sciences*. Brooks/Cole, Monterey, CA.

Draper, N.R. and Smith, H. (1981) *Applied Regression Analysis*, Second Edition. Wiley, New York.

Dunn, O.J. and Clark, V.A. (1987) *Applied Statistics: Analysis of Variance and Regression Analysis*, Second Edition. Wiley, New York.

Friendly, M. (1991) *SAS System for Statistical Graphics*, First Edition. SAS Institute Inc., Cary, NC.

Henderson, C.R., Kempthorne, O., Searle, S.R., and von Krosigk, C.N. (1959) Estimation of environmental and genetic trends from records subject to culling. *Biometrics*, 15, 192-218.

Kenward, M.G. and Roger, J.H. (1997) Small sample inference for fixed effects from restricted maximum likelihood. *Biometrics*, 53, 983-997.

Kirk, R.E. (1982) *Experimental Design*, Second Edition. Brooks/Cole, Monterey, CA.

Koopmans, L.H. (1987) *Introduction to Contemporary Statistical Methods*, Second Edition. Duxbury, Boston, MA.

Kuehl, R.O. (2000) *Design of Experiments: Statistical Principles of Research Design and Analysis*. Brooks/Cole, Pacific Grove, CA.

Kutner, M.H., Nachtsheim, C.J., and Neter, J. (2004) *Applied Linear Regression Models*, Fourth Edition. McGraw-Hill/Irwin, Chicago.

Kutner, M.H., Nachtsheim, C.J., Neter, J., and Li, W. (2005) *Applied Linear Statistical Models*, Fifth Edition. McGraw-Hill/Irwin, Chicago.

Littell R.C., Freund, R.J., and Spector, P.C. (1991) *SAS System for Linear Models*, Third Edition. SAS Institute Inc., Cary, NC.

Lund, R.E. (1975) Tables for an approximate test for outliers in linear models. *Technometrics*, 17, 473-476.

Mason, R.L., Gunst, R.F., and Hess, J.L. (1989) *Statistical Design & Analysis of Experiments*. Wiley, New York.

McClave, J.T., Benson, G.P., and Sincich T.L. (2000) *Statistics for Business and Economics*, Eighth Edition. Prentice Hall Inc., Englewood Cliffs, NJ.

McDonald, G.C. and Schwing, R.C. (1973) Instabilities of regression estimates relating air pollution to mortality. *Technometrics*, 15, 463-482.

Milliken, G.A. and Johnson, D.E. (2001) *Analysis of Messy Data, Volume III: Analysis of Covariance*. Chapman & Hall/CRC, Boca Raton, FL.

Montgomery, D.C. (1991) *The Design and Analysis of Experiments*, Third Edition. Wiley, New York.

Morrison, D.F. (1983) *Applied Linear Statistical Methods*, Prentice Hall Inc., Englewood Cliffs, NJ.

Ostle, B. (1963) *Statistics in Research*, Second Edition. Iowa State University Press, Ames.

Ott, R.L., Larson, R.F., and Mendenhall, W. (1987) *Statistics: A Tool for the Social Sciences*, Fourth Edition. Duxbury, Boston, MA.

Ott, R.L. and Longnecker, M. (2001) *An Introduction to Statistical Methods and Data Analysis*, Fifth Edition. Duxbury, Pacific Grove, CA.

Rice, J.A. (1988) *Mathematical Statistics and Data Analysis*. Wadsworth & Brooks/Cole, Pacific Grove, CA.

Sahai, H. and Ageel, M.I. (2000) *The Analysis of Variance*. Birkhäuser, Boston, MA.

SAS/GRAPH 9.1 Reference, Volumes 1 and 2. SAS Institute Inc., Cary, NC.

Schlotzhauer, S.D. and Littell, R.C. (1997) *SAS System for Elementary Statistical Analysis*, Second Edition. SAS Institute Inc., Cary, NC.

Searle, S.R. (1971) *Linear Models*. Wiley, New York.

Searle, S.R., Casella, G., and McCulloch, C.E. (1992) *Variance Components*. Wiley, New York.

Simpson, J., Olsen, A., and Eden, J. (1975) A Bayesian analysis of a multiplicative treatment effect in weather modification. *Technometrics*, 17, 161-166.

Snedecor, G.W. and Cochran, W.G. (1989) *Statistical Methods*, Eighth Edition. Iowa State University Press, Ames.

Sokal, R.R. and Rohlf, J.F. (1995) *Biometry: The Principles and Practice of Statistics in Biological Research*, Third Edition. Freeman, New York.

Tukey, J.W. (1949) One degree of freedom for nonadditivity. *Biometrics*, 5, 232-242.

Weisberg, S. (1985) *Applied Linear Regression Analysis*, Second Edition. Wiley, New York.

Index

#n, 38
FREQ, 77
_N_, 24
TYPE, 77

a =, 140
a priori comparisons, 283, 286, 339
added-variable plot, 225
adjust =, 319
adjust = (proc glm), 432
adjust = (proc mixed), 437
adjusted means, 311
adjusted R^2, 251
AIC criterion, 240
all-subsets method, 238
alpha =, 82, 293
Anderson-Darling test, 82
annotate =, 133, 166
annotate data set, 165
areas =, 133
array, 22, 23, 27, 28
assignment statements, 16
asycov (proc mixed), 403
asymmetric lambda, 97, 102
at means, 317
at option, 317
attributes, 6, 45
axis, 140
axis options, 130

backward elimination, 243
bar chart, 114, 116, 117, 169
bartlett, 296
best =, 242, 248
best estimates, 281
best linear unbiased predictor, 400
bias, 239
block, 170
block chart, 114
block effects, 364
blocking, 358
blom, 156
BLUP, 400, 403, 405, 407, 429, 430, 437, 447, 449, 451, 453, 461
Bonferroni adjustment, 465
Bonferroni method, 200, 202, 219, 290, 465
box plot, 82, 85, 104
boxanno macro, 165
boxstyle =, 179
bwidth =, 144
by statement, 42

c =, 140, 142
case statistics, 217
catalog entry, 73
caxes=, 207, 225, 248
cell frequency, 91, 93, 99

cell means, 328, 331, 332
cellchi2, 93
centered variables, 259
cfill =, 176
chart variable, 169
chartype, 139
chi-square statistic, 89, 90, 93
chi-square test, 89, 99
chisq, 93
chref =, 150
cibasic, 82, 87
cl (proc mixed), 403, 461
class, 75, 83
class level information, 362
clb, 190, 207, 226
cldiff, 332
cli, 207
clm, 207
co =, 142
coefficient of determination, 188
collin, 226
collinearity diagnostics, 226
column input, 14
comparing slopes, 253
comparison operators, 18
concatenation, 59, 60
conditional execution, 18
constrained parameters (CP) model, 442

contrast, 318
contrast (proc glm), 297, 301, 343, 344, 350, 353, 355, 366
Cook's D, 197, 204, 220
covariance analysis, 309
covariate, 311, 313
cp, 243
Cp statistic, 239, 251
Cramer's V, 94, 98
Cramer's V, 94
ctext=, 207, 225, 248
cv =, 142
cvref =, 150

data =, 67, 81, 114
data set, 2, 3, 5
data step, 2
data step programming, 16
delimiter, 56
delimiter =, 58
design matrix, 275, 314, 392, 427, 429
details, 69
details = all (proc reg), 243
deviance, 402
device =, 167
diagnostic statistics, 217
discrete, 134
divisor =, 302, 308, 357
dlm =, 58
do loop, 20, 28, 29, 159
drop, 7, 27, 29, 81
dsd, 58

effects model, 282, 332, 343, 347
enhanced scatter plot, 166
error contrasts, 393
estimable functions, 278, 350, 353
estimate (proc glm), 297, 307, 345, 350
exact, 90, 94
exact test, 99
expected, 93

expected mean squares, 394, 397
experimentwise error, 290, 295
experimentwise error rate, 465
externally studentized residuals, 219

f =, 138
F-to-delete, 237
F-to-enter, 236, 243
F-to-remove, 237
filename, 57, 64, 79
fileref, 56, 64
firstobs =, 58
fisher, 93
Fisher's exact test, 90, 94, 99
fitted values, 217
footnote, 139, 159
format, 4, 45, 69, 73, 81
format(proc step), 48
formatted input, 3, 12
formatted-value, 70
formchar =, 104, 114, 119
forward selection, 236
fuzz =, 71

Gamma, 96, 101, 103
goodness-of-fit test, 89
goptions, 138, 167
graphic catalogs, 169
group =, 136, 171
groupnames = (proc reg), 251
gsasfile, 181

h =, 138, 140, 142
Hat Diag, 197, 200
hat matrix, 218
haxis =, 132
hbar, 133, 170
hierarchical, 419, 453
higher-order terms, 251
histogram, 169
histogram(univariate), 176
hm =, 161

homogeneity, 89
homogeneity of variance, 296
hovtest =, 296
hsize =, 162

i =, 140, 142, 143
if-then, 18
if-then/else, 21
infile, 56, 79
influence, 197, 199, 220
informat, 5, 12, 14, 69, 71, 73
informatted-value, 70
input, 3, 7, 8, 11, 12, 14–16, 21, 24, 26–28, 31, 33, 35, 45
input buffer, 24, 27, 33
inset location, 177
inset(univariate), 176
interaction, 209, 336, 338
interaction comparisons, 339
interaction plot, 159, 332
interaction test, 329
intraclass correlation, 396
invalue, 71
IQR, 151
iterative procedure, 391

j =, 138

Kendall's tau-b, 97, 101, 103
Kenward-Roger, 450
keylabel, 119
Kolmogorov-Smirnov test, 82
kurtosis, 75

l =, 142, 143
label, 45, 141
label =, 141
label option, 48
labeling statements, 19
lack of fit, 196
least squares, 187
least squares method, 209, 281

Index   555

legend, 159, 161, 162, 207
legend =, 132
length, 45, 71
levene, 296
leverage, 197, 220
lhref =, 150
libname, 65–67, 79, 81, 97, 106, 114
library =, 69
libref, 65, 66, 68
likelihood, 391
line pointer control, 38
linear trend, 303, 305, 481
list input, 12
log page, 3
log-likelihood, 391
logical operators, 18
LSD, 288, 295, 299, 364
lsmeans (proc glm), 313, 317, 350
lspace =, 139
lvref =, 150

main effects, 329
Mallows' Cp, 239
Mantel-Haenszel, 90
maxdec =(proc means), 76
maximum likelihood, 281
maximum likelihood estimates, 391
maximum likelihood method, 402
maxis =, 136
maxr, 251
means (proc glm), 300
means model, 282, 342
measures, 93
memtype =, 69
merge, 58, 64
method = type3 (proc mixed), 403
method of moments, 390, 395
method of moments estimates, 399, 403
midpoint =, 173

midpoints =, 134, 146, 171
minr, 251
missing values, 8
missover, 58
MIVQUE(0), 401
mixed model equations, 429
ML estimates, 393
MLE, 391
model(proc anova), 292
modifier :, 35, 37
move =, 139, 162
mu0 =, 87
multicollinearity, 226, 228
multinomial probabilities, 89
multiple comparisons, 289
multiple correlation coefficient, 239
multiway tables, 91

n =(infile statement), 38
nested do loops, 30
nested factor, 453, 466
nonadditive, 331, 367
nonadditivity, 367
noprint, 93
normal, 82
normal equations, 209, 277
normal probability plot, 82, 85, 86, 104, 155, 194, 224

obs =, 58
ODS, 229
one-way classification, 282
options in reg, 200
order –, 83, 140, 363
origin=, 162
orthogonal polynomials, 302, 481
output, 63, 75, 79
output (proc glm), 336
Output Delivery System, 229

output(data step), 25, 27, 29
overlay, 132, 140

pairwise comparisons, 288
parameter estimates, 278
partial, 226
partial regression residual plot, 225
partial slope, 225
partial sums of squares, 233, 243
partitioning SS, 339
pattern, 146, 171, 173
pctldef =, 82
pctlpre =, 88
pctlpts =, 88
pdiff (proc glm), 317
PDV, 24, 27, 33, 34
Pearson's correlation, 89, 97, 102, 103
per-comparison error rate, 288
percent = (gchart), 174
permanent data set, 5
pfill =, 176
pie chart, 114, 169
plot-request, 130
plot2, 158
plots, 82
pointer, 14, 31
pointer control, 14, 21, 30, 38
polynomial models, 259
position =, 177
power terms, 251
power transformation, 337
precedence rules, 17
predictable functions, 429, 471
predicted values, 204, 217
prediction, 209
prediction interval, 205
preplanned comparisons, 283, 286, 339, 361
probplot(univariate), 176
proc anova, 292, 293

556   Index

proc chart, 114, 116
proc contents, 69
proc corr, 74, 103
proc datasets, 69
proc format, 70, 73, 170
proc freq, 89, 90, 93, 95, 98
proc gchart, 133, 169
proc gplot, 130, 137
proc greplay, 169
proc means, 75, 76
proc plot, 104, 106, 107, 110
proc sort, 42, 46
proc statement options, 40
proc step, 40
proc tabulate, 119
proc univariate, 81, 82, 87
procedure information statements, 41
product terms, 251
profile log-likelihood, 392
profile plot, 159, 332
program data vector, 24, 27

Q option (proc glm), 432
qqplot, 151, 154
quadratic form, 390, 432, 433
quadratic trend, 303
quantile plot, 148
quantiles, 146

R^2, 210
rank, 150, 157
raxis =, 136
RCBD, 358
reduction notation, 231
reference line, 308
reference lines, 133
refpoint =, 177
REML estimates, 393
REML method, 402
reps, 358
residual plots, 222, 308
retain, 36

rsquare (proc reg), 242
RStudent, 200, 202, 204

SAS/GRAPH, 136
Satterthwaite, 395, 403, 417, 440, 445, 450, 459, 463, 483, 484, 486
scatmat macro, 167
scatter plot, 163
scatter plot matrix, 163, 166
Scheffé procedure, 290
Scheffé's method, 290
selection =, 241
sequential sums of squares, 233
set, 58, 60–63
Shapiro-Wilk test, 82
side-by-side box plots, 178
skewness, 75, 86
sle =, 249
slice = (gchart), 174
sls =, 249
Somers' D, 97
Spearman's correlation, 97, 102, 103
squared terms, 251
standard normal quantiles, 308
start =, 242, 248
statistic keyword, 75
stepwise, 237, 249
stname1, 71
stop =, 242, 248
Stuart's tau c, 97, 103
Studentized range, 290, 441
studentized residuals, 194, 219
subgroup =, 136, 173
subscripting, 22
subset selection, 235
subsetting, 9
subsetting if, 58
summary variable, 169
sumvar =, 135, 171
sumvar = (gchart), 174

symbol, 140, 144, 161, 177, 207

table, 119, 121, 122
tables, 91, 93
targetdevice, 138
temporary data set, 5
test, 90
title, 138
title2, 139
trailing at symbol, 159, 298, 362
transformation, 336
trim =, 82, 87
trimmed mean, 82, 88
Tukey procedure, 290, 295
Tukey's method, 290
Tukey's test, 367
two-level data set names, 65–67, 85, 106, 116
two-way factorial, 328
type =, 74, 171, 173
type = (gchart), 174
Type I, 233, 348
Type II, 233, 243, 348
Type III, 299, 348
Type III E(MS), 397, 405, 413–415, 424, 425, 433, 437, 448, 457
Type IV, 348
types, 78
Types of sums of squares, 233

unadjusted means, 350
unconstrained parameters (UP) model, 442
unequal sample sizes, 347
univariate, 176
unqual sample sizes, 405, 450
unweighted means, 350

v =, 140, 142
value =, 140
value = (gchart), 174

var, 75
vardef =, 82
variable attribute
    statements, 45
variance components,
    392, 397, 401
variance inflation factor,
    226
variance-covariance
    matrix, 392
vaxis =, 132
vbar, 146, 170
vif, 226, 230, 259
vsize =, 162

Wald statistic, 417, 438
ways, 78

where, 11
Wilcoxon signed rank
    test, 82
work.gseg, 170

X'X matrix, 277
x-outlier, 197, 220

y-outlier, 197, 219

springer.com

Time Series Analysis with Applications in R

Jonathan D. Cryer and Kung-Sik Chan

Time Series Analysis With Applications in R, Second Ed., presents an accessible approach to understanding time series models and their applications. Although the emphasis is on time domain ARIMA models and their analysis, the new edition devotes two chapters to the frequency domain and three to time series regression models, models for heteroscedasticty, and threshold models. All of the ideas and methods are illustrated with both real and simulated data sets. A unique feature of this edition is its integration with the R computing environment.

2008. 2nd Ed., 494 pp. (Springer Texts in Statistics) Hardcover
ISBN 0-387-75958-6

Statistical Learning from a Regression Perspective

Richard A. Berk

This book considers statistical learning applications when interest centers on the conditional distribution of the response variable, given a set of predictors, and when it is important to characterize how the predictors are related to the response. As a first approximation, this is can be seen as an extension of nonparametric regression. Among the statistical learning procedures examined are bagging, random forests, boosting, and support vector machines. Response variables may be quantitative or categorical.

2008. Approx. 370 pp. (Springer Series in Statistics) Hardcover
ISBN 978-0-387-77500-5

Modern Multivariate Statistical Techniques
Regression, Classification, and Manifold Learning

Allen Julian Izenman

This book is for advanced undergraduate students, graduate students, and researchers in statistics, computer science, artificial intelligence, psychology, cognitive sciences, business, medicine, bioinformatics, and engineering. Familiarity with multivariable calculus, linear algebra, and probability and statistics is required. The book presents a carefully-integrated mixture of theory and applications, and of classical and modern multivariate statistical techniques, including Bayesian methods.

2008. Approx 760 pp. (Springer Texts in Statistics) Hardcover
ISBN 978-0-387-78188-4

Easy Ways to Order▶ Call: Toll-Free 1-800-SPRINGER • E-mail: orders-ny@springer.com • Write: Springer, Dept. S8123, PO Box 2485, Secaucus, NJ 07096-2485 • Visit: Your local scientific bookstore or urge your librarian to order.

Printed in the United States of America